2006 Bipolar/BiCMOS Circuits and Technology Meeting Proceedings

Maastricht, Netherlands
8-10 October 2006

IEEE Catalog Number: 06CH37813
ISBN: 1-4244-0458-4

Copyright © 2006 by The Institute of Electrical and Electronics Engineers, Inc.
All Rights Reserved

Copyright and Reprint Permissions: Abstracting is permitted with credit to the source. Libraries are permitted to photocopy beyond the limit of U.S. copyright law for private use of patrons those articles in this volume that carry a code at the bottom of the first page, provided the per-copy fee indicated in the code is paid through Copyright Clearance Center, 222 Rosewood Drive, Danvers, MA 01923.

For other copying, reprint or republications permission, write to IEEE Copyrights Manager, IEEE Operations Center, 445 Hoes Lane, Piscataway, New Jersey USA 08854. All rights reserved.

IEEE Catalog Number: 06CH37813

ISBN: 1-4244-0458-4

ISSN: 1088-9299

Additional Copies of This Publication Are Available from:

IEEE Service Center
445 Hoes Lane
Piscataway, NJ 08854
IEEE Service Center
445 Hoes Lane
Piscataway, NJ 08854
Phone: (800) 678-IEEE
              (732) 981-1393
Fax: (732) 981-9667
E-mail: customer-service@ieee.org

Table of Contents

Linearization Techniques at the Device and Circuit Level 1
L.C.N. de Vreede and M.P. van der Heijden

Linear Efficiency Load-pull Measurements on GaAs HBT versus Si BJT for EDGE and OFDM Modulation 9
L.C.M van den Oever, R.M. Heeres, H.F.F. Jos

Analysis of Temperature Modulation on a SiGe Power Amplifier Non Linearity 13
L. Leyssenne, J.M. Pham, P. Jarry, E. Kerhervé, and Daniel Saias

MEMS-Based Reconfigurable Multi-band BiCMOS Power Amplifier 17
A.J.M. de Graauw, P.G. Steeneken, C. Chanlo, J. Dijkhuis, S. Pramm, A. van Bezooijen, H.K.J. ten Dolle, F. van Straten, P. Lok

Miniaturized Quad-Band Front-End Module for GSM using Si BiCMOS and Passive Integration Technologies 21
A.J.M. de Graauw, A. van Bezooijen, C. Chanlo, A. den Dekker, J. Dijkhuis, S. Pramm, and H.K.J. ten Dolle

Physical Description of the Mixed-Mode Degradation Mechanism for High Performance Bipolar Transistors 25
T. Vanhoucke, G.A.M. Hurkx, D. Panko, R. Campos, A. Piontek, P. Palestri, L. Selmi

Analysis of Factors Contributing to Common-Base Avalanche Instabilities in Advanced SiGe HBTs 29
Curtis M. Grens, John D. Cressler, and Alvin J. Joseph

Electrothermal Phenomena in Bipolar Transistors and ICs: Analysis, Modeling, and Simulation 33
N. Rinaldi, V. D'Alessandro, and F. M. De Paola

Reliability Issues in SiGe HBTs Fabricated on CMOS-Compatible Thin-Film SOI 41
Marco Bellini, Tianbing Chen, Chendong Zhu, John D. Cressler, and Jin Cai

Unified Analysis of Degraded Base Current in SiGe:C HBTs after Reverse and Forward Reliability Stress 45
M. Ruat, N. Revil, G. Pananakakis and G. Ghibaudo

A BiCMOS Technology Featuring a 300/330 GHz (fT / fmax) SiGe HBT for Millimeter Wave Applications 49
B. A. Orner, M. Dahlström, A. Pothiawala, R. M. Rassel, Q. Liu, H. Ding, M. Khater, D. Ahlgren, A. Joseph, J. Dunn

A 0.13μm thin SOI CMOS technology with low-cost SiGe:C HBTs and complementary high-voltage LDMOS 53
L. Boissonnet, F. Judong, B. Vandelle, L. Rubaldo, P. Bouillon, D. Dutartre, A. Perrotin, G. Avenier, P. Chevalier, A. Chantre and B. Rauber

A 205/275GHz fT/fmax Airgap Isolated 0.13μm BiCMOS Technology featuring on-chip High Quality Passives 57
S. Van Huylenbroeck, L. J. Choi, A. Sibaja-Hernandez, A. Piontek, D. Linten, M. Dehan, O. Dupuis, G. Carchon, F. Vleugels, E. Kunnen, P. Leray, K. Devriendt, X. P. Shi, R. Loo, E. Hijzen and S. Decoutere

Simultaneus Integration of SiGe High Speed Transistors and High Voltage Transistors 61
R.K. Vytla, T.F. Meister, K. Aufinger, D. Lukashevich, S. Boguth, H. Knapp, J. Bock, H. Schafer and R. Lachner

RF, Analog and Mixed Signal Technologies for Communication ICs - An ITRS Perspective 65
W. Margaret Huang, Herbert S. Bennett, Julio Costa, Peter Cottrell, Anthony A. Immorlica Jr., Jan-Erik Mueller, Marco Racanelli, Hisashi Shichijo, Charles E. Weitzel, and Bin Zhao

A High-Slew Rate SiGe BiCMOS Operational Amplifier for Operation Down to Deep Cryogenic Temperatures 72
Ramkumar Krithivasan, Yuan Lu, Laleh Najafizadeh, Chendong Zhu, John D. Cressler, Suheng Chen, Chandradevi Ulaganathan, and Benjamin J. Blalock

86dB. 10Gb/s SiGe Transimpedance Amplifier Using Photodiode Capacitance Neutralization and Vertical Threshold Adjustment 76
Adrian Maxim

A 70 MHz - 4.1 GHz 5th-Order Elliptic gm-C Low-Pass Filter in Complementary SiGe Technology 80
Yuan Lu, Ramkumar Krithivasan, Wei-Min Lance Kuo, Xiangtao Li, John D. Cressler, Hans Gustat, and Bernd Heinemann

Table of Contents

A 3.5GHz Low Power Programmable Transversal Filter RFIC Implemented in 47GHz SiGe Technology................84
Vasanth Kakani, Xuefeng Yu, Foster F. Dai, Richard C. Jaeger

SiGe BiCMOS Precision Voltage References for Extreme Temperature Range Electronics 88
Laleh Najafizadeh, Chendong Zhu, Ramkumar Krithivasan, John D. Cressler, Yan Cui, Guofu Niu, Suheng Chen, Chandradevi Ulaganathan, Benjamin J. Blalock, and Alvin J. Joseph

Quantitative Analysis of Errors in On-Wafer S-Parameter Deembedding Techniques for High Frequency Device Modeling................92
Rob Groves, Jing Wang, Lawrence Wagner and Ava Wan

BJT Base and Emitter Resistance Extraction from DC Data................96
Colin C. McAndrew

Simultaneous Extraction of Collector and Substrate Series Resistance by Simple DC Measurement................100
Tianbing Chen, Tracey L. Krakowski, Andy Strachan, Yun Liu, Alexei Sadovnikov, and Je Babcock

Joint Extraction of the Base and Collector Resistances with the Base-Collector Capacitance Split of HBT/BJT Transistors................104
Z. Huszka, E. Seebacher and W. Pflanzl

Compact Modeling of GaAs Heterojunction Bipolar Transistors using the new Mextram 3500 model................108
R. van der Toorn, J.C.J. Paasschens, J.J. Dohmen, R.M.T. Pijper

Substrate Current Injection and Latchup in Complementary Vertical Bipolar Processes................112
Andy Strachan

10A 12V 1 chip DC/DC converter IC using bump technology................118
Kazutoshi Nakamura, Kenichi Matsushita, Norio Yasuhara, Koichi Endo, Fumito Suzuki, Morio Takahashi and Akio Nakagawa

High RF performances asymmetric spacer NLDMOS integration in a 0.25μm SiGe:C BiCMOS Technology................122
Bertrand Szelag, Dorothée Muller, Jocelyne Mourier, Caroline Arnaud, Halim Bilgen, Fabienne Judong, Alexandre Giry, Denis Pache, Agustin Monroy

A Modular 0.18 um Analog / RFCMOS Technology Comprising 32 GHz FT RF-LDMOS and 40V Complementary MOSFET Devices................126
Zachary Lee, Robert Zwingman, Jie Zheng, Will Cai, Paul Hurwitz, and Marco Racanelli

A Concurrent Fully-Integrated BiFET LNA for W-CDMA/IEEE 802.11a Applications................130
C. P. Moreira, E. Kerherve, P. Jarry, D. Belot

A 77 GHz (W-band) SiGe LNA with a 6.2 dB Noise Figure and Gain Adjustable to 33 dB................134
Ralf Reuter, Yi Yin

A 5GHz Series Coupled BiCMOS Quadrature VCO with Wide Tuning Range................138
Vasanth Kakani, Foster F. Dai and Richard C. Jaeger

Design and Scaling of SiGe BiCMOS VCOs Above 100GHz................142
S. T. Nicolson, K.H.K Yau, K.A. Tang, P. Chevalier, A. Chantre, B. Sautreuil, and S. P. Voinigescu

Turn-On Voltage Control in BSCR and LDMOS-SCR by the Local Blocking Junction Connection................146
V.A. Vashchenko and P.J. Hopper

Deep Trench NPN Transistor for Low-RON ESD Protection of High-Voltage I/Os in Advanced Smart Power Technology................150
A. Gendron, C. Salamero, N. Nolhier, M. Bafleur, P. Renaud, P. Besse

GaAs HBT ESD Diode Layout and its Relationship to Human Body Model Rating................154
Douglas A. Teeter, Kathy Muhonen, David Widay, Mike Fresina

Human Body Model ESD Protection of RF Bipolar Circuits................158
Paul Davis and Brian Horton

Characterization and Modeling of Intermodulation Linearity in a 200 GHz SiGe HBT Technology................162
Guofu Niu, Ying Li, Zhiming Feng, Jun Pan, and David C. Sheridan

Table of Contents

Impact of Collector-Base Space Charge Region on RF Noise in Bipolar Transistors 166
Kejun Xia and Guofu Niu

SiGe Profile Optimization for Improved Cryogenic Operation at High Injection 170
Yan Cui, Guofu Niu, Yun Shi, Chendong Zhu, Laleh Najafizadeh and John D. Cressler, and Alvin Joseph

Influence of the Ge Profile on VBE and Current Gain Mismatch in Advanced SiGe BICMOS NPN HBT with 200 GHz fT 174
M. Dahlstrom, K. Walter, S. Von Bruns, R.M. Malladi, Kim M. Newton and A.J. Joseph

Substrate Transfer: an Enabling Technology for System-in-Package Solutions 178
R. Dekker

Above IC RF MEMS and BAW filters: fact or fiction? 186
P. Ancey

Lithography for the 32-nm Node and Beyond 191
J. N. Burghartz, M. Irmscher, F. Letzkus, J. Kretz, and D. Resnick

Applications of SiGe Material for CMOS and Related Processing 196
S.Monfray, T.Skotnicki, P.Coronel, S.Harrison, D.Chanemougame, F.Payet, D.Dutartre, A.Talbot, S.Borel

Design and Modeling of mm-Wave Monolithic Transformers 203
Tak Shun D. Cheung and John R. Long

Lumped Modeling of Differentially Driven Symmetric Inductors for RF IC Design 207
E. Ragonese, A. Scuderi, T. Biondi, A. Italia, and G. Palmisano

Demonstration of three-dimensional 35nF/mm² MIM Capacitor integrated in BiCMOS Circuits 211
JC. Giraudin, F. Badets, JP. Blanc, E. Chataigner, A. Bajolet, T. Jagueneau, C. Rossato and P. Delpech

Integrated BiCMOS 10 GHz S-Parameter Module 215
Jangsup Yoon, Robert M. Fox, and William R. Eisenstadt

RF Switch on Standard SiGe-CMOS Technology for System-on-a-chip Radio Transceivers 219
Domenico Zito and Bruno Neri

SiGe BiCMOS for Analog, High-Speed Digital and Millimetre-Wave Applications Beyond 50 GHz 223
S.P. Voinigescu, T. Chalvatzis, K.H.K. Yau, A. Hazneci, A. Garg, S. Shahramian, T. Yao, M. Gordon, T.O. Dickson, E. Laskin, S.T. Nicolson, A.C. Carusone, L. Tchoketch-Kebir, O. Yuryevich, G. Ng, B. Lai, and P. Liu

A 64-bit High-Speed Read-Write Look-Up Table Memory Implemented in GaAs HBT 231
Andre G. Metzger and Peter M. Asbeck

Low-Power, Low-Phase Noise SiGe HBT Static Frequency Divider Topologies up to 100 GHz 235
E. Laskin, S. T. Nicolson, P. Chevalier, A. Chantre, B. Sautreuil, S. P. Voinigescu

A Low Power 12.5Gb/s SiGe Limiting Amplifier Using a Feed-forward Adjustable Threshold Loss-Of-Signal Detector 239
A. Maxim, D. Smith

250-GHz self-aligned Si/SiGeC HBT featuring an all-implanted collector 243
P. Chevalier, C. Raya, B. Geynet, F. Pourchon, F. Judong, F. Saguin, T. Schwartzmann, R. Pantel, B. Vandelle, L. Rubaldo, G. Avenier, B. Barbalat, and A. Chantre

Development of a Cost-Effective, Selective-Epi, SiGe:C HBT Module for 77GHz Automotive Radar 247
Jay P. John, Jim Kirchgessner, Matt Menner, Hernan Rueda, Francis Chai, Dave Morgan, Jill Hildreth, Morgan Dawdy, Ralf Reuter, and Hao Li

Experimental Study of Metallic Emitter SiGeC HBTs 251
B. Barbalat, F. Judong, L. Rubaldo, P. Chevalier, M. Proust, C. Richard, G. Borot, B. Vandelle, F. Saguin, D. Dutartre, N. Zerounian, F. Aniel and A. Chantre

Schottky Barrier Diodes for Millimeter Wave SiGe BiCMOS Applications 255
R.M. Rassel, J.B. Johnson, B.A. Orner, S.K. Reynolds, M.E. Dahlström, J.S. Rascoe, A.J. Joseph, B.P. Gaucher, J.S. Dunn, S.A. St. Onge

Table of Contents

High Performances 3D Damascene MIM Capacitors Integrated in Copper Back-End Technologies...........259
S. Crémer, C. Richard, D. Benoit, C. Besset, J.-P. Manceau, A. Farcy, C. Perrot, N. Segura, M. Marin, S. Bécu, S. Boret, M. Thomas, S. Guillaumet, A. Bonnard, P. Delpech and S. Bruyère

5.8 GHz RF Transceiver LSI Including On-Chip Matching Circuits...........263
Minoru Nagata, Hideaki Masuoka, Shin-ichi Fukase, Makoto Kikuta, Makoto Morita, Nobuyuki Itoh

5-GHz WLAN Standards Compliant Image Reject Radio Receiver on Low-cost SiGe-CMOS Technology...........267
Domenico Zito and Bruno Neri

A SiGe BiCMOS 9.75/10.6GHz Frequency Synthesizer for DBS Satellite LNB Down-Converters Using Half-Rate Oscillators...........271
A. Maxim, M. Gheorghe, C. Turinici

Ruggedness Improvement by Protection...........275
André van Bezooijen, Anton de Graauw, Lennart Ruijs, Skule Pramm, Christophe Chanlo, Henk Jan ten Dolle, Freek van Straten, Reza Mahmoudi, and Arthur H.M. van Roermund

Compact Modeling of High Frequency Correlated Noise in HBTs...........279
P.Sakalas, J.Herricht, A.Chakravorty, M.Schroter

Modeling of Intrinsic Base Majority Carrier Thermal Noise for SiGe HBTs Including Fringe BE Junction Effect...........283
Kejun Xia and Guofu Niu

Fully coupled dynamic self heating model for power SOI Lateral Insulated Gate Bipolar Transistors...........287
S. Gamage, V. Pathirana and F. Udrea

A Broadband and Scalable On-Silicon-Chip Inductor Model for Varying Substrate Resistivities...........291
J. C. Guo and T. Y. Tan

"Linearization Techniques at the Device and Circuit Level" (Invited)

L.C.N. de Vreede and M.P. van der Heijden[1],

Department of Microelectronics, Laboratory of Electronic Components, Technology & Materials,
Delft University of Technology, Feldmannweg 17, 2628 CT, Delft, The Netherlands.
[1]Philips Research Laboratories, Integrated Transceivers Group, High Tech Campus 5, 5656 AE,
Eindhoven, the Netherlands

Abstract—**Modern telecommunication applications set high demands on transceiver technology and design in terms of speed, noise, linearity and power consumption. This paper provides an overview of techniques, at the device and circuit level, which improve the classical DC power consumption / linearity trade-off found in (bipolar) telecommunication circuits.**

Index Terms—**Nonlinear distortion, bipolar, CMOS linearization, current-mode, feedback, transconductance, out-of-band termination, IM3 cancellation, power amplifier, harmonic load-pull, device optimization.**

I. INTRODUCTION

Linearity is one of the key parameters in the circuit design of modern wireless communication systems. The continuous search to improve on the classical trade-off between DC power consumption and linearity has resulted in a wide variety of linearization techniques, each with its own benefits and drawbacks. This paper gives an overview of the most relevant principles and techniques to improve device / circuit linearity. The focus is on bipolar devices and circuits, but on some points the view is extended to CMOS based designs. We start with the basic distortion mechanisms present in bipolar and CMOS devices. Using this knowledge, we are able to classify the different linearization principles and make conclusions about their effectiveness. Special attention will be given to the use of out-of-band linearization techniques, where we utilize the distortion cancellation in the active device itself to improve the overall linearity. We conclude our discussion with the latest innovations / trends in direct down conversion receivers and device optimization for linearity.

II. DOMINANT DEVICE NON-LINEARITIES

Active devices can be modeled by nonlinear current and charge sources, which depend on the voltages of the (internal) device terminals. When driven with a modulated signal, these nonlinear sources will give rise to distortion. For not too large signal amplitudes, this distortion can be studied through analytical techniques like the Volterra series method [1]. Using this method, each of the nonlinear sources is approximated by a Taylor series, representing the 1st, 2nd, 3rd,

4th etc. order dependency of the source quantity on the nodal voltages. In this description, the 1st order terms give the desired linear device transfer, while the higher-order nonlinearities of the device give rise to (intermodulation) distortion that corrupts the desired signal transfer. It is for this reason that we first determine the dominant device nonlinearities.

A. Bipolar devices

Although bipolar devices exhibit an almost perfect linear current-current transfer, most applications will suffer from the exponential nature of the device. Consequently, when considering the equivalent schematic (Fig. 1a) the most dominant contributors to nonlinear distortion are the transconductance and the base-emitter capacitance, which both include an exponential dependency on the base-emitter voltage. [2]. Other elements in the equivalent schematic of a bipolar transistor can be, in a first order approximation, considered to be linear for most applications with the exclusion of power amplifiers.

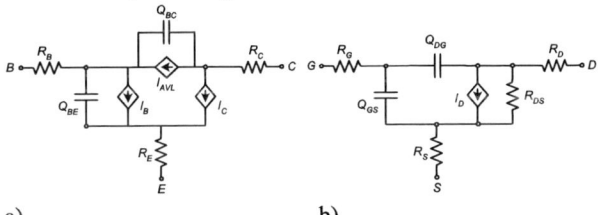

Fig. 1 The dominant nonlinear current and charge sources indicated in the equivalent schematics of a (a) bipolar device and a (b) CMOS device.

B. FET devices

When considering FETs (e.g. CMOS) devices, again the nonlinear transconductance is one of the most dominant nonlinearities, while the other elements in their equivalent schematic (Fig. 1b) can be assumed linear for most applications. Although the transconductance of FET devices is also nonlinear, short gate-length CMOS devices have a more linear reputation than bipolar transistors when operated in the saturation region, since their transconductance tends to be less a function of bias [3], more specifically: $g_m = \partial I_D / \partial V_{gs} \approx \mu_n C_{ox} W.E_{sat}/2$, in which μ_n, C_{ox}, W and

E_{sat} are respectively: the electron mobility, the gate-oxide capacitance per unit gate width, the gate width and the electrical field strength at which the carrier velocity has dropped to half the value extrapolated from the low field mobility (~4.10⁶V/m) [3]. Unfortunately, CMOS devices in practical analogue circuits do not exhibit a perfectly constant g_m, since they are often operated below the saturation region for reasons of DC power consumption and an ever-decreasing supply voltage. As a result, also for these devices linearization techniques are required to meet the stringent specifications of wireless applications.

Based on these rather general device considerations we are able to categorize the various linearization concepts in literature and give their principle of operation. Independent of whether one is dealing with bipolar or CMOS design, a key point in these discussions is how to reduce / compensate the impact of the nonlinear transconductance without violating other important circuit specifications like gain, noise or power consumption.

III. THE DESIGN OF LINEAR LOW -POWER RF CIRCUITS

A. Current-mode design

When considering the linear collector-base current relation of bipolar devices one of the most logical ways to improve for linearity is the use of current-mode or translinear operation [4]. At RF frequencies however, device parasitics will affect the intended linear transfer. For this reason in Fig. 2 we evaluate the bandwidth constrains for current-mode operation of a CE-stage (Fig. 2a) and a CB-stage (Fig. 2b) using as reference transistor, the 70 GHz SiGe QUBiC technology [6]. In these schematics C_{jE} and C_{DE} refer to the depletion and diffusion capacitance, respectively. As measure for the linearity in current-mode operation, we use the OIP3 in dBmA (relative to 1mA) of the collector current as function of operating frequency [5]. Essential in these experiments is the value of the source resistance in relation to the input impedance of the transistor stage. Note that a high source impedance / input impedance ratio dictates current-mode operation, while a very low ratio results in a voltage driven condition. Due to the much lower input impedance of the CB-stage, current-mode operation is easier to establish at RF frequencies. This is also visible from Fig. 2c where current-mode operation of a CB-stage gives a linearity advantage over the voltage mode (R_s=0) up to the cut-off frequency. A typical circuit example of current-mode operation at RF frequencies is found in the Gilbert cell mixer, where the upper switching "CB-stages" are current driven by the driver stage (Fig. 11). In most practical mixer designs it is the nonlinear transconductance of this driver stage that acts as the limiting factor for the IIP3 mixer linearity. For this reason many design techniques are focused on improving the IIP3 of the driver stage.

B. Local and overall feedback techniques

Overall feedback [7][8] and local feedback [9] are commonly used techniques to improve for linearity. Especially local feedback based on inductive source or emitter degeneration of the active device is very attractive for RF design, since it can provide (in addition to an improved linearity) also simultaneous noise and impedance matching conditions [10], explaining its popularity in RF design. In [9] it was shown that, when aiming for linearity, inductive degeneration is much more effective than resistive degeneration at radio frequencies. This advantage can be explained by "partial" IM3 cancellation effects, which occur for inductive degeneration. This phenomenon has been analyzed for two-tone test conditions using the circuit of Fig. 3 (assuming Z_b=0), yielding the following relation for the IM3 distortion [9]:

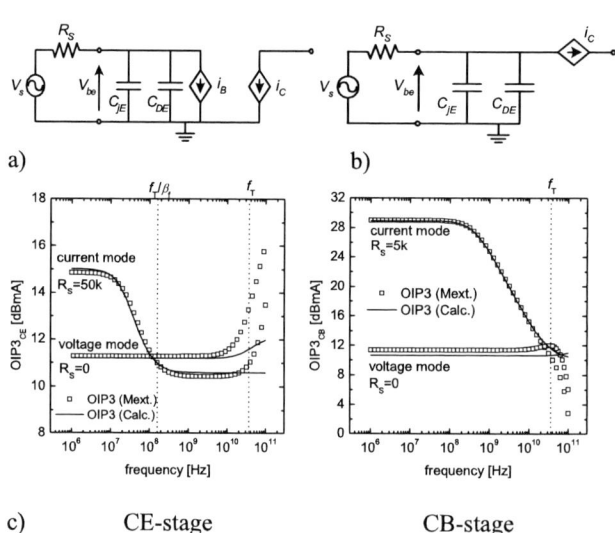

Fig. 2   Current mode operation of a 70 GHz SiGe QUBiC transistor [6] analyzed using Volterra series as well simulated with the full Mextram model. a) Schematic used to analyze a CE-stage b) Schematic used to analyze the CB-stage. c) The resulting OIP3 in dBmA for the CE-stage and CB-stage.

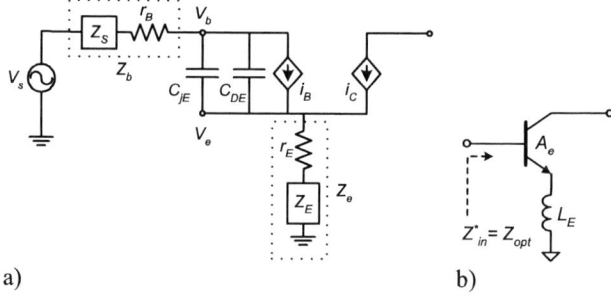

Fig. 3   a) Large-signal model of a CE-stage for studying the effects of series feedback. b) CE-stage with inductive emitter degeneration

$$IM3 = \frac{3}{4}\frac{\hat{V}_s^2}{V_T^2}\left|H_1(s)\right|^3 \cdot \left|T(s)\right| \cdot \left\|\left(\frac{1}{2}H_1(2s)T(2s) + H_1(\Delta s)T(\Delta s) - 1\right)\right\| \quad (1)$$

in which V_s is the source signal voltage, V_T the thermal voltage and H_1 represents the transfer of the voltage source to internal base-emitter junction. In this equation the function T is defined as

$$T(s) = 1 + sC_{JE}Z_e(s) \quad (2)$$

Note that when Z_e is inductive, in theory, perfect IM3 cancellation can be achieved by resonating C_{JE} and Z_e at the design frequency, yielding a zero in T and consequently in the IM3. However, closer inspection shows that such a dimensioning would result in too high values of the inductance and/or active device area and is therefore incompatible with achieving simultaneous noise and impedance match conditions. For this reason, in LNA or mixer designs it is more appropriate to achieve "partial" IM3 cancellation through manipulation of the terms $T(2s)$ and $T(\Delta s)$, which relate to the circuit response at the sum and difference frequencies, respectively, and which are also known as the out-of-band frequencies. In practical designs, the response at the difference or baseband frequency Δs is defined by the bias network, which can be utilized to manipulate the linearity. This was addressed in literature [11] but not always fully recognized in practical implementations. To support the above we consider in Fig. 4 a 5 GHz noise and impedance matched bipolar cascode LNA, which is tuned for its linearity by the value of the bias network resistance (R_B). From Fig. 4 it can be observed that the best linearity is achieved when R_B is set to 30 ohms, while no other changes are made to the circuit. Note that the change in R_B does not affect the bias conditions, since V_{B1} is swept and the linearity is plotted versus collector current. From the results it is observed that the use of a higher tone spacing causes asymmetry in the lower and upper IM3 bands. This is a consequence of the fact that the IM3 cancellation in this example is only effective at one single frequency. In section D we will discuss techniques that overcome this limitation.

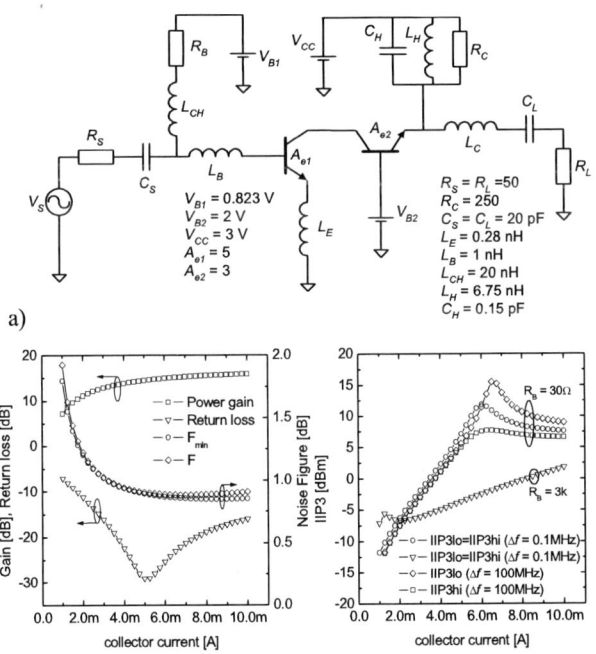

b)

Fig. 4   a) A 5 GHz cascode LNA with inductive emitter degeneration. The bias networks are utilized to tune the inband linearity. b) The resulting LNA linearity in IIP3 as function of the bias current. Note that R_B=30 ohm yields very high linearity compared to R_B=3k ohm at identical bias currents.

C. Shaping the transconductance

An alternative for improving linearity by feedback is to aim for a linearization of the non-linear transfer function itself by shaping the transconductance of the active device.

1) Bipolar devices

A well-known example in bipolar design is the multi-tanh doublet, which provides, compared to a conventional differential pair, a flattened transconductance versus input voltage [12]. Since the multi-tanh doublet is focused on improving the linearity of the transconductance function itself, it is most beneficial at lower frequencies with relative large bandwidths, while narrow-band applications operating at RF / microwave frequencies seem to benefit more from inductive emitter degeneration.

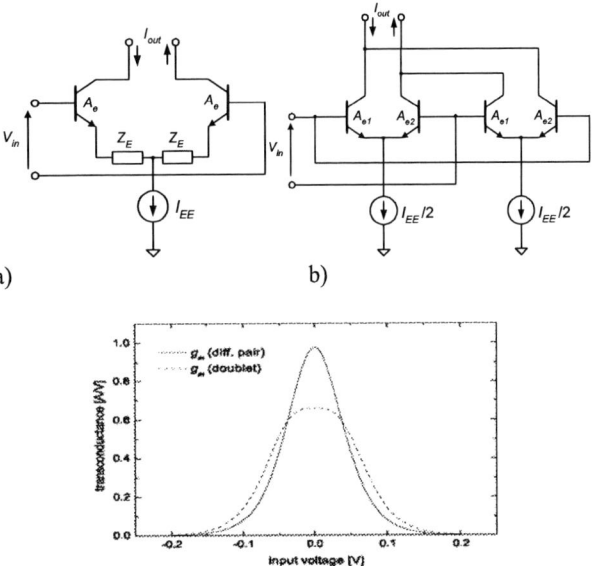

c)

Fig. 5   (a) Emitter degenerated differential pair stage and (b) Multi-tanh doublet. (c) The normalized transconductance of a differential pair and a multi-tanh

In order to illustrate this we provide a linearity comparison of the multi-tanh doublet with a resistive and inductive degenerated differential pair [5]. For this experiment we assume that the different configurations are driven with a two-tone voltage source, each having a current budget of 10mA, while the design frequency is 5 GHz. Next we take the transconductance of the multi-tanh doublet as reference, and dimension the inductive and resistive degeneration of the differential pairs in such a way that they provide the same transconductance at 5 GHz. In addition, the active emitter area of the inductive degenerated differential pair is scaled to the point where

$$\omega^2 C_{JE} L_E \approx 1 . \tag{3}$$

Note that this choice yields "partial" cancellation of the IM3 through the fundamental term $T(s)$ of equation (1). It must be mentioned that full IM3 cancellation does not occur due to the presence of the device series parasitics, like base and emitter

3

resistance. When comparing the linearity versus frequency of the various circuits (Fig. 6), a very pronounced improvement is found for the inductive degenerated differential pair at the design frequency of 5 GHz, outperforming the linearity of the multi-tanh doublet, while the resistively degenerated differential pair performs the worst of all three circuits. When considering the linearity versus power at the design frequency, we observe that the linearity improvement is effective over a large power range. The above confirms the linearity performance of the multi-tanh doublet for wide-band applications, while inductive degeneration is the best choice for "narrowband" wireless applications.

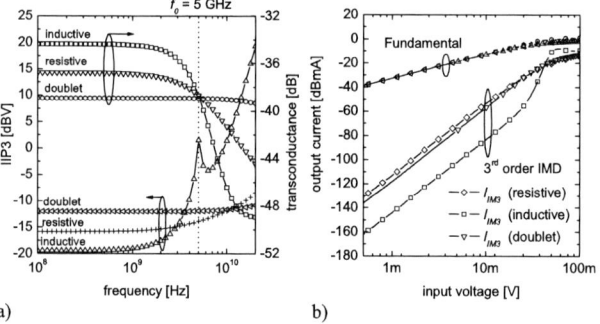

a)   b)

Fig. 6   The fundamental and third-order intermodulation collector current in dBmA for the resistive / inductive emitter degenerated differential pairs and the multi-tanh doublet: a) as function of frequency, b) as function of input voltage @ 5 GHz.

2) FET devices

Also for FET devices (e.g. CMOS), techniques are present that focus on the shaping of the transfer characteristic. The most well known is the derivative superposition technique [13]. This technique is based on the drain current dependence versus V_{gs}. When considering the related 3rd order Taylor coefficient of this current (gm_3), which represents the third-order nonlinearity, this bias dependence results in magnitude and phase reversals for gm_3 in the threshold region. The generalized circuit topology to be used with derivative superposition is shown in Fig. 7.

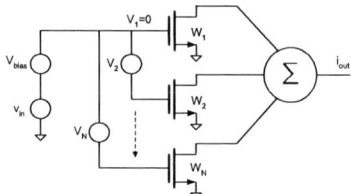

Fig. 7   The derivative superposition approach. The overall linearized transconductance is composed out of several gate bias shifted common source FET devices, which are current summed at their outputs.

By shifting the bias point and superposing the device characteristics the third-order nonlinearity can be cancelled over a limited input voltage range. There are two bias operation points that can be focused upon, namely class-AB operation and Class-A operation. One should be aware that when aiming for class-AB operation, the 2nd order frequency components (base-band and 2nd harmonic) must be short circuited to avoid secondary mixing of these products with the fundamental, which again contributes to third-order intermodulation distortion. Since this technique is primarily based on low-frequency considerations, additional measures need to be taken at RF frequencies. An example of such a situation is found in [14]. In this work, a modified LNA topology is given that utilizes a tapped inductor to avoid linearity degradation at higher frequencies (Fig. 8).

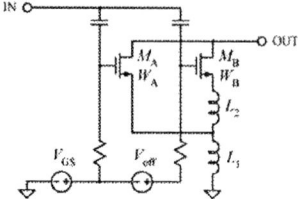

Fig. 8   Modified Derivative Superposition method applied in a CMOS LNA with inductive degeneration (From [14]) © IEEE 2005.

The resulting reported LNA linearity [14] exhibits a fifth-order dependency, illustrating the effective cancellation of the third-order intermodulation distortion. The LNA achieved +22-dBm IIP3 with 15.5-dB gain, 1.65-dB NF, and 9.3mA@2.6-V power consumption. This result indicates the potential of derivative superposition techniques for modern wireless receivers implemented in CMOS technology.

D. Out-of-band linearization techniques

Although the influence of out-of-band terminations was always present in any circuit realization, it is only recently that the full extend of its importance has been recognized [15]. The principle of out-of-band linearization is illustrated in Fig. 9, which shows the typical output spectrum of a nonlinear device driven by a two-tone signal.

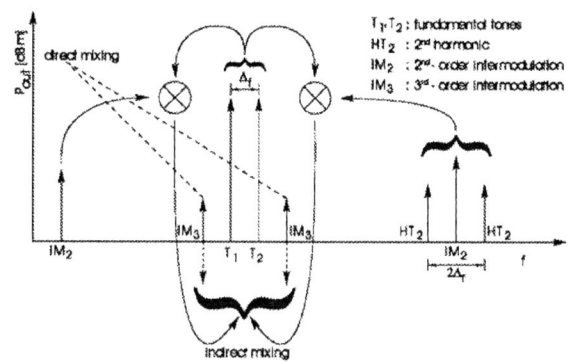

Fig. 9   Basic principle of the out-of-band linearity optimization. By controlling the out-of-band terminations at the baseband and 2nd harmonic frequencies. The IM3 products of the indirect mixing (fundamental with 2nd order products) over the 2nd order device nonlinearities are used to cancel out the direct IM$_3$ products, which result directly from the 3rd order device nonlinearities.

The interpretation of Fig. 9 is as follows: if an active device is driven with a modulated (e.g. two-tone) signal the device nonlinearities will give rise to intermodulation distortion. The IM3 distortion resulting directly from the 3rd-order device non-linearities is the so-called 'direct' IM3 product. When applying the *out-of-band linearization technique*, we aim to

4

cancel these 'direct' IM3 components by making use of the "indirect" IM3 products, which are present anyway. The latter distortion products result from the secondary mixing of the fundamental tones with the IM2 products (base-band and 2nd harmonic frequencies), over the nonlinear junctions of the device. By controlling the base-band and 2nd harmonic terminations at the in- and output of the device, the magnitude and phase of these indirect components can be adjusted. Consequently, the indirect IM3 components can be used, under some specific conditions, to cancel the direct ones. This method can provide excellent linearity over a large power range up to the compression point [18].

In this overview we focus on the out-of-band linearization of bipolar devices, which provide high linearity at low DC-current levels. In [16] it was shown that for a given bipolar technology there is an optimum combination of base-band & 2nd harmonic input termination that result in perfect IM3 cancellation over an unrestricted bandwidth. Based on the large-signal model in Fig. 3 under the constraint of short-circuited out-of-band terminations at the output of the active device, while neglecting the parasitic base and emitter resistance, the input out-of-band conditions for a CE-stage for frequency independent IM3 cancellation are given by:

$$Z_{b,BB} = Z_{b,2nd} = \frac{\beta_F}{2g_m} \qquad (4)$$

$$I_{cq,opt} = V_T \frac{C_{jE}}{2\tau_F} \qquad (5)$$

where

$Z_{b,BB}$ is the base-band impedance in the base path,
$Z_{b,2nd}$ the 2nd harmonic impedance in the base path,
$I_{cq,opt}$ the optimum quiescent collector current,
g_m the transconductance,
V_T the thermal voltage,
β_F the forward current gain of the transistor,
C_{jE} the emitter-base depletion capacitance in the operating point,
τ_F the forward delay time in the operating point.

Satisfying the conditions in equation (4) and (5) yields large improvements in linearity over a very wide bandwidth for both the lower and upper third-order intermodulation sidebands. Although in theory this IM3 cancellation is frequency independent, this is not always easy to achieve in practical circuits. For this reason we compare base and emitter IM3 compensated designs and give their implementation sensitivity versus bandwidth.

For this experiment we consider the circuit of Fig. 3. For the base compensated design we set $Z_{b,BB}$ and $Z_{b,2nd}$ according to (4) while Z_e is set to zero. For the emitter out-of-band linearization we assume the out-of-band terminations in the base path equal to zero ($Z_{b,BB} = Z_{b,2nd} = 0$), but now we use the appropriate (ohmic) out-of-band conditions in the emitter path $Z_{e,BB} = Z_{e,2nd} = 1/(2g_m)$ [5]. In all practical circuit realizations the intended IM3 wideband cancellation only occurs when (5) is also satisfied. Deviation from the condition

in (5) results in linearity degradation at higher frequencies. Fig. 10a shows the simulation results when the area of the active device in both cases is off by 30%, yielding a violation of the condition in (5) From the results one can conclude that base compensated designs are more sensitive to deviations from (5). Note that this is a situation comparable to that previously found for current-mode operation. If the condition in (5) is also fulfilled (e.g. by proper device scaling), this leads to nearly unrestricted wide-band IM3 cancellation. This is shown in Fig. 10b for the emitter out-of-band compensated stage. Note that in all these experiments the out-of-band terminations at the output of the active device (2nd harmonic and base-band components) are short-circuited. This is essential to avoid feedback of harmonic components at the output to the device input, which would result in a drastic change of the initial IM3 cancellation conditions.

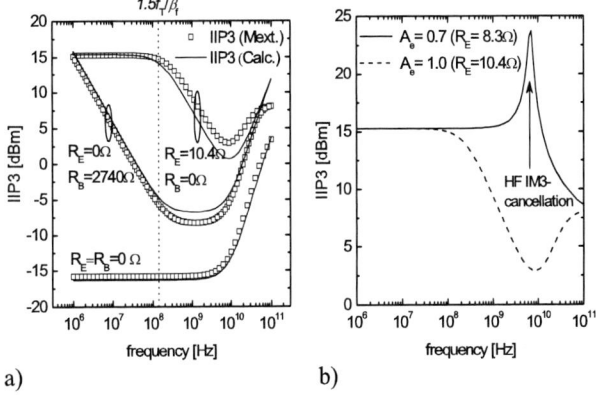

a)   b)

Fig. 10   IIP3 versus frequency with the 70 GHz f_T SiGe QUBiC reference transistor using out-of-band compensation a) Only the real part of the IM3 cancellation condition is fulfilled (imaginary part off by 30%) b) Both the real as well the imagery part of the IM3 cancellation condition are fulfilled by proper device scaling. Note that the IM3 cancellation at higher frequencies is less sensitive for emitter out-of-band compensated designs.

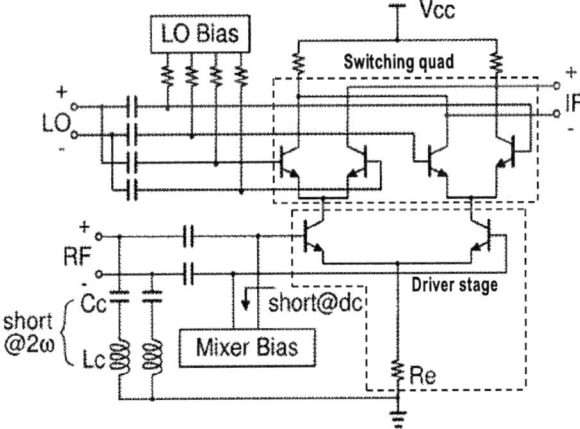

Fig. 11   Mixer with emitter out-of-band IM3 compensation implemented through the correctly valued resistor R_e and the 2nd harmonic shorts in the base [20]. The resulting linearized driver stage is current mode coupled to the switching quad to yield a highly linear mixer © IEEE 2003.

Practical examples of circuits that utilize out-of-band termination are given in [15][16][17]. Note that the best reported performance of LNA implementations in terms of linearity, noise and DC power consumption are based on the use of out-of-band IM3 cancellation techniques [17], [19], indicating the high potential of this linearization method. A nice example of a mixer stage [20] utilizing emitter out-of-band compensation is given in Fig. 11. Here, the drive stage is optimized using emitter out-of-band compensation for improved linearity without any penalty in DC power consumption. The drive stage in turn is connected in current-mode operation to the switching quad yielding a highly linear low-power mixer.

IV. DIRECT CONVERSION RECEIVER LINEARITY REQUIREMENTS

Currently, direct conversion receivers (DCR) are receiving much attention in industry due to their better suitability for integration than heterodyne receivers. In contrast to heterodyne receivers, DCR concepts do not have high demands only on the IIP3, but also on the IIP2. This latter requirement is essential to suppress the undesired down conversion of interfering signals to the base band [21]. Depending on the implementation of the DRC, if there is no filter present between the LNA and the down converting Mixer, the IIP2 requirements on the mixer can be as high as 60dBm or more. It is common knowledge that high IIP2 can be achieved by balancing of the circuitry [21][22]. In theory perfect balancing can result in a very high common mode rejection factor and consequently in a very high IIP2. Unfortunately, in practical implementations, component spread corrupts the balancing to some extend, significantly lowering the IIP2. To suppress these effects, common-mode feedback loops are used to enforce a better balancing of the mixer [23]. Special attention requires the capacitive loading of the common terminal of the switching stages. Reducing this capacitive loading results almost directly in an IIP2 improvement, since a better balancing of the switching pair is enforced by the higher impedance of the common terminal. In [24] an inductor between the points A and A' in Fig. 12 is proposed. By tuning this inductor the capacitive loading is resonated out. Although positive effects have been reported in [24], this configuration does not offer any advantage for improving the node impedances of A and A' for common mode signals. In [25] consequently, an extension to this technique is proposed by including a shunting capacitor between the point B and the ground. By choosing the component values such that they resonate at the local oscillator frequency, the capacitive loading of the nodes A and A' are now also neutralized for common-mode signals. This increase in the effective impedance seen by the switching pair, yields an additional improvement in CMRR and consequently in IIP2. IIP2 values above 78 dBm have been reported using this technique [25], indicating the feasibility of fully-integrated direct-conversion receivers without inter-stage filter.

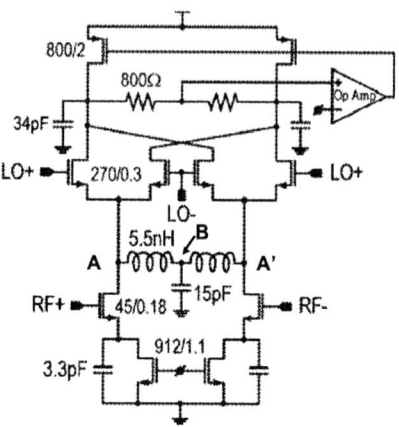

Fig. 12   Mixer topology with improved IIP2 for use in direct conversion receivers [25] © IEEE 2006.

V. POWER AMPLIFIER LINEARIZATION

Until now, the focus was placed on the linearity of "small signal" circuits, which operate at relative low signal powers. When considering bipolar PA applications the need to optimize simultaneously linearity and efficiency is even more stringent then for low-power circuits. Consequently, to address these specific requirements in power amplifiers, modification of the out-of-band termination technique is required to establish an IM3 sweet spot close to the gain compression point. Note that such an approach will improve the linearity at higher power levels rather then focusing on linearity at large power back-off conditions. The IM3 sweet spot can be created by adjusting the out-of-band solutions and the quiescent current [26]. An illustration of what can be achieved is given in Fig. 13 which gives the measured linearity for a Class-AB biased GaAs Skyworks HBT [27]. The fundamental frequency at the device input is impedance matched while the fundamental at the output is matched for maximum power. The out-of-band terminations at the output are short circuited to establish class-AB operation. The input out-of-band terminations are first set to zero (classical class-AB case) yielding a poor linearity, which is used here as reference. Next the out-of-band terminations and quiescent current are set to the low power IM3 cancellation condition of eq. (4) and (5) (for this particular device $Z_{b,BB}=Z_{b,2nd}=17.5\Omega$, $I_{qs}=9.8mA$) yielding IM3 cancellation and consequently a 5th order dependence of the IM3 on the input power. With these impedances, one can create an IM3 sweet spot by increasing the quiescent current, which can be positioned close to the compression point of the device [26]. The efficiency can be improved even more by increasing the baseband impedance, which reduces the required quiescent current but now demands a complex 2nd harmonic input termination [27]. As a result the IM3 spot is even closer to the compression yielding a very high linearity at high efficiency for bipolar PA's in class-AB operation.

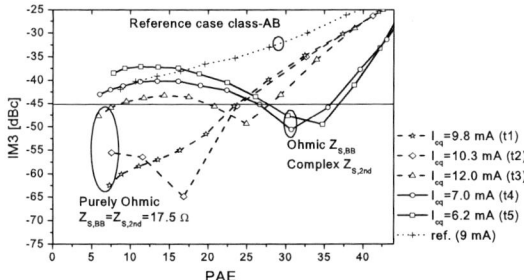

Fig. 13 Improved PA class-AB linearity by manipulation of the input out-of-band terminations and quiescent current using a GaAs Skyworks HBT.

A. Device characterization for linear PA's

So far, mainly analytical considerations formed the basis for the various bipolar linearization concepts. At higher power levels, however, analytical considerations become impractical due to increased influence of the higher-order terms and device parasitics making it difficult to predict the optimum device terminations. To overcome this problem, load-pull measurement techniques are generally used. However, in order to obtain meaningful characterization of active devices, one must be able to provide the optimum reflection coefficients for the various frequency components. In view of this, passive harmonic load-pull setups fail to provide the high (on-wafer) reflection coefficients for the out-of-band terminations. In turn, active load-pull systems are in general not suited for a reliable device linearity optimization when aiming for wide-band modulated signal testing. In order to overcome these limitations, a custom Active Harmonic load-pull setup has been developed at the TU Delft that facilitates the simultaneous control of the baseband, fundamental and 2nd harmonic impedances at both in- and output of the DUT. Special measures have been taken to make the system compatible, in terms of dynamic range and bandwidth, with realistic applications. An early version of this system is described in [28]. The modified version of this system has been used for the experimental results of Fig. 13. In fact, the optimum complex 2nd harmonic loading conditions for the source has been found by a source-pull sweep [27]. Due to the fact that the system is capable to provide optimum loading conditions at the baseband fundamental and 2nd harmonic, it proves to be also a very valuable tool for in the on-wafer linearity evaluation of new device generations. Examples for the linearity optimization of LDMOS devices are found in [29].

1) Device linearity optimization for PA applications

Higher quiescent current levels result in a more linear transconductance of the active device [5], consequently other device non-linearities come into play. In view of this the base-collector or gate-drain capacitance is generally considered as an important limitation for linearity. This can be understood by considering the active device as a system with feedback. If the feedback behaves non-linearly, also the total transfer will be nonlinear. This is illustrated in Fig. 14, which gives the resulting linearity of our QUBiC reference transistor applied

in a typical bench-mark circuit for linearity. The linearity is given as lines of constant 3rd order Output Intercept Point (OIP3) in the bias output plane of the active device under test using a low power two-tone signal as input. Inspection of this OIP3 data allows identification of the dominant sources of distortion. E.g. at low current levels the linearity depends primary on the current level and not on the collector-emitter voltage, so clearly here the exponential dependency of the bipolar device dominates. At low V_{ce} or high current levels when the "lightly" doped collectors is no longer fully depleted, the base-collector charge proves to be the limiting factor for linearity. When quasi-saturation (q.s) or the so-called Kirk effect set in, a strong (nonlinear) increase in the base-collector charge occurs, thus limiting the device linearity. The onset of q.s. is also visible by the drop in the cut-off frequency of the bipolar device. To support the above, in Fig. 14 the bias conditions for both the peak f_T and the point where the f_T is reduced to the half maximum value due to high current effects are indicated. Finally, at (very) high V_{ce} values the useful / linear operating region of the transistor is limited by (weak) avalanche effects.

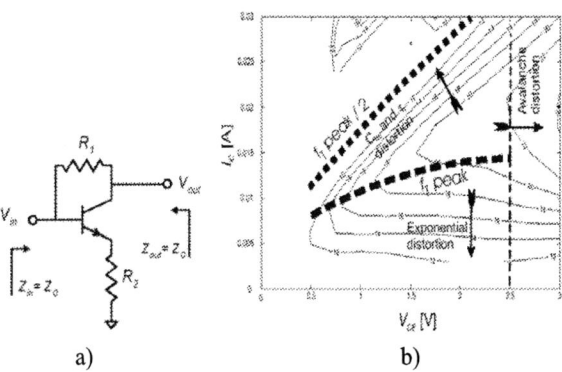

a) b)

Fig. 14 a) AC-schematic of a typical benchmark circuit for two-tone testing of an bipolar device for linearity. b) Constant OIP3-contours in dBm on the I_C - V_{CE} plane for the 70GHz QUBiC4 SiGe ref device (f_0 = 1GHz and Δf = 1MHz).

Although demonstrated here for a relativly small device, similar phenomena occur for large devices. Since the non-linear collector-base charge is an important limitation for linear device operation, many works have been reported in literature to make the base-collector charge behave more linearly, e.g. by optimizing the profile of the lightly doped collector [31][32]. In view of the above, III-V HBT devices are claimed to behave more linearly than silicon devices, due to the different material properties yielding a more linear bias dependence of the base-collector charge [33]. This is one of the main reasons why III-V HBT devices currently dominate the PA market for handset devices.

Feedback by device parasitics also result in the injection of harmonics present at the device output back to its input, yielding secondary mixing phenomena, which give rise to additional intermodulation components. This phenomenon can play a dominant role in degrading as well as improving the device linearity [30] and is quite difficult to analyze. In

7

general, less feedback of harmonics results in a more predictable and therefore more easy to use linear device operation. This can be accomplished by providing short-circuited out-of-band terminations or by lowering the feedback capacitance through unilateralization.

VI. FUTURE TRENDS / CONCLUSIONS

In this work we have given an overview of the most relevant design techniques for improved linearity. Besides the more classical methods like current-mode and feedback, attention has been given to the out-of-band linearization technique, which utilize IM3 cancellation effects. Although perfect IM3 cancellation is difficult to implement, sub-optimum solutions can already provide significant linearity improvements. Currently, the best reported values for the Dynamic Range Figure of Merit [17] are designs utilizing some kind of IM3 cancellation. In spite of this progress in linear circuit design, most concepts are still "narrow" band in nature, since their actual implementation is based on the use of reactive elements to obtain the desired harmonic terminations, noise matching, or the compensation of capacitive loading effects. Consequently, in view of the future demands of multi-band transceivers, or "pseudo" software-defined radios, new wide-band circuit techniques are still needed to successfully realize multi-octave highly linear LNAs and mixers.

When considering linear PA applications, also here significant progress has been made at both the circuit level as well at the device level, yielding excellent linearity-efficiency performance for class-AB or inverse class-AB operation, while being simple and cost effective. The continuous drive to even higher efficiencies, however, will lead to the adaptation of more exotic amplifier concepts like Doherty, EER, polar, etc. , which in their final implementation will most likely utilize digital pre-distortion to meet the linearity specifications without any compromise on the efficiency. For handsets the suitability for multi-band, multi-mode operation will determine the final PA candidate.

ACKNOWLEDGMENT

The authors acknowledges Philips Semiconductors, Philips Research, Infineon, Skyworks, Agilent, Maury, BSW and especially all their colleagues and fellow students from the Delft University of Technology for their support.

REFERENCES

[1] P Wambacq, W Sansen. *Distortion Analysis of Analog Integrated Circuits.* Dordrecht: Kluwer Academic Publishers, 1998.

[2] P Antognetti, G Massobrio. *Semiconductor device modeling with SPICE.* New York: McGraw-Hill, 1987.

[3] Thomas Lee, The design of CMOS Radio-Frequency Integrated Circuits, Cambridge University Press, ISBN 0-521-63922-0, 1998

[4] C. Toumazou, *et. al., Analogue IC design: the current-mode approach.* IEE Circuits and Systems series 2. London: Perigrinus, 1990.

[5] L.C.N. de Vreede and M.P. van der Heijden, *"Chapter 9.4. of the: Silicon Heterostructure Handbook,"* Edited by J.D. Cressler, 2005.

[6] P. Deixler, *et. al,* "QUBiC4G: a f_T/f_{max} = 70/100GHz 0.25μm low power SiGe-BiCMOS production technology with high quality passives for 12.5 Gb/s optical networking and emerging wireless applications up to 20GHz," *Proc. BCTM2002*, pp. 201-204, Oct. 2002.

[7] EH Nordholt. *Design of High-Performance Negative Feedback Amplifiers. Elsevier,* 1983.

[8] C.J.M. Verhoeven, *et al. Structured Electronic Design,* Dordrecht: Kluwer Academic Publishers, 2003.

[9] K. L. Fong, R.G. Meyer, "High-Frequency Nonlinearity Analysis of Common-Emitter and Differential-Pair Transconductance stages," *IEEE J. Solid-State Circ.* Vol. 33, No. 4, pp. 548-555, April 1998.

[10] S.P. Voinigescu, *et al.,* "A scalable high-frequency noise model for bipolar transistors with application to optimal transistor sizing for low-noise amplifier design," *IEEE JSSC*, pp.1430-1439, Sept. 1997.

[11] K. L. Fong, *et al.,*" High-Frequency Analysis of Linearity Improvement Technique of Common-Emitter Transconductance Stage Using a Low-Frequency-Trap Network," *IEEE J. Solid-State Circ.* Vol. 35, No. 8, pp. 1249-1252, Aug. 2000.

[12] B. Gilbert, "The Multi-tanh Principle: A Tutorial Overview," *IEEE J. Solid-State Circ.* Vol. 33, No. 1, pp. 2-17, Jan. 1998.

[13] D. Webster, *et al.,* "Control of Circuit Distortion by the Derivative Superposition Method," *IEEE Microwave And Guided Wave Lett.,* Vol. 6, No. 3, pp. 123-125, March 1996.

[14] V. Aparin, L.E. Larson, "Modified derivative superposition method for linearizing FET low-noise amplifiers," IEEE Trans. Microwave Theory and Techn., Vol. 53, No. 2, pp. 571–581, Feb. 2005.

[15] V. Aparin, C Persico, "Effect of out-of-band terminations on intermodulation distortion in common-emitter circuits," *in IEEE MTT-S Digest*, vol.3, pp. 977-980, June 1999.

[16] M.P. van der Heijden, *et al.,* "A Novel Frequency-Independent Third-Order Intermodulation Distortion cancellation Technique for BJT Amplifiers," *IEEE J. Solid-State Circ.,* pp. 1176-1183, Sept. 2002.

[17] V. Aparin, *et al.* "Highly Linear SiGe BiCMOS LNA and Mixer for Cellular CDMA/AMPS Applications," *Proc. RFIC2002*, June 2002.

[18] M. van der Heijden, *et al.,* "A 2 GHz high-gain differential InGaP HBT driver amplifier matched for high IP3," *in IEEE MTT-S Digest*, vol. 1, pp. 235-238, June 2003.

[19] M. van der Heijden, *et. al.,* "A High Performance Unilateral 900 MHz LNA with Simultaneous Noise, Impedance, and IP3 Match," *Proc. BCTM2003*, pp. 45-48, Sept. 2003.

[20] L. Sheng, *et al.,* "An Si-SiGe BiCMOS Direct-Conversion Mixer with Second-Order and Third-Order Nonlinearity Cancellation for WCDMA Applications," *IEEE Trans. MTT.,* pp. 2211-2220, Nov. 2003.

[21] A. A. Abidi, "Direct conversion radio transceivers for digital communications, " *IEEE JSSC.,* pp. 1399-1410, Dec. 1995.

[22] B. Razavi, "Design considerations for direct-conversion receivers," IEEE Trans. Circuits Syst. II, vol. 44, pp. 428-435, Dec. 1997.

[23] T. Furukawa, J. Arisawa, "Common-mode rejection of the differential operational amplifier with shunt common-mode feedback," *IEEE J. Solid-State Circ.,* Vol. 10, No. 6, pp. 539-540, Dec. 1975.

[24] T.A. Phan, *et al.,* "Low noise and high gain CMOS down conversion mixer," ICCCAS2004., pp.1191–1194, June 2004

[25] M. Brandolini, *et al.,* "A +78 dBm IIP2 CMOS direct downconversion mixer for fully integrated UMTS receivers" *IEEE JSSC,* March 2006.

[26] M.P. van der Heijden, *et al.,* "On the Optimum Biasing and Input Out-of-band Terminations of Linear and Power Efficient Class-AB Bipolar RF Amplifiers," *in proc. BCTM2004*, Sept. 2004.

[27] M. Spirito, *et al.,* "Experimental procedure to optimize out-of-band terminations for highly linear and power efficient bipolar class-AB RF amplifiers," *Proc. BCTM 2005*, pp. 112-115, Oct. 2005.

[28] M. Spirito, *et al.,* "A novel active harmonic load-pull setup for on-wafer device linearity characterization," IMS 2004, p 1217, June 2004.

[29] *L.C.N. de Vreede, Topical Workshop on* Power Amplifiers for Wireless Communications, San Diego, January 2006

[30] D.M.H. Hartskeerl, *et al.,* "On the optimum 2nd harmonic source and load impedances for the efficiency-linearity trade-off of RF LDMOS power amplifiers," *Digest of RFIC 2005,* pp. 447-450. June 2005.

[31] L.C.N. de Vreede, *et. al,* "Bipolar transistor epilayer design using the MAIDS mixed-level simulator," *IEEE JSSC,* p. 1331, Sept. 1999.

[32] W.D. van Noort, *et. al.,* "Reduction of UHF power transistor distortion with a nonuniform collector doping profile," *IEEE JSSC,* Sept. 2001.

[33] L.C.N de Vreede, *et al.,* "The effect of non-saturated electron drift velocity on bipolar device linearity," *Proc. BCTM2002*, Sept. 2002.

2006 Bipolar/BoCMOS Circuits and Technology Meeting Proceedings

Linear Efficiency Load-pull Measurements on GaAs HBT versus Si BJT for EDGE and OFDM Modulation

L.C.M van den Oever, R.M. Heeres, H.F.F. Jos

Philips Semiconductors, Innovation Center RF, Gerstweg 2, 6534 AE Nijmegen, The Netherlands

Abstract — **We measured the large-signal performance of GaAs HBT and Si BJT for 1-tone, EDGE, and OFDM excitation. We optimized loading and biasing for linear efficiency. For EDGE GaAs HBT shows a slightly higher linear efficiency than Si BJT. For OFDM both technologies reach comparable performance. This makes silicon a good alternative for linear power amplifiers in mobile communication systems.**

Index terms — **GaAs HBT, silicon bipolar transistor, linearity, efficiency, RF power transistors.**

I. INTRODUCTION

Today's market for power amplifiers (PAs) for mobile communication systems is dominated by GaAs HBT, because of its high power capability, gain, efficiency, linearity, and breakdown voltage. But GaAs is more expensive and more limited with respect to integration than Si. Si, on the other hand, has lower breakdown. Recently it was shown that with protection circuitry Si BJT is capable to fulfill the breakdown and ruggedness requirements [1]. This makes Si an interesting option to replace GaAs.

For Si good RF power performance has been published with high peak efficiencies for saturated PAs [2]. Applications for higher data rates such as EDGE, WCDMA, and OFDM, use linear PAs and have more severe requirements on linearity. Knowing that GaAs and Si have different material properties and device structures [3], it is important to compare GaAs and Si with respect to the efficiency-linearity trade-off, and particularly in terms of the so-called linear efficiency at the point where linearity meets the spec limit. Therefore in this paper we present load-pull measurement results of the linear efficiency for GaAs HBT versus Si BJT for EDGE and OFDM modulation.

II. LINEAR EFFICIENCY

Consider the power sweep in Figure 1. We define linear efficiency as the (collector) efficiency where linearity equals the spec limit, e.g. IMD, ACP, or EVM. Linear output power is defined similarly, and power back-off (PBO) is the difference between the 1dB compression point (P_{1dB}) and the linear output power.

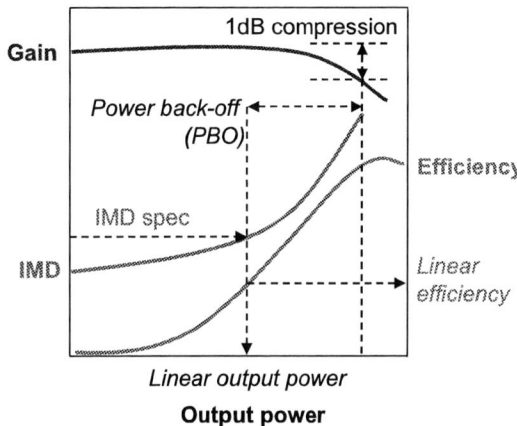

Figure 1: Definition of linear efficiency.

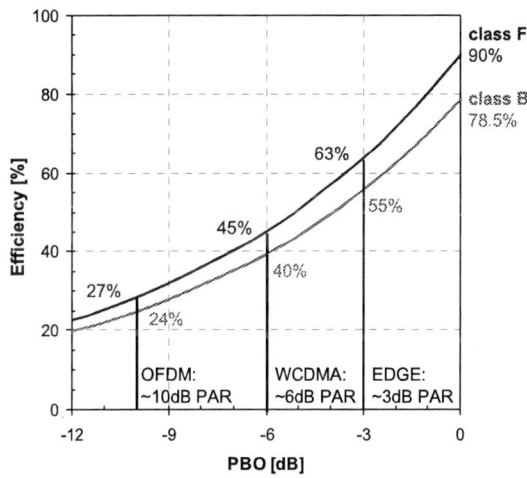

Figure 2: Theoretical linear efficiency versus PBO and peak efficiency.

In general maximum linear efficiency requires minimum power back-off and maximum saturated efficiency. For modulated signals the required PBO is strongly related, though not identical, to the peak-to-average ratio (PAR). The PBO together with the peak efficiency set an upper limit to the linear efficiency that can be reached. This is illustrated in Figure 2 for

1-4244-0458-4/06/$20.00 ©2006 IEEE

EDGE, WCDMA, and OFDM with a PAR of 3, 6, and 10 dB, respectively, and for class-B and F with a peak efficiency of 78.5% and 90%, respectively.

III. EXPERIMENTAL SET-UP

For this experiment we used GaAs HBTs and Si BJTs from commercial processes with comparable small-signal performance, comparable emitter area, and integrated bias circuitry. The GaAs HBTs were designed in TriQuint's HBT3 process [4]. They had substrate vias for the emitter and bond wires for the base and collector, and were mounted on laminate substrates. The Si BJTs were designed in QUBIC4+ BiCMOS and flip-chipped on a PASSI™ (passive integration on Si) die, which was mounted on laminate [5]. The laminates contain pre- and post-matching.

We used a Maury fundamental mechanical load-pull system to perform power sweeps for excitation with 1-tone and with complex modulation and measured output power (Pout), transducer gain (Gt), collector efficiency (CollEff), and linearity. For EDGE we used the adjacent channel power ratio (ACPR) in a 30kHz band at 200kHz offset. For OFDM (IEEE802.16e derivative) we used the adjacent channel leakage ratio (ACLR) in a bandwidth of 8.447MHz at 9 and 13.5MHz offset [6]. Special attention was paid to decoupling and averaging to get symmetric and reproducible results for ACPR/ACLR. We measured the devices at different load states and at different quiescent currents, keeping the source impedance unchanged. Both the EDGE and the OFDM measurements were performed at 1710 MHz so that we could use the same samples.

IV. OFDM RESULTS

The GaAs HBT was biased at Icq=100mA, Vce=3.5V. Figure 3 shows the load-pull measurement results for the ACLR for OFDM. For meeting the linearity spec, ACLR at 9MHz (upper traces) appeared to be slightly more critical than ACLR at 13.5MHz (lower traces). For each load state (each trace) we extracted the linear efficiency, linear output power, and PBO. Figure 4 shows the load-pull power sweeps for 1-tone and OFDM excitation. We found a maximum linear collector-efficiency of 35% at 24.8dBm output power and 6.3dB PBO.

In the same way we measured Si BJT at Icq=100mA, Vce=3.5V. The results are shown in Figure 5. We found a maximum linear efficiency of 34% at 24.5dBm output power and 5.9dB PBO, which is comparable to the result for GaAs HBT.

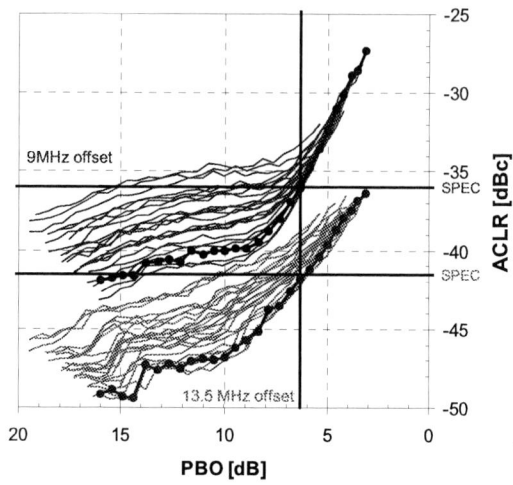

Figure 3: ACLR load-pull results for OFDM for GaAs HBT at Icq=100mA. The upper/lower traces are for different frequency offsets and the specs are indicated by the lines. The different traces are for different load states. The trace with markers corresponds to the load with the highest linear efficiency.

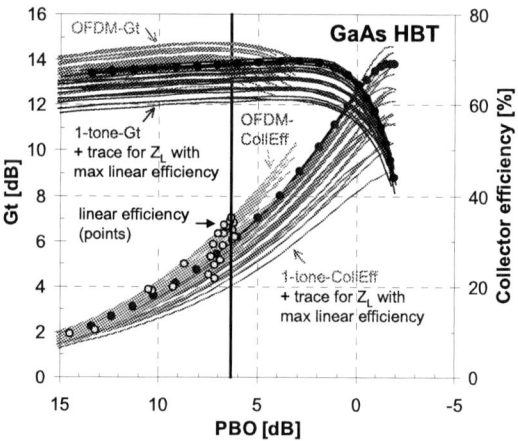

Figure 4: Load-pull results for 1-tone and OFDM excitation for GaAs HBT at Icq=100mA. The green points give the linear efficiency where ACLR equals spec. The trace with markers corresponds to the load with the highest linear efficiency (Z_L=3.4-j1.1Ω).

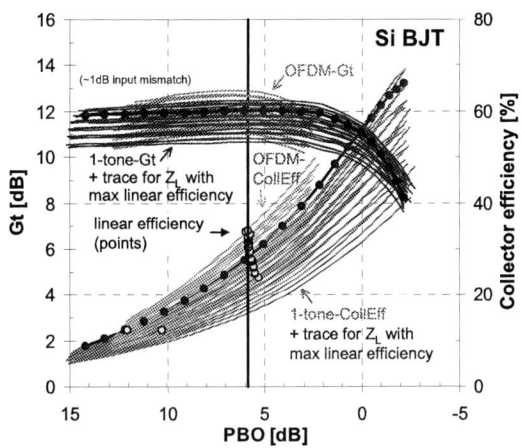

Figure 5: Load-pull results for 1-tone and OFDM excitation for Si BJT at Icq=100mA. The green points give the linear efficiency where ACLR equals spec. The trace with markers corresponds to the load with the highest linear efficiency (Z_L=4.1-j1.7Ω).

We also measured both devices at Icq=200 and 300mA. The results for the linear efficiency, PBO, and linear output power are summarized in Figure 6 and in Table 1. Note that the points in Figure 6 for Icq=100mA correspond to the points in Figure 4 and Figure 5.

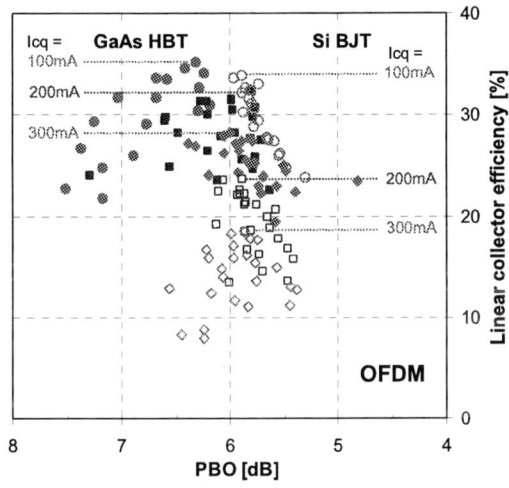

Figure 6: Linear efficiency vs PBO for OFDM for GaAs HBT (filled markers) and Si BJT (open markers) at different bias levels (circles for 100mA, squares for 200mA, diamonds for 300mA), Vce=3.5V. The points are for different load states.

For both devices the best linear efficiency was achieved for Icq=100mA. Higher Icq yielded a higher linear output power at a slightly lower linear efficiency for GaAs HBT. For Si BJT higher Icq resulted in lower linear efficiency and lower linear output power.

Table 1: Summary of linear efficiency load-pull results for OFDM (IEEE802.16e) for ACLR(9MHz)=-36dBc and ACLR(13.5MHz)=-41.5dBc at 1710MHz, Vce=3.5V.

OFDM	GaAs HBT			Si BJT		
Icq [mA]	100	200	300	100	200	300
Linear collector efficiency [%]	35	32	28	34	24	18
Power back-off (PBO) [dB]	6.3	5.8	6.0	5.9	5.9	5.9
Linear output power [dBm]	24.8	25.4	26.2	24.5	23.6	23.3

V. EDGE RESULTS

We used the same devices and followed the same procedure to measure the linear efficiency for EDGE[1] at Icq=180mA, Vce=3.5V, at 1710MHz.

The load-pull measurement results are summarized in Figure 7 and in Table 2. For GaAs HBT we found 45% linear efficiency at 28.0dBm linear output power and 2.8dB PBO (Z_L=3.5+j0 Ω). For Si BJT we found 41% linear efficiency at 28.3dBm linear output power and 2.1dB PBO (Z_L=5.4-j1.4 Ω). In spite of the larger PBO we measured a 5% higher linear efficiency for GaAs HBT. We attribute this difference to a difference in the steepness of the efficiency vs output power characteristics (1-tone). For GaAs HBT we measured 5% higher saturated peak efficiency than for Si BJT. And the difference between the output power at peak efficiency and P_{1dB} was 1.3dB less for GaAs HBT compared to Si BJT. This saturation behavior is more relevant in case of low PBO as for EDGE, because the effect is less flattened by averaging over the dynamic range for output power.

EVM was 2.7% for GaAs HBT and 3.7% for Si BJT, and Gt was 14.4 and 13.9dB, respectively.

---

[1] In these measurements the load impedance was optimized for EDGE. In commercial PAs EDGE is combined with saturated GSM, and the load is usually optimized for saturated GSM and not for EDGE.

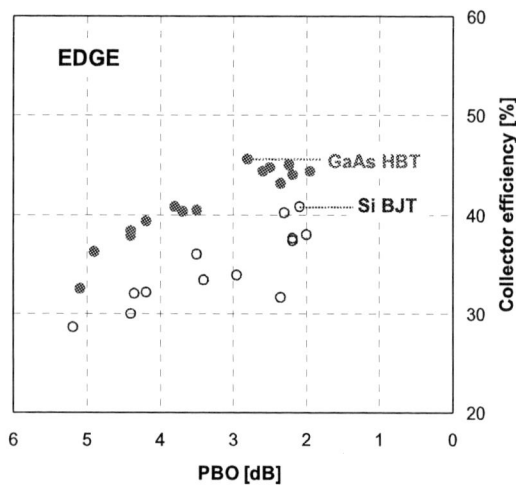

Figure 7: Linear efficiency vs PBO for EDGE for GaAs HBT (filled markers) and Si BJT (open markers) at Icq=180mA, Vce=3.5V. The points are for different load states.

Table 2: Summary of linear efficiency load-pull results for EDGE for ACPR(200kHz)=-57dBc at 1710MHz, Vce=3.5V.

EDGE	GaAs HBT	Si BJT
Icq [mA]	180	180
Linear collector efficiency [%]	46	41
Power back-off (PBO) [dB]	2.8	2.1
Linear output power [dBm]	28.0	28.3

VI. CONCLUSIONS

The linear efficiencies of GaAs HBT and Si BJT were compared using load-pull measurements. We found that for low power-back-off as for EDGE, GaAs HBT shows slightly better performance than Si BJT. For high power-back-off as for OFDM, GaAs and Si can reach comparable performance. From this we conclude that Si BJT can be a good alternative for GaAs HBT in linear power amplifiers for mobile communication systems.

REFERENCES

[1] A. van Bezooijen, F. van Straten, R. Mahmoudi, and A.H.M. van Roermund, "Avalanche breakdown protection by adaptive output power control", *2006 IEEE Radio and Wireless Symp. Dig.*, pp. 519-522, January 2006.

[2] D.M.H. Hartskeerl, H.G.A. Huizing, P. Deixler, W. van Noort, and P.H.C. Magnée, "High performance SiGeC HBT integrated into a 0.25μm BiCMOS technology featuring record 88% power-added eficiency", *2004 IEEE MTT-S Int. Microwave Symp. Dig.*, pp. 979-982, June 2004.

[3] L.C.N. de Vreede, H.C. de Graaff, B. Rejeai, "The effect of non-saturated electron drift velocity on bipolar device linearity", *Proc. BCTM 2002*, pp. 88-91, 2002.

[4] O. Berger, "GaAs HBT for power applications", *Proc. BCTM 2004*, pp. 52-55, 2004.

[5] A.J.M. de Graauw, A. van Bezooijen, C. Chanlo, A. den Dekker, J. Dijkhuis, S. Pramm, and H.K.J. ten Dolle, "Miniaturized quad-band front-end module for GSM using Si BiCMOS and passive integration technologies", *Submitted to BCTM 2006.*

[6] O. Kuijken *et al.*, "Optimization of PA linearity and efficiency through loadpull measurements and simulations using modulated signals", *IEEE MTT Wireless Week (PA Topical Workshop)*, January 2006.

Analysis of Temperature Modulation on a SiGe Power Amplifier Non Linearity

L. Leyssenne[1], J.M. Pham[1], P. Jarry[1], E. Kerhervé[1], and Daniel Saias[2]

[1] IXL Laboratory, CNRS UMR 5818, Bordeaux, 33000 France

[2] ST Microelectronics, Crolles, 38926 France

Abstract — **The following discussion presents the design and measurements of a power amplifier for a WCDMA handset application on silicon with a STM BiCMOS 0.25µm process. An analysis of the impact of temperature on linearity especially on adjacent channel power regrowth is proposed**

Index Terms — **BiCMOS process technology, SiGe simulation, RF power amplifier, physical thermal dependency.**

I. INTRODUCTION

This work presents a W-CDMA differential double stage PA (1920 – 1980 MHz), including the functionality of power stage bypass, at low power operation, for battery saving. It was implemented on silicon with the SiGe BiCMOS7RF STM technology. Though SiGe HBTs generally feature a smaller $fT*BV_{CE0}$ product than their GaAs counterparts [1] and present higher dielectric losses via the substrate, they make possible the integration of the power amplifier with control or digital blocks. Designing a Power Amplifier (PA) for W-CDMA implies to fulfill severe requirements in terms of linearity, under the constraint of efficiency conservation. For a single carrier RF signal, the adjacent channel must be rejected 35dBc under the main channel. The integrity of an RF signal with a 3.84MHz power dynamic range can be damaged by non-linearities in electrical parameters, but also by low frequency memory effects. SiGe HBTs present a higher f_T (~ 35GHz for a high voltage npn) than MOS or BJT transistors; therefore the maximum power gain level per stage is maximized and the contribution in the overall non linearity of preamplifier stages is reduced. Meanwhile, at high current densities, Kirk effect and the quasi-saturation region appearing in the characteristic $I_{CC}(V_{CE})$ compel the use of HBT below the optimum f_T bias. Moreover, a collector current collapse can be noted at high T° in low base-load conditions, especially for multi-finger-HBT. This remark stresses the impact of temperature on PA linearity. In order to understand the issue of thermal stability, it is more appropriate to begin with the DC thermal modeling of the DUT

II. HBT LOW FREQUENCY THERMAL PROPERTIES

Some works have demonstrated the importance of base and emitter degeneration on HBT's collector current thermal stability [2]. Base ballast is made profitable against thermal runaway by the current gain negative dependency with temperature [1], [3]. Let us apply the Kirchhoff law to our power stage fig.1. Though it is differential, the power stage is represented as a single-ended one, all transistors being in parallel from a thermal point of view.

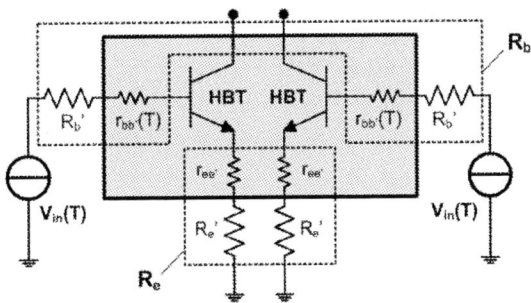

Fig. 1. DC degenerated HBT schematic

The obtained expression can be differentiated, thus giving:

$$\frac{1}{I_c} \cdot \frac{\partial I_{CC}}{\partial T} \approx \frac{(\alpha_1 + \alpha_1 + \alpha_3 \cdot g_m \cdot R_b)}{1 + g_m \cdot \left(\frac{R_b}{\beta_F} + R_e\right)} \tag{1}$$

With:

$$\alpha_1 = \frac{1}{U_t} \cdot \frac{\partial V_{in}}{\partial T} \tag{2}$$

$$\alpha_2 = \frac{\partial(\ln(I_s))}{\partial T} - \frac{1}{U_t}\frac{\partial U_t}{\partial T} \cdot \ln\left(\frac{I_c}{I_s}\right) \tag{3}$$

$$\approx \frac{\left(E_{g0} - \frac{q \cdot V_{be}}{m}\right)}{kT^2} + \frac{\alpha \cdot \beta}{k} \cdot \frac{1}{(T+\beta)^2} + \frac{4-n}{T}$$

$$\alpha_3 = \frac{1}{\beta_F^2} \cdot \frac{\partial \beta_F}{\partial T} < 0 \tag{4}$$

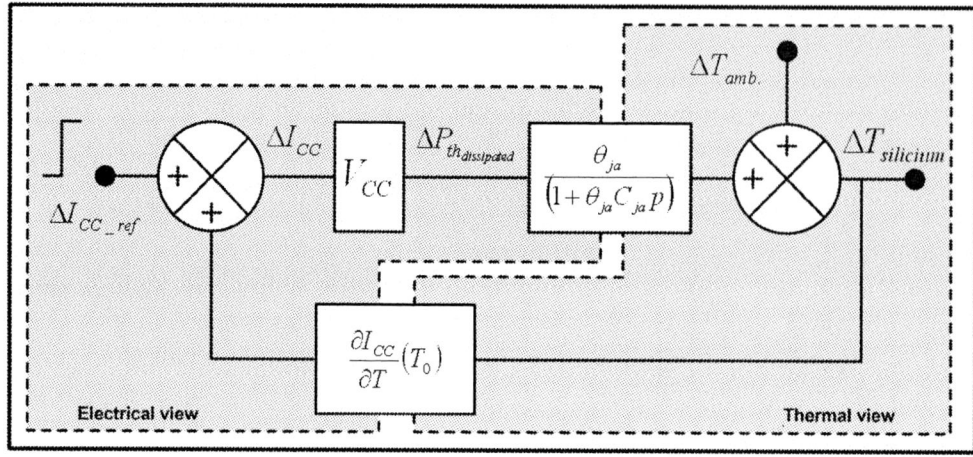

Fig. 2. Electro-thermal loop governing HBT junction temperature

Thermal stability of collector current can be understood via the electro-thermal loop fig. 2. Assumptions were made of a 1st order thermal Laplace response for the die and the package, and of a linear collector current dependency with temperature. The stability condition is thus given by:

$$\forall T \in \left[-40°C, 125°C\right],$$
$$\left(1 - \frac{\partial I_{CC}}{\partial T}(T) \cdot V_{CC} \cdot \theta_{ja}\right) > 0 \qquad (5)$$

Considering the worse case from the point of view of thermal runaway, when the negative term α_3 is negligible compared to the other terms, the condition (5) becomes:

$$\frac{R_b}{\beta_F} + R_e > U_t \cdot \left[\frac{1}{I_{CC\_th}} - \frac{1}{I_{CC}}\right] \qquad (6)$$

With:

$$I_{CC\_th} = \left(V_{CC} \cdot \theta_{ja} \cdot \left(\alpha_1(T) + \alpha_2(T)\right)\right)^{-1} \qquad (7)$$

For a typical SiGe technology, with a 3.3V V_{CC}, and a 0.25A collector bias current, we obtain at 125°C the condition to prevent thermal runaway:

$$\frac{R_b}{\beta_F} + R_e > 0.038 \cdot \left[0.15^{-1} - 0.25^{-1}\right] \approx 0.1\Omega$$

For an overall $1250\mu m^2$ emitter area (fig. 3), intrinsic base and emitter resistors are below 0.01Ω, which justifies the use of distributed ballast resistors.

III. SIGE POWER AMPLIFIER MEASUREMENTS

The "power mode" features a 20.4dB transducer gain working in the linear class A regime, a 23.3dBm

output CP_1, and a 26.6dBm maximum output power, whereas the "bypass mode" features a 0.3 dB gain, and a 6.6dBm output CP_1, as depicted on the continuous-wave measurements (fig.4).

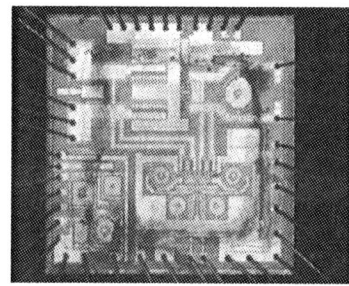

Fig. 3. 2.2mm X 2.3mm Power Amplifier photograph.

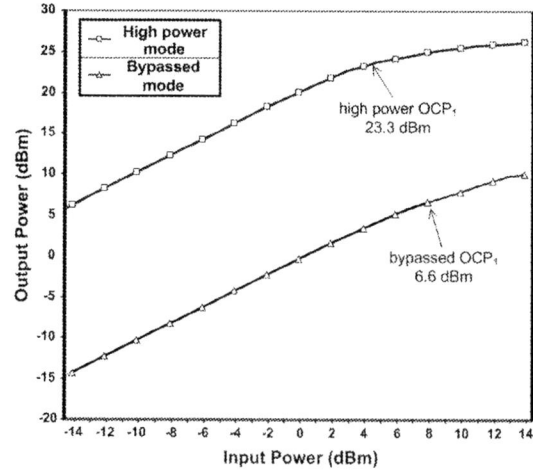

Fig. 4. Output power vs. Input power, both in high and low power modes.

This CW power performance is close to the 24dBm specified for class1 W-CDMA. The measurements obtained with a modulated signal are depicted on fig.5. Degradation in adjacent channel power rejection is monitored on fig.6. With a 0.5 roll-off factor, the output channel power CP_1 is this time close to 18dBm, whereas we would have expected (8):

$$OCP_{1\_channel} \approx OCP_{1\_CW} - PAR_{QPSK}\left(\alpha = 0.5\right)$$
$$\approx 23.3 - 3.3 = 20dBm > 18dBm \qquad (8)$$

In a CW test, power gain compression is only determined by 3rd degree non linearities, whereas in dual-tone or modulated tests, the channel power compression is also sensitive to 2nd degree terms.

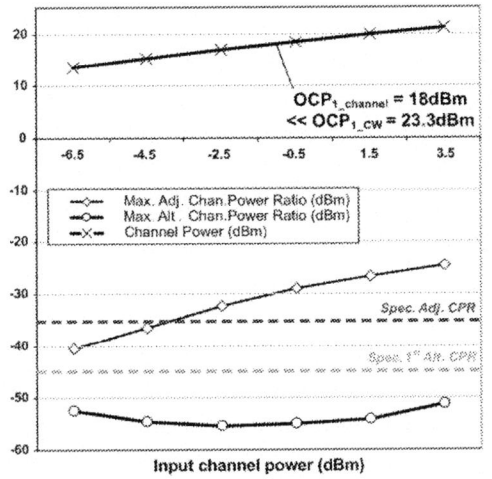

Fig. 5. Measured output power, ACLR vs. Input power for a 0.5 roll-off factor

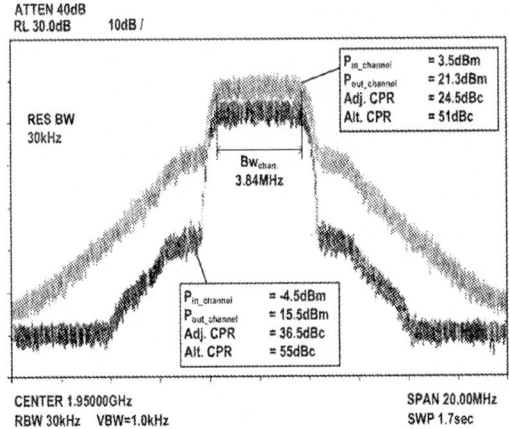

Fig. 6.   Measured channel power and spectral regrowth.

The differential topology cancels 2nd degree distortion in theory. In practice, however, positive and negative output collector accesses are not loaded the same on the broad range of candidate frequencies for intermodulation, thus generating a common-mode signal that degrades the PA sensitivity to 2nd degree non linearities. Nevertheless, this cannot completely explain this strong adjacent channel regrowth. The following development proposes an explanation for this excessive gain compression.

IV. EFFECT OF TEMPERATURE MODULATION ON PA LINEARITY

Continuous-wave power variations at high frequency have little impact on temperature. Nevertheless, QPSK modulation presents low frequency envelope power variations that can no longer be neglected from a thermal point of view [4]. The 3.84MHz W-CDMA channel bandwidth is typically within the DUT thermal low-pass region. When considering such a modulated signal, the thermal loop is more complex than in the DC case (fig.7). The instantaneous power that is dissipated in the HBT (base current neglected) is given by the equation (9):

$$P_{HBT\_inst} = \underbrace{V_{CC} \cdot I_{CC\,0}}_{DC} - \underbrace{\hat{I}_{CC}^{\,2} \cdot R_{load}}_{BB\_\&\_2f_C}$$
$$+ \underbrace{\left(V_{CC} - I_{CC\,0} \cdot R_{load}\right) \cdot \hat{I}_{CC}}_{f_C} \qquad (9)$$

The output collector current spectrum self-mixes to give a base-band and a $2f_C$ dissipated power spectra, both with a $2.BW_{channel}$ bandwidth. Whereas the latter is filtered by the thermal cut-off frequency, the former is converted into a base-band temperature spectrum that modulates the power stage parameters (f_T, g_m...). The input signal is therefore mixed with this memory effect spectrum [5] via the HBT transconductance gain, thus generating spectral regrowth in the channel vicinity. The differential topology cannot make the circuit linearity insensitive to this 2nd degree phenomenon since both semi-circuits are "in parallel" from a thermal point of view. The thermal time constant is determined by the active devices on silicon but also by the packaging, the assembly technique and offers little margin to decrease this memory effect. A solution to this problem is suggested by (9) and depicted fig.8. It consists in a fast voltage regulator providing a variation of V_{CC}, properly synchronized, likely to compensate for the temperature modulation, without degrading the differential power gain. This architecture has already been used to manage the PA efficiency as a function of the input envelope power [6].

15

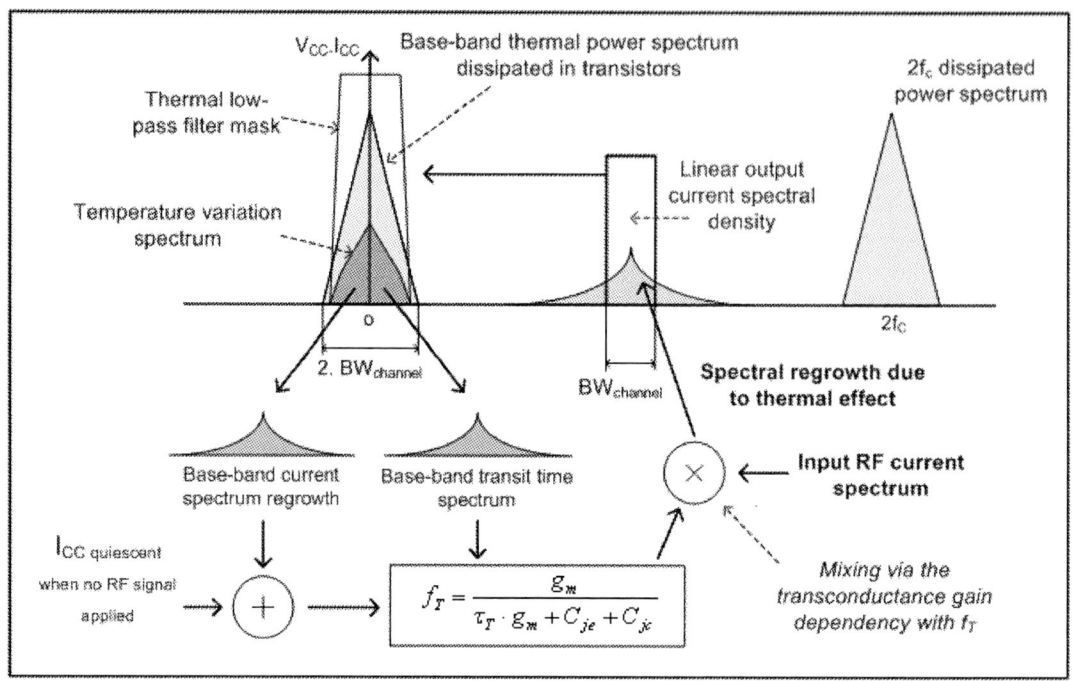

Fig. 7.   Dynamic thermal/electrical loop (RF + Base-Band spectra)

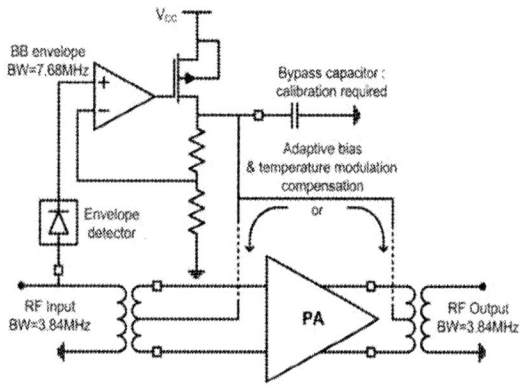

Fig 8: Architecture for temperature modulation compensation

VII. CONCLUSION

We have presented an analysis of the temperature modulation effect on channel compression and adjacent channel power regrowth. It consists in a second degree thermal phenomenon related to base-band dissipated power variations that cannot be removed by a differential topology and that is all the more sensitive as the DUT thermal time constant is short. We presented a temperature compensation architecture that is compliant with the need for the average efficiency optimization.

ACKNOWLEDGEMENT

This work was carried out in collaboration with ST Microelectronics and supported by the French Ministry for Research, in the frame of the *Asturies* project (financial support ref.: N°04510)

REFERENCES

[1] K. Nellis, K. Choi, N.S. Cheng, P. Zampardi, M.F. Chang, "A Comparison of BJT, SiGe HBT, and GaAs HBT Technologies for Linear Handset PA Applications", *IEEE Topical Workshop on Power Amplifiers for Wireless Communications*, September 2002.

[2] P. D. Tseng &Al, "A 3-V Monolithic SiGe HBT Power Amplifier for Dual-Mode (CDMA/AMPS) Cellular Handset Applications", *IEEE Journal of Solid-State Circuits*, Vol. 35, N°9, Sept. 2000.

[3] J. Paaschens, W. Kloosterman, Ramses van der Toorn, "Mextram (level 504) the Philips Model for Bipolar Transistors", *FSA Modeling Workshop*, September 2002.

[4] N. Nenadovic, "Extraction and Modeling of Self-Heating and Mutual Thermal Coupling Impedance of Bipolar Transistor", *IEEE Journal of Solid-State Circuits*, Vol. 39, N°10, October 2004.

[5] J.H.K. Vuolevi, T. Rahkonen, "Measurement Technique for Characterizing Memory Effects in RF Power Amplifiers" *IEEE Transaction on Microwave and Techniques*, Vol. 49, N°8, August 2001.

[6] B. Sahu,"Integrated Dynamically Adaptive Supplies for Linear RF Power Amplifiers in Portable Applications", thesis for the degree of Doctor, November 2004.

MEMS-Based Reconfigurable Multi-band BiCMOS Power Amplifier

A.J.M. de Graauw, P.G. Steeneken*, C. Chanlo, J. Dijkhuis, S. Pramm, A. van Bezooijen, H.K.J. ten Dolle, F. van Straten, P. Lok.

Philips Semiconductors, ICRF, Gerstweg 2, 6534AE Nijmegen, The Netherlands
*Philips Research Laboratories, Prof. Holstlaan 4, 5656AA Eindhoven, The Netherlands

Abstract — **This paper presents a small dual-band 0.9GHz /1.8GHz inverse class F power amplifier with load-switch functionality using a single BiCMOS amplifier line-up with a MEMS based reconfigurable matching network. The realized prototype measures 40mm², offers 31dBm with 40% efficiency at 0.9GHz and 30dBm with 34% at 1.8GHz. The load-switch provides up to 10% efficiency improvement at 0.9GHz for reduced power levels.**

Index Terms — **RF power amplifier, Reconfigurable Matching Networks, BiCMOS, MEMS.**

I. INTRODUCTION

Future mobile handsets are expected to offer increased functionality by providing access to multiple cellular-, connectivity- and broadcast-wireless networks. The diversity of used frequency bands and modes will result in a dramatic increase in the complexity of the RF functionality of the phone.

A critical function in the RF section is the power amplifier (PA). The PA should be efficient as it directly affects the talk-time of the phone. In addition it has to be sufficiently linear to meet the spectral requirements of the modulation standards. Finally, it has to be small in order to fit in the continuously shrinking size of the RF part of the phone.

The conventional concept for a multi-band/mode PA is the use of a dedicated PA for each service which is optimized with respect to the band and mode under consideration. A disadvantage of this approach is that operation in n-frequency bands requires n-amplifier line-ups resulting in a cost and size of the PA function that increases linear with the number of bands. The concept presented here uses a single PA line-up in combination with a reconfigurable output matching network to cover multiple frequency bands/modes. This approach results in size and cost advantages compared to the conventional approach due to the re-use of circuits as shown in Fig. 1.

The use of PA's with reconfigurable RF circuits to meet the multi-band/mode challenges is well known in the wireless industry and is currently an important research and development topic. Recent work has demonstrated the use of reconfigurable matching networks in PA's for frequency-band switching [1,2] as well as load-line adaptation [3].

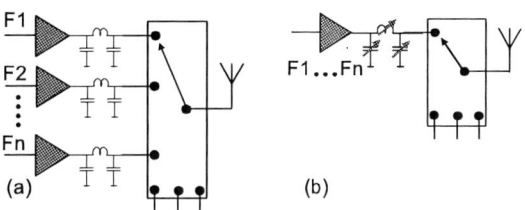

Fig. 1. (a) Conventional multi-band PA, (b) Single Line-Up reconfigurable multi-band PA .

The RF performance is however limited in those concepts to optimization of the PA load impedance at the fundamental frequency. High efficiency PA operation classes like (inverse) class D,E and F require in addition optimum loading at higher harmonics and proper suppression of these harmonics at the output to meet the spectral requirements.

This work presents a concept that offers optimization of efficiency and linearity for each frequency band by reconfiguration of the output matching network using switchable RF-MEMS capacitors. The new output matching network topology offers highly efficient class F^{-1} amplifier operation in two frequency bands by optimizing both the fundamental and harmonic impedance levels in combination with strong suppression of those harmonics at the output of the amplifier. In addition, the network also offers the possibility to optimize the fundamental load impedance as function of the modulation type and power level. The concept was demonstrated by a 40mm² prototype which offers 31dBm with 40% efficiency at 0.9GHz and 30dBm with 34% efficiency at 1.8GHz. A build-in load-switch provides up to 10% efficiency improvement at 0.9GHz for reduced power levels.

II. RECONFIGURABLE CLASS F^{-1} AMPLIFIER

The optimum load impedance for the output stage for class F^{-1} operation is given by [4]:

$$Zl = \frac{(\pi.Vc)^2}{(8.Psat)} \quad for \quad F = Fo \quad (1)$$

$$Zl \longrightarrow \infty \quad for\ all\ even\ harmonics$$

$$Zl \longrightarrow 0 \quad for\ all\ odd\ harmonics$$

where Zl is the collector load impedance, Vc the DC collector voltage and Psat the maximum RF output power. The conditions stated in (1) result in a square wave collector current waveform and a half sinusoidal collector voltage waveform without overlap in the time domain corresponding to no power dissipation in the transistor. The theoretical peak efficiency of class F^{-1} operation of 100% is considerable better than the theoretical value of 78.5% for class B operation but requires a more complex network. An efficiency of 90% can however be achieved by providing the optimum impedance only at the second and third harmonics, fig. 2 shows the wave-shapes for that case.

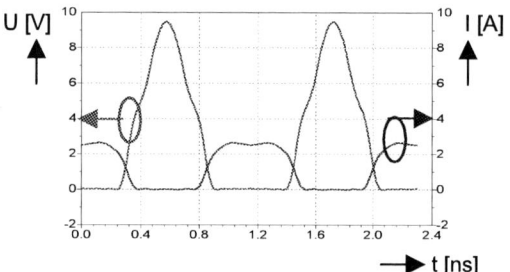

Fig.2.   Simulated Class F^{-1} collector voltage and current wave-shape with optimum H2, H3 termination (η_c=90%).

The highly reflective terminations at the second and third harmonic frequency bands result in a low level of the transfer of harmonic power to the antenna terminal and is desired for a clean output spectrum.
The collector efficiency of a class F^{-1} stage is a function of the output power level and is given by:

$$\eta_C = \sqrt{\frac{Pload}{Psat}} \qquad (2)$$

where Pload is the actual power and Psat the maximum available power, both with Zl given by (1).

(2) shows that the efficiency drops with the square root of the actual load power. Figure 3 shows the relation between efficiency, power and load resistance for a typical 0.9GHz GSM PA which is designed to offer a maximum power of 35.5dBm at Vcc=3.0V using a nominal load-line of 3Ohms. The efficiency at reduced power levels can be restored to the peak value by adjusting the load impedance such that it doubles for every 3dB reduction in power.

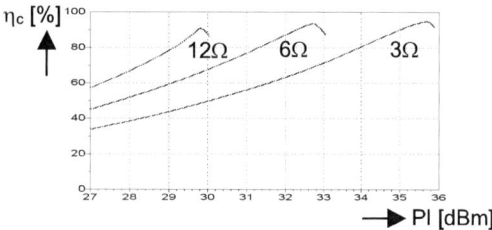

Fig.3. Simulated relation between collector efficiency and power level as function of load impedance level.

III. RECONFIGURABLE OUTPUT MATCHING NETWORKS

The requirements on the load impedance which has to be realized by the reconfigurable output matching network for dual-band class F^{-1} operation can be summarized as:
a.   Real valued load impedance at the fundamental frequency which depends upon the RF power.
b.   Open circuit at second harmonic- and short circuit at third harmonic-frequency band.

Fig. 4 shows a basic network topology in which a single switchable capacitor is used to switch the fundamental load impedance between two real values.

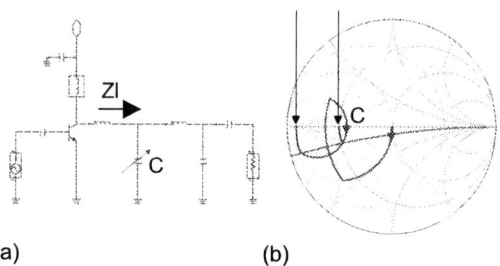

Fig.4. (a) Basic load-switch network topology. (b) Simulated Load-impedance (Zo=50Ω).

Fig. 5 shows a basic circuit topology for the realization of the desired harmonic terminations. Band switching is obtained by a single switched capacitor which tunes the harmonic trap L,C between the second and third harmonic of the low frequency band.

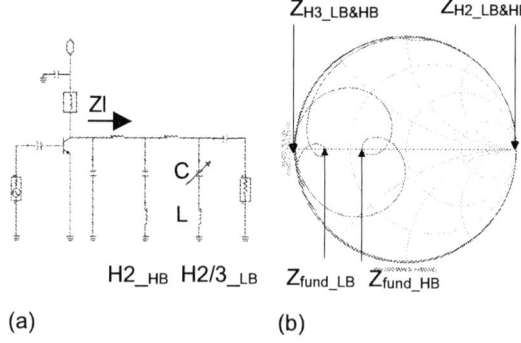

Fig.5. (a) Basic network topology for dual-band harmonic loading. (b) Simulated Load-impedance (Zo=10Ω).

A combination of the topologies shown in fig. 4 and 5 can be used to meet both requirements a. and b. mentioned above in a dual-band 0.9GHz/1.8GHz matching network for class F^{-1} operation including 0.9GHz load-switch functionality. Figure 6 shows the overall network topology with 3 MEMS used for band switching and 1 for loadline switching.

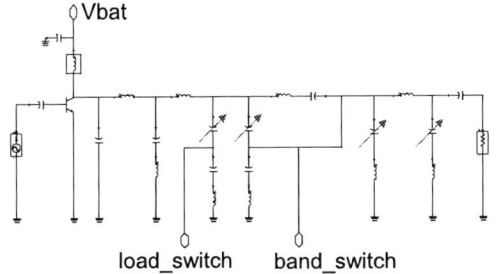

Fig.6. Dual-band reconfigurable output matching network with optimized second and third harmonic loading and loadswitch functionality.

Table 1 shows the three operating states of the network: low-band maximum power, low-band reduced power and high-band.

State	Zl	load_switch	band_switch
LB, 33dBm	3Ω	high	high
LB LS, 30dBm	6Ω	low	high
HB, 30dBm	5Ω	low	low

Table 1. Operating states of reconfigurable output matching network.

IV. IMPLEMENTATION

The amplifier was implemented as a System in a Package (SiP) to allow for an optimum partitioning of the functionality with respect to cost, size and performance. The SiP measures 40mm^2 and consists of a stack of an organic laminate and a high-Ohmic silicon substrate (PASSI) with flipped BiCMOS (QUBiC) and MEMS on top as shown in Figure 7.

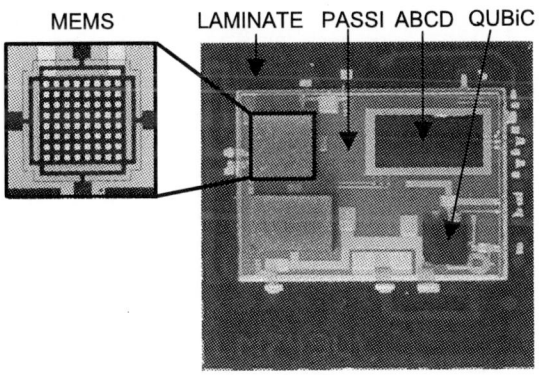

Fig.7. RF-SiP using stack of laminate, high-Ohmic Silicon and flipped BiCMOS and MEMS dies.

The laminate serves as a module carrier and as a heat spreader/interposer between the Passive die and the phone-board. It offers the interconnect to the phone board by means of an Land Grid Array (LGA) on the backside. The laminate is also used for integration of high Q inductors and coarse interconnects using two 25um thick Copper layers.

The high Ohmic Silicon crystal serves as a carrier for the flipped crystals on top and as a heat spreader /interposer between the flipped dies and the laminate converting the fine pitch (90um) of the dies to the coarse pitches (>300um) used on laminate. The silicon is also used for integration of high Q MIM capacitors (C=145pF/mm^2), a high density MIS capacitor (C=25nF/mm^2) and Poly-silicon resistors (R=10Ohm/square).
The RF LineUp is realized in a 0.25um BiCMOS technology with special High Voltage NPN device for PA applications (Ft=26GHz, BV$_{CBO}$=18V).
The MEMS switched capacitors are made in a low cost thin film passive integration technology on high Ohmic Silicon [5]. This technology offers capacitors with a high switching ratio (Con/Coff >10) in combination with low parasitic losses (ESR < 0.4 Ω). The charge pump used for actuation of the MEMS is realized in a 1.2um SOI process (ABCD) with a High Voltage (BV> 60V) DMOSFET device.

V. MEASUREMENTS

The output match was evaluated separately as well as combined with the BiCMOS active line-up. Fig. 8 shows the input impedance of the matching network for the network states defined in Table 1. The low-band results show that the load-impedance in the pass-band is doubled from 4Ω to 8Ω if the network is switched from the nominal to the load-switch state. The impedance in both low-band states is high at the second and low at the third harmonic frequency band. The high-band results show a nominal pass-band impedance of 4Ω again in combination with the desired second- and third-harmonic termination for class F^{-1} amplifier operation.

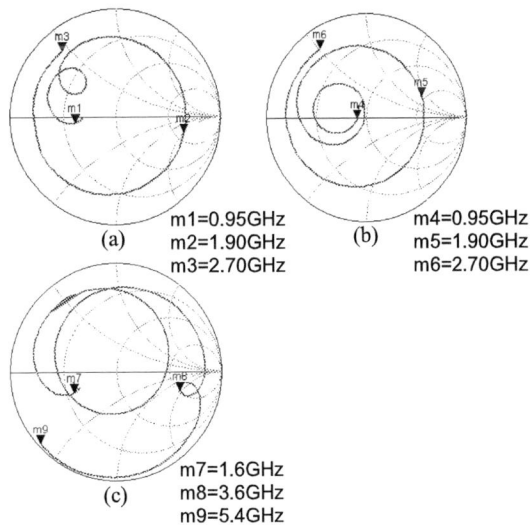

Fig.8. Measured input impedance of output matching network (Zo=10Ω):
(a) LB, 33dBm, (b) LB LS, 30dBm, (c) HB, 30dBm.

The pass-band losses and harmonic filtering properties of the matching network were evaluated by measuring the frequency transfer using a source impedance equal to the conjugated input impedance. The results are shown in Fig. 9. The low-band results shows about 1.5dB pass-band loss with over 40dB attenuation in the second and third harmonic frequency band. The high-band pass-band loss is about 1.5dB with over 30dB second and third harmonic attenuation.

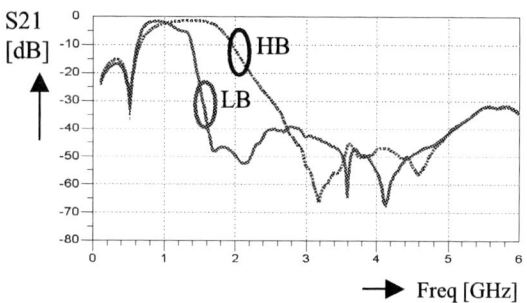

Fig.9. Measured pass-band loss and harmonic rejection of matching network in LB- and HB-state with matched source at input terminal.

The linearity of the matching network has been evaluated by exciting the network with an 8PSK (EDGE) signal from a conjugated matched source and measuring the EVM and ACPR of the output signal. The low-band results are shown in Fig. 10. Both EVM and ACPR are well below the typical PA requirements of 4% and -57dBc at power levels up to 33dBm. This demonstrates the good linearity of the applied MEMS-based switched capacitors.

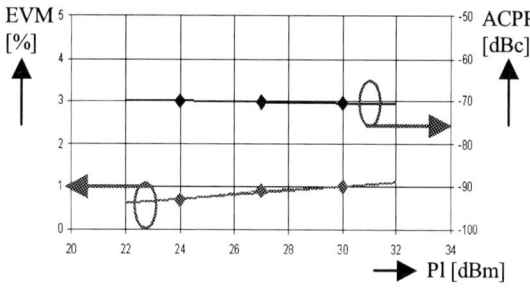

Fig.10. Measured EVM and ACPR (400kHz) of matching network in LB-state when excited at 0.9GHz with matched 8PSK (EDGE) source at the input terminal.

The full amplifier consisting of the BiCMOS active line-up connected to the MEMS-based re-configurable matching network has been evaluated with respect to output power and efficiency. Fig. 11 shows the overall efficiency versus output power for the three states defined in Table 1. The 0.9GHz results shows a maximum power level of 31dBm with 40% efficiency while at 1.8GHz a maximum power of 30dBm has been achieved with 34% efficiency. The build-in load-switch provides up to 10%

efficiency improvement at 0.9GHz for reduced power levels.

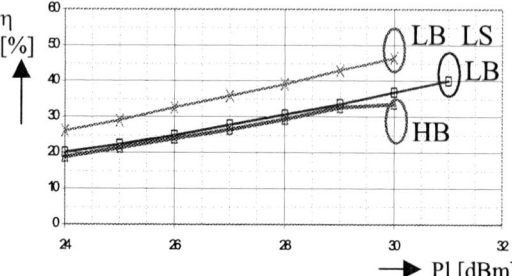

Fig 11. Measured power efficiency versus output power as function of network state.

VI. CONCLUSION

This work demonstrates the feasibility of the use of a MEMS-capacitor based reconfigurable output matching network for efficient dual-band operation of a single BiCMOS amplifier line-up with EDGE compliant linearity. In particular it was demonstrated that a new reconfigurable matching network topology offers proper wideband termination for efficient class F^{-1} amplifier operation in both bands in combination with low-band load-switch functionality.

The demonstrated amplifier concept offers cost, size and performance advantages for multi-band/mode handset PA's, a advantage that increases with the number of bands/modes covered.

ACKNOWLEDGEMENT

The authors wish to acknowledge the support of Philips Research and the ICRF Concept and Technology development teams of Philips Semiconductors.

REFERENCES

[1] A. Fukada, "Novel 900MHz/1.9GHz Dual-Mode Power Amplifier Employing MEMS Switches for Optimum Matching", *IEEE Microwave and Wireless components Letters, Vol 14, No 3, 3 March 2004 pp.121-123.*

[2] Y. Lu, "A MEMS Reconfigurable Matching Network for Class AB Amplifier", *IEEE Microwave and Wireless components Letters, Vol 13, no 10, Oct 2003 pp.437-439.*

[3] X. Liu, Y. Lin, W.C.E.Neo, L.C.N. de Vreede et al, "Improved Power Amplifier Efficiency Using Varactor-Based Tunable Matching Networks", *Proceedings BCTM 2006.*

[4] F.H. Raab, "Maximum Efficiency and Output of class F amplifiers", *IEEE Transactions on Microwave Theory and Techniques, Vol. 49, no 6, June 2001 pp.1162- 1166.*

[5] J.T.M. van Beek et al, "High-Q Integrated RF Passives and micro-mechanical Capacitors on Silicon", *Proceedings BCTM 2003, pp.147-150.*

Miniaturized Quad-Band Front-End Module for GSM using Si BiCMOS and passive integration technologies

A.J.M. de Graauw, A. van Bezooijen, C. Chanlo, A. den Dekker, J. Dijkhuis, S. Pramm, and H.K.J. ten Dolle

Philips Semiconductors, Innovation Center RF, Nijmegen, 6534AE, The Netherlands

Abstract — We present a highly miniaturized front end module (FEM) for GSM. The module consists of a pure silicon BiCMOS RF power amplifier (PA) chip, a BiCMOS die for bias/control functions and a pHEMT switch die, all flipped on a passive integration die, which is mounted on an organic laminate substrate. All passives are integrated. Partitioning of passives is optimized to achieve minimum size (6x6x1 mm^3) and low cost at acceptable performance. A Power Added Efficiency (PAE) of 35% is demonstrated in the low band (870 MHz) for an output power of 32 dBm with the second harmonic (H2) below -40 dBm and third harmonic (H3) below -45 dBm.

Index Terms — Silicon bipolar, BiCMOS, passive integration, power devices, analog circuits, RF circuits.

I. INTRODUCTION

Since the introduction of cellular phones, there has been a strong trend to make cheap small phones with more and more functionality, requiring the same from the phone's components.

In this work we show a miniaturized FEM of 6x6x1 mm^3 using a new passive integration technology as well as a new assembly platform. The achieved product area reduction is 44% compared to a product with the same functionality in our previous assembly platform. Additionally, we present the FEM performance, particularly for 900MHz band.

II. PASSIVE INTEGRATION TECHNOLOGY

The development of the passive integration technology Passi™ [1] was the first step for Philips on the road of passive integration in Silicon, with a limited functionality available in the process. Now, a new passive integration technology (Passi4) is developed together with a new assembly platform, in order to integrate more functionality in a limited space. The new assembly platform consists of a Passi4 die mounted on an organic laminate substrate, with multiple active dies in different technologies flipped on the Passi4 die (Fig.1).

Passi4 allows the integration of resistors (R=10Ω/sq.), coils (Q>25), small accurate capacitors (C=145pF/mm^2, ESR<100mΩ) suitable for RF matching, as well as large capacitors (C=25nF/mm^2) suitable for DC decoupling.

Fig. 1. FEM in new assembly platform: two silicon BiCMOS dies and a pHEMT switch are flipped on a Passi4 die on top of 4-layer organic laminate substrate.

Availability of via holes (L=20pH, R=10mΩ) enables a Passi4 die to act as heat spreading interposer between fine-pitch (85μm) flipped active dies and the laminate (Fig.2), thus realizing excellent thermal behavior of the total stack.

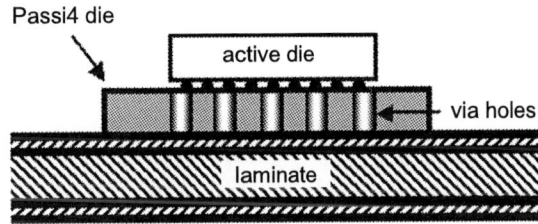

Fig. 2   Schematic cross section of the assembly stack.

The organic laminate substrate has four metal layers. Thickness of the copper top metal is 30 μm. Several passive elements (capacitors, coils) can be made in the laminate, in Passi4 or in the active die, making it possible to optimize the design with respect to cost, size and performance by carefully selecting where these elements are realized. To allow this optimization and make simulation of the stack possible, a multi technology design environment has been developed.

III. ACTIVE DEVICE TECHNOLOGY

Because of integration possibilities and to keep cost down, Philips pure silicon BiCMOS technology QUBiC4plus [2] is used to implement both the RF PA and the bias and control functions.

Dedicated protection circuits, based on output power adaptation [3-4], are included in the RF PA chip to avoid destructive breakdown of the power transistor under extreme conditions. Over-voltage, over-temperature and over-current protection are included, giving the possibility to maintain save operation relatively close to the breakdown limits of the technology and thus realize enhanced performance.

In order to predict performance of the QUBiC4plus power transistor in the new assembly platform, load pull measurements were done at 900MHz and at 1710MHz. Fig.3 shows the actual device under test. With use of proper EM simulation and dedicated de-embedding structures, the load and source impedances seen by the transistor are determined.

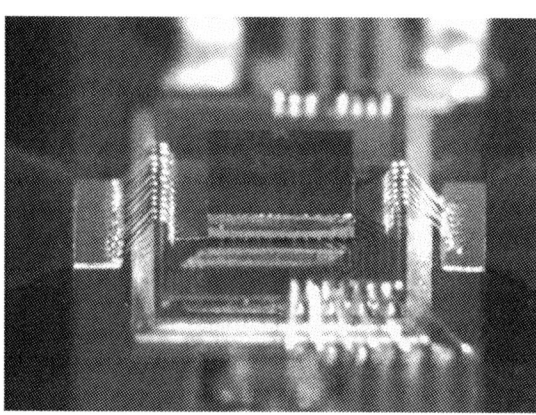

Fig. 3. Device under test for load pull: QUBiC4plus end-stage die flipped on Passi4 wire bonded to laminate.

Fig.4 shows the results of sweeping the end-stage load impedance over a specific region of the Smith-chart while measuring the PAE and gain (Gt), with a constant source impedance. The best trade-offs for each band are those highlighted at the upper side of the cloud of loads. At the chosen trade-off point for the low band, PAE is 75% with 16dB gain. At the chosen trade-off point for the high band, PAE is 65% with 12.5dB gain.

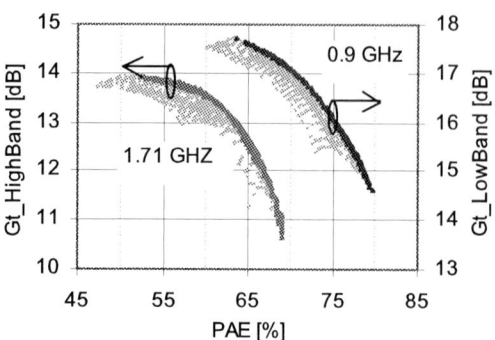

Fig. 4 Load pull measurement result: Trade-off between PAE and Gt for different load impedances at 35dBm output power (0.9 GHz) and 32.5dBm (1.71 GHz)

IV. MODULE DESIGN

The design of a reliable, well performing RF module suitable for high volume production requires modeling and (co-) simulation of electrical, thermal and mechanical aspects. A new multi-technology module design environment has been set-up for design definition, simulation and mask-generation. The electrical performance is evaluated in a combined 2.5D EM simulation of the entire passive stack with a harmonic balance nonlinear simulation of the QUBiC4plus RF PA die.

The thermal resistance and mechanical stress parameters of the total assembly stack was simulated in a 3D finite element simulator. The results were used to optimize the flip-chip bump interconnect design.

A. Functional Partitioning

The module consists of the following functional blocks: two 3-stage RF PA's for high and low frequency bands, output match and harmonic filters, SP6T antenna switch, power control loop, module control logic, antenna ESD protection inductor and protection circuits (Fig. 5.)

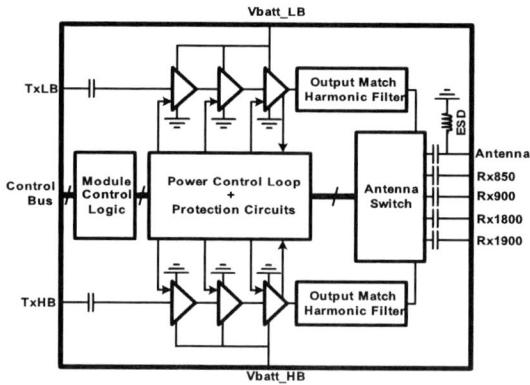

Fig. 5. Functional block diagram.

Fig. 1 shows the FEM on an application board. Table I shows the mapping of functions over the different technologies for optimized cost, size and performance.

TABLE I
FUNCTIONALITY TECHNOLOGY MAPPING

Technology: Functionality	LAMP4	PASSI4	RFdie Qubic4+	GaAs pHEMT	Controller Qubic4+
Input Match	x	x			
RF PA stages with bias mirrors			x		
Inter-stage matches		x	x		
Output Match & Harmonic Filter	x	x			
Antenna Switch				x	
Antenna ESD-protection L		x			
Bias circuits					x
PowerControl loop			detector		actuator
Control logic					x
Protection circuits			det/act		

The following measures were taken to avoid coupling and stability problems due to extreme miniaturization: cross-coupled supply connections, damping resistance in interstage matches, emitter degeneration and separate grounding of RF stages.

B. The Output Match

To achieve a small module size, the output match and the low pass filter are combined in a single network, which is critical for the FEM performance because:

1) it determines the final stage operation class;
2) it affects peak voltages, thus ruggedness;
3) it determines the harmonic suppression of the module;
4) low loss is critical for module efficiency;
5) the impedance below the working frequency has a large influence on stability.

Inverse class F was chosen as operating class for the final stages of both low- and high-band amplifiers, in order to give a good compromise between efficiency, ruggedness and size. The characteristic high impedance for the second harmonic (H2) and low impedance for the third harmonic (H3) at the internal collector of the transistors is realized by the network in Fig. 6. Capacitor C_1 is in series resonance with L_1 at H3 and in parallel resonance with the inductive part of the rest of the network at H2.

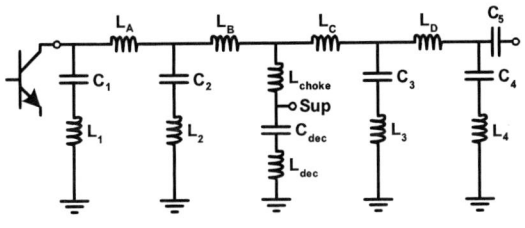

Fig. 6.   Output match and harmonic filter network.

High harmonic suppression is obtained by the realized resonances in the network at frequencies close to H2 and H3. There are two resonances near H2 and one at H3. The first L- matching section is made entirely on the Passi4 die to prevent coupling via bond wires. As it is common to both the RF and the DC path, it also filters H2 out of the DC path. The other inductors are integrated into the laminate to reduce loss and cost.

Dedicated module sub-structures have been made to evaluate the impedance seen when looking from the final stage into the load. Fig.7 illustrates the measurement result for the low frequency band.

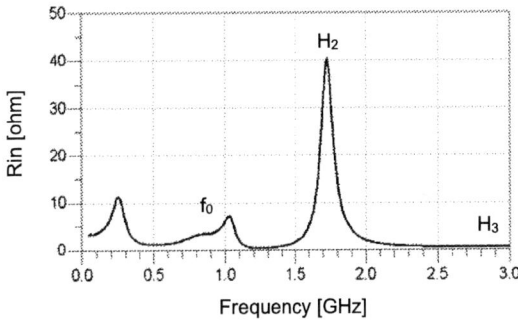

Fig. 7. Real part of measured impedance at the collector of the final stage for the low band network. Note that the impedance at H2 is high and at H3 is low. The impedance at the fundamental is around 3 ohm.

V. FEM EVALUATION RESULTS

According measurements of the FEM in the low frequency band, the module delivers an output power of 32 dBm, with PAE of 35%. The power control curve of the module is smooth and monotone (Fig.8.).

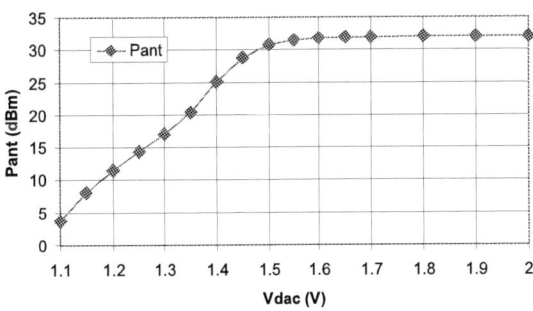

Fig. 8.   Antenna power vs. control voltage.

Further analysis shows that final stage collector efficiency at an output power of 32 dBm is approximately 70%, indicating potential to further improve the PAE of the entire FEM. Fig. 9 shows the measured PAE of the FEM versus the output power together with the determined collector efficiency of the final stage.

23

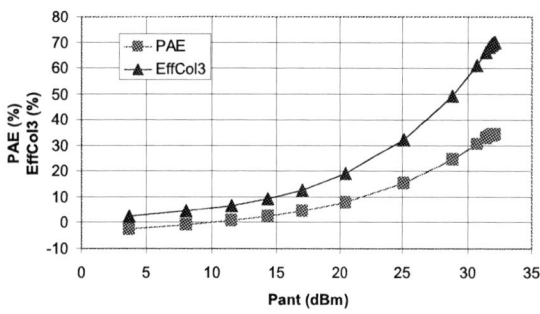

Fig. 9.   PAE of the FEM and collector efficiency of the final stage (EffCol3).

Measured harmonic levels (Fig. 10) are well within specification. Worst-case levels are -40 dBm for H2 and -45 dBm for H3, indicating proper operation of the miniature sized combined output match / low pass filter network.

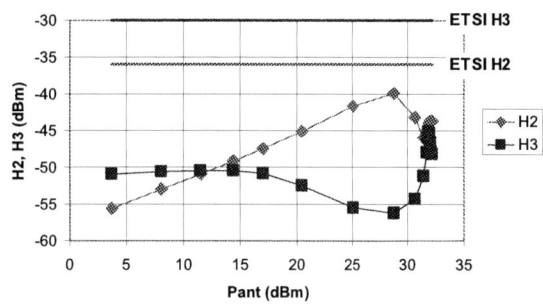

Fig. 10.   Harmonic levels indicate proper operation of combined output match / lowpass filter.

The module is stable under VSWR 10:1 load mismatch at nominal conditions (870 MHz, supply voltage Vsup=3.5 V, Pin = 10 dBm).

Measured insertion loss for the Rx part of the FEM is shown in Table II together with Tx-Rx isolation, determined at 32 dBm output power. Both are sufficiently good in all frequency bands.

TABLE II
PHEMT INSERTION LOSS & TX-RX ISOLATION

Transfer	Insertion Loss
Ant->Rx850	0.84 dB
Ant->Rx900	0.84 dB
Ant->Rx1800	1.16 dB
Ant->Rx1900	1.26 dB
Transfer	Tx-Rx Isolation
TxLB->Rx850	32.8 dB
TxLB->Rx900	36.8 dB
TxLB->Rx1800	40.6 dB
TxLB->Rx1900	43.4 dB

VI. CONCLUSION

This work demonstrates substantial miniaturization of low cost RF-SiP modules for cellular handsets by using passive integration. The achieved product area reduction is 44% for the demonstrated FEM, while maintaining sufficient performance.

Passive integration eliminates the need for smd components and reduces process spread, which improves high volume production yield and therefore leads to further cost reductions.

The design of a miniaturized multi-technology module can be successfully managed by integral simulation of electrical, thermal and mechanical aspects. This approach reduces the development time, which is essential in order to cope with the short lifetime of current products.

ACKNOWLEDGEMENT

We would like to thank our colleagues at Philips Semiconductors in Caen and Nijmegen for their support and contribution to the achieved results.

REFERENCES

[1] J.T.M. van Beek, M.H.W.M. van Delden, A. van Dijken, P. van Eerd, M. van Grootel, A.B.M. Jansman, A.L.A.M. Kemmeren, Th.G.S.M. Rijks, P.G. Steeneken, J. den Toonder, M. Ulenaers, A. den Dekker, P. Lok, N. Pulsford, F. van Straten, L. van Teeffelen, J. de Coster, R. Peurs, "High-Q integrated RF passives and micro-mechanical capacitors on silicon," *2003 IEEE Bipolar/BiCMOS Circuits and Technology Meeting, Proceedings*, pp. 147-150, September 2003.

[2] P. Deixler, T. Letavic, T. Mahatdejkul, Y. Bouttement, R. Brock, P.C. Tan, V. Saikumar, A. Rodriguez, R. Colclaser, P. Kellowan, H. Sun, N. Bell, A. Yao, R. van Langevelde, T. Vanhoucke, W.D. van Noort, G.A.M. Hurkx, D. Crespo, C. Biard, S. Bardy, and J.W. Slotboom, "QUBiC4plus: A Cost-effective BiCMOS Manufacturing Technology with Elite Passive Enhancements Optimized for 'Silicon-based' RF-System-in-Package Environment," *2005 IEEE Bipolar/BiCMOS Circuits and Technology Meeting, Proceedings,* pp. 272-275, October 2005.

[3] A. van Bezooijen, F. van Straten, R. Mahmoudi, and A.H.M. van Roermund, "Avalanche breakdown protection by adaptive output power control," IEEE RSW Proceedings, pp. 519-522, Jan. 2006.

[4] L. Ruijs, A. van Bezooijen, R. Mahmoudi, and A.H.M. van Roermund, "Novel voltage limiting concept for avalanche breakdown protection," to be published, *2006 IEEE MTT-S Int. Microwave Symp. Dig.*

Physical Description of the Mixed-Mode Degradation Mechanism for High Performance Bipolar Transistors

T. Vanhoucke, G.A.M. Hurkx, D. Panko[1], R. Campos[1], A. Piontek[2], P. Palestri[3], L. Selmi[3]

Philips Research Leuven, Kapeldreef 75, 3001 Leuven, Belgium
[1] Philips Semiconductors Fishkill, PO Box1279, Hopewell Junction, New York, 12533-1279
[2] IMEC, Kapeldreef 75, 3001 Leuven, Belgium
[3] DIEGM, University of Udine, Udine 33100, Italy

Abstract — **We study the mixed-mode stress regime for bipolar transistors. From Monte Carlo simulations we found that high-energy avalanche generated holes dominate the degradation and we experimentally investigate the energy dependence of the hot carriers on the device degradation including the pinch-in effect.**

Index Terms — **Reliability Modeling, Mixed-Mode stress, Bipolar Transistor, Hot Carriers, SiGe HBT.**

I. INTRODUCTION

The introduction of SiGe HBTs has opened new perspectives with respect to high-speed applications. However, device scaling in combination with higher current densities has consequences for the reliability under extreme operation conditions [1]. The trend of decrease in breakdown voltage (BV_{CEO}) makes negative base current operation an important issue in the reliability and instability of HBTs [2].

Recently, Zhang *et al.* [3] introduced the mixed-mode reliability regime where the transistor is biased with constant emitter current (I_E) and high collector-base voltage (V_{CB}) such that $V_{CE} > BV_{CEO}$ with V_{CE} the collector-emitter voltage. It has been found [3,4] that under these conditions, higher V_{CB} results in more device degradation and that the mechanism can be modeled by the time integrated avalanche current [4]. Yang *et al.* [4] have shown that the device damage occurs in the dielectric regions away from the bulk silicon junction and that the degradation scales with the emitter perimeter. Zhu *et al.* [5] have simulated carrier distributions showing that hot-electrons are the dominant high-energy carriers injected into the base-emitter (BE) spacer.

In this work we investigate the physical mechanism of the mixed-mode stress by systematically varying different stress parameters.

II. STRESS MEASUREMENTS

The wafer level measurements are performed at constant forward I_E and high V_{CB} for several ambient temperatures (T_0) from 25 to 150 °C using the Philips QUBiC4G/X process [7]. Stress times of 1 hour are used but similar results are obtained for longer times [6]. Several devices with $BV_{CEO} \approx 2.2$ V are studied

with J_C at peak f_T ranging from 1.0 to 4.0 mA/μm². In this study we focus on forward gummel plots taken before and after the stress at $V_{CB} = 0$ and $T_0 = 25$ °C to quantify the device degradation.

No influence of stress on the collector current (I_C) is observed and the decrease of the current gain (h_{FE}) is attributed to the increase of the non-ideal component of the base current (I_B) (see later also). To quantify the device degradation we define ΔI_B at a given V_{BE} as the ratio of I_B after and before stress.

III. RESULTS AND DISCUSSION

The avalanche related degradation mechanism relevant for mixed-mode stress is shown schematically in the inset of Fig. 1. The primary electrons enter the collector-base (CB) depletion region and gain enough energy by the electric field to generate electron-hole pairs due to impact ionization ($V_{CE} > BV_{CEO}$). The generated secondary holes and electrons drift towards the base and collector region respectively. For sufficiently high V_{CB}, however, the secondary carriers generate tertiary carriers. This avalanche multiplication process ultimately leads to avalanche breakdown [8] experimentally observed by an increase of I_C (electrons) and reversal of I_B (holes). Some carriers might possess enough energy to cross the base layer, and reach the Si-SiO₂ BE-spacer interface to create traps. The traps are located within the BE-depletion region and they affect the non-ideal I_B by Shockley-Read-Hall recombination [3]-[5].

Measurements have shown that trap creation not only depends on the amount of carriers but also on their energy [4]. Therefore, increasing V_{CB} is expected to enhance device degradation because, firstly, more carriers are generated and, secondly, the average energy of the carriers is higher. This is presented in Fig. 1 showing the V_{CB} dependence of ΔI_B and of the avalanche current (I_{AV}). I_{AV} is defined as the increase of I_C due to impact ionization and can be determined from the I_B - V_{CB} relation [8]. Figure 1 indicates that the degradation is related to the avalanche mechanism.

To further investigate the effect of hot carriers, we study the influence of temperature on the degradation.

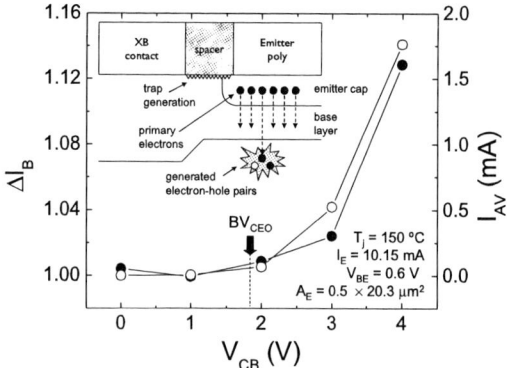

Fig. 1. V_{CB} dependence of ΔI_B and I_{AV} together with a schematic view (inset) of electron-hole pair generation by the primary electrons in absence of pinch-in.

A. Junction Temperature Dependence

We have justified that not only the amount of hot carriers but also the carrier energy plays a role in the device degradation. For high-energy carriers traveling from the CB-depletion region towards the BE-spacer, however, energy can be lost due to phonon scattering. It is well known that the scattering mechanism strongly depends on lattice temperature and therefore less devices degradation is expected for higher junction temperature (T_j) [9]. This is presented in Fig. 2 showing ΔI_B as function of T_j while keeping $I_{AV} = 500\ \mu A$ and $V_{CB} = 3.0\ V$. T_j has been increased by changing T_0 during the stress using $T_j \approx T_0 + R_{TH}I_EV_{CE}$ with R_{TH} the thermal resistance. Note that for higher T_0, higher stress I_E has been used to keep I_{AV} constant and therefore the dependence of T_j on the avalanche mechanism is included. We found that higher T_j results in lower ΔI_B in agreement with the enhanced phonon scattering rate [9].

Next we show that phonon scattering not only determines the temperature dependence of ΔI_B, but also the influence of the lateral emitter dimensions.

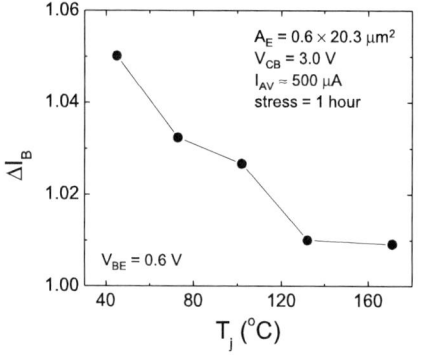

Fig. 2. Influence of junction temperature on degradation.

B. Emitter Width Dependence and Pinch-in

Hot carriers generated far from the spacer will not contribute to the degradation because they will have lost their energy before reaching the spacer. This means that the lateral distribution of the generated

electron-holes pairs is important for the degradation. We now study the influence of emitter width (W_E) for constant length (L_E), I_{AV} and V_{CB} (Fig. 3). Since R_{TH} depends on W_E, we adapt T_0 to keep $T_j = 150\ °C$.

The decrease of ΔI_B for wider devices is a result of two effects. Firstly, since we keep I_{AV} constant for all widths, the avalanche current density decreases for increasing W_E. Assuming a uniform carrier distribution, this reduces the available amount of hot carriers close to the spacer. Secondly, the non-zero base resistance causes the avalanche current to be concentrated in the middle of the device. Since $I_B < 0$ during mixed-mode stress ($V_{CE} > BV_{CEO}$), the internal V_{BE} is higher in the centre of the device (two base contacts) due to the base pinch resistance $R_{pinch} > 0$. This effect is known as pinch-in [10] and forces the device to operate in the centre. The pinch-in effect effectively increases the distance between the avalanche carrier generation and the BE-spacer.

In order to separate the effect of the reduced avalanche current density from pinch-in when increasing W_E, we have simulated the lateral I_{AV} distribution with (high R_{pinch}) and without (low R_{pinch}) pinch-in effect using MEDICI for a constant $I_{AV} = 25\ \mu A$ per micron emitter length. No thermal effects are included in the simulations and therefore we exclude the influence of T_j on ΔI_B. By integrating the simulated current under the BE-spacer, the amount of electron-hole pairs has been extracted from the simulations to quantify the degradation as function of W_E. From the inset of Fig. 3 it can be seen that pinch-in effects play a role in the device degradation and that the degradation does not scale with the device perimeter in contrast to Ref. [4].

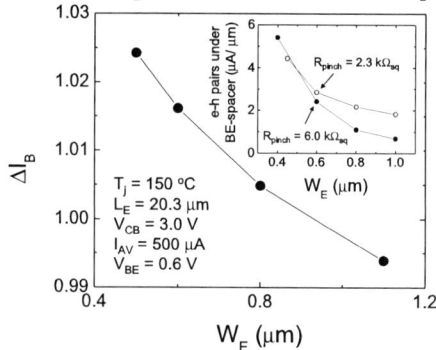

Fig. 3. Scaling of ΔI_B and simulated electron-hole pairs under the BE-spacer (see text).

C. Primary and Secondary Carriers

We have shown that ΔI_B increases with increasing I_{AV}. Now we investigate whether the primary electrons or the avalanche generated carriers are responsible for the degradation. Figure 4 shows the relation between I_{AV} and I_E for several values of V_{CB}. At low I_E, I_{AV} increases with increasing emitter current at given V_{CB} as a result of increasing impact ionization. At high I_E, however, further raising I_E

results in a decrease of I_{AV} mainly caused by high injection effects in the collector. Note that for all currents, $I_{AV} \ll I_E$ meaning that there are more primary electrons than generated electron-hole pairs.

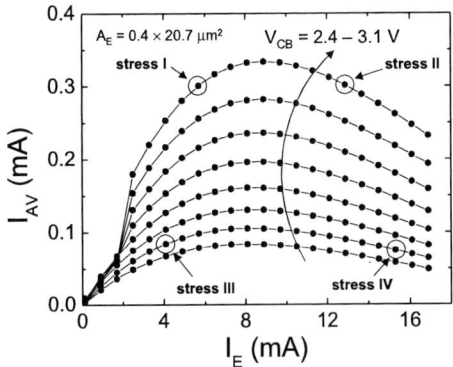

Fig. 4. Emitter current dependence of I_{AV} for different V_{CB}. The arrows indicate the stress conditions used in Fig. 5.

We have performed degradation measurements using the I_{AV} - I_E combinations indicated by the arrows in Fig. 4. Since stress point I and II have identical I_{AV} and V_{CB} (but different I_E), the degradation is expected to be similar according to the avalanche related mechanism in Fig. 1. T_j, however, is about twice in stress point II which reduces the degradation as demonstrated in Fig. 2. Therefore, stress measurement I has been repeated at higher T_0 to increase T_j estimated from R_{TH}. Results in Fig. 5, however, still show a difference in ΔI_B between I and II. This may be due to a change in R_{TH} at very high I_E resulting in different T_j. For very low I_{AV} (III and IV), no influence of I_E is found demonstrating once again that the degradation mechanism is avalanche related.

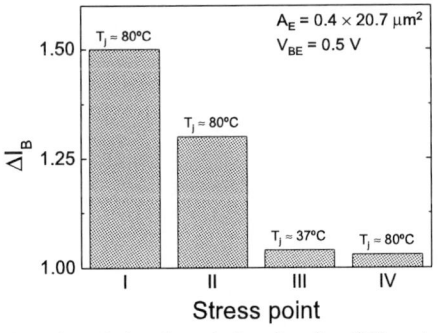

Fig. 5. Results of the degradation for the different stress points indicated in Fig. 4 together with T_j estimated from the low current thermal resistance.

The previous analysis indicates that the primary electrons are not responsible for the degradation. At this stage, however, it is not clear whether the avalanche generated electrons or holes cause the degradation. For both carriers, the analysis given above is valid. To distinguish between the high-energy electrons and the holes we have performed Monte Carlo simulations.

IV. MONTE CARLO SIMULATIONS

We used the Full-Band Monte Carlo (MC) simulator [11] to calculate the hot-electron and hot-hole populations during mixed-mode stress. A representative 1D cut of the device (*i.e.* large area device) and doping profile has been considered (inset Fig. 6). The MC model accounts for phonon and impact ionization (II) scattering. As for its transport properties, the SiGe layer is treated as pure Si and alloy scattering is neglected [12]. The II scattering rate has been calibrated to reproduce the ionization coefficients of [13]. As shown in Fig. 6, the simulator reproduces the $M_n - 1$ curve.

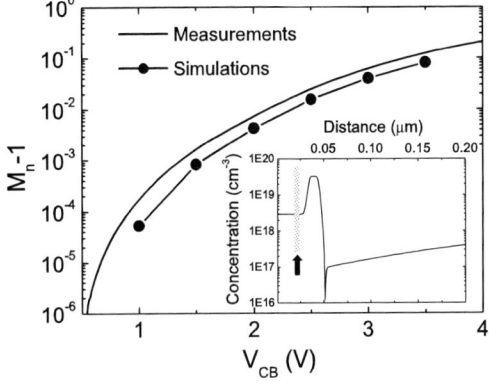

Fig. 6. Monte Carlo simulated (symbols) and measured (line) $M_n - 1$ values and (inset) used doping profile.

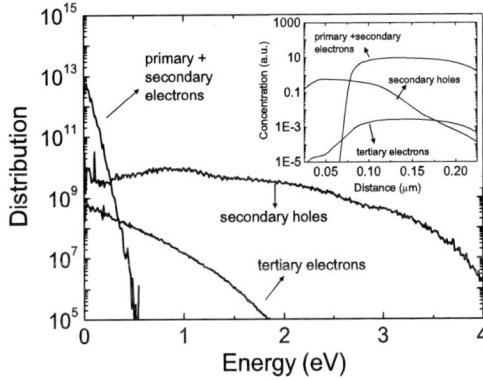

Fig. 7. Simulated distribution of avalanche carriers in the region indicated by the arrow in Fig. 6. The inset shows the carrier population with energy higher than 1 eV.

To separate the populations of primary and secondary electrons, secondary holes and tertiary electrons we have followed a three step simulation approach [14]. During the first step, electrons are injected at the beginning of the neutral base and moved according to the conduction band profile $E_C(x)$ given by drift-diffusion simulations and the II generation rate $G_{II,n}(x)$ is recorded. During the second step, secondary holes generated according to $G_{II,n}(x)$ are moved according to the valence band profile $E_V(x)$ and their impact ionization generation rate $G_{II,p}(x)$ is recorded. Note that $E_C(x)$ and $E_V(x)$ include the conduction and valence band shifts due to the Ge grading respectively. Finally, we simulate tertiary electrons generated according to $G_{II,p}(x)$ and moved according to $E_C(x)$. This approach provides the energy distributions of primary and secondary electrons (step 1), secondary holes (step 2) and tertiary electrons (step 3) at different positions inside the device.

Figure 7 reports these distributions at the beginning of the base (see arrow in Fig. 6) for $V_{CB} = 3.0$ V. It is clear that the distributions are far from heated Maxwellians with a constant effective temperature (as implicitly assumed by energy balance and lucky-electron models as those used in Ref. [5]) and that the population of high energy holes is orders of magnitude higher than that of hot electrons. Moreover, since the carriers receive energy from the electric field, we observed that the population of hot primary electrons increases moving from the neutral base towards the collector region. On the other hand, the population of secondary hot-holes increases moving from the collector side of the depletion region (*i.e.* where the majority of secondary holes is generated) toward the base. Tertiary electrons are generated in proximity of the base region, but due to the electric field, their concentration is higher in the high-field region than in the neutral base. For this reason, in the neutral base the population of hot-holes is much larger than the population of hot-electrons (Fig. 7). These results strongly suggest that trap generation at the BE-spacer is caused by avalanche generated hot-holes and not by hot-electrons as previously suggested based on energy balance simulations and the lucky-electron models [5].

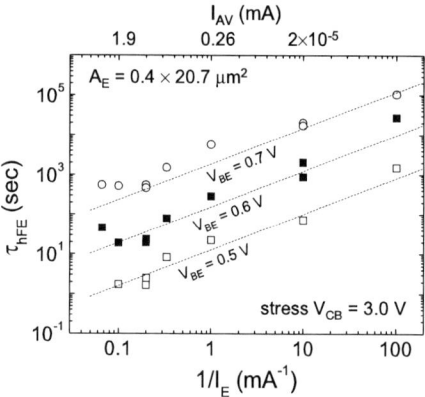

Fig. 8. Example of the life-time model showing I_E and I_{AV} dependence of the life-time τ_{hFE} (see text). Symbols are measurements and lines are model results [6].

V. PRACTICAL LIFE-TIME MODEL

Our understanding of the degradation mechanism has been presented in a new mixed-mode reliability time-to-fail model [6]. It allows the designer to estimate the device life-time (τ_{hFE}) for given operation conditions (I_E, V_{CB}) using, for example, 10% h_{FE} reduction. The model also includes the V_{BE} readout of h_{FE} and has been tested for stress times exceeding 12 hours including statistics over the wafer. From the relationship between I_E and I_{AV} in Fig. 4, stress parameters I_E and V_{CB} together with the device geometry are used in the model as a practical set of parameters. A detailed model description can be found in Ref. [6] but some results are repeated in Fig. 8 as an example.

VI. CONCLUSION

We have studied the mixed-mode stress regime and have shown that the degradation mechanism is related to the avalanche generated carriers in the CB-depletion region. Temperature measurements have shown that higher T_j leads to less degradation as expected by the reduction of high-energy carrier population. The influence of device width shows that wider devices results in less degradation for constant I_{AV} and V_{CB}. From 2D simulations, the influence of the pinch-in effect is shown. From 1D Full-Band Monte Carlo simulations, we have distinguished between hot-electrons and hot-holes and found that the number of hot-holes entering the BE-spacer is orders of magnitude higher than hot-electrons. Our findings are collected in a practical degradation model which enables the designer to estimate the device life-time under mixed-mode conditions.

ACKNOWLEDGEMENT

The authors thank all BiCMOS project members. The work was supported by the IWT EXAP project.

REFERENCES

[1] A. Piontek *et al.*, *Proc. ISTDM 2006*, pp. 242-243.
[2] T. Vanhoucke *et al.*, *Proc. BCTM 2005*, pp. 37-40.
[3] G. Zhang *et al.*, *IEEE Trans. Electron Devices*, Vol. 49, No. 12, pp. 2151-2156, Dec. 2002.
[4] Z. Yang *et al.*, *Proc. IRPS 2003*, pp. 339-343.
[5] C. Zhu *et al.*, *IEEE Trans. On Device and Materials Reliability*, Vol. 5, No. 1, pp. 142-149, March 2005.
[6] D. Panko *et al.*, *Proc. IRPS 2006*, pp. 512-515.
[7] P. Deixler *et al.*, *Proc. BCTM 2004*, pp. 223-236.
[8] A. Y-K Su *et al.*, *Solid-State Electronics 45*, pp. 761-765, 2001
[9] U. Gogineni *et al.*, *IEEE Trans. Electron Devices*, Vol. 47, No. 7, pp. 1440-1448, July 2000.
[10] M. Rickelt *et al.*, *Solid State Circuits*, Vol. 37, No. 9, pp. 1184-1197, Sept. 2002.
[11] P. Palestri *et al.*, *IEEE Trans. Electron Devices*, Vol. 47, No. 5, pp. 1044-1051; idem, *Proc. SISPAD 2000*, pp. 38-41.
[12] P.Palestri *et al.*, *IEEE Elect. Dev. Lett.*, Vol. 22, No. 11, pp. 533-535, November 2001.
[13] P. Palestri *et al.*, *Proc. IEDM 1998*, pp. 885-888.
[14] D. Esseni *et al.*, *IEEE Trans. Electron Devices*, Vol. 47, No. 11, pp. 2194-2200.

Analysis of Factors Contributing to Common-Base Avalanche Instabilities in Advanced SiGe HBTs

Curtis M. Grens, John D. Cressler, and Alvin J. Joseph [1]

School of Electrical and Computer Engineering, 85 Fifth Street, N.W.
Georgia Institute of Technology, Atlanta, GA, 30308, USA
[1]IBM Microelectronics, Essex Junction, VT, 05452, USA

Abstract — **Factors contributing to the observed bias-dependent features in common-base (CB) dc instability characteristics are examined for advanced SiGe HBTs. Parameters relevant to CB instabilities are identified and extracted from data to yield improved physical insight, a straightforward estimation methodology, and simple comparison between samples.**

Index Terms — **avalanche breakdown, bipolar transistors, electric variables measurement, impact ionization, safe-operating-area, SiGe HBT BiCMOS technology.**

I. INTRODUCTION

SiGe HBTs continue to make rapid in-roads into commercial RF and mixed-signal circuit applications, and through careful lateral scaling and vertical profile optimization, can achieve impressive performance (currently in excess of 350 GHz peak f_T) [1], while maintaining high yield and compatibility with best-of-breed CMOS. As SiGe HBTs are optimized to obtain higher peak f_T, increased collector doping is required to suppress Kirk effect and heterojunction barrier effects, and thereby allow operation at higher current density. As a consequence, the impact-ionization rate at the collector-base junction increases [2]. Thus, an inherent (and well known) tradeoff exists between peak f_T and breakdown voltage in SiGe HBT device design [3].

As breakdown voltages continue to decrease with technology scaling in SiGe HBTs (depicted in Fig. 1), circuit designers must be attentive to the increasingly rigorous (and often complex) upper-voltage limits for reliable device biasing for dynamic operation – the so-called safe-operating-area (SOA) [1], [4], [5]. The common-emitter (CE) electrothermal behavior of bipolar devices under constant base-emitter voltage drive were examined in [6], and operating voltage constraints were studied in [7] across several generations of SiGe HBT technology, focusing on device and circuit reliability implications.

This paper presents a detailed analysis of the bias dependence of common-base (CB) collector voltage stability thresholds, with emphasis on unusual characteristics observed at low currents (down to approximately $5\mu A/\mu m^2$), and including unreported statistical variations. We identify and analyze key parameters that characterize stability limits across bias

in high-speed SiGe HBTs. In addition to providing useful metrics for technology or device comparisons, these parameters translate into simple and useful guidelines for circuit designers and device modelers to ensure safe and stable device operation.

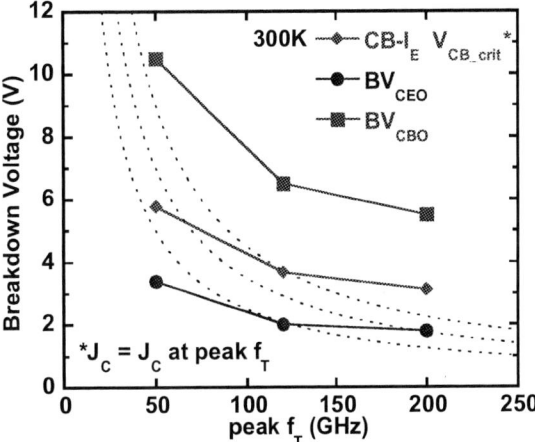

Fig. 1. Voltage constraints as a function of technology generation (peak f_T) for high performance SiGe HBTs.

II. CB BREAKDOWN OVERVIEW

The CB bias configuration, driven by a constant emitter current (labeled here as CB-I_E), lends itself to useful circuit topologies including cascode configurations and differential pairs for numerous practical applications (e.g., amplifier output stages). Since the base current is not fixed, excess hole current from impact ionization is allowed to freely exit the base terminal, and at $V_{CE} \cong BV_{CEO}$ the product of current gain (β) and avalanche multiplication factor ($M-1$) exceeds unity and the base current reverses sign [8], [9].

The voltage drop of the reverse base current across the distributed base resistance can result in central current crowding and bias instabilities at higher V_{CB} (in-between BV_{CEO} and BV_{CBO}), often referred to as "pinch-in" effects [4], [5], [7], which dictate the maximum voltage limits of safe device operation. Analysis in [4] demonstrates that in CB-I_E operation the lateral current collapse is 2-dimensional and is

1-4244-0458-4/06/$20.00 ©2006 IEEE

actually initiated along the *length*-axis of the emitter stripe.

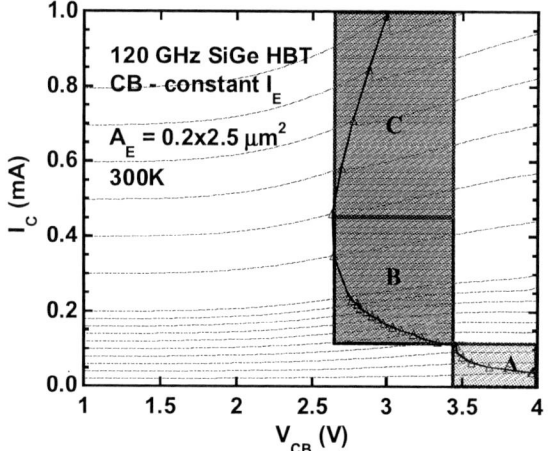

Fig. 2. CB-I_E output characteristics with highlighted regions showing different $V_{CB-crit}$ behavior across bias.

In Fig. 2, *dc* I-V curves for CB-I_E drive are plotted, with the CB instability threshold ($V_{CB-crit}$ – the triangles) indicated across bias. Three distinct regions of differing breakdown behavior are identified and discussed below.

A. Strong Pinch-in

At lowest current (region *A* in Fig. 2), $V_{CB-crit}$ is characterized by a well-defined pinch-in and decreases (from approx. BV_{CBO}) as I_E increases. Thus it can be inferred that the critical avalanche multiplication factor $(M-1)_{crit}$ likewise decreases with increasing I_E. The magnitude of the critical reverse base current ($|I_{B-rvs-crit}|$) increases with I_E in a near-linear fashion. The variation across multiple samples is small and hence predictable.

B. Weak Quasi-Pinch-in

With increased current (region *B* in Fig. 2), however, $V_{CB-crit}$ becomes less well-defined, undergoing either "soft" pinch-in (in which a clear discontinuity in the terminal characteristics is not apparent) or a random fluctuation between "pinched" and "stable" states. This region represents a balancing of the positive and negative feedback mechanisms, such that neither influence is dominant to clearly determine the threshold of device stability. Consequently, the conditions for quasi-instability are less severe compared to strong pinch-in, resulting in an apparent branch in the $V_{CB-crit}$ characteristic vs. I_C. In this region $|I_{B-rvs-crit}|$ is approx. constant vs. I_E and both $(M-1)_{crit}$ and $V_{CB-crit}$ decrease as I_E increases. Because the conditions for breakdown are defined vaguely here, they are more prone to random fluctuations, and hence are less predictable. Thus, variation across multiple samples is larger in this region.

C. Strong Quasi-Pinch-in

As the collector current is increased further (region *C* in Fig. 2) the quasi-pinch-in branch begins to exhibit strong pinch-in characteristics, and $V_{CB-crit}$ is once again well-defined. $|I_{B-rvs-crit}|$ increases with increasing bias and regains an approx. linear relationship vs. I_E, and $(M-1)_{crit}$ approaches a constant value. As self-heating effects become more prominent at higher currents, $(M-1)$ for a given V_{CB} is reduced due to phonon scattering, and as a result $V_{CB-crit}$ increases with I_E.

III. GENERAL RELATIONS

For our analysis of $V_{CB-crit}$ as a function of bias for CB-I_E operation, we use the following simple (but general) relations, as derived in [4] and explored in [7]:

$$|I_{B-rvs-crit}| = (v_o + r_e" I_E)/r_b'. \qquad (1)$$

Therefore,

$$(M-1)_{crit} \cong |I_{B-rvs-crit}|/I_E = (v_o/I_E + r_e')/r_b' \qquad (2)$$

and

$$\Delta V_{B-crit} = |I_{B-rvs-crit}| \cdot r_b' = v_o + r_e" I_E. \qquad (3)$$

The simple ratio approximation for $(M-1)$ in (2) relies on the assumption that the product $(M-1)\cdot\beta >> 1$. In all cases the subscript "crit" denotes that a parameter correlates to $V_{CB-crit}$ and represents the critical value to induce pinch-in at the given bias. The factor r_b' represents the destabilizing effect due to the distributed (extrinsic and intrinsic) nature of the base resistance during base current reversal. The product of I_{B-rvs} and r_b' (ΔV_B) denotes the increase in the average intrinsic base potential, which corresponds to an increase in the peak base potential at the center of the device. Similarly, ΔV_{B-crit} signifies the upper-limit of intrinsic base voltage offset before pinch-in occurs.

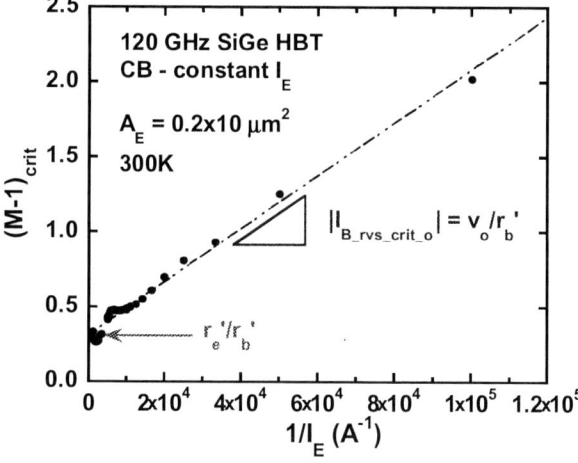

Fig. 3. $(M-1)_{crit}$ vs. $1/I_E$ data from a 120 GHz SiGe HBT with fitting using (2): slope = v_o/r_b', y-intercept = r_e'/r_b'.

The factor r_e' physically represents the degree to which the emitter resistance ballasts the intrinsic B-E voltage distribution to prevent the accumulation of junction potential at the center of the device. The factor v_o physically represents the $\Delta V_{B\text{-crit}}$ in the case of $r_e' = 0$, and provides the bias dependence of $(M-1)_{crit}$ that is observed empirically.

For example, in Fig. 3 the avalanche instability threshold $(M-1)_{crit}$ is extracted from CB-I_E output characteristics for a 120 GHz SiGe HBT ($A_E = 0.2 \times 10 \mu m^2$) and plotted as a function of $1/I_E$ to take the form shown in (2). The bias dependence of $(M-1)_{crit}$ is clearly observable and is related to the slope term v_o/r_b', which represents the "primary" critical reverse base current $|I_{B\text{-rvs-crit-o}}|$ in the absence of the stabilizing influence of the emitter ballast. The y-intercept term shows the useful approximation that $(M-1)_{crit}$ asymptotically approaches the ratio r_e'/r_b' at high currents. Both terms can serve as useful metrics for comparison of the resultant CB avalanche instability characteristics in different SiGe technologies.

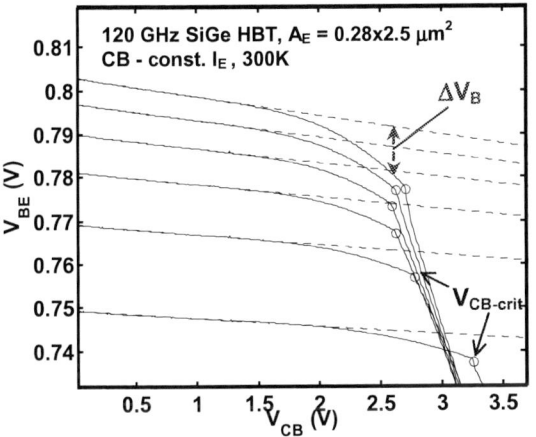

Fig. 4.   Extraction of ΔV_B and $\Delta V_{B\text{-crit}}$ from CB-I_E V_{BE} vs. V_{CB} characteristics.

IV. ANALYSIS

To obtain further insight into the nature of CB instabilities, the parameters ΔV_B and $\Delta V_{B\text{-crit}}$ are examined in detail. Using an approach similar to the one described in [10], the V_{BE} vs. V_{CB} characteristic from a CB-I_E measurement is used to infer the increase in base voltage that results from the potential drop of $I_{B\text{-rvs}}$ across the base resistance. As shown in Fig. 4, V_{BE} vs. V_{CB} is plotted for fixed I_E with a logarithmic fit of V_{BE} extrapolated from the region $0 < V_{CB} < BV_{CEO}$ to account for the Early effect and self-heating influences on V_{BE} as V_{CB} increases. The difference between V_{BE} and the extrapolated fit determines ΔV_B (and thus $\Delta V_{B\text{-crit}}$, as indicated in Fig. 4). By plotting ΔV_B vs. $I_{B\text{-rvs}}$ (not shown) an estimate for r_b' can be easily determined.

Fig. 5 shows the average value and standard deviation of $\Delta V_{B\text{-crit}}$ as a function of I_E for a sample size of 6 devices ($A_E = 0.2 \times 2.5 \mu m^2$). As shown by this plot, $\Delta V_{B\text{-crit}}$ deviates from its linear dependence on I_E (as does $I_{B\text{-rvs-crit}}$) at the transition between strong pinch-in and weak quasi-pinch-in regions (at I_E of approx. 100μA in this case). At this transition, an increase in the statistical variance of $\Delta V_{B\text{-crit}}$ is also clearly evident.

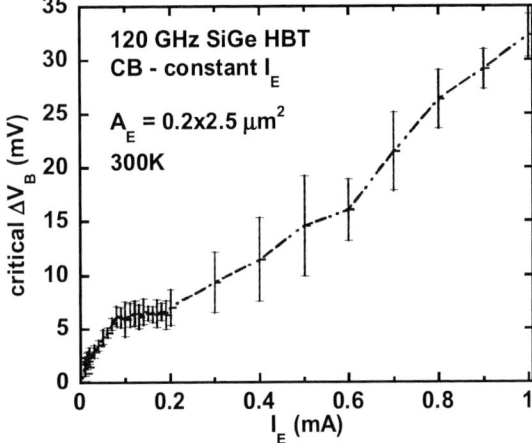

Fig. 5. $\Delta V_{B\text{-crit}}$ vs. I_E: mean and standard deviation for 6 devices.

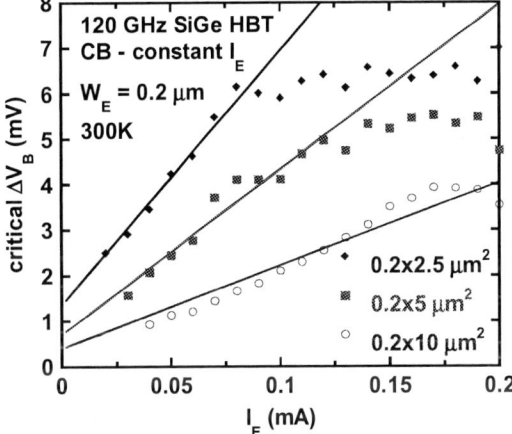

Fig. 6.   Average $\Delta V_{B\text{-crit}}$ vs. I_E (low injection) for 3 emitter lengths.

Fig. 6 shows the low current (strong pinch-in) $\Delta V_{B\text{-crit}}$ characteristics (averaged from 12 devices across 4 die) vs. I_E, for three emitter lengths. For each L_E, a least-squares fit was applied in the linear region to estimate r_e' and v_o (slope and y-intercept, respectively, as indicated in (3)). To contrast the analysis given in [4], which theoretically concludes v_o to be the thermal voltage V_T (26mV at 300K), our data clearly indicates that stability in practical SiGe HBTs is far more delicate, requiring a substantially lower base voltage increase to achieve pinch-in (on the order of 1 or 2 mV, 10-20x lower that in [4]). Our analysis also indicates that even at very low emitter currents the contribution of $r_e'' I_E$ is significant compared to v_o, so the ballasting influence of the emitter resistance on CB stability warrants consideration at all practical current levels.

V. GEOMETRY DEPENDENCE OF CB STABILITY

As shown in Fig. 6, a clear geometry dependence of the parameters related to the CB instability threshold (particularly the slope component r_e', which scales with L_E as expected) can be deduced, and analysis at higher currents (up to $I_E = 1$ mA) indicates that these trends persist.

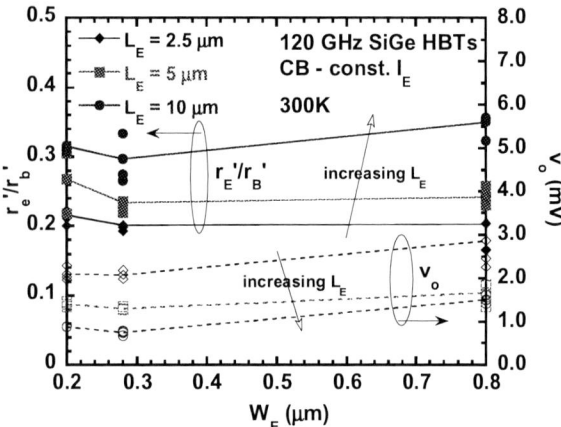

Fig. 7.   Extracted r_e'/r_b' and v_o for various device sizes.

As an example of the geometrical analysis of CB instability thresholds, the extracted parameters r_e'/r_b' (which represents the $(M-1)_{crit}$ behavior at the high current limit) and v_o (which represents the $(M-1)_{crit}$ behavior at the low current limit) of 120 GHz SiGe HBTs are plotted in Fig. 7 as functions of emitter width (W_E), for 3 different emitter stripe lengths (L_E). A total of 9 device geometries (4 samples of each to monitor variations and ensure a representative set of data) were measured.

Fig. 7 shows that with increasing L_E (from 2.5μm to 10μm) v_o decreases on the order of 1 mV (evidence of the importance of the emitter-length dimension in initiating pinch-in), while the r_e'/r_b' factor increased from approx. 0.2 to 0.3, implying that the reduction in r_e' with L_E is countered by a greater reduction in r_b' for 'enhanced' device stability, for a given I_E. Interestingly, these data show only small (and non-monatomic) variations of r_e'/r_b' and v_o as functions of W_E, suggesting that the stability of the intrinsic base potential distribution is far more sensitive to variations of L_E than W_E. This is a potentially important point for circuit designers.

When the results from Fig. 7 are used to generate fitted curves for the $(M-1)_{crit}$ vs. I_E characteristics (not shown), only a small overall variation is observed among all device geometries. However, when viewed instead as a function of emitter current *density* (J_E), a more pronounced geometry dependence becomes

apparent, as Fig. 8 demonstrates with the plot of (mean and standard deviation of) $V_{CB\text{-}crit}$ vs. I_E for the 3 emitter lengths measured. This variation is most clearly observed in the strong pinch-in region at low currents, exhibiting as much as 1V reduction in $V_{CB\text{-}crit}$ for a given J_E as L_E increases from 2.5μm to 10μm, a clearly significant point for circuit applications.

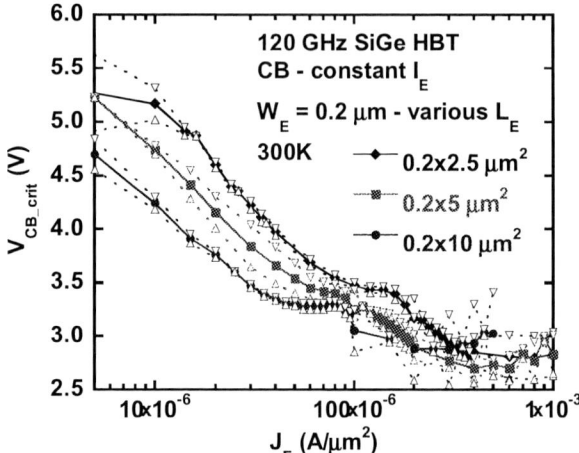

Fig. 8.   Mean (solid curves) and standard deviation (dashed curves) of $V_{CB\text{-}crit}$ vs. J_E for 3 emitter lengths.

VI. SUMMARY

We present a detailed analysis of the bias dependence of common-base (CB) collector voltage stability thresholds, highlighting distinct regions of operation, examining new statistical variations, and identifying and analyzing the key parameters v_o and r_e'/r_b'. These factors adequately capture complex stability limits across bias, and can be used to provide simple design and modeling guidelines, physical insight, and useful metrics for device or technology comparisons.

ACKNOWLEDGEMENT

This work was supported by IBM and the Georgia Electronic Design Center at Georgia Tech. The authors are grateful to D. Harame, D. Ahlgren, G. Freeman, J.-S Rieh, D. Herman, B. Meyerson, and the IBM SiGe team for their contributions.

REFERENCES

[1]   G. Freeman, *et. al., Proc. IEEE IRPS*, p. 332, 2003.
[2]   P. Lu *et.al., IEEE Trans. Elec. Dev.*, 36, p. 1182, 1989.
[3]   J. S. Rieh, *et. al., Solid State Elec.*, 48, p. 339, 2003.
[4]   J. D. Cressler, *IEEE Trans. Dev. Mat. Rel.*, 4, p. 222, 2004.
[5]   M. Rickelt, *et. al., IEEE Trans. Elec. Dev.*, 48, p. 774, 2001.
[6]   N. Rinaldi *et al., IEEE Trans. Elec. Dev.*, 52, p. 2009, 2005.
[7]   C. M. Grens, *et. al., Proc. IEEE IRPS*, p. 409, 2005.
[8]   J. D. Hayden, *et. al., IEEE Elec. Dev. Lett.*, 12, p. 407, 1991.
[9]   M. R. Shaheed, C. M. Maziar, *IEEE BCTM*, p. 42, 1992.
[10]   L. Vendrame *et.al., IEEE Tran. Elec. Dev.*, 42, p.1636, 1995.

2006 Bipolar/BoCMOS Circuits and Technology Meeting Proceedings

Electrothermal Phenomena in Bipolar Transistors and ICs: Analysis, Modeling, and Simulation

N. Rinaldi, V. d'Alessandro, and F. M. De Paola

Department of Electronics and Telecommunications Engineering, University of Naples
"Federico II", via Claudio 21, 80125, Naples, Italy, E-mail: nirinald@unina.it

Abstract — This paper addresses some relevant issues and recent developments concerning the analysis and simulation of electrothermal effects in bipolar devices and circuits.

Index Terms — Bipolar Junction Transistor (BJT), electrothermal simulation, thermal instability, impact ionization, bipolar modeling and simulation, analog circuits, device physics.

I. INTRODUCTION

It has been repeatedly stated that several current technology trends (namely, the increase in operation speed and integration density, the use of oxide-based isolation schemes and thermally resistive substrates such as GaAs, SOI, and silicon-on-glass) have made thermal management a crucial aspect in the design of RF/analog circuits. Thermal constraints impose a major change in the design process in that it is not possible to dissociate electrical and thermal phenomena, so that device temperature cannot be regarded as a mere parameter but rather as a state variable to be determined concurrently with the electrical analysis.

There are three fundamental requirements for a successful thermal-aware design:
1) A clear understanding of how thermal effects influence and limit the electronic operation at device level.
2) A clear understanding of the impact of electrothermal phenomena on basic electronic circuits.
3) The identification of the most significant issues and optimized methodologies to include electrothermal effects in commercial CAD tools, thus enabling fully automated and reliable electrothermal circuit simulation to become a common step in the design flow.

The aim of this paper is not only to review some recent developments in this field, but also to present some novel results. In Section II, we examine the limits to the operation of the bipolar transistor imposed by both self-heating (SH) and impact ionization (II). Optimal ballasting strategies are also discussed.

In Section III, we investigate the impact of SH effects on the operation of two basic building analog blocks: the current mirror and the differential pair. By means of experimental results, theoretical analysis

and numerical simulations it is shown that SH effects can significantly modify circuit operation with respect to the isothermal case. Theoretical conditions for the correct operation of these circuits are given for the first time. Both numerical simulations and theoretical results demonstrate that under severe self-heating conditions the behavior of the differential pair is profoundly modified, giving rise to a hysteresis phenomenon.

In Section IV we critically review the concept of thermal resistance in connection to the modeling of the distributed thermal behavior of devices. A novel compact expression for the thermal resistance of single-finger bulk devices is derived and its scaling properties are discussed.

Finally, some important issues concerning the development of electrothermal simulation codes are discussed in Section V, and a practical example is illustrated.

II. INSTABILITY EFFECTS IN BIPOLAR TRANSISTORS

The operation of BJTs can be limited by either SH, II of by the interplay between these effects. Although related to entirely different physical mechanisms, SH and II share a common feature in that they introduce a positive feedback action in the operation of the bipolar transistor. As a result, they introduce the same kind of singularities in the I-V behavior, which limit the operating area of the device. The effect of both SH and II can be studied by means of the simplified model illustrated in Fig. 1. Here E', B', and C' represent the device terminals and r_E, r_B, and r_C are the device intrinsic parasitic resistances. Also included are the external (ballast) resistances R_{Ex}, R_{Bx}, and R_{Cx}, so that the overall resistances applied to the device are $R_E = r_E + R_{Ex}$, $R_B = r_B + R_{Bx}$, $R_C = r_C + R_{Cx}$. V_{T0} is the thermal voltage at room temperature, I_T represents the transport current flowing across the neutral base region, $I_b = I_T/\beta_F$ denotes the ideal base current component, $I_{av} = (M-1)I_T$ represents the avalanche current, and M is the multiplication factor.

As well known, the positive action of SH may be viewed as a decrease of the base-emitter voltage needed to sustain a given value of I_C: at high collector voltages the dissipated power $P = I_B V_{B'E'} + I_C V_{CE'}$ causes an increase of the junction temperature $\Delta T = R_{TH}P$, and hence V_{BE} must decrease by a quantity $\phi\Delta T$ (ϕ is

1-4244-0458-4/06/$20.00 ©2006 IEEE

assumed positive) in order to maintain a constant value of I_C. The positive feedback introduced by II is quite similar: an increase of V_{CE} leads to an increase of the avalanche current I_{av}, and hence a decrease in the voltage drop $R_B I_B = R_B(I_b - I_{av})$ across the base resistance. As a result, the external voltage must decrease in order to keep I_C constant. Indeed, it can be shown that the internal voltage applied to the B-E junction is given by $V_{BEi} = V_{BE} - R_{eq}I_C$, being $R_{eq} = (R_B + R_E)/M\alpha_F - R_B$ an equivalent feedback resistance [1]. Note that R_{eq} decreases monotonically with increasing V_{CB} from $R_{EB} = R_E/\alpha_F + R_B/\beta_F$ for $V_{CB} \to 0$ (M=1) to $-R_B$ for $V_{CB} \to BV_{CBO}$ (M→∞). Since R_{eq} becomes negative at high V_{CB} values, the internal voltage V_{BEi} becomes higher than V_{BE}, which makes the device prone to instability. This suggests that the base resistance plays an important role in II-related instability.

From the simplified model of Fig. 1 it follows that the collector current can be approximated as

$$I_C = \frac{MI_S}{1 - V_{CE}/V_A} e^{\frac{V_{BE}+I_C\left[\phi R_{TH}\left(V_{eq}-R_x I_C\right)-R_{eq}\right]}{\eta V_{T0}}} \qquad (1)$$

where $V_{eq} = V_{CE} + V_{BE}/M\alpha_F$ is an "equivalent collector voltage" [1]. The "equivalent external resistance" $R_x = R_{Cx} + R_{Ex}/(M\alpha_F)^2 + R_{Bx}(1/M\alpha_F - 1)^2$ [2] takes into account the effect that the power dissipated on the external resistances does not contribute to SH.

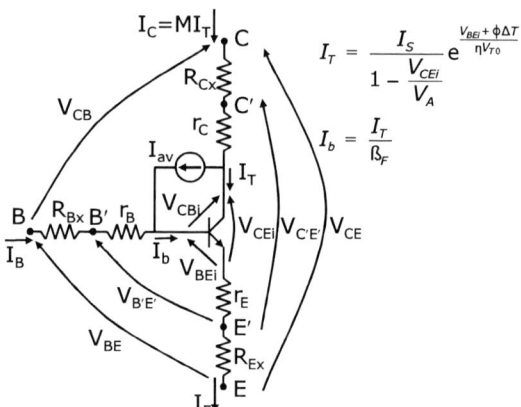

Fig. 1. Simplified bipolar transistor model including SH and II.

This model can be used to derive the condition of the onset of SH/II instability for a device driven with a constant V_{BE}. Fig. 2 shows the I_C-V_{CE} characteristics corresponding to different values of the base-emitter voltage. The results shown in this figure were obtained by means of the ATLAS 2-D simulation code with both SH and II included. Transport parameters and doping profiles were calibrated to fit experimental results relative to a single-finger 1×20 μm^2 Si *npn* BJT fabricated by means of a back-wafer contacted silicon-on-glass bipolar process [3]. Due to the low thermal conductivity of the glass substrate this device has an extremely high thermal resistance

(about 19×10^3 K/W). As can be seen, the output characteristics show a "flyback" behavior: the current increases monotonically with increasing V_{CE} until a "flyback" or "snapback" point is encountered where a negative resistance branch is originated. Note that this behavior occurs if either II or SH (or both) are activated. Also shown in Fig. 2 is the "flyback" locus, i.e., the collection of flyback points in the V_{CE}-I_C plane, which represents one of the boundaries of the safe operating area of the device (often referred to as "2nd breakdown" boundary). Note that the open-base breakdown voltage BV_{CEO} (also indicated in Fig. 2) is completely uncorrelated with respect to the instability onset for these driving conditions. From (1) the flyback condition can be derived as [2]

$$I_C = \frac{\eta V_{T0}}{\phi R_{TH}\left(V_{eq} - 2R_x I_C\right) - R_{eq}} \qquad (2)$$

Condition (2) includes SH, II, the effect of external resistances, but not high current effects [2]. If we neglect II then $V_{eq} \cong V_{CE}$ and $R_{eq} = R_{EB}$, while if we assume $R_x = 0$ (2) reduces to a result first obtained in [1]. On the other hand, if SH effects are negligible ($R_{TH}=0$), then (2) reduces to $I_C = -\eta V_{T0}/R_{eq}$, which indicates that II-induced instability can only occur if $R_{eq}<0$. This II-related flyback locus is also shown in Fig. 2. It can be seen that the considered device is mostly thermally limited. Generally speaking, (2) indicates that the SOA boundary tends to be limited by II at high V_{CE} values, and by SH at low V_{CE} and high I_C values. In order to enhance the SOA, we can i) reduce R_{TH}; ii) increase the ballasting resistances R_{Ex} and R_{Bx}. While both base and emitter ballasting can be used to improve thermal stability, it has been claimed that base ballasting should be preferred in HBTs, due to the negative temperature coefficient of the current gain β_F, which leads to an increase of R_{EB}, and hence of the ballasting action, at high dissipated power [4]. Note, however, that this picture is true only for a thermally limited device. If the SOA boundaries are also limited by II, then base ballasting leads to a widening of the SOA at low V_{CE} values but an SOA shrinkage at high voltages. This can be inferred from the fact that the limit value of the flyback locus for $V_{CB} \to BV_{CBO}$ depends only on R_B (see Fig. 2), so that an increase in R_B leads to a shift downwards of the flyback locus. In this case emitter ballasting should be preferred. Fig. 3 shows a comparison between different ballasting strategies. Also shown is the SOA boundary imposed by the maximum temperature specification. All results were obtained from numerical simulations of a silicon-on-glass BJT.

An additional undesirable thermally driven effect in bipolar transistors is the "current hogging" or "current crunch", which consists in a sudden focusing of the current density when a critical operating condition is reached.

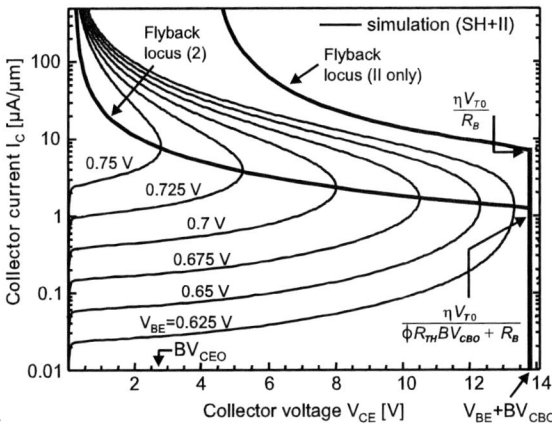

Fig. 2. I_C-V_{CE} characteristics under constant V_{BE} driving conditions.

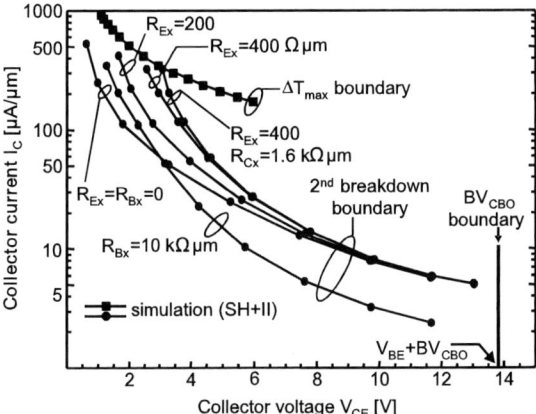

Fig. 3. Bipolar transistor SOA boundaries.

This effect can be activated by both SH and II, and, although not necessarily destructive, it should be avoided since is accompanied by hot spot formation and uneven active area utilization. To illustrate this phenomenon, let us consider the case of a bipolar transistor composed of two identical fingers biased with a constant emitter current $I_E=I_{E1}+I_{E2}$ forced into the common emitter input node, while a voltage V_{CB} is applied to the common collector node (Fig. 4). As can be seen, at low values of V_{CB} the two fingers show an identical behavior and carry an identical current $I_{C1}=I_{C2}$, as expected. As V_{CB} is increased, a critical "bifurcation" condition is eventually reached in which one device carries most of the current (see e.g., [4]-[6]), while the other finger is gradually turned off. In GaAs-based HBTs the onset of current bifurcation is accompanied by a decrease of current gain due to the negative temperature coefficient of the current gain in HBTs [4], [6]. It has been shown that this behavior must be attributed to the nonlinearity introduced by SH that causes the existence of multiple solutions [5]: after the bifurcation condition is reached, in addition to the "symmetric" operation mode ($I_{C1}=I_{C2}$), an "asymmetric" operation mode is allowed, in which one finger becomes hotter and hence carries most of the current ($I_{C1}<I_{C2}$), while

keeping the overall current $I_{E1}+I_{E2}$ constant. In practice, due to the inherent asymmetry in real devices, the symmetric operation mode ($I_{C1}=I_{C2}$) cannot occur, and only the asymmetric operation is observed. Including II, the bifurcation condition can be derived as [2]

$$I_C = \frac{\eta V_{T0}}{\phi\left(R - R_m\right)\left(V_{eq} - 2R_x I_C\right) - R_{eq}} \quad (3)$$

where R and R_m represent the self-heating and mutual thermal resistances of the fingers, respectively. This bifurcation locus is also shown in Fig. 4, which refers to the same device considered in Fig. 2. Equation (3) indicates that thermal coupling can substantially improve the immunity to current bifurcation. Indeed, if we increase R_m by e.g., reducing the spacing between the two fingers, the bifurcation current (3) increases. To understand this phenomenon, we note that when current bifurcation is triggered, one finger carries most of the current and becomes progressively hotter, while the other device is turned off and cools down. Thermal coupling tends to counteract this behavior, since the hotter device tends to heat up the cooler device, and vice versa. It is important to note that current bifurcation occurs independently of the driving condition at the input port (V_{BE}=const, I_B=const, or I_E=const) [5]. Finally, we remark that the flyback current is always lower than the bifurcation current, and therefore represents a more conservative condition for a safe device operation.

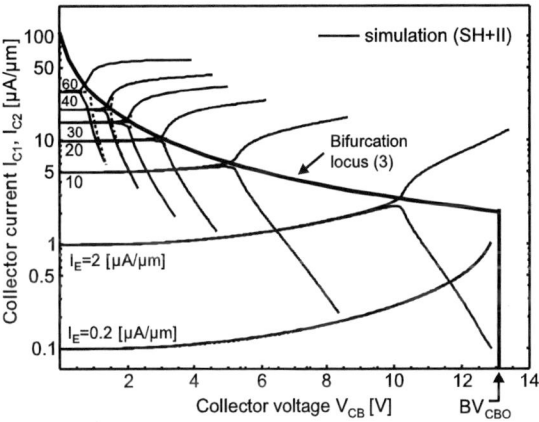

Fig. 4. Common-base output characteristics of a two-finger bipolar transistor.

III. ELECTROTHERMAL PHENOMENA IN BASIC ANALOG BUILDING BLOCKS

The analysis of the influence of electrothermal effects on the operation of analog circuits has received relatively less attention in the literature, although some simulation and experimental studies have been published. In the following we briefly sketch the results of a theoretical analysis of the current mirror and of the differential pair.

A. Analysis of the current mirror

The operation of the current mirror may be strongly modified by electrothermal effects. Indeed, it has been suggested that the nonisothermal operation of current mirror can be used as a means to investigate the relevance of SH effects in SOI BJTs [7].

Fig. 5 shows the measured output characteristics (solid lines) of a current mirror fabricated in GaAs technology with an emitter area of 56 μm^2 for the reference device (Q_1) and a mirror ratio of 2. The characteristic "flyback" behavior typical of SH operation is clearly noticeable, which makes the current mirror operation strongly nonideal. The different curves refer to I_{REF} values ranging from 2 mA to 10 mA with a step of 1 mA. Dashed lines were obtained by electrothermal circuit simulation (see Section V). For the lowest characteristic also indicated are: the reference current I_{REF}=2 mA, the ideal value of the output current $2I_{REF}$=4 mA, and the collector current of the reference transistor I_{C1}. Note that the flyback condition for the output transistor corresponds to a sudden decrease of I_{C1}, a behavior which can be assimilated to current bifurcation. Using (1) to model the collector current, and evaluating the current ratio I_{C2}/I_{C1}, it is possible to derive the following equation for the onset of the current bifurcation

$$I_{OUT} = \frac{\eta V_{T0}}{\phi \left(R_{22} - R_{12} \right)\left(V_{OUT} - 2R_{Ex}I_{OUT} \right) - R_{EB2}} \quad (4)$$

where avalanche effects have been neglected ($V_{eq} \cong V_{CE}$, $R_{eq} = R_{EB}$). Equation (4) is formally similar to condition (3) obtained for the two-finger device. In this case, however, due to the difference in active area, the thermal resistances and feedback resistances cannot be assumed equal ($R_{11} \neq R_{22}$, $R_{12} \neq R_{21}$, $R_{EB1} \neq R_{EB2}$). Also shown in Fig. 5 is the bifurcation locus as given by (4) (with R_{Ex}=0 for the considered devices). Again, (4) indicates that thermal coupling should be optimized in order to improve the SH behavior, a concept well known to analog designers.

B. Analysis of the differential pair

A theoretical analysis of the differential pair (see inset in Fig. 6) can be carried out on the basis of the simplified model given by (1). It can be shown that when SH effects are taken into account, the transfer characteristics of the differential pair are governed by the following equation:

$$x = \frac{I_{C1} - I_{C2}}{I_{CC}} = \tanh \left(v + \kappa x \right)$$
$$\kappa = \frac{\phi \left(R - R_m \right)\left(V_{CC} - R_L I_{CC} + V_{BE2} \right) - R_{EB}}{\eta V_{T0}} \frac{I_{CC}}{2} \quad (5)$$

where $x = (I_{C1} - I_{C2})/I_{CC}$ is the normalized current difference (with $I_{CC} = \alpha_F I_{EE}$) and $v = V_I/(2\eta V_{T0})$ is a normalized input voltage. Note that for negligible thermal effects (κ=0), (5) reduces to the classical

result x=tanh(v). As we will see shortly, the effect of the additional term κx due to electrothermal effects in the argument of the hyperbolic tangent in (5) is not only a distortion of the *i-v* characteristics of the emitter coupled pair (compared to the isothermal case), but also the introduction of multiple solution branches in the *i-v* behavior, which lead to a hysteresis phenomenon. Fig. 6 shows the normalized collector current $i_{C1} = I_{C1}/I_{CC}$ as a function of the normalized input voltage *v* for different values of the parameter κ (a complementary behavior is observed for i_{C2}). Solid lines were obtained from ATLAS simulations with all parameters extracted with reference to a silicon-on-glass process [3], while dashed lines refer to the above model. The curve for κ=0 was obtained from an isothermal analysis, while the curves for κ=0.55 and κ=2.26 were obtained by including SH effects, and considering two values of the supply voltage V_{CC}. It can be seen that for κ<1, SH introduces a simple distortion in the transfer characteristics. On the other hand, for κ>1 a completely different behavior is observed resulting in a hysteresis phenomenon, which can be illustrated as follows. Assume that the input voltage is initially negative, corresponding to point P_1. If we increase *v*, the operating point moves along the branch P_1-P_2-P_3. After point P_3 the operating point jumps to the branch P'_3-P'_4. If we now decrease the input voltage starting from point P'_4, the operating point moves leftwards along the branch P'_4-P'_3-P'_2, and then jumps to the branch P_2-P_1. Again, it can be recognized that the critical condition κ=1 corresponds to a bifurcation condition.

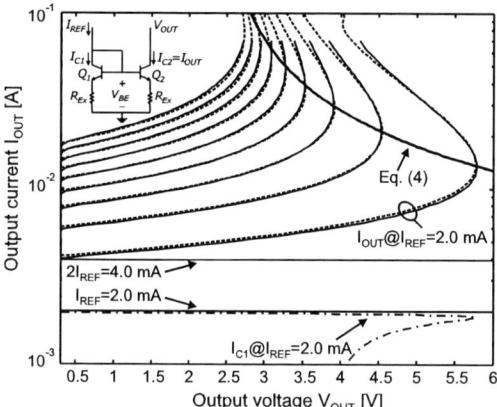

Fig. 5. I_{OUT}-V_{OUT} characteristics of a current mirror for different values of I_{REF}.

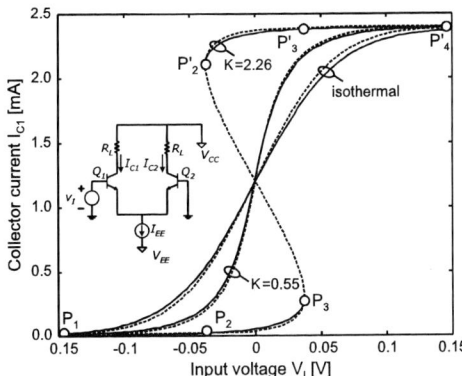

Fig. 6. Transfer characteristics of a differential pair for different values of parameter κ.

IV. R_{TH} Modeling and Scaling Properties in Single and Multifinger Devices

The temperature increase of a device is traditionally evaluated from the thermal resistance R_{TH}, which is defined as the ratio of the "junction temperature" to the dissipated power. In this definition it is implicitly assumed that the temperature variation across the device is negligible, so that distributed effects do not occur. In real devices, however, the temperature actually varies across the device [8], leading to a nonuniform current density distribution. Therefore, to accurately model a device one should in principle discretize the active area into elementary sections where the temperature and current density can be assumed to be uniform. The current of each elementary device must be then calculated by taking into account thermal interactions, i.e., by solving a rather complex electrothermal problem. This "distributed model" is indeed the most rigorous approach to model the electrothermal behavior of multifinger devices, where the temperature variation between fingers may lead to a strongly uneven current distribution. In most practical cases, however, a simplified "lumped approach" is used, where a "junction temperature" increase ΔT is assigned to the device, and calculated from the thermal resistance as $\Delta T = R_{TH} P$. In the following, we wish to analyze under which conditions the "lumped model" approach is accurate, and how the thermal resistance of the lumped model should be defined to account for distributed effects, at least in a first-order approximation. It is also important to note that the thermal resistance involved in the "lumped model" approach is the quantity that is obtained by conventional R_{TH} measurement methods based on a temperature sensitive electrical quantity (e.g., V_{BE}).

Let us consider a semiconductor chip with a heat dissipating device. Neglecting initially nonlinear effects, the temperature field will be proportional to the dissipated power P: $\Delta T(\mathbf{r}) = R(\mathbf{r})P$, where the function R depends on the position inside the chip $\mathbf{r}=(x,y,z)$ in which the temperature is being evaluated,

as well as on the geometry of the heat source and on material and technology properties of the substrate. In bipolar transistors the heat is generated in the base-collector space-charge region, so that the heat source can be approximated by a thin rectangle of dimensions $W_E \times L_E$ located at a depth $z=z_{HS}$ beneath the surface, where the heat source depth is inside the BC SCR. In some studies, the thermal resistance is defined by evaluating $R(\mathbf{r})$ at a reference point, e.g., the top center of the device [9], [10] where the temperature profile reaches a maximum, thereby leading to an overestimation of R_{TH}. A more reasonable approach suitable for modeling purposes is to define R_{TH} as an average value given by

$$R_{TH} = \frac{1}{W_E L_E} \iint_{A_E} R(x, y, z_{ref}) f(x, y) \, dx \, dy \quad (6)$$

where $f(x,y)$ is a weight function. Note that the integral should be evaluated over the active (emitter) area $A_E = W_E L_E$ at a depth $z=z_{ref}$ beneath the chip surface equal to the base-emitter junction $z_{ref}=X_{je}$ [11]. This definition can be straightforwardly extended to the case of the mutual resistance R_{THm} between two devices, by simply evaluating the integral over the active area of the neighboring device. Although in a more refined choice the collector current density $J_C(x,y)$ should be used as the weight function [11], in the following we assume f=1 for the sake of simplicity.

It is possible to derive a closed-form analytical expression for the "average" thermal resistance of a single-finger device, as defined by (6). To this end, we need a model for the temperature field function $R(\mathbf{r})$. For bulk devices (i.e., non trenched) this can be obtained from the analysis presented in [9]. By evaluating the double integration in (6) with f=1 and $z_{ref}=0$, we obtain after considerable manipulation

$$R_{TH} = \frac{1}{4k\sqrt{W_E L_E}} \chi(a, \zeta) \quad (7)$$

being k the thermal conductivity of the substrate and χ a correction factor which depends on the aspect ratio $a=L_E/W_E$ and normalized heat source depth $\zeta = z_{HS}/\sqrt{W_E L_E}$. This factor is expressed as $\chi = [\varphi(a,\zeta) + \varphi(1/a,\zeta)]2/\pi$ where the function φ is given by

$$\varphi(a, \zeta) = \frac{2}{3}\zeta^3 - 2\zeta \arctan\left(\frac{1}{\zeta\Delta}\right) + \zeta^2\left[\frac{4}{3}\left(\frac{\Delta}{2} - \delta\right) + \right.$$
$$\left. +\sqrt{a}\ln\left(\frac{-\sqrt{a}+\Delta}{\sqrt{a}+\Delta}\frac{\sqrt{a}+\delta}{-\sqrt{a}+\delta}\right)\right] + \frac{2}{3}a(\delta-\Delta) + \frac{1}{\sqrt{a}}\ln\frac{\sqrt{a}+\Delta}{-\sqrt{a}+\Delta} \quad (8)$$

being $\Delta = (\zeta^2 + a + 1/a)^{1/2}$ and $\delta = (\zeta^2 + a)^{1/2}$. Note that this result is exact as long as the device size is small compared to the chip size and thickness, so that the effect of chip boundaries on the temperature field can be neglected [9]. In [12] an empirical correction term $\chi = [1 + (\pi\zeta/2)^2]^{-1/2}$ is introduced, which does not include the dependence on the aspect ratio.

Figure 7 shows R_{TH} as a function of the emitter length L_E with a constant emitter width $W_E = 0.28 \, \mu m$. Two cases are considered: 1) a single-finger device

($N_f=1$) with two different values of the heat source depth z_{HS}; 2) a three-finger device ($N_f=3$) with $z_{HS}=0.2$ µm and different values of the center-to-center spacing S. The thermal resistance calculated from (7)-(8) (solid lines) for the single-finger device is compared to experimental results for bulk Si devices reported by Pacelli *et al.* [12] (dots) as well as to the results obtained by the empirical approximation also proposed in [12] (dashed lines). It can be noted that the present model predicts a significantly more marked dependence on the heat source depth compared to the approximation of [12]. This implies that vertical scaling of bipolar transistors may adversely affect the thermal behavior. It can also be seen that including a non zero value for z_{HS} gives rise to a weaker scaling behavior of R_{TH} with respect to the emitter length [11]. Another implication of such dependence is that high current effects (Kirk effect) cause a variation of the BC SCR, leading to an increase of the heat source depth at high current levels and hence a decrease of R_{TH} [13].

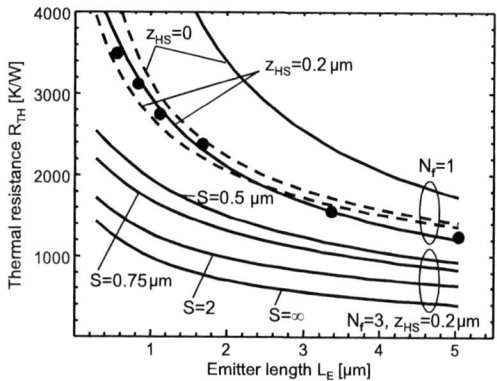

Fig. 7. R_{TH} vs. L_E for 1-finger [calculated from (7), (8)] and 3-finger [calculated from (9)] bulk-Si transistors.

The lumped model approach can be extended to multifinger devices. As discussed before, a "distributed" analysis of a multifinger transistor requires each finger (or parts of each finger) to be treated as a single device interacting thermally and electrically with the neighboring fingers. For such analysis the self-heating thermal resistance R of each finger, as well as the mutual thermal resistance R_{ij} between the i-th and j-th finger, is needed. Instead, in a "lumped model" approach the device is treated as a single device characterized by an overall thermal resistance R_{TH} which can be calculated by means of the average resistance method outlined above as

$$R_{TH} = \frac{1}{N_f^2} \sum_{i,j=1}^{N_f} R_{ij} \qquad (9)$$

where N_f is the number of fingers. In many cases, assuming identical fingers, the thermal resistance matrix is symmetrical: $R_{ii}=R$ and $R_{ij}=R_{ji}$. Since the mutual resistance R_{ij} is always lower than the self-heating resistance R, it can be seen that R_{TH} is lower for a multifinger device compared to a single-finger device ($R_{TH}=R$). As shown in Fig. 7, the thermal resistance decreases when the spacing S between adjacent fingers increases. It is interesting to note that for a large spacing, thermal coupling effects become negligible ($R_{ij}=0$) and R_{TH} reaches the lower bound $R_{TH}=R/N_f$ [10], as also indicated in Fig. 7 (S=∞).

Several simulations have been carried out to compare the lumped and distributed models in multifinger devices. Fig. 8 shows the I_C-V_{CE} characteristics of a three-finger silicon-on-glass BJT biased with a constant V_{BE}. As can be seen, the lumped model gives accurate results before the onset of the flyback. Other simulations carried out by varying the number of fingers gave similar results, showing that the lumped approximation is fairly accurate, particularly for strongly thermally coupled fingers. This has an important implication in that the locus of the onset of the flyback phenomenon in a multifinger device can be predicted using the simple equation (2), by considering the "average" thermal resistance given by (9). After the onset of the flyback, the model becomes less accurate since the temperature and current distribution become less uniform, as the central finger becomes hotter and thus draws more current. The "lumped model" cannot obviously account for current hogging effects, as those shown in Fig. 8, and its validity is therefore limited to the operation region where these effects do not occur. Concluding, as a rule of the thumb, the lumped model approximation remains accurate provided that the operation point remains inside the SOA limits dictated by either flyback or current bifurcation. When there boundaries are exceeded, the current and temperature distributions become nonuniform, and the lumped approximation no longer applies.

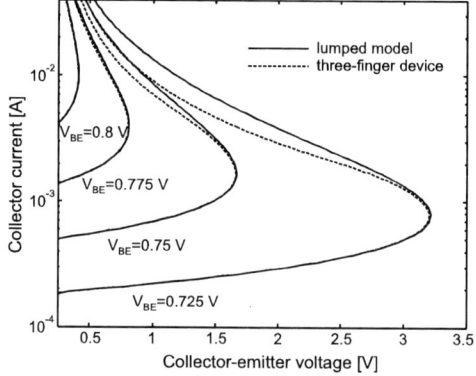

Fig. 8. Comparison between distributed and lumped models for a 3-finger BJT.

V. ELECTROTHERMAL SIMULATION

In this section, the most important approaches to device/circuit electrothermal simulation will be reviewed, and some important issues and limitations will be discussed.

A. Physical device simulation

Presently available device simulation tools offer the possibility to perform a fully consistent SH simulation by solving the lattice heat flow equation jointly with the semiconductor equations. However, it should be noted that device simulation tools have limited capabilities for electrothermal analysis:
i) number of mesh nodes constraints limit the structure complexity;
ii) thermal boundary conditions are over simplified;
iii) while 2-D simulators provide a realistic description of the electrical behavior in most practical cases, this is not the case for the thermal behavior [13]. Some approaches to overcome these limitations, at least partially, are discussed in [13].

B. Circuit simulation

Electrothermal circuit simulators are circuit simulators with two added features: i) devices are modeled by compact models provided by a thermal node [14]; ii) some thermal model is implemented for the evaluation of the junction temperatures on the basis of the knowledge of the power dissipated by the devices. There are two possible implementations of the thermal model:
a) A numerical solution of the heat flow equation in the chip at each iteration step of the electrical analysis. This approach has the main advantage that all heat transfer modes (conduction, convection, and radiation) as well as nonlinear effects can be included. On the other hand, from the CPU and memory requirements point of view, it is practically impossible to perform a numerical simulation in parallel to a circuit simulation.
b) In view of the above limitations, reduced thermal models have to be used to represent the thermal behavior of electronic devices and/or systems. In the most general form, the steady-state thermal behavior of any electronic system components can be described in terms of the thermal resistance matrix $\underline{R}_{TH}$ [15]. In this regard, some important considerations are in order:
b1) although widely used, thermal equivalent networks must be regarded simply as a possible implementation of the thermal resistance matrix of the system [15];
b2) the use of compact thermal models can be applied to any complex electronics system composed of different parts exchanging heat with each other (e.g., MMICs, packaged ICs, multichip modules, etc.). In a reduced model representation, each part of the system is modeled by discretizing its surface in a small set of portions (thermal nodes) where the temperature and heat flux can be assumed to be uniform. The heat fluxes and temperatures at thermal nodes are related by a set of linear equations, which can be generally described by means of a thermal resistance matrix [15]. Such a model is said to be

boundary condition independent, because no value is assigned to the temperature or heat flux at the boundary nodes. Instead, these must be determined when the system parts are "put together" by imposing the temperature and heat flux continuity conditions;
b3) the thermal model representation in terms of a thermal resistance (or impedance in the case of time or frequency domain analysis) is fully compatible with circuit simulation engines, by interpretation of thermal impedance matrices in terms of generalized multiport network parameters, so that the synthesis of equivalent RC thermal networks can be avoided.

It should be kept in mind, however, that linear compact models such as the thermal resistance representation (or equivalent thermal networks) do not include nonlinearities, such as those due to the temperature dependence of material parameters [16] (thermal conductivity and diffusivity), or to nonlinear boundary conditions (heat transfer by convection or radiation). Although these effects can be approximately modeled by averaged model parameters, it is possible to extend the validity of linear models by means of suitable variable transformations. For example, it is well known that the temperature dependence of k makes the steady-state heat flow equation nonlinear. This nonlinear equation can be linearized by applying the Kirchhoff transform [14], [17], so that the thermal problem can be treated as a linear problem (and the corresponding thermal resistance matrix can be determined). Once the linearized temperature is evaluated, the "real" temperature can be found by an analytical inverse transform. However, it should be noted that, strictly speaking, the Kirchhoff transform cannot be applied to non-homogeneous systems, such as multilayer structures [17]. In addition, in the time dependent case, the heat flow equation is not linearized upon application of the Kirchhoff transform, although a time transformation can be introduced to improve the accuracy of a linear approximation [14].

As pointed out above, the most general approach to include electrothermal interactions in CAD tools developed for RF/analog circuit design is through the thermal resistance matrix $\underline{R}_{TH}$ (or the thermal impedance matrix for time or frequency domain simulation). The thermal resistance matrix can be directly obtained from measurements (e.g., [18]), or from the solution of the heat flow equation by means of either analytical or numerical methods. In principle, 3-D numerical solutions such as FEM are to be regarded as the most accurate and general methods, as they allow to treat complex geometry structures (trench/STI isolation, metal layers, etc.). Scaling laws can be extracted from numerical simulations (or measurements), so that the thermal resistance matrix can be automatically evaluated from layout data. This approach, however, is more difficult to apply for the calculation of mutual resistances R_{ij}, which depend on the relative position of the devices, and on the distance from chip boundaries.

Several analytical methods (Green's functions, e.g., [9], [13], Fourier series [14], etc.) can be used to solve the heat flow equation. Although these methods are developed for simple geometries, it is also possible to treat arbitrary geometries (e.g., trenched devices [19]), by discretization of the domain into a set of subvolumes with a simple shape that can be modeled using a thermal matrix resistance representation [14].

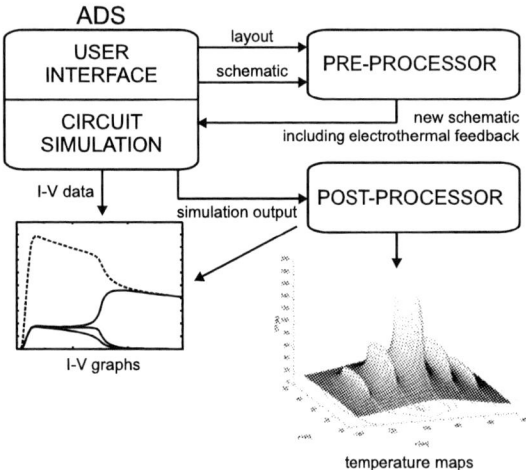

Fig. 9. Scheme of an electrothermal simulation code.

Fig. 9 shows a possible scheme of an electro-thermal simulation tool based on a commercial code (Agilent ADS) [20]. This scheme consists of three basic blocks: a pre-processing routine, the main ADS core, and a post-processing code. If not already available, the thermal resistance matrix $\underline{R}_{TH}$ is automatically evaluated from the layout file associated to the circuit as follows. First, the user provides information about depth and thickness of the heat sources along with the thermal properties of the substrate. Subsequently, the code scans the layout file in order to detect the coordinates of the emitter windows. Lastly, the thermal resistance matrix is computed by means of the closed-form analytical formulation proposed in [9]. As a second step, the pre-processing routine creates a new circuit component that performs the calculation of the power dissipated by each transistor (on the basis of the thermal resistance matrix) and the corresponding junction temperature increase. The circuit schematic is then automatically modified by adding this electrothermal feedback block, which can be implemented in ADS by means of a Verilog-A behavioral description, or using SDD (Symbolically

Defined Device) parts. These are multi-port components specified by user-defined equations. Note, however, that in the case of transient analysis the dissipated power must be properly calculated by excluding reactive elements that do not contribute to heating [21]. After the electrothermal analysis is completed, a post-processing routine can be invoked to plot the temperature field for assigned bias conditions.

ACKNOWLEDGEMENT

The authors wish to thank Dr. P. Zampardi for providing test structures and technology data.

REFERENCES

[1] T. Vanhoucke and G. A. M. Hurkx, *Proc. IEEE BCTM*, pp. 37-40, 2005.
[2] N. Rinaldi and V. d'Alessandro, *IEEE Trans. El. Dev.*, vol. 53, no. 7, pp. 1683-1697, 2005.
[3] L. Nanver *et al.*, *IEEE Trans. El. Dev.*, vol. 51, no. 1, pp. 42-50. 2004.
[4] W. Liu *et al.*, *IEEE Trans. El. Dev.*, vol. 43, no. 2, pp. 245-251, 1996.
[5] N. Rinaldi and V. d'Alessandro, *IEEE Trans. El. Dev.*, vol. 52, no. 9, pp. 2009-2033, 2005.
[6] W. Liu, *IEEE Trans. El. Dev.*, vol. 42, no. 6, pp. 1033-1038, 1995.
[7] J. Kim *et al.*, *Proc. IEEE BCTM*, pp. 305-308, 2004.
[8] J. Andrews *et al.*, *Proc. IEEE BCTM*, pp. 50-53, 2005.
[9] N. Rinaldi, *Solid-State Electron.*, vol. 44, no. 10, pp. 1789-1798, 2000.
[10] D. Walkey *et al.*, *IEEE Trans. El. Dev.*, vol. 49, no. 8, pp. 1375-1383, 2002.
[11] A. Reid *et al.*, *Proc. IEEE BCTM*, pp. 130-133, 2000.
[12] A. Pacelli *et al.*, *IEEE Trans. El. Dev.* vol. 49, no. 6, pp. 1027-1033, 2002.
[13] V. d'Alessandro and N. Rinaldi, *Solid-State Electron.*, vol. 46, no. 4, pp. 487-496, 2002.
[14] W. Batty *et al.*, *IEEE Trans. Comp. and Packag. Tech.*, vol. 24, no. 4, pp. 566-590, 2001.
[15] E. G. T. Bosch, *IEEE Trans. Comp. and Packag. Tech.*, vol. 26, no. 1, pp. 173-178, 2003.
[16] J. C. J. Paasschens *et al.*, *Proc. IEEE BCTM*, pp. 96-99, 2004.
[17] F. Bonani and G. Ghione, *Solid-State Electron.*, vol. 38, no. 7, pp. 1409-1412, 1995.
[18] N. Nenadović *et al.*, *IEEE J. Solid-State Circ.*, vol. 39, no. 10, pp. 1764-1772, 2004.
[19] J.-S. Rieh *et al.*, *IEEE Trans. El. Dev.*, vol. 52, no. 12, pp. 2744-2752, 2005.
[20] F. M. De Paola *et al.*, *Proc. IEEE ISPSD*, pp. 41-44, 2006.
[21] C. C. McAndrew, *Proc. IEEE BCTM*, pp. 200-203, 1992.

2006 Bipolar/BoCMOS Circuits and Technology Meeting Proceedings

Reliability Issues in SiGe HBTs Fabricated on CMOS-Compatible Thin-Film SOI

Marco Bellini, Tianbing Chen, Chendong Zhu, John D. Cressler, and Jin Cai [1]

School of Electrical and Computer Engineering, 777 Atlantic Drive, N.W.
Georgia Institute of Technology, Atlanta, GA 30332-0250, USA (bellini@ece.gatech.edu)

[1] IBM Thomas Watson Research Center, P.O. Box 218, Yorktown Heights, NY 10598 USA

Abstract — **A comprehensive investigation of reliability issues in both fully-depleted and partially-depleted SiGe HBTs-on-SOI is presented. The devices were subjected to "mixed-mode" stress at 300 K and at 77 K, and we have analyzed the changes in base current I_B, collector resistance R_C, M-1, and AC performance. A comparison of mixed-mode stress to conventional reverse EB bias stress is also made, and significant differences are observed. The thermal resistance R_{th} of the devices is extracted over the 50 K – 300 K range.**

I. INTRODUCTION

Silicon-Germanium Heterojunction Bipolar Transistors (SiGe HBTs) fabricated on Silicon-on-Insulator (SOI) substrates have recently received increasing attention [1]-[3]. These devices combine the well-known high gain, low noise, high output resistance, and high transconductance of conventional bulk SiGe HBTs, with the many advantages of SOI, which include: reduction of device parasitics and signal cross-talk, capability for high temperature operation, decreased vulnerability to radiation-induced soft errors (a significant problem in bulk SiGe HBTs), significant reduction of C_{CS}, and elimination of latch up. From a radiation perspective, SiGe HBTs-on-SOI share the established built-in radiation tolerance of bulk SiGe HBTs and potentially an additional reduced vulnerability to single event upset due to the thin SOI layer. These characteristics, coupled with the fact that SiGe HBTs-on-SOI show capability of operation in the temperature range between 20 K and 473 K make this device potentially an ideal candidate for extreme environment applications such as space electronics.

Clearly, before this new technology can be used for such applications, it must be proven reliable. The effect of substrate bias during room temperature stress on fully-depleted SiGe HBTs-on-SOI was discussed in [4], showing that a grounded substrate appears to be the worst case stress condition. In the present work, for the first time, SiGe HBTs-on-SOI are stressed at 77 K and compared with stress applied at 300 K, noting the effects on I_B, current gain, and avalanche multiplication. The effects of stress on the AC characteristics are also

investigated, and comparisons are made to more conventional reliability burn-in techniques.

II. DEVICE STRUCTURE AND PRIOR STUDIES

Fabricating a SiGe HBT-on-SOI requires a fundamentally different structure from a bulk SiGe HBT, because the silicon layer is too thin to accommodate the (thick) heavily doped sub-collector needed in high-speed transistors. Recently, a new "folded" SiGe HBT structure has been proposed [1]. While the emitter and base doping are comparable to a second-generation SiGe HBT bulk device, the sub-collector is replaced here by either a fully or partially-depleted collector (by changing the doping), as shown in Fig. 1. The absence of the sub-collector results in a 2D current flow, which can be significantly altered by the substrate voltage V_S [4]. The devices investigated in this work share the same substrate as a 130 nm SOI CMOS technology, featuring a 120 nm silicon layer on top of a 140 nm buried oxide layer.

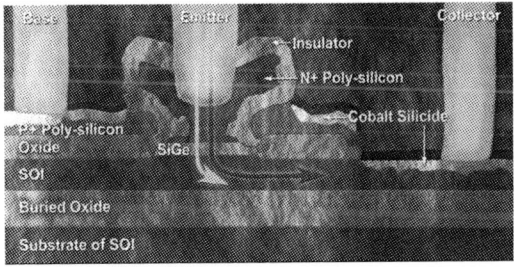

Fig. 1. Cross-sectional SEM of the SiGe HBT-on-SOI. The current flow can be separated into a vertical path directly under the emitter and a lateral path parallel to the SOI/Buried Oxide interface, as indicated by the arrows.

Since collector doping also significantly affects the DC and AC performance of the device, a fully-depleted transistor with a collector doping N_C of 1.5 x 10^{17} cm^{-3} and a partially-depleted transistor with N_C of 4.8 x 10^{17} cm^{-3} are both investigated here in order to assess any impact of collector doping on reliability. As shown in prior experiments, the substrate voltage V_S affects the

1-4244-0458-4/06/$20.00 ©2006 IEEE

current flow and the electric field, altering f_T, f_{max}, R_C, and M-1 [4]-[5].

III. EXPERIMENT

The devices were subjected to a "mixed-mode" stress. [6]-[8]. This technique is based on simultaneous application of both high current density and collector-base voltage, and is thought to emulate actual mixed-signal circuit operation in a better way than more traditional stress techniques [9]-[10]. Measurements were performed with an Agilent 4155 Semiconductor Parameter Analyzer and a closed-cycle cryogenic test system.

A. DC performance

All devices showed negligible degradation until a current as high as I_E=4 mA – which is about 20 times the peak f_T current – was applied. Overstressing was applied also in reliability studies of the bulk SiGe HBTs that share the same emitter and base profiles, but in that case the stress current needed was only 4-5 times the peak f_T current [8]. Fig. 2 shows the effect of "mixed-mode" stress at I_E=4 mA and V_{CB}=3 V for the fully-depleted device at 300 K and at 77 K.

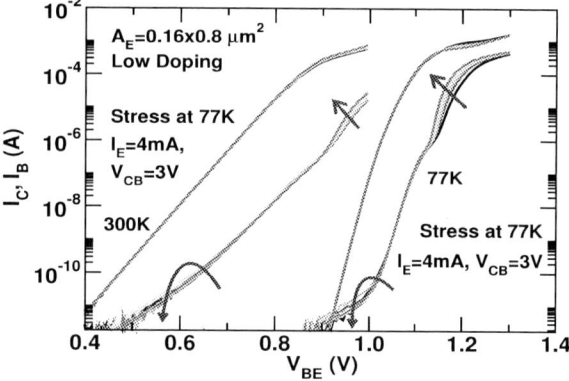

Fig. 2. Effects of stress on the Gummel characteristics of a fully-depleted SiGe HBT-on-SOI after a 4 hour stress sequence, at both 300 K and at 77 K.

The forward and inverse mode Gummel characteristics were measured at 40 logarithmically-spaced points over a total stress time of 4 hours. The data show a small increase in the base current at low V_{BE} and the presence of significant spontaneous damage recovery. The other most notable effect is the increase in the quasi-saturation effect at high V_{BE}, resulting from an increase in collector resistance R_C. This degradation is especially relevant for the fully-depleted SiGe HBT-on-SOI device which is characterized by a larger R_C than a comparable bulk device because of lower collector doping. This damage shows no sign of recovery and becomes more significant at low temperatures due to

additional collector freeze-out. The stress-induced base current of the partially-depleted device behaves similarly; however, the increase in R_C appears more modest, due to its higher collector doping.

The presence of spontaneous damage recovery has been consistently observed in multiple devices at different temperatures. Yet, interestingly, the amount of recovery is shown to vary substantially between samples. Although the exact amount of I_B degradation cannot be reliably predicted (this is in itself a potential reliability issue), the amount of leakage current is modest and confined to a low-current region rarely employed in actual circuit design. The presence of fluctuations in the excess base current such as those shown in Fig. 3 was reported in [9] and attributed to a process of bond breaking and creation of a dangling silicon bond at the Si-SiO$_2$ interface, followed by re-passivation of such dangling bonds. The high junction temperature reached during "mixed-mode" stress due to the large R_{th} of the device could enhance the re-passivation process, as suggested in [9].

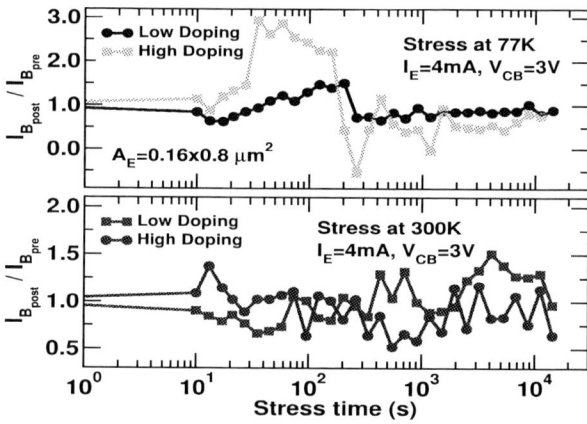

Fig. 3. Ratio of post-stress and pre-stress base current extracted at a voltage V_{BE} such that $I_{B,pre}$ is 20 pA.

As further argued in [9], fluctuations would be observed especially if the number of interface traps is small and the rates of generation and annealing are slow. The inverse mode operation is similarly affected by both degradation and recovery. However, since the inverse mode Gummel characteristics already show a high base leakage current – possibly due to interface defects introduced by the SOI fabrication process – the relative degradation introduced by stress is negligible.

Stress also significantly affects the avalanche multiplication characteristics (unlike in bulk devices), changing M-1 without showing any sign of recovery during stress. As it is shown in Fig. 4, the effects on the fully-depleted device at 77 K include a marked reduction of M-1 at lower collector-base voltages V_{CB}, and a subsequent increase of M-1 at higher collector voltages.

M-1 in the low V_{CB} range is associated with avalanche multiplication in the vertical current flow path, as observed in [4] and confirmed here by simulations, while M-1 in the high V_{CB} range is due to multiplication in the lateral current flow path. At 300 K, the effects of stress are different for $V_S=0$ V and $V_S=20$V, possibly as a result of the different current flow path in the device. BV_{CE0} was extracted from the base current inversion point and is shown in Table 1. The results are consistent with Fig. 4 if one observes the variation in M-1 at a fixed V_{CB} close to the original BV_{CE0}. Interestingly, "mixed-mode" stress results in an increase of BV_{CE0} at 77 K and a small decrease at 300 K.

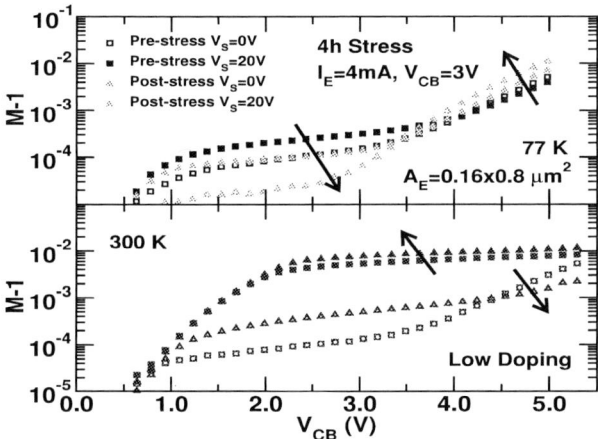

Fig. 4. Effects of mixed-mode stress on M-1 of a fully-depleted SiGe HBT-on-SOI.

Conversely, the M-1 of the partially-depleted device is hardly affected by substrate voltage V_S or by stress, both at 300 K and 77 K. Consequently, the impact of stress and of V_S on BV_{CE0} is minimal - the values of 1.35 V at 300 K and of 0.75 V at 77 K remain unchanged. It was also observed that substrate bias V_S and x-ray (or proton) irradiation have little effect on M-1 of this device.

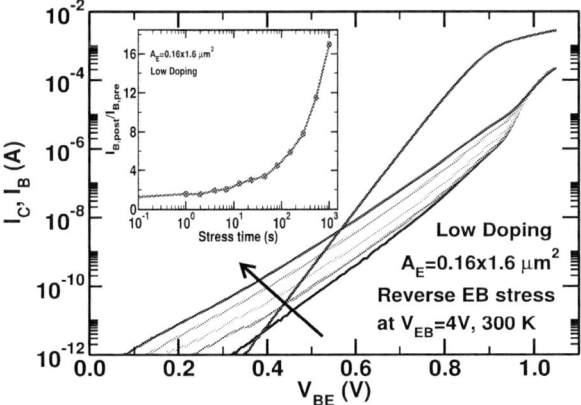

Fig. 5. Excess base-current of a fully-depleted SiGe HBT-on-SOI stressed for 1000 seconds with $V_{EB}=4$ V.

TABLE I
BV_{CE0} OF A FULLY-DEPLETED SiGe HBT-ON-SOI

Temperature	$V_S=0$ V	$V_S=20$ V
77 K pre-stress	3.18 V	1.53 V
77 K post-stress	3.23 V	2.73 V
300 K pre-stress	4.59 V	1.69 V
300 K post-stress	4.44 V	1.59 V

Given the limited impact of "mixed-mode" stress on this device (clearly good news from a reliability point of view) more traditional reverse emitter-base bias stress, as described in [10], was also applied.

By using an increasing V_{EB} the damage threshold to reverse EB bias stress was determined to be between 3 and 4 V, as shown in Fig. 5. In direct contrast to "mixed-mode" stress, this stress mechanism results in large degradation without any sign of recovery, possibly because self-heating effects are less important during reverse emitter-base bias stress. Unlike for "mixed-mode" stress, no change in the collector resistance was observed. "Mixed-mode" stress was applied after reverse EB bias stress, resulting in a sharp reduction of the non-ideal I_B component and confirming the presence of an underlying self-annealing mechanism.

Calibrated TCAD simulations were used to show that due to the reduced collector doping and to the nature of 2-D doping profile of the folded collector, the peak of the electric field is situated far from the emitter-base spacer in the case of "mixed-mode" stress, as shown in Fig. 6, while it is obviously adjacent to the EB spacer for the reverse EB bias stress, consistent with our data.

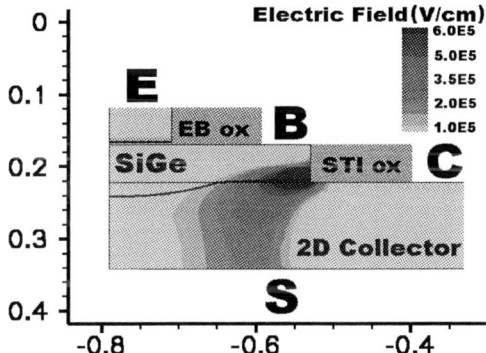

Fig. 6. Electric field contour of a fully-depleted SiGe HBT-on-SOI biased at $I_E=1$ mA and $V_{CB}=3$ V.

Comparing simulations of these SiGe HBTs-on-SOI with those of a corresponding bulk device, as described in [8], one can easily observe that the electric field distribution and the position of its peak are significantly altered by the differences in the collector profiles. Since mixed-mode" stress combines high current operation with large collector-base bias, it is not surprising that the collector doping affects the extent of the stress damage in a fundamentally different manner.

B. Thermal resistance

Although self-heating is obviously a concern for all SOI devices, at low temperatures the thermal conductivities of both silicon and SiO_2 naturally increase. As shown in Fig. 7, the thermal resistance R_{th}, extracted using the technique detailed in [11], decreases significantly with cooling. Even though R_{th} is about 5 times that of a comparable bulk device, we can expect some mitigation of any self-heating triggered reliability issues at low temperatures, clearly good news for cryogenic applications.

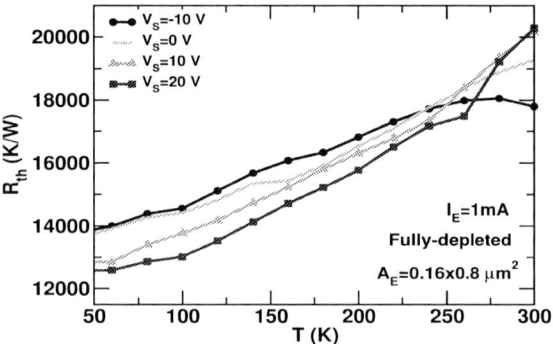

Fig. 7. Thermal resistance of a fully-depleted SiGe HBT-on-SOI in the temperature range between 50 K and 300 K.

C. AC Performance

The effects of stress on AC performance have also been investigated under "mixed-mode" stress, using a fixed stress time of 1,000 seconds. As shown in Fig. 8, an emitter current I_E of 8 mA (about 4 times the peak f_T current) and V_{CB} of 2 V result in negligible changes in

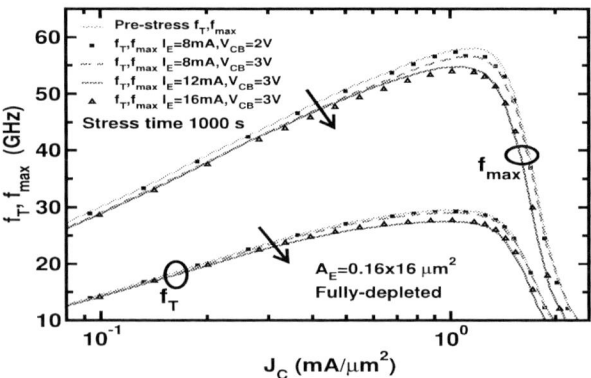

Fig. 8. Effects of mixed-mode stress on f_T and f_{max} of a fully-depleted SiGe HBT-on-SOI.

both f_T and f_{max}. The increase of V_{CB} to 3 V with I_E=8 mA results in a small but noticeable decrease in f_{max}, suggesting that the degradation process has a threshold situated between 2 V and 3 V. The increase of the emitter current to higher values results in only a slightly lower f_T and in a more noticeable reduction in f_{max}, indicating negligible damage in the intrinsic device [8]. The degradation of f_{max} for low J_C could be due to stress-induced changes in the base resistance r_{bb}, either in the extrinsic or intrinsic base region (or both). The slight change in f_{max} at high J_C is due to the observed increase in the collector resistance R_C. The partially-depleted device shows a similar behavior, with small decreases in f_T and f_{max}. Due to the higher doping of the collector, peak f_T is larger and occurs at a higher current density.

IV. SUMMARY

Both fully-depleted and the partially-depleted SiGe HBTs-on-SOI show little degradation under applied "mixed-mode" stress. The observed larger degradation shown by I_B under reverse emitter-base bias stress is expected, but is of less concern because this operational mode is seldom encountered in normal circuit operation. The effects of "mixed-mode" stress on M-1 and on AC performance were also discussed. Finally, the thermal resistance of the device was extracted in the range of temperatures between 50 K and 300 K.

ACKNOWLEDGEMENT

This work was supported by the Georgia Electronic Design Center at Georgia Tech, IBM, NASA NEPP, and DTRA, under a contract from NAVSEA Crane. The authors are grateful to T. Ning, L. Cohn, A. Clark, D. Platteter, and K. LaBel, and the SiGe group at Georgia Tech.

REFERENCES

[1] J. Cai *et al., 2005 Proceedings of the IEEE BCTM*, pp. 256-259, October 2005.

[2] G. Avenier *et al., 2005 Proceedings of the IEEE BCTM*, pp. 128-131, October 2005.

[3] I. Z. Mitrovic *et al., Solid State Electronics,* vol. 49, no. 9, pp. 1556-1567, January 2005.

[4] T. Chen *et al., 2005 Proceedings of the IEEE BCTM*, pp. 256-259, October 2005.

[5] T. Chen *et al., IEEE Trans. on Nuclear Science.,* vol. 52, no. 6, pp. 2353-2357, December 2005.

[6] G. Zang *et al., IEEE Trans. on Electron Devices,* vol. 49, no. 12, pp. 2151-2156, December 2002.

[7] C. Zhu *et al., 2005 Proceedings of the IEEE BCTM*, pp. 41-44, October 2005.

[8] C. Zhu *et al., IEEE Transactions on Device and Materials Reliability,* vol. 5, no. 1, pp. 142-149, March 2005.

[9] R. A. Wachnik *et al., Journal of Applied Physics,* vol. 63, no. 9, pp. 4734-4740, May 1988.

[10] A. Neugroschel *et al., IEEE Transactions on Electron Devices,* vol. 43, no. 8, pp. 1286-1290, August 1996.

[11] T. Vanhoucke *et al., IEEE Electron Device Letters,* vol. 25, no. 3, pp. 150-152, October 2005.

2006 Bipolar/BoCMOS Circuits and Technology Meeting Proceedings

Unified Analysis of Degraded Base Current in SiGe:C HBTs after Reverse and Forward Reliability Stress

M. Ruat[1,2], N. Revil[1], G. Pananakakis[2] and G. Ghibaudo[2]

[1]STMicroelectronics, 850 rue Jean Monnet, F-38926 Crolles Cedex, France
[2]IMEP/INPG, 28 Avenue des Martyrs, F-38000 Grenoble, France

Abstract — A 150GHz f_T/f_{max} SiGe:C NPN Heterojunction Bipolar Transistor was subjected to reverse reliability stress, high current forward stress and Mixed-Mode stress. A base current degradation comparison puts in evidence a Shockley-Read-Hall leakage mechanism for all the degradation modes, giving a unified basis for device degradation physics study. Temperature and recovery behavior, acceleration factors and defect location studies add to discussion on aging models in SiGe:C HBTs.

Index Terms — BiCMOS technology, DC characteristics degradation, Heterojunction Bipolar Transistor, Reliability stress, Temperature effect.

I. INTRODUCTION

Advanced SiGe and SiGe:C Heterojunction Bipolar Transistors (HBTs) are very attractive today due to their high speed, high transconductance, low noise, and excellent power performance. They are particularly relevant for high-speed digital as well as analog/mixed-signal applications for communication systems [1]. Performance improvement was most achieved through the vertical scaling of the device, which is generally accompanied by an increased electric fields and operation current density. Consequently, device robustness and reliability studies must be updated. Three main types of reliability degradation modes are reported today: reverse mode, high current forward mode and mixed-mode. Reverse stress has been extensively described in the literature over the last 30 years and is well known to create mid-gap defects at SiO_2/Si lateral spacer interface [2]-[8]. These traps give rise to Generation-Recombination base leakage current, which is observable at low monitoring bias. High current forward stress has also been reported in the literature and associated degradation has been shown to either originate from electromigration effect, Hot-Carriers created by Auger recombination or from a Recombination Enhanced Impurity Diffusion (REID) phenomenon [9]-[17]. Increasing operation current density recently led to a specific interest in the use of the device in avalanche regime. In 2002, Zhang *et al.* first described "Mixed-Mode", a degradation mode which is triggered when devices are simultaneously polarized at very high current and high Base-Collector bias. In this mode, Hot-Carriers generated by enhanced impact ionization create defects at both spacers and STI interfaces [18]-[21].

Up to now, these three degradation modes were distinctly studied. In this work, they are all assessed in a unified way based on similar electrical response to stress, despite subtle differences at physics level. Experimentally observed identical Shockley-Read-Hall (SRH) generation-recombination mechanism in the Space Charge Region (SCR) of the base-emitter (BE) junction led to a unique base current degradation model. This model is developed and enhanced including temperature activation of degradation. At the end, a similar equation will allow evaluating base current degradation whatever the reliability stress type, temperature and device geometry. Stress acceleration factors as well as defect recovery and location were also studied for discussion on degradation device physics.

II. DEVICE DESCRIPTION

The measurements were performed on a 150GHz f_T/f_{max}, 1.7V BV_{CEO} NPN Heterojunction Bipolar transistor from a 0.13μm BiCMOS technology [22]. Its schematic structure is shown in Fig.1. This device combines a non selective epitaxial base and a double polysilicon, quasi self-aligned base-emitter structure. Nitride and Oxide spacers are used to separate the emitter from the extrinsic base region. Shallow Trench Isolation (STI) separates base from collector regions. $Si_{1-x}Ge_x$:C base profile was optimized using a graded Ge concentration across the base, with a maximum content x=25% on collector side. Carbon was introduced to suppress Boron diffusion during processing. Typical DC Gummel plot is shown in Fig.2, for an emitter size of 0.3x5.74μm². It exhibits low leakage current and a maximum current gain of 600.

III. AGING PROCEDURE AND MEASUREMENT

Reliability stress has been carried out in reverse mode, forward mode and mixed-mode. Base and collector currents were measured before, during and after reliability stress, and Gummel plots of base-emitter junction were obtained. Typical Gummel plot aging after any type of reliability stress is presented in Fig.3. Base current was shown to increase with stress time, especially at low monitoring V_{BE} bias, while collector current remained fairly constant. These measurements were used to monitor the base current shift as a function of stress time, assuming

$$\Delta I_B(t) = I_B(t) - I_{B0} \qquad (1)$$

where I_{B0} is the virgin base current and $I_B(t)$ the measured base current at stress time t. Insignificant

1-4244-0458-4/06/$20.00 ©2006 IEEE

collector current shift was monitored and will not be discussed in this work. Reverse stress was carried out for a reverse Base-Emitter bias $V_{EB}=2.5V$, with collector left open. During high current forward stress, the emitter current density is usually forced at its nominal level (here $J_E=8.5mA/\mu m^2$) under its nominal Collector-Emitter bias ($V_{CE}=1.5V$ for studied HBT). However, wafer level experiments were conducted for $J_E=8.5$, 19, 26 and $36mA/\mu m^2$, whereas keeping $V_{CE}=1.5V$. Resulting base current shift is shown in Fig.4, where degradation curves have been subjected to a time scale translation. Such a method was presented in [16]. It allows demonstrating that degradation mode is not changed in the studied stress current range. In the following paragraphs, we will focus on results obtained for stress $J_E=36mA/\mu m^2$.

Mixed-Mode stress was triggered with $J_E=36mA/\mu m^2$ and $V_{CE}=3.5V>>BV_{CE0}$.

IV. BASE CURRENT IDEALITY FACTOR ANALYSIS

Unified basic degradation equation

I_{B0} and $\Delta I_B(t)$ characteristics were measured before and at every stress time interruption of reverse degradation experiment. They are shown in Fig.5. While I_{B0} characteristic exhibits an ideality factor of 1, $\Delta I_B(t)$ characteristic ideality factor is close to 2 for any stress time. So $\Delta I_B(t)$ characteristics are shifted in a parallel fashion as stress time increases, giving a "set of degradation curves". It is thus demonstrated that virgin base current I_{B0} is not impacted by the stress while a stress induced excess base current with an ideality factor close to 2 appears as soon as stress is applied. This can be explained assuming that a Shockley-Read-Hall recombination mechanism in the SCR of the base-emitter junction is responsible for observed leakage [3].

From (1) and usual bipolar operation equations, $I_B(t)$ under reverse stress can be expressed as :

$$I_B(t) = I_{B0} + \Delta I_B = I_S \cdot \exp(\frac{q \cdot V_{BE}}{kT}) + \frac{P}{A} \cdot I_S' \cdot t^a \cdot \exp(\frac{q \cdot V_{BE}}{2 \cdot kT}) \quad (2)$$

where t is the stress time, k the Boltzmann constant, T the temperature, V_{BE} the base-emitter bias, I_S and I_S' the saturation currents of respectively I_{B0} and $\Delta I_B(t)$. P is the transistor perimeter, A its area. P/A ratio allows taking into account the peripheral localization of degradation, which is discussed is Section V. Time acceleration of reverse degradation is empirically described with a power law of coefficient a. This acceleration is identical and observable over the whole monitoring bias range, which has not been previously shown.

The same analysis was repeated after high forward current stress and Mixed-Mode stress. It demonstrated a similar leakage mechanism. Ideality factors of 1.7 and 1.8 were respectively exhibited for high current forward stress and mixed-mode stress, as shown in Fig. 6 and 7.

Reverse, forward and mixed degradation modes in SiGe:C HBTs are thus characterized by the same SRH recombination mechanism giving rise to a non-ideal stress induced base leakage current. A unified basis for bipolar transistors degradation study is here provided.

Stress acceleration factors

At the moment the model does not take into account stress acceleration factors such as reverse or forward current, and stress bias. These acceleration factors can be distinct according to reliability stress type. Reverse stress was shown to accelerate as a power law of the reverse stress current [3][4]. This was verified on our transistors. From recent experiments, we tend to show that forward and Mixed-Mode stress are parts of a same degradation mode as far as acceleration factors are concerned. In Fig.8 $\Delta I_B(1000s)$ is presented base as a function of stress current after forward stress ($V_C=0.5V$) and mixed-mode stress ($V_C=2.5V$). We can see that only one exponential acceleration factor of 0.25/A is exhibited, despite a degradation level shift. Two other experiments were then conducted for a fixed stress current of $36mA/\mu m^2$ while V_{CB} was varying from 0.15 to 4V. $\Delta I_B(1000s)$ is presented in Fig.9 and demonstrate for a prevailing exponential acceleration factor of 1.2/V. Acceleration factors' implementation into (2) will allow having a complete base current shift model.

V. DEFECT LOCATION

The degraded base current analysis also demonstrates that defects are universally located in the base-emitter Space Charge Region. However, whereas reverse stress induced defects are known to be located at base-emitter spacers interface [3][4][7], mixed-mode stress induced defects are created at both spacer and STI interfaces within base-collector junction space charge region [19]-[21]. Base-collector junction Gummel plots measurements were then conducted on our devices during reverse, high current and mixed-mode stress. Fig.10 shows that base-collector junction is only degraded after a mixed-mode stress, corroborating previous works. Directly comparing the three degradation modes allows exhibiting differences at physical level. However, (2) will not be impacted as all defects behave as SRH recombination centers.

VI. TEMPERATURE ACTIVATION ANALYSIS

Temperature experiments

Temperature acceleration of stress must also be studied for the determination of worst reliability case. A few studies discuss temperature effects on bipolar reliability stress results [16][23][24]. Reliability stress in reverse, forward and mixed-mode was repeated at room temperature, 75°C and 125°C. A sufficiently long hold time was respected before stress start to take into account device temperature stabilization. During stress interruptions, Gummel plots were obtained keeping device at the same temperature. Initial DC characteristics of bipolar transistors depend on temperature, so careful attention must be paid for a correct determination of the temperature activation energy associated to a failure mode. In this work, parameters from (2) that are intrinsically dependent on temperature were isolated from stress acceleration calculation. Extracted activation energies are thus physically representative of temperature impact on defects creation. This analysis can be extended to the study of either reverse, forward or mixed-mode reliability stress.

First, I_{B0} was expressed as :

$$I_{B0} = I_S(T) \cdot \exp(\frac{q \cdot V_{BE}}{k \cdot T}) \qquad (3)$$

An empirical law was extracted for $I_S(T)$ and implemented in a simulator. Fig.11 shows simulated values for reverse stress (Hci), high current forward stress (On) and mixed-mode stress (MM).

Then $\Delta I_B(t)$ was then expressed as :

$$\Delta I_B(t) = \frac{P}{A} \cdot I_S'(T,t) \cdot (\frac{t}{t0})^a \exp(\frac{q \cdot V_{BE}}{n(T) \cdot k \cdot T}) \qquad (4)$$

where n is the ideality factor, which was also revealed to depend on temperature. Initial stress time t0 was introduced to normalize time dependency. As for I_S, an empirical law was extracted for $I_S'(T)$ (at a given stress time) and implemented in a simulator. n(T) obtained values are shown in Fig.12.

At t0, I_S, $I_{S'}$ and n will be dependent on temperature. Finally, the only relevant parameter for describing *temperature acceleration of degradation* is the stress time power law coefficient a. It represents the width of the set of ΔI_B curves, which will vary with temperature. a is modeled assuming an Arrhenius law:

$$a(T) = a0 \cdot \exp(-\frac{Ea}{k \cdot T}) \qquad (5)$$

Experimental a0 and Ea values have been used in Fig.13 which plots ln(a) as a function of 1/kT. Straight lines are obtained here and Arrhenius model is verified. It is also demonstrated in Fig.13 that degradation decreases as temperature increases. It was expected since Hot-Carriers loose energy due to phonon scattering enhancement at high temperature. From this study we found Ea values of 0.05eV for reverse stress, and 0.1eV for high current forward stress and mixed-mode stress.

Using (2) enhanced with temperature model, an accurate prediction of degradation is made, for any temperature and stress type. However it only describes degradation at one stress current and/or bias. In Fig.14 is shown experimental and modeled base current at t0 and t=70000s under reverse stress. Results are presented at 25°C and 125°C. A perfect agreement between experiment and model can be achieved.

Recovery experiments

After stress stops, devices were annealed at different temperatures. Degradation from reverse stress exhibited good recovery at low temperature (250°C). However, defects created by forward stress and mixed-mode stress must be annealed at temperatures as high as 400°C to be recovered. This suggests that different nature of Si/SiO_2 interface defect are created during reverse stress and

forward/mixed-mode stress. It can originate in the fact that different hot carriers are involved in the degradation, as defects were shown to be created by hot holes in reverse stress with open collector [7][8], whereas hot electrons were simulated to be responsible for degradation during mixed-mode stress [21].

VII. CONCLUSION

This work provides a basis for a unified study of either reverse, high current forward or mixed degradation modes in SiGe:C bipolar transistors. Defects created are shown to give rise to SRH recombination mechanism in all cases. Temperature activation model was developed for the accurate determination of degradation acceleration with temperature. This model can be extended to all degradation modes. However, degradation modes in bipolar transistors were shown to be distinct as far as acceleration factors, recovery behavior and defects location are concerned. Defect nature is also different. Acceleration factors must be included in this model for a complete description of reliability stress impact on electrical characteristics of advanced bipolar transistors.

REFERENCES

[1] J.D. Cressler, *IEEE Trans. On Device Materials and Reliability*, Vol.4, n°2, pp. 222-236, June 2004.
[2] S. A. Petersen and G.P. Li, *IEDM Digest 1985*, pp. 22–25.
[3] J.D. Burnett and C. Hu, *IEEE Trans. Electron Devices*, Vol.35, n°12, pp. 2238-2244, Dec. 1988.
[4] J.D. Burnett and C. Hu, *Proc. IRPS 1990*, pp. 164-169.
[5] C.J. Huang *et al.*, *IEEE Trans. Electron Devices*, vol.40, n°.9, pp. 1669-1674, Sep. 1993.
[6] A. Neugroschel *et al.*, *IEEE Trans. Electron Devices*, vol.42, n°.7, pp. 1380-1383, July 1995.
[7] A. Neugroschel *et al.*, *IEEE Trans. Electron Devices*, vol.43, n°.8, pp. 1286-1290, Aug. 1996.
[8] U. Gogineni *et al.*, *IEEE Trans. Electron Devices*, vol.47, n°7, pp. 1440-1448, July 2000.
[9] S.R. Sheng, *et al.*, *IEEE ISSDR 2001*.
[10] R.S. Hemmert *et al.*, *Journal of Applied Physics*, Vol.53, n°6, pp. 4456-4462, June 1982.
[11] R.A. Wachnik *et al.*, *Journal of Applied Physics*, Vol.63 n°9 pp. 4734-4740, May 1988.
[12] D.D.-L. Tang *et al.*, *IEEE Trans. Electron Devices*, Vol.37, n°7, pp. 1698-1706, July 1990.
[13] J. Zhao *et al.*, *IEEE Electron. Dev. Letters*, Vol.14, n°5, pp. 252-255, May 1993.
[14] M.S. Carroll *et al.*, *IEEE Trans. Electron Devices*, Vol.44, n°1, pp. 110-117, Jan. 1997.
[15] K. Hofmann *et al.*, *Proc. IRW 2002*, pp. 79-82.
[16] J.-S. Rieh *et al.*, *IEEE Trans. On Device Materials and Reliability*, Vol.3, n°2, pp. 31-38, June 2003.
[17] P.A. Rosenthal *et al.*, *Proc. RCSW 2004*.
[18] G. Zhang *et al.*, *Proc. BCTM 2002*, pp. 32-35.
[19] C. Zhu *et al.*, *IEDM Technical Diges 2003*.
[20] Z. Yang *et al.*, *Proc. IRPS 2003*.
[21] C. Zhu *et al.*, *IEEE Transactions on Device Materials and Reliability*, Vol.5, n°1, pp. 142-149, March 2005.
[22] M. Laurens *et al.*, *Proc. BCTM 2003*, pp. 199-202.
[23] J.D. Burnett and C. Hu, *IEEE Trans. Electron Devices*, Vol.37, n°4, pp. 1171-1173, April 1990.
[24] H.S. Momose and H. Iwai, *IEEE Trans. Electron Devices*, vol.41, no.6, pp. 978-987, June 1994.

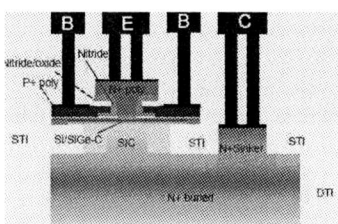

Fig. 1. Schematic cross-section of studied NPN HBT.

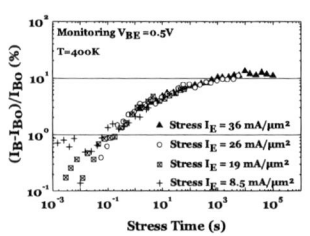

Fig. 2. Typical Gummel plot of studied NPN HBT for a device area of 0.3*5.74µm².

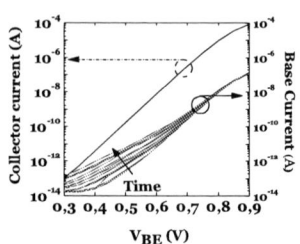

Fig. 3. Typical Gummel plot aging after any type of reliability stress for studied NPN HBT

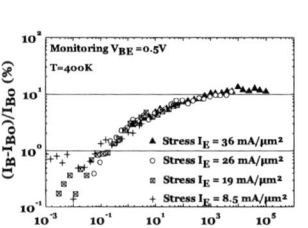

Fig. 4. Base current relative shift monitored at V_{BE}=0.5V after forward stress under I_E=8.5 (nominal), 19, 26 and 36mA/µm², with V_{CE}=1.5V. T=400K.

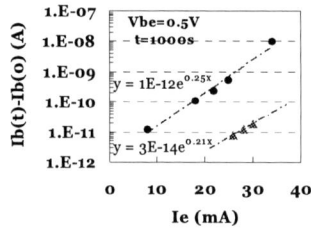

Fig. 5. I_{Bo} and $\Delta I_B(t)$ characteristics measured before and at every *reverse* stress time interruption. Extraction of ideality factors.

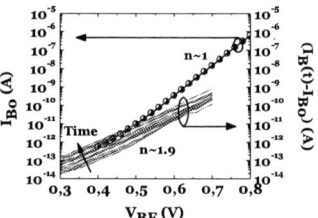

Fig. 6. I_{Bo} and $\Delta I_B(t)$ characteristics measured before and at every *high current forward* stress time interruption. Extraction of ideality factors.

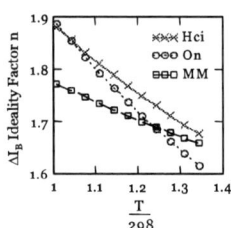

Fig. 7. I_{Bo} and $\Delta I_B(t)$ characteristics measured before and at every *mixed-mode* stress time interruption. Extraction of ideality factors.

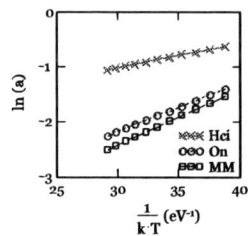

Fig. 8 ΔI_B as a function of stress current for high current forward stress (V_C=0.5V) and mixed-mode stress (V_C=2.5V) at t=1000s and V_{BE}=0.5V.

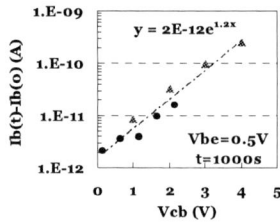

Fig. 9 ΔI_B as a function of stress bias V_{CB} for high current forward stress and mixed-mode stress (I_E=36mA/µm²) at t=1000s and V_{BE}=0.5V.

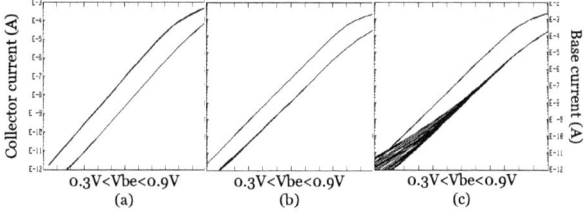

Fig. 10 Base-collector junction Gummel plots measured during (a) reverse stress, (b) high current forward stress, (c) mixed-mode stress.

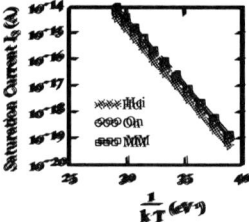

Fig. 11. I_s as a function of 1/kT for reverse stress (Hci), high current forward stress (On) and mixed-mode stress (MM).

Fig. 12 n as a function of T/298K for reverse stress (Hci), high current forward stress (On) and mixed-mode stress (MM).

Fig. 13. ln (a) as a function of 1/kT for reverse stress (Hci), high current forward stress (On) and mixed-mode stress (MM).

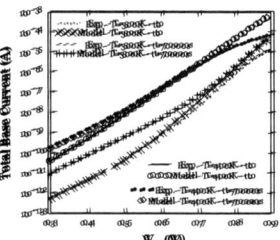

Fig. 14 Experimental base currents before (t0) and after (t=70000s) reverse stress at 300K and 400K. Comparison with model.

48

2006 Bipolar/BiCMOS Circuits and Technology Meeting Proceedings

A BiCMOS Technology Featuring a 300/330 GHz (f_T / f_{max}) SiGe HBT for Millimeter Wave Applications

B. A. Orner, M. Dahlström, A. Pothiawala, R. M. Rassel, Q. Liu, H. Ding, M. Khater*, D. Ahlgren**, A. Joseph, J. Dunn

IBM Microelectronics Division, IBM, Mailstop 972D, Essex Junction, VT 05452, 802-769-5139, email: ornerb@us.ibm.com
*IBM T. J. Watson Research Center, Yorktown Heights, NY 10598
**IBM Microelectronics Division, Hopewell Junction, NY 12533

Abstract — We present a 0.13 μm SiGe BiCMOS technology for millimeter wave applications. This technology features a high performance HBT (f_T = 300 GHz /f_{max} = 330 GHz) along with various newly developed millimeter wave features, such varactor, Schottky and p-i-n diodes and other back end of line passives.

Index Terms — Silicon bipolar/BiCMOS process technology, Millimeter wave bipolar integrated circuits, Millimeter wave devices, Heterojunction bipolar transistors, Silicon Alloys.

I. INTRODUCTION

The speed of the SiGe HBT has continued to increase over time, resulting in the application of SiGe process technologies to ever increasing frequency regimes. There have been reports of HBT transistors with f_{max} = 350 GHz [1]. Such transistors allow SiGe technologies to be extended well into the millimeter wave (MMW) band, previously the domain of III-V materials. Doing so allows the realization of MMW mixed signal circuits at a high level of integration [2].

The performance of the SiGe HBT is no longer the limit to integrating a MMW transceiver, but rather the on-chip passives and their accurate characterization. For example, a high frequency Schottky barrier diode (SBD) can be used as an ideal mixer, and optimized p-i-n diode can be used as a transmit/receive switch, and back end passives can be used as matching and tuning components in the MMW frequencies.

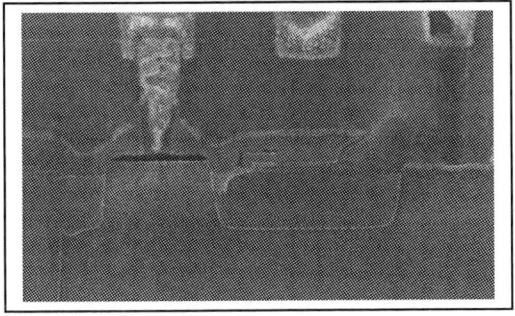

Fig. 1   SEM cross-section of HBT

Thus, the MMW SiGe BiCMOS technology demonstrated here includes an HBT (f_T=300 GHz/f_{max}=330 GHz), a Schottky barrier diode (SBD) (Fc = 1.1 THz), a p-i-n diode (isolation = 16 dB and insertion loss = 1 dB at 60 GHz), back end passives (couplers, bends, tees), and standard 0.13 μm CMOS, and other typical devices.

II. PROCESS OUTLINE

The process integration is similar to BiCMOS8HP, which is a 200 GHz/0.13 μm high performance SiGe BiCMOS technology [3]. This was chosen for easy

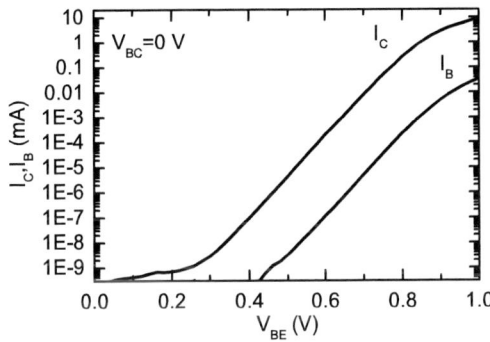

Fig. 2   HBT Gummel plot, A_E = 0.12 x 2.5 μm^2

migration of designs between technologies. We have retained logic library compatibility, maintained a common back end, and have retained several resistors and capacitors.

The process begins with a high resistivity silicon substrate, on which buried n+ layers are formed by implantation and nominally intrinsic epitaxial growth. This is followed by deep and shallow trench isolation formation. A series of block level masks are employed to create patterned implants which support the FET wells, reach through implants, and implants to support passive devices such as the junction varactor and SBD. The CMOS gates, extensions, and spacers are formed next, followed by the HBT SiGe

1-4244-0458-4/06/$20.00 ©2006 IEEE   49

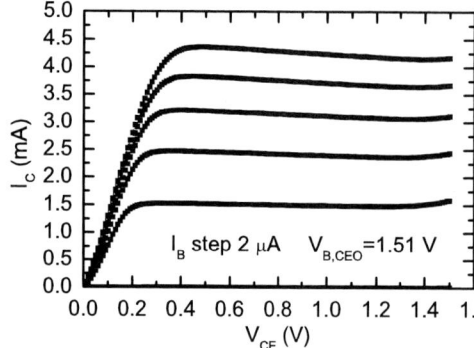

Fig. 3   HBT output characteristics $A_E = 0.12 \times 2.5$ μm^2

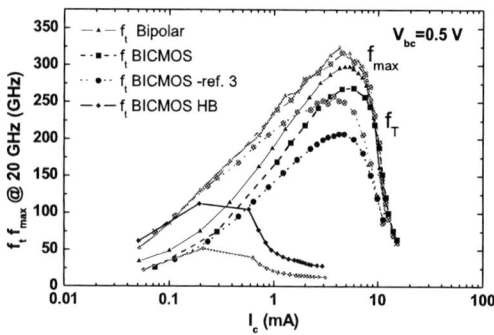

Fig. 4   HBT AC characteristics.   Bipolar and BiCMOS wafer have equivalent performance.

base epitaxy.   A second epitaxy forms the raised extrinsic base. An emitter opening is etched through a dielectric and the raised extrinsic base, stopping on the intrinsic base. After emitter poly is deposited, the HBT films are selectively removed from the other devices. The FET sources and drains are formed and the wafers are silicided.

Tungsten stud contacts are formed by established methods, and several layers of copper metal and vias are formed by a dual damascene method. The final two metal layers are aluminum metals with tungsten vias.

III. HBT

The HBT (Fig. 1) is of a raised extrinsic base design.   Device f_T has been increased relative to previous technologies through vertical scaling to reduce the collector and base transit time.   The increase in f_T is a component of the f_{max} increase; further components of the f_{max} increase include processing   improvements   which   reduce   base resistance.

DC gain (Fig. 2) for a 0.12x2.5 μm HBT was between 1200 and 1700 due to the aggressive Ge profile and the low Gummel number. The output characteristics (Fig. 3), illustrate that BV_{CEO} was over 1.4 V. AC measurements were done using open-short calibration at 1-110 GHz using ground-signal-ground probes. Both f_T and f_{max} were extracted at 20 GHz with a -20dB/dec slope. The peak f_T observed was greater than 300 GHz while the value in Fig. 4 is the average of several wafers. f_{max} was extracted from Masons Unilateral Gain (U).   Peak f_{max} was estimated extraction of U necessitated a conservative estimate. We noted the noise in U decreases if the device is biased past the Kirk threshold, suggesting that pad and wiring capacitance is on the order of the base-collector capacitance.   Future work is expected to investigate   more   advanced   de-embedding methodologies.

An optional high breakdown HBT, built along with the 300 GHz device, exhibited an $f_T = 60$ GHz, $f_{max} = 110$ GHz, and $BV_{ceo} = 3.0$ V

IV. P-I-N DIODE

A p-i-n diode [4] is being developed for use as an RF switch. A cross-sectional sketch of the integrated p-i-n diode is shown in Fig. 5. The n+ cathode layer is formed by a high dose implant, the i-layer is formed by a nominally intrinsic epitaxial Si growth, and the p+ anode is formed by the same p+ epitaxial film that is used for the extrinsic base of the HBT. The buried layer is contacted with an n+ reach-through implant, and deep trench isolation provides perimeter isolation from the substrate.

Often the bipolar subcollector is used as the p-i-n diode cathode. The subcollector of the HBT used in this technology is too shallow to be shared with a p-i-n diode in this case; the intrinsic layer was not wide enough to give the desired isolation.   To permit integration   of   these   two   devices   and   allow

Table I: FET comparison. The bipolar processing did not perturb the CMOS devices.

Parameter	Nominal Value of base CMOS technology	BiCMOS experimental results
Lpoly	88 nm	85 nm
NFET Vtsat	355 mV	340 mV
NFET Ioff	0.3 nA/μm	0.45 nA/μm
NFET Vtlin	170 mV	182.6 mV
PFET Vtsat	-330 mV	-343 mV
PFET Ioff	-0.45 nA/μm	-0.12 nA/μm
PFET Vtlin	-240 mV	-237 mV

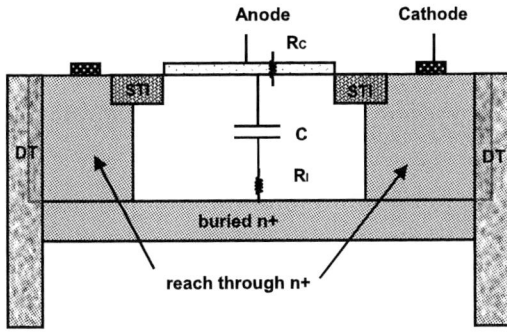

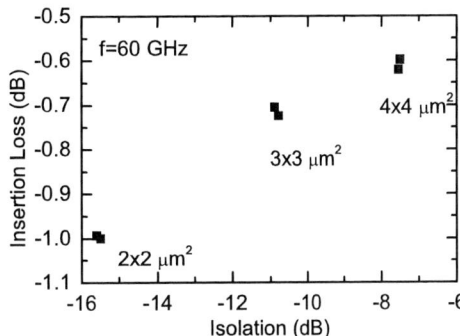

Fig. 5   Integrated p-i-n diode

Fig. 6   Isolation vs. insertion loss for various pin diode layouts. The designer may improve one parameter at the cost of the other by varying the anode area

independent optimization of each, we have introduced a second, deeper n+ layer which provides a wider intrinsic region.

A wider intrinsic region generally improves the isolation in the "off" state, but this must be balanced against degradation of insertion loss in the "on" state. The intrinsic region is fixed by the process, but we have found the designer has latitude to balance insertion loss and isolation by varying the anode size (Fig. 6). The model and parameterized cell support this scaling.

V. SCHOTTKY BARRIER DIODE

The Schottky barrier diode (SBD) is similar in structure to the p-i-n diode and shares much of the processing. In particular, both devices have the same n+ buried layer cathode. Though there have been reports of devices which perform well using an NW-like implant [5], we have found the n+ layer enhances performance. Additional implants tailor the dopants profile near the junction and form a perimeter guard

ring.

Electrical results for zero volt bias conditions are shown in Fig. 7. The cutoff frequency is calculated from Ron and Cj extracted from s-parameter measurements averaged over a frequency of 10-100 GHz. The cutoff frequency increases as anode size decreases. For the optimized device, the Fc = 1.1 THz. Moreover, the device shows good DC characteristics, with reverse bias leakage below 10^{-6} mA/μm^2.

VI. CMOS

Compatibility with a standard logic library is an important feature of a BiCMOS technology. Therefore, we require that all devices in the technology be integrated into the technology in a manner which allows design ground rule compatibility and allows key electrical device

Table II: Device Library, with nominal device parameters indicated. A full complement of active and passive devices is available.

NPN	High Performance	f_{MAX} = 330 GHz, f_T = 300 GHz, BV$_{CEO}$ = 1.5 V
	High Breakdown	f_{MAX} = 110 GHz, f_T = 60 GHz, BV$_{CEO}$ = 3.0 V
FET	1.2V NFET	Lp >= 0.092, Tox = 2.2 nm
	1.2V PFET	Lp >= 0.092, Tox = 2.2 nm
	2.5V NFET	Leff >= 0.24, Tox = 5.2 nm
	2.5V PFET	Leff >= 0.24, Tox = 5.2 nm
	1.2V Isolated NFET	Lp >= 0.092, Tox = 2.2 nm
	2.5V Isolated NFET	Leff >= 0.24, Tox = 5.2 nm
Resistors	p-polysilicon	300 Ω/sq, 1450 Ω/sq
	Sucollector	8.8 Ω/sq
	n-diffusion	73 Ω/sq
	BEOL	50 Ω/sq
Capacitors	MIM	upto 4 fF/μm^2
	MOS Varactor (Tox = 2.2)	11 fF/μm^2, tunability 5:1
	MOS Varactor (Tox =5.2)	5.8 fF/μm^2, tunability 3.5:1
	HA Junction Varactor	2.5 fF/μm^2, tunability 3:1
Diodes	p-i-n diode	I.L. = 1 dB, isolation = 16 dB at 60 GHz
	SBD	Fc = 1.1 THz
BEOL Passives		inductors, bends, tees, inverters dividers

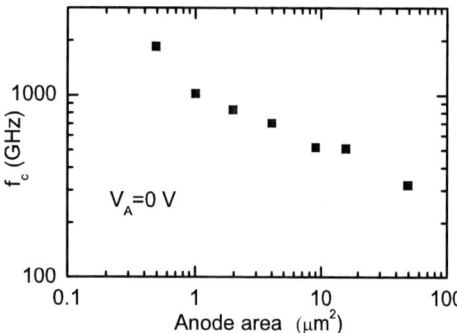
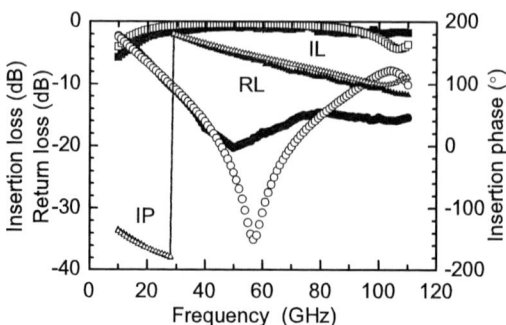

Fig. 8   Measured   v.   simulated   phase   inverter characteristics

parameters to be matched. Doing so requires consideration of the thermal processing impact as well as structural impacts such as the ability to planarize the surface when required or the ability to deposit and remove films without damaging underlying features. As performance demands increase each device has less processing latitude, making this a formidable challenge. Table I compares nominal values for key FET parameters for a CMOS-only 0.13 μm technology to FETs on wafers which have been processed through the HBT processing steps used to fabricate the HBT described above. This demonstrates the feasibility of integrating this collection of high performance bipolar and FET transistors. Similarly, the SBD and p-i-n diodes have been integrated into an existing BiCMOS technology.

VII. OTHER FEATURES

Additional devices (table II) include a full component of resistors, varactor diodes, and MIM capacitors fabricated by established methods [3]. The first several levels of metal are copper and are ground rule compatible with other technologies in IBM's 0.13 μm CMOS family. The uppermost metal levels are thick aluminum analog metal levels. The processing of these metal levels is not novel, but we are enhancing the design environment by including parameterized layout cells and models for a variety of passive devices fabricated in the back end of the line. These include transmission lines with bends and tees, couplers, and power dividers. An example is given in Fig. 8. Integrating these into the design kit as parameterized devices automates this part of the design process and increases the chances of first time right design.

VIII. CONCLUSION

We have described a next generation 0.13 μm BiCMOS technology. The technology has a high performance active and passive device set and is suitable for millimeter wave communication, radar, and other applications.

ACKNOWLEDGEMENT

This work is supported in part by funding from the Defense Advanced Research Projects Agency (DARPA) under contract number N-66001-05-C-8013.

REFERENCES

[1] M. Khater, *et al., Technical Digest of the IEEE International Electron Devices Meeting*, 247 – 250, (2004).

[2] B. Gaucher *et al., Proceedings of the International Silicon Germanium Technology and Device Meeting*, in press.

[3] B. A. Orner *et al., Proceedings of the2003 Bipolar/BiCMOS circuits and Technology Meeting*, 203-206 (2003).

[4] B. A. Orner, *et al., Proceedings of the International Silicon Germanium Technology and Device Meeting*, in press.

[5] S. Sankaran, K. K. O, *IEEE Electron Device Letters*, **26** (7), 492-494 (2005).

2006 Bipolar/BoCMOS Circuits and Technology Meeting Proceedings

A 0.13μm thin SOI CMOS technology with low-cost SiGe:C HBTs and complementary high-voltage LDMOS

L. Boissonnet, F. Judong, B. Vandelle, L. Rubaldo, P. Bouillon, D. Dutartre, A. Perrotin, G. Avenier, P. Chevalier, A. Chantre and B. Rauber

STMicroelectronics, 850 rue Jean Monnet, F-38926 Crolles Cedex, France

Abstract — We demonstrate the integration, in 0.13μm thin SOI CMOS technology, of low-cost high-performance high-voltage LDMOS and HBT transistors. These specific devices are obtained, without affecting CMOS core process devices. Static and dynamic characteristics for both type of transistors are presented, showing state of the art devices suitable for RF/analog/digital System On Chip integration.

Index Terms — High-voltage MOS, bipolar transistors, heterojunction, SOI, analog/RF, process integration, 130nm

I. INTRODUCTION

Compared to bulk, thin Silicon-On-Insulator (SOI) high-resistivity substrate presents advantages, such as complete isolation between actives or blocks, latch-up immunity, reduced substrate coupling and losses, improved passive device performances and improved figure of merit in digital circuits. To address mixed RF/analog application needs (precise analog amplifiers for instance) specific devices like high-voltage MOS and NPN HBT are needed. In this paper, we present the integration of these devices in ST 0.13μm thin SOI technology, keeping all the CMOS and passives advantages. The fabrication of low-cost fully-self-aligned NPN HBT, using only 4 dedicated masks, and of N&P Lateral DMOS is described. Their DC and AC performances are largely detailed. Specific studies on collector doping and on NPN bipolar layouts are also reported. It is shown that these high-performance specific devices are obtained, without affecting CMOS core process devices.

II. FABRICATION PROCESS

Our complementary high-voltage LDMOS and NPN HBT fabrication process is fully compatible with conventional 0.13μm CMOS ST technology. Compared to the core process, only four lithography masks are added for complementary N&P HV LDMOS. The HBT, which consists in a double-polysilicon fully self-aligned structure, is also implemented in a low-cost way with only four specific masks.

A high-resistivity (>1kOhm.cm) Unibond SOI substrate, composed of 160nm silicon over 400nm buried oxide, required for advanced 0.13μm SOI

CMOS [1], is used in our technology. The process integration is presented schematically in Fig. 1.

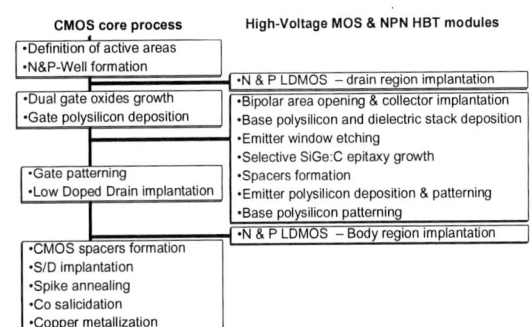

Fig. 1. Schematic process integration.

The fabrication starts with the Shallow Trench Isolation formation and CMOS well implantations. The low doped drift regions for N & P LDMOS are then implanted using two specific masks. Dual gate oxide (20Å and 50Å) is grown and the gate polysilicon is deposited. At this point, the HBT module is inserted. Integration of bipolar transistors starts with the bipolar area opening and the collector implantation with the first added mask. Next, an oxide/implanted base polysilicon/oxide/nitride stack is deposited. The 0.28μm-width emitter window is patterned and etched (2nd mask). Nitride spacers are formed in order to protect the base polysilicon sidewalls during selective epitaxy. A wet de-oxidation is then used to etch the pedestal oxide and to form the base cavity. SiGe:C selective epitaxy is grown inside this cavity and oxide/nitride inner spacers are fabricated, leading to a final effective emitter width of 0.15μm. Arsenic in-situ doped polysilicon is deposited and etched (3rd mask) to form the emitter. The last step of this module consists in the base polysilicon patterning (4th mask). Then, the CMOS core process resumes with the polysilicon gate patterning and the Low Doped Drain implantations. The two last specific implantations for N & P LDMOS channels are performed, with two specific masks, just before CMOS spacer formation to provide gate self-alignment. The end of the process is the core CMOS one, with the S/D implantations (the N+ implant is also used for the HBT collector contacts), the spike annealing, the salicidation, and the 6 Cu-metal level back-end.

1-4244-0458-4/06/$20.00 ©2006 IEEE   53

The main advantage of this integration scheme is that core process devices are not affected by the added bipolar thermal budget since the bipolar module is inserted prior to the formation of CMOS source/drain regions. Fig. 2 presents the trade-off saturation current versus leakage current for N&P MOS devices (W/L=10/0.13μm) without or with the integration of the bipolar process, and shows that electrical parameters are equivalent.

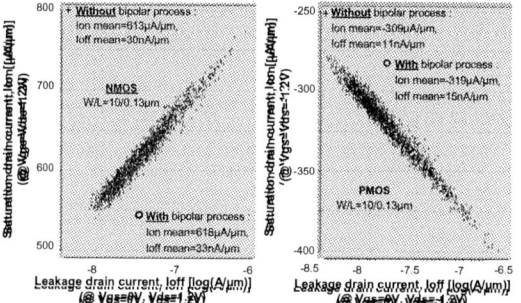

Fig. 2. Ion vs Ioff for N&P MOS with or without bipolar process.

On the other hand, one critical point of this integration is that the bipolar height has to be small enough, in order to be protected during critical CMOS lithography steps (CMOS polysilicon gate patterning in particular). Compared to the bipolar structure on SOI presented in [2], the thickness reduction of the different layers has required a significant process optimization (change of the inside spacer module from L-shape to D-shape, for instance). The TEM pictures shown in Fig. 3 clearly illustrate the vertical scaling which has been applied to the HBT structure while moving from the HBT-only to the BiCMOS process.

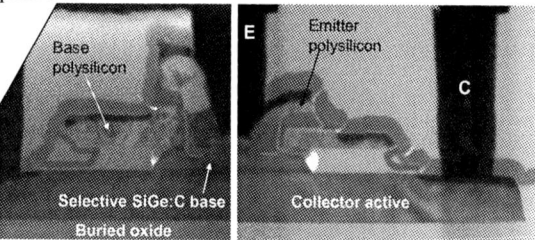

Fig. 3. TEM cross-sections of the SOI SiGe HBT: (left) HBT-only process, (right) BiCMOS process.

III. HBT ELECTRICAL RESULTS

A. HBT Electrical Characteristics

The NPN transistor used in this process has a double-polysilicon fully self-aligned structure. This self-alignment is obtained thanks to a selective epitaxial base. Carbon is incorporated in the SiGe in order to minimize boron thermal diffusion. Process parameters (thicknesses, boron dose, carbon and germanium contents) have been optimized to improve static-dynamic-analog characteristics trade-off. The

reduction of the emitter width, thanks to inner spacers, allows to minimize current crowding effects and to reduce intrinsic base resistance. Since SOI substrate is used, base active area and collector contacts have to be in the same active. The collector doping concentration has an important influence on the NPN behaviour [2]. A low doped collector (fully-depleted, $Nc\sim10^{17}cm^{-3}$) is chosen in order to reach high breakdown voltages. On the contrary, a highly doped collector (partially depleted) allows to achieve high transition frequency. The high-speed bipolar is realized at no extra cost by adding a CMOS core process implantation into the collector.

Gummel characteristics of HBTs with drawn emitter area of $0.28\times1.42\mu m^2$ are shown in Fig. 4, for High-Voltage (HV) and High-Speed (HS) NPN transistors. Ideal plots are obtained down to Vbe=0.4V. The maximum current gain is close to 200 and constant over more than 5 decades of current. HV and HS transistors present an open-base E/C breakdown voltage of 5.5V and 2.3V respectively. The B/C breakdown voltage reaches 10V and 8.4V for HV and HS transistors. For both devices, the Early voltage, extracted from the output characteristics at fixed Vbe=0.7V is larger than 300V. Forced-Ib common-emitter output characteristics are presented in Fig. 5 for different Ib values, and for HV and HS devices. For high current level, a negative slope on Ic curves is observed, which is due to the self-heating effect. As the heat cannot be dissipated through the substrate due to the dielectric isolation, the self-heating effect has to be taken into account in models and safe-operating-area definition. Comparison of bulk and SOI substrates and layout impact on self-heating are described in [3].

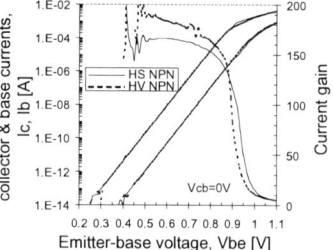

Fig. 4. Gummel characteristics for HV & HS HBTs.

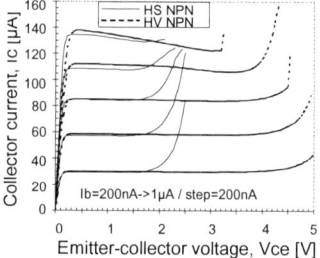

Fig. 5. Output characteristics for HV & HS NPN HBTs.

The cut-off frequency f_T and the maximum oscillation frequency f_{max} are extracted from s-parameters measurements at Vcb=0.5V. Fig. 6 shows

f_T versus Ic curves for the two types of NPN. Measurements have been made on 0.28×4.86µm² emitter drawn area structures made of three 0.28×1.42µm² emitter area cells. HV NPN presents a peak f_T value of 34GHz, while high peak f_T value of 70GHz is achieved for HS transistor. These performances are reached at a collector current density of about 1mA/µm² and 2.2mA/µm² respectively. A Mason gain, U, of 20dB is obtained at a frequency of 14GHz for HV and HS transistors. This corresponds to a peak f_{max} value of 140GHz, assuming a -20dB/decade roll-off of Mason gain versus frequency.

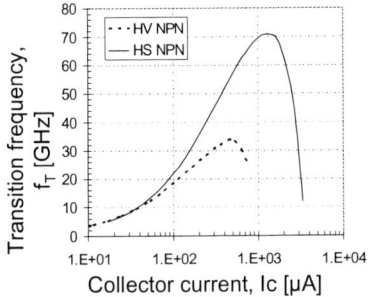

Fig. 6. f_T- Ic for HV & HS NPN HBTs.

B. HBT Layout Study

Unlike bulk architecture, the current flow in the SOI HBT is lateral and thus a short distance between intrinsic E/B region and collector contacts is needed to achieve high-speed performances. A C_BE^BC layout, described in [3], is used to reduce the collector resistance (for partially depleted collector), the base/collector capacitance and the base/collector transit time (for fully depleted collector). In this C_BE^BC structure, the base contacts are located in a perpendicular plane and thus a cellular layout is preferred to minimize extrinsic base resistance, see Fig. 7. C_BE^BC layouts, used for HV and HS devices presented in the previous section, lead to high RF performances. Here, we investigate the impact of the layout on static and dynamic behaviour. Fig. 7 shows the two different considered layouts : the C_BE^BC one, with a short emitter/collector distance, and the more classical CBEBC one, with the base-emitter-collector contacts, all in the same plane.

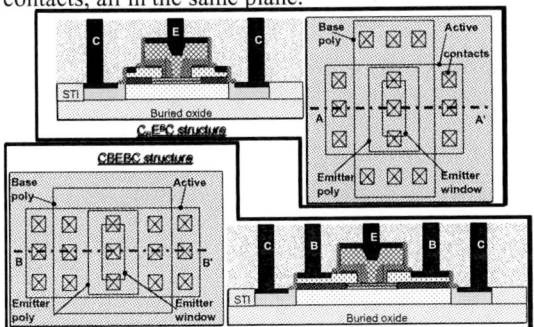

Fig. 7.   Top-view and cross-section of C_BE^BC and CBEBC layouts.

Apart from breakdown voltages, HV CBEBC and HV C_BE^BC devices have similar static parameters. Thanks to the larger distance between emitter and collector, the CBEBC layout presents higher breakdown voltages. The open-base E/C breakdown voltage reaches 7.3V and the B/C breakdown voltage is 16V, compared to 5.5V and 10V for HV C_BE^BC device. But, as expected, this breakdown voltages increase is at the expense of the cut-off frequency f_T. Fig. 8 shows f_T versus Ic for the two types of layouts. Measurements have been made on 0.28×4.86µm² emitter drawn area structures made of three 0.28×1.42µm² emitter area cells. The CBEBC f_T falls down to 27GHz, due to the larger base/collector transit time. Concerning f_{max}, the two layouts exhibit the same value, 140GHz. For the CBEBC layout, the decrease of f_T is counterbalanced by the lower base resistance. These two types of layout allow to address different applications.

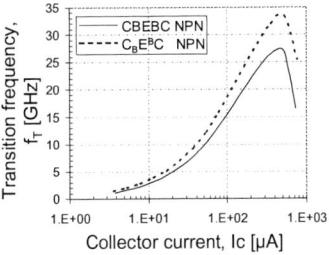

Fig. 8. f_T - Ic for C_BE^BC and CBEBC NPN HBTs.

IV. HIGH-VOLTAGE LDMOS ELECTRICAL RESULTS

As mentioned before, complementary high-voltage LDMOS are inserted in 0.13µm SOI CMOS technology thanks to two specific masks for each device only. In order to sustain 2.5V on the gate, a 50Å gate oxide is used for these devices. High voltage can be applied on the drain thanks to a low doped drift region ($\sim 10^{17}$cm^{-3}), which is unsalicided (protected from the salicidation with a dielectric stack). This N or P drift region is critical to reach the specified breakdown voltage and its doping and length are optimized to achieve an off-state BVds up to 15V. In order to achieve high performances, the channel is implanted with a specific mask just before the CMOS spacer formation. The body implants (Boron for NLDMOS and Phosphorus for PLDMOS) are therefore self-aligned to the gate and then diffused according to the core process thermal budget, thus leading to very short channels.

Fig. 9 shows a schematic top-view and two cross-sections of the NLDMOS. PLDMOS architecture is similar, replacing N-type by P-type dopants. Whereas floating body is an interesting effect for high-speed digital circuits, the body has to be contacted in case of these analog high-voltage devices. One of the major challenges is the efficiency of the body contact in order to avoid kink effect. Unlike bulk substrate, the thin SOI substrate prevents to take the body contact with a link under the source [4]. In the case of NLDMOS structure, the layout presents regularly placed P+ body contacts instead of N+ source. N+ source and P+ body are shorted by the salicidation, which provides a higher layout density. Such a technique allows to obtain a perfect control of the body potential [5].

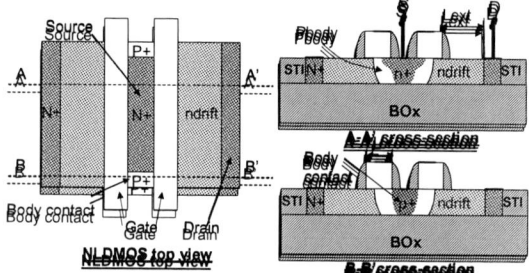

Fig. 9.   Schematic NLDMOS top-view and cross-sections

Electrical results presented hereafter concern typical N&P LDMOS structures with Lpoly=0.5μm, Lext=0.6μm, W=2×5μm and with two body contacts. Fig.10 shows output characteristics for N&P LDMOS. High off-state breakdown voltages (BVds) are reached, up to 18V for NLDMOS and 16V for PLDMOS. Despite the high drain resistance due to the thin SOI substrate, a large saturation current of 360μA/μm (Ion measured at Vgs=2.5V and Vds=5V) is obtained for N-type LDMOS and nearly 250μA/μm for P-type one. Thanks to efficient body contacts, no floating effect ("kink" effect) can be observed for the two structures. The threshold voltages are 0.65V and - 0.60V for N & P LDMOS respectively. A good optimization of the channel and the drain regions allows to obtain a low value for the product silicon area × On-resistance, S.Ron, which is a main figure of merit for this kind of device. S.Ron values smaller than 9mΩ.mm² and 22mΩ.mm² are reached for N & PLDMOS respectively. At high Vgs, the self-heating effect is at the origin of the drain current decrease.

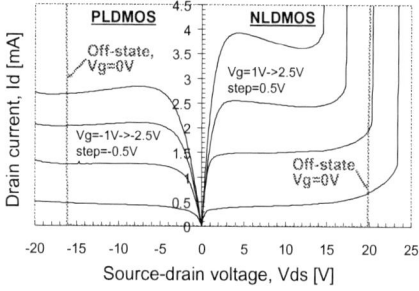

Fig. 10.   Output characteristics for W/L=10/0.5 N & P LDMOS.

The cut-off frequency f_T and the maximum oscillation frequency f_{max} are extracted from s-parameters measurements at Vds=4V, for W/L=250/0.5 devices. A peak f_T value of 12GHz and a peak f_{max} of 20GHz are achieved for PLDMOS.

Higher dynamic performances are obtained for NLDMOS with a peak f_T value of 19GHz and a peak f_{max} of 32GHz.

V. CONCLUSION

In this paper, we have demonstrated the integration, in 0.13μm thin SOI CMOS technology, of low-cost high-performance high-voltage LDMOS and HBT transistors. These specific devices are obtained, without affecting CMOS core process devices. Trade-off and optimization of HV LDMOS has been presented. Specific studies on collector doping and on NPN bipolar layouts have been largely described. The main electrical parameters, summarized in Tables 1 and 2, demonstrate state-of-the-art devices suitable for RF/analog/digital System On Chip integration.

TABLE I
NPN HBT electrical parameters
(Ae cell=0.28×1.42μm² - 1 cell for DC and 3 cells for RF)

	$C_B E^B C$ HS	$C_B E^B C$ HV	CBEBC HV	Conditions
gain	194	190	181	Vbe=0.75V, Vbc=0V
BV_{ceo} [V]	2.3	5.5	7.3	Vbe=0.7V, Ib=0A
BV_{cbo} [V]	8.4	10	16	
V_{Early} [V]	320	900	971	Vbe=0.7V
F_t [GHz]	70	34	27	Vcb=0.5V
F_{max} [GHz]	140	140	140	Vcb=0.5V

TABLE II
N & P LDMOS electrical parameters
(W/L=10/0.5 for DC and W/L=250/0.5 for RF results)

	NLDMOS	PLDMOS	Conditions
V_t [V]	0.65	-0.6	Vds=0.1V
I_{on} [μA/μm]	360	-250	Vgs=2.5V, Vds=5V
$S.R_{on}$ [mΩ.mm²]	8.8	22	Vds=0.1V, Vgs=2.5V
BV_{ds} [V]	18	-16	Vgs=0V
I_{off} [pA/μm]	3	-4	Vgs=0V, Vds=5V
F_t [GHz]	19	12	Vds=4V, W/L=250/0.5
F_{max} [GHz]	32	20	Vds=4V, W/L=250/0.5

REFERENCES

[1] C. Raynaud et al., "Is SOI CMOS a promising technology for SOCs in high frequency range?", *Proceedings of the 207th ECS Meeting - 12th International Symposium on Silicon-on-Insulator Technology and Devices*, May 15 - May 20, 2005, pp. 331-344

[2] G. Avenier et al., "A self-aligned vertical HBT for thin SOI SiGeC BiCMOS", *Proceedings of the 2005 Bipolar/BiCMOS Circuits and Technology*, pp 128-131

[3] P. Chevalier et al., "Low-Cost Self-Aligned SiGeC HBT Module for High-Performance Bulk and SOI RFCMOS Platforms", *IEDM Tech. Dig.*, 2005, pp. 983-986.

[4] S. Matsumoto et al., "RF performance of a state-of-the art 0.5μm-rule thin-film SOI power MOSFET", *IEEE Electron Devices*, vol. 48, n°6, June 2001, pp 1251-1255

[5] O. Bon et al., "High voltage devices added to a 0.13μm high resistivity thin SOI CMOS process for mixed analog-RF circuits", *2005 IEEE International SOI Conference* (QSIC 2005), pp 171

A 205/275GHz f_T/f_{max} Airgap Isolated 0.13 m BiCMOS Technology featuring on-chip High Quality Passives

S. Van Huylenbroeck, L. J. Choi, A. Sibaja-Hernandez[1], A. Piontek, D. Linten, M. Dehan, O. Dupuis, G. Carchon, F. Vleugels, E. Kunnen, P. Leray, K. Devriendt, X. P. Shi, R. Loo, E. Hijzen[2] and S. Decoutere

IMEC, Kapeldreef 75, B-3001 Leuven, Belgium
[1]K. U. Leuven, ESAT, Kasteelpark Arenberg 10, B-3001 Leuven, Belgium
[2]Philips Research Leuven, Kapeldreef 75, B-3001 Leuven, Belgium

Abstract — A QSA airgap isolated HBT module, embedded in a 0.13 m BiCMOS technology, reaches f_T/f_{max} values of 205/275GHz and a 3.5ps CML gate-delay. A 17GHz LNA using high quality passives sustains above 8kV HBM ESD stress.

Index Terms — BiCMOS process technology, BiCMOS integrated circuits, Current mode logic, Heterojunction bipolar transistors, Silicon Germanium.

I. INTRODUCTION

SiGe:C HBTs have proven their capability to support large bandwidth and high data rates for high-speed communication systems. Continuous improvement of the transistor performance is enabled by reducing the device parasitics and has shown impressive f_T/f_{max} values. The low complexity quasi self-aligned architecture (QSA) has been often criticized not to be suitable for the fabrication of such high-speed HBT devices that will be used in the next generation high-end RF applications like 60GHz UWB and 77GHz car radar. The main disadvantages of this architecture are said to be e.g. the lack of scalability, high parasitic values and the difficulty to minimize at the same time the external base-collector capacitance and the base resistance.

In this work however, we introduce a QSA BiCMOS technology featuring 205GHz f_T and 275GHz f_{max} HBT devices isolated by airgap deep trenches. The various process optimizations, required in order to reduce device parasitics, will be explained. This technology is embedded into a 0.13 m RF-CMOS baseline process and is further equipped with on-chip high quality passive components. Yield data and circuit performance demonstrates the manufacturability of this technology.

II. BiCMOS TECHNOLOGY DESCRIPTION

A low complexity architecture yielding f_T and f_{max} values of 200GHz and 160GHz respectively has been described in [1]. It implements a non-selective SiGe:C epitaxial base, grown in an ASM Epsilon 2000 production epi reactor, and a single poly quasi self-

aligned (QSA) emitter structure into a minimal topography HBT. In order to further improve the RF figure-of-merit for power gain f_{max}, the following modifications to the process and to the layout rules, aiming at a reduction of the device parasitic resistance and capacitance values, have been realized:

i. The sheet resistance of the n-type buried layer is reduced from 22 to 6.5Ω/sq by increasing the arsenic implantation dose with a factor of four. The new buried layer module has been verified to be compatible with a defect free deposition of a 0.4µm n-type collector epitaxial layer. Together with the reduction of the lateral spacing between the base active and the collector active region from 1 m to 0.5 m, the constant part of the collector resistance R_{cc} has been reduced from 11Ω/ m to 4.5Ω/ m.

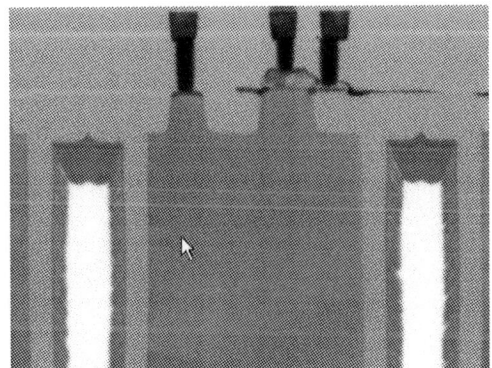

Fig. 1. Cross-section of this paper s HBT surrounded by airgap deep trenches.

ii. The deep trench module as presented in [1] is implementing 1 m wide trenches that are partially filled with oxide and partially with polysilicon. Compared to the previous DTI module, the oxide/polysilicon trench filling has been replaced by an airgap encapsulated in an oxide plug (figure 1). The integration scheme is fully compatible with standard STI processing and does not require any complex process steps [2]. Since air is the best electrical insulator, the peripheral collector-substrate capacitance $C_{cs,p}$ is further reduced from 0.11fF/ m to a record value of 0.02fF/µm. The thermal resistance is only marginally increased compared to the previous DTI module as the heat flux for both isolation schemes basically flows vertically into the substrate.

iii. A reduction of the emitter window width is appropriate since it will reduce the pinched base resistance and the intrinsic base emitter junction capacitance. The use of 130nm wide emitter window trenches, compared to the 200nm emitter windows for the reference HBT in [1], is made possible through the introduction of 193nm DUV lithography. Figure 2 shows the measured emitter window CD as a function of the emitter length. The 130nm wide emitter trenches are stable down to emitter lengths of 0.2µm. An end-of-line shortening of 75nm per side, irrespective of the emitter length, has been verified experimentally. This constant correction at the line-ends is therefore applied to the final emitter window mask.

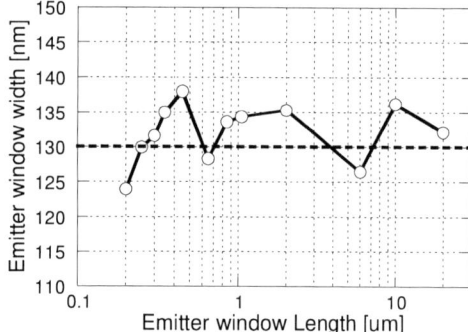

Fig. 2. Emitter window width vs. emitter window length, 193nm litho.

iv. The poly-Si emitter module consists of an arsenic in-situ doped layer deposited in an A400 ASM-E furnace. An optimization of the cleaning procedure prior to this poly emitter deposition, needed to remove a 20nm oxide layer inside the 130nm wide emitter window trenches, reduced the remaining interfacial oxide between the mono silicon base epitaxial layer and the poly silicon emitter layer. Together with the use of a nickel silicide module, we were able to further reduce the specific emitter resistance from a previously reported $8.5\Omega\mu m^2$ [1] to a value of $3.5\Omega\mu m^2$.

v. The lateral emitter window enclosure by the poly-Si emitter (PE-EW) and the poly-Si emitter enclosure by the active area (AA-PE) have been scaled down to nominal values of 0.13µm and 0.03µm respectively. The active area width of an HBT device is therefore reduced to 0.45µm. A drastic scaling of the AA-PE enclosure, it used to be 0.15µm on previous generation, has to be accompanied with an optimized external base strategy. The external base implantation is done after the poly-emitter patterning and before poly-emitter resist strip. The boron dopants have to cover

the nearby poly/mono interface of the non-selectively grown base epitaxial layer in order to prevent base-collector leakage. At the same time they should not diffuse to deep as this would increase the base-collector capacitance C_{bc}. The small AA-PE enclosure we have now prevents a deep diffusion of the external base dopants since a large part of the dopants is implanted on top of field regions. An increase of the implant energy becomes therefore feasible with the additional advantage of obtaining a lower base resistance R_b.

Fig. 3. Cross-section of this paper s HBT (PE-EW=0.13µm, AA-PE=0.03µm).

A TEM cross-section of the new reference HBT device can be seen in figure 3. The HBT module is embedded into a 0.13µm RF-CMOS baseline process with 65nm physical gate length devices and a heavily nitrided DPN gate oxide with EOT equal to 1.5nm. Dopants activation of both the CMOS and HBT devices is done during an 1100 C spike anneal. The minimized thermal budget of the bipolar module and the nitridation of the gate oxide make it possible to put the complete CMOS processing before the bipolar module.

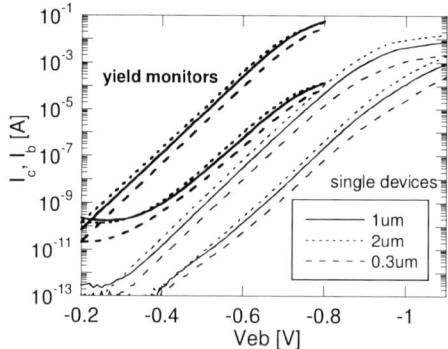

Fig. 4. Gummel plots, 0.13x0.3µm² (9.6K), 0.13x1.0µm² (8K), 0.13x2.0µm² (6.4K), single device and yield arrays.

III. HBT DEVICE PERFORMANCE

The dopant profiles of the internal device have not been changed compared to [1]. Because of the 2-step Ge-profile (8% and 21% respectively), we can limit the current gain to 500 and obtain a BV_{ceo} value of 1.85V.

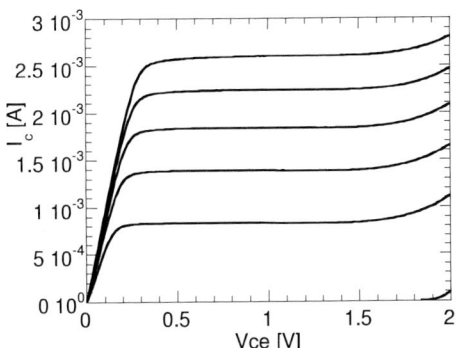

Fig. 5. I_b forced I_c-V_{ce} output curves, $0.13 \times 2.0 \mu m^2$ single device.

Excellent base and collector current ideality, especially at low Vbe bias, is demonstrated on single devices in figure 4. I_c-V_{ce} curves (I_b forced) are presented in figure 5, showing the excellent early voltage for collector currents close to peak f_T. The RF-performance of this HBT device is shown in figure 6 and is compared with data from [1].

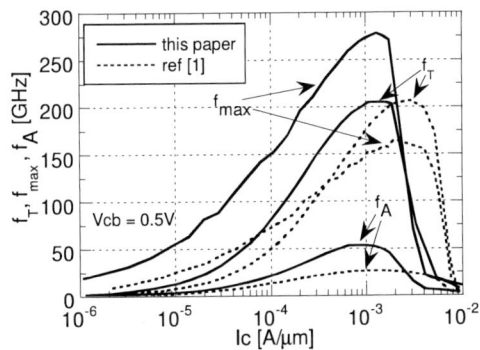

Fig. 6. f_T, f_{max}, f_A versus I_c for this paper s HBT device and ref [1].

The peak f_T value is unchanged but the current at this peak f_T is reduced with a factor of two as we scaled the emitter window width from 200nm to 130nm. The f_{max} value is increased from 160 to 275GHz because of the lateral scaling, demonstrating the serious reduction of the base resistance R_b and of the C_{be} and C_{bc} capacitance. A more realistic demonstration of the device RF capabilities is the f_A figure-of-merit [3] as it demonstrates the device performance in typical circuit configurations. For a DC gain of 10, we now obtain a f_A above 50GHz, more than double the value of what the devices of ref [1] reached. The serious reduction of the collector-substrate C_{cs} capacitance due to the introduction of the airgap deep trench module contributes to this high f_A value. Table 1 summarizes the principle electrical performance parameters of a $0.13 \times 2 \mu m^2$ HBT device.

A selective collector implantation (SIC) balances the occurrence of base push-out and C_{bc} for the high-speed devices. Omitting this implant leads to a device with much lower f_T and f_{max} (70 and 170GHz respectively). However, due to the lower C_{bc} for the non-SIC HBT, this device outperforms the high-speed device at low currents, as can be seen on the f_A versus I_c curve (figure 7), making it an interesting HBT device for low power applications.

Device parameter	Value
Peak f_T/f_{max} [GHz]	205/275
Current Gain [-]	500
BV_{ceo} [V]	1.85
BV_{cbo} [V]	6.2
V_A [V]	> 100
C_{be} [fF] (0V)	4.3
C_{bc} [fF] (0V)	3.2
C_{cs} [fF] (1V reverse)	0.7
R_e [Ω]	11
R_b [Ω] (peak f_T)	40

Table. 1. Electrical parameters for a $0.13 \times 2.0 \mu m^2$ HBT.

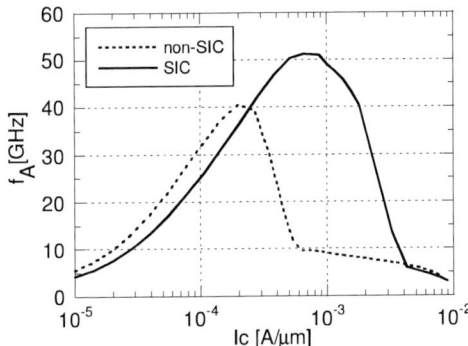

Fig. 7. f_A versus I_c of the high speed SIC device and the non-SIC device.

The manufacturability of this technology has been assessed by measuring gummel plots on yield arrays with up to 9600 parallel connected devices. Excellent base and collector ideality is obtained on device arrays with emitter window lengths varying from 0.3 m to 2 m (figure 4).

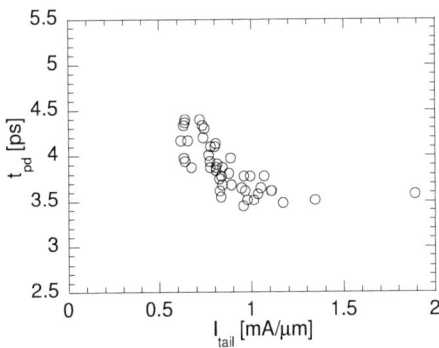

Fig. 8. Measured gate delay t_{pd} vs. tail current I_{tail} for the CML ring oscillators.

Ring oscillators based on a CML topography for maximum speed have been measured as well. The gates were designed to operate in the 0.5-1.5mA/μm range where they are expected to give peak performance. The number of stages is 15 and the logic swing equals 250mV. Figure 8 shows the measured gate delay versus tail current. A gate delay per stage down to 3.5ps is achieved, which is among the best reported values for full BiCMOS technologies.

IV. ON-CHIP PASSIVE COMPONENTS

The planar MIM capacitor module uses TaN layers for the plates and an ONO dielectric optimized for excellent voltage linearity and targeting a 1.1fF/ m2 specific capacitance. The TaN bottom layer is shunted with the underlying Cu metal layer in order to minimize series resistance. A Q-factor of 20 is obtained at 50GHz for a 30fF unit cell capacitor (figure 9). The TaN bottom layer can also be used as a back-end resistor with a 40Ω/sq sheet resistance and a low temperature coefficient.

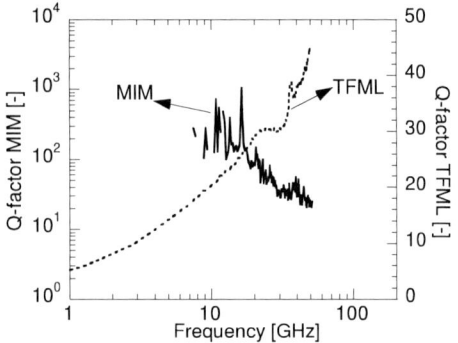

Fig. 9. Q-factor vs. freq. of the MIM capacitor (30fF unit cell) and the thin film microstrip line (50Ω impedance).

Above-IC inductors using a 5 m thick Cu layer are processed on top of the passivation and a 16 m thick BCB (benzo-cyclobutene, K=2.65) layer. The thick above-IC dielectrics drastically lower the parasitic coupling to the ground substrate while the thick Cu-wiring lowers the series resistance. In this way, high Q-factors for inductors and transmission lines can be obtained. Figure 9 for instance shows the Q-factor of a thin film microstrip line with 50Ω characteristic impedance realized in the above-IC Cu-wiring. The ground is formed by the parallel connection of three standard back-end-of-line metal layers. A Q-factor above 30 is obtained for frequencies above 25GHz.

A 17 GHz ESD protected LNA was designed in this BiCMOS technology (figure 10, 11). Because of the high Q-factor of the above-IC inductor used in the ESD input protection [4], this LNA can sustain ESD stress events of more than 8 kV HBM. The non ESD protected LNA reference holds less than 100V HBM stress.

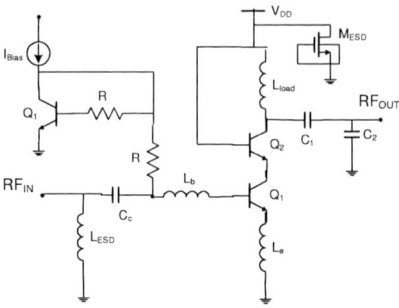

Fig. 10. Schematic of the 17 GHz ESD protected LNA Q-factor vs. freq.

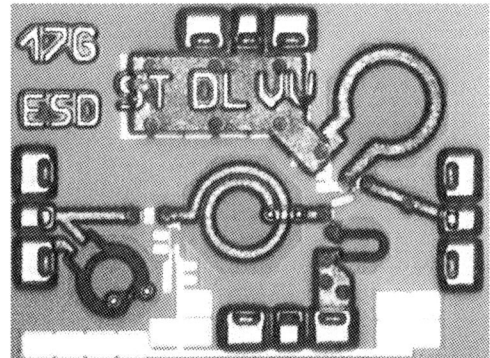

Fig. 11. Micrograph of the 17 GHz ESD protected LNA.

V. CONCLUSION

An HBT device processed in a low complexity quasi self-aligned architecture features f_T/f_{max} values of 205/275GHz and is implemented into a 0.13 m BiCMOS technology. Lateral scaling of the layout rules and the introduction of airgap deep trench isolation allowed a serious reduction of the device parasitics. A CML gate-delay of 3.5ps is obtained, among the best reported values for BiCMOS technologies. With the availability of high quality passives, a 17GHz LNA is demonstrated that is protected against ESD stress events of 8kV HBM and more.

REFERENCES

[1] S. Van Huylenbroeck, et al., Lateral and Vertical Scaling of a QSA HBT for a 0.13 m 200GHz SiGe:C BiCMOS Technology , BCTM, pp 229-232, 2004.

[2] L. J. Choi, et al., A Novel Deep Trench Isolation featuring Airgaps for a High-Speed 0.13 m SiGe:C BiCMOS Technology , VLSI-TSA, pp 88-89, 2006.

[3] G.A.M. Hurkx, et al., RF figures-of-merit for process optimization , IEEE Transactions on Electron Devices, volume 51, pp. 2121-2128, 2004.

[4] D. Linten, et al. "A 5 GHz fully integrated ESD-protected low-noise amplifier in 90 nm RF CMOS" IEEE Journal of Solid-State Circuits Special Issue on the 2004 European Solid State Circuits Conference (ESSCIRC), volume 40, no. 7, pp. 1434-1442, 2005.

2006 Bipolar/BoCMOS Circuits and Technology Meeting Proceedings

Simultaneous Integration of SiGe High Speed Transistors and High Voltage Transistors

R. K. Vytla,[1,2] T. F. Meister,[1] K. Aufinger,[1] D. Lukashevich,[1] S. Boguth,[1]
H. Knapp,[1] J. Böck,[1] H. Schäfer,[1] and R. Lachner[1]

[1]Infineon Technologies, Am Campeon 1-12, 85579 Neubiberg, Germany
Phone: +49 89 234-25942, Fax: +49 89 234-9554929, email: Rajeev.Vytla@infineon.com
[2]University of Bremen, Otto-Hahn-Allee NW 1, 28359 Bremen, Germany

*Abstract—*Integration of high voltage transistors and varactors with high tuning range into high frequency SiGe bipolar technologies is challenging due to the requirement of a shallow collector for the high speed transistor. In this paper we present a high speed SiGe bipolar technology using a novel concept with two epitaxial layers for the simultaneous integration of high speed transistors, high voltage transistors, and varactors. Using this concept high speed transistors with 209 GHz cut-off frequency and 3.3 ps gate delay have been combined with high voltage transistors providing an emitter-collector breakdown voltage of 5 V. Additionally in this concept a varactor has been developed and optimized to achieve a high tuning range of 13 GHz and low phase noise for a 77 GHz VCO.

Index Terms – SiGe HBT, Double epi, High voltage npn, BV_{CEO}, Varactor.

I. INTRODUCTION

Future application fields like wireless communication systems in the 60 GHz range and automotive radar around 77 GHz require improved device performance for the operation at high frequencies. SiGe HBTs with cut-off frequencies in excess of 200 GHz have been achieved ([1]–[3]) by optimizing the vertical doping profiles and using thin and highly doped collector layers.

For circuit applications additional devices like npn transistors with high breakdown voltages (e.g. for ESD protection or driver stages) are required. Therefore, these technologies typically offer a high voltage npn in addition to the high speed npn by using a lower dose pedestal collector implant. This increases the breakdown voltage at the expense of lower transit frequency. But due to the requirement of the shallow collector for 200 GHz transistors the range for breakdown voltage optimization by different collector implantations is small due to the limitation of the maximum space charge layer width by the buried layer.

Some approaches to overcome this limitation have been suggested in the past. In [4] several subcollector design options with varying implantation doses and materials (As, Sb, P) have been investigated in addition to the pedestal collector doping variations. In [5] the effects of different subcollector layout options on the f_T-BV_{CEO} trade-off have been studied. Though these methods increase the collector-emitter breakdown voltage of the high voltage npn up to a certain extent the achievable BV_{CEO} still suffers from the shallow epitaxial collector required for the high speed npn. In this work, we present a new concept with two epitaxial layers ("double epi concept") that enables

the integration of high speed transistors with a cut-off frequency of 209 GHz and BV_{CEO} of 1.8 V together with high voltage transistors with f_T of 52 GHz and BV_{CEO} of 5 V on the same chip.

The additional degree of freedom given by the formation of two epitaxial collector layers has also great advantages for the integration of further devices. It will be demonstrated that the "double epi concept" also allows the integration of varactors with high quality factors and high capacitance tuning ratios into high frequency SiGe bipolar technologies for enabling high frequency voltage controlled oscillators (VCOs) with large tuning range and low phase noise.

II. DEVICE STRUCTURE AND PROCESS CONCEPT

As compared to our previous work [3] several changes have been made in the process flow as shown in fig. 1. The collector region for the high voltage (HV) transistor and the cathode region for the varactor are formed by implanting the buried subcollector layer and the first epitaxial layer (Epi 1) is grown subsequently (fig. 1(a)). Then the collector region for the high speed (HS) transistor is formed by the implantation of the HS buried layer followed by the subsequent growth of an additional thin epitaxial layer (Epi 2). In the HV transistor and the varactor the HS buried layer is also used to provide a contact to the HV buried layer (fig. 1(b)). The final thicknesses of the active collector regions are adjusted to 60 nm in the HS transistor and to 350 nm in the HV transistor.

Then the deep/shallow trench transistor isolation is formed, resulting in a completely planar transistor isolation with different thicknesses of the active collector regions in the high speed and high voltage transistors.

After the fabrication of the deep/shallow trench transistor isolation a CVD oxide with a thickness of 60 nm is deposited. By using a resist mask the CVD oxide in the varactor regions and in the substrate contact regions (not shown in fig. 1(c)) is removed. Then the phosphorus implantations for the varactor profile are performed (fig. 1(c)). Next a stack consisting of the p^+ polysilicon base electrodes, a CVD oxide layer and a nitride layer is deposited and patterned for forming the emitter window and the anode of the varactor (fig. 1(d)).

As described in [6], emitter/base processing of the double polysilicon self aligned (DPSA) SiGe HBTs is continued

1-4244-0458-4/06/$20.00 ©2006 IEEE

by forming nitride spacers inside the emitter window, by performing the self-aligned HS and HV transistor collector implantations and by removing the CVD-oxide on the active transistor regions (fig. 1(e)). Now the SiGe base is integrated by selective epitaxial growth (SEG) and a mono crystalline emitter is formed (fig. 1(f)).

This approach has the following advantages:

1) The thicknesses of collector regions of the high voltage and high speed transistors can be adjusted completely independent from each other. Therefore, this double epi concept provides the maximum degree of freedom for realizing transistors with different f_T - BV_{CE0} trade-offs on one chip.

2) With a minimum number of additional process steps (one additional mask), varactors are easily integrated with this concept into a high frequency SiGe HBT technology. By using this approach, varactor epi thickness and varactor doping profile can be optimized for the required application field without imposing any restrictions on the active collector thickness and high frequency properties of the high speed transistors.

III. ELECTRICAL CHARACTERISTICS

In this section the electrical results of the high speed transistor, high voltage transistor, and varactor are reported. All measurements have been performed at room temperature. Fig. 2 and fig. 3 show the output characteristics of the high speed and high voltage transistors, respectively. Due to the SiGe base a high Early voltage is obtained (> 100 V). BV_{CE0} for the high speed transistor is 1.8 V (fig. 2) and for the high voltage transistor 5 V (fig. 3). The base-collector breakdown voltages (BV_{CB0}) are 6.0 V and 17 V for the high speed and high voltage transistors, respectively.

On-wafer S-parameter measurements using OPEN and SHORT de-embedding structures are used to evaluate the high-frequency performance of transistors with an emitter area of 0.14×2.6 μm^2. At a collector-base voltage of $V_{CB} = 0$ V a transit frequency of $f_T = 209$ GHz for the high speed transistor (fig. 4) and at a collector-base voltage of $V_{CB} = 2$ V a transit frequency of $f_T = 52$ GHz for the high voltage transistor (fig. 5) are obtained.

These values of cut-off frequency have been obtained at current densities of 11 mA/μm^2 and 0.85 mA/μm^2 for the high speed and high voltage transistors, respectively. The dependency of maximum oscillation frequency on collector current for the high voltage npn is shown in fig. 6. For this device a maximum oscillation frequency of 174 GHz has been achieved at a collector-base voltage of $V_{CB} = 2$ V.

Table I summarizes the most important transistor parameters which have been measured on devices with an effective emitter area of 0.14×2.6 μm^2. The high speed and high voltage transistors exhibit maximum oscillation frequencies of 237 GHz at a collector-base voltage of $V_{CB} = 1$ V and 174 GHz at a base-collector voltage of $V_{CB} = 2$ V, respectively. State-of-the-art CML ring oscillator gate delay performance of 3.3 ps has been obtained in the double polysilicon self-aligned SiGe bipolar technology.

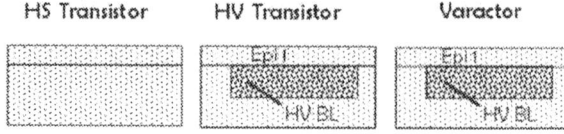

(a) Implantation of HV buried layer and growth of epi 1.

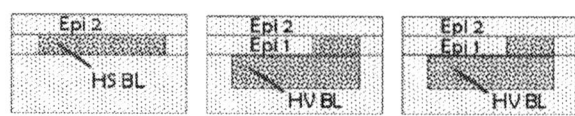

(b) Implantation of HS buried layer and growth of epi 2.

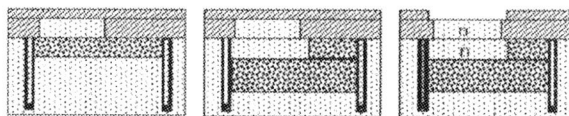

(c) Formation of deep/shallow trench isolation, CVD oxide deposition, and phosphorus implantations for varactor.

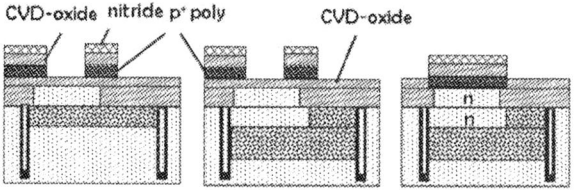

(d) Deposition of p$^+$ polysilicon/CVD-oxide/nitride stack and formation of emitter window.

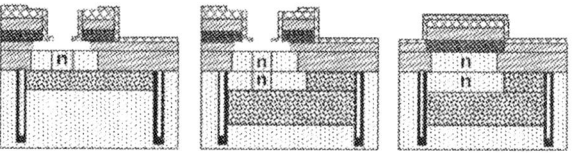

(e) Formation of nitride spacers, collector implantations, and removal of CVD oxide on active transistor regions.

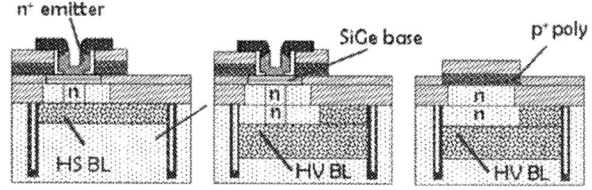

(f) Selective SiGe deposition, nitride removal, formation of emitter-base spacer, and emitter processing.

Fig. 1. Fabrication of the high speed npn, high voltage npn, and varactor on the same chip (a)-(f).

In comparison with our previous single epi approach these results show that by using the double epi concept an increase of BV_{CE0} from 3 V to 5 V and of BV_{CB0} from 10 V to 17 V can be obtained for the high voltage transistor while maintaining the performance of the high speed transistor.

Fig. 7 shows the capacitance-voltage characteristics of the varactor. The measured varactor is a structure with 5 stripes and an effective area of $5 \times 1 \times 10$ μm^2. The capacitance tuning ratio between 0 V and -5 V is about 2.3.

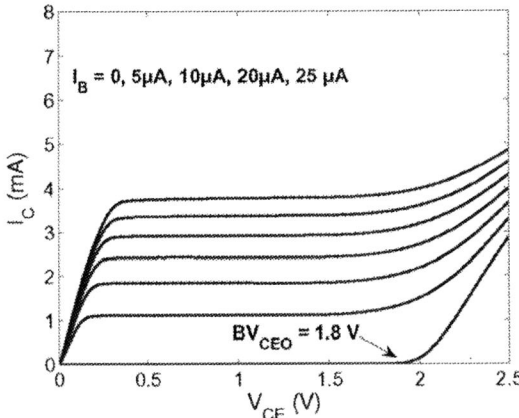

Fig. 2. Output characteristics of high speed npn ($A_E = 0.14 \times 2.6 \ \mu m^2$).

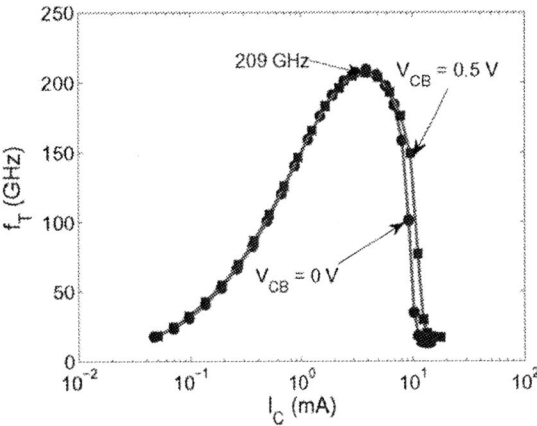

Fig. 4. Cut-off frequency f_T vs. collector current I_c for high speed npn ($A_E = 0.14 \times 2.6 \ \mu m^2$).

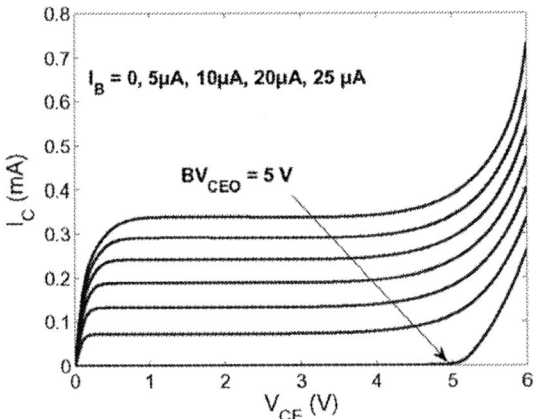

Fig. 3. Output characteristics of high voltage npn ($A_E = 0.14 \times 2.6 \ \mu m^2$).

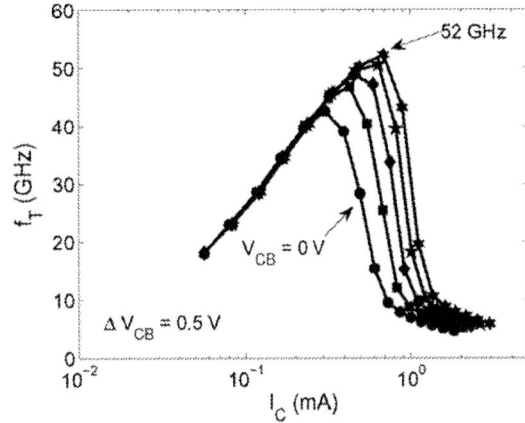

Fig. 5. Cut-off frequency f_T vs. collector current I_c for high voltage npn ($A_E = 0.14 \times 2.6 \ \mu m^2$).

Fig. 8 illustrates the measured quality factor versus junction voltage at frequencies of 20, 35, and 50 GHz, respectively. For example at 50 GHz the quality factor is about 9.5 at a junction voltage of -1 V and about 16 at a junction voltage of -5 V.

IV. CIRCUIT RESULTS

By using the double epitaxial collector layer concept a 77 GHz VCO suited for automotive radar applications has been fabricated. The VCO is similar to that described in [7] and includes an integrated output buffer. The buffer has two outputs and provides a total signal output power of 16 dBm. In this circuit the new varactor has been integrated for realizing high tuning range. Fig. 9 shows the tuning and phase noise characteristics of the fabricated VCOs.

For comparison, the figure also shows the tuning and phase noise characteristics of our previous VCOs fabricated in the conventional single epitaxial layer approach, which has used the base-collector junction of the high speed transistors for frequency tuning. The integration of the varactor by the double epitaxial layer concept has significantly

improved the VCO tuning range from 7 to 13 GHz while maintaining nearly the same phase noise performance.

V. CONCLUSIONS

By using a new concept with two epitaxial layers we have developed a SiGe bipolar technology which provides the maximum degree of freedom for realizing transistors with different f_T - BV_{CE0} trade-offs on one chip. The high speed transistors have a cut-off frequency f_T of 209 GHz, a maximum oscillation frequency f_{max} of 237 GHz, and an emitter-collector breakdown voltage BV_{CE0} of 1.8 V.

The high voltage transistors provide an f_T of 52 GHz and an f_{max} of 174 GHz which have been combined with a high emitter-collector breakdown voltage of 5.0 V. Additionally a new varactor has been developed and integrated in a 77 GHz VCO suited for automotive radar applications. As compared to our previous single epitaxial layer approach the varactor has increased the tuning range of the VCO from 7 to 13 GHz.

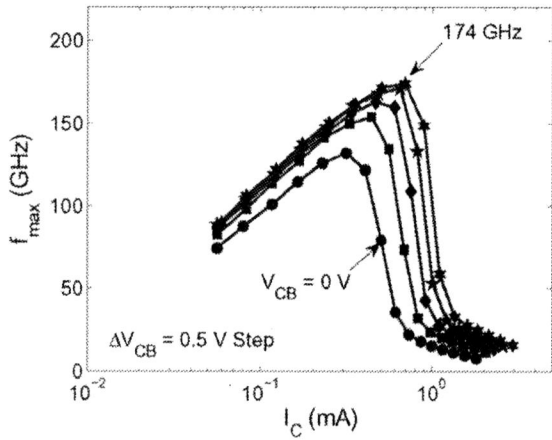

Fig. 6. Maximum oscillation frequency f_{max} vs. collector current I_c for high voltage npn ($A_E = 0.14 \times 2.6\ \mu m^2$).

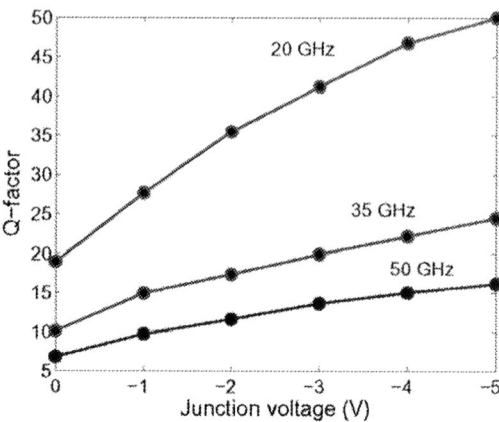

Fig. 8. Quality factor of the varactor vs. junction voltage measured at different frequencies ($A_E = 5 \times 1 \times 10\ \mu m^2$).

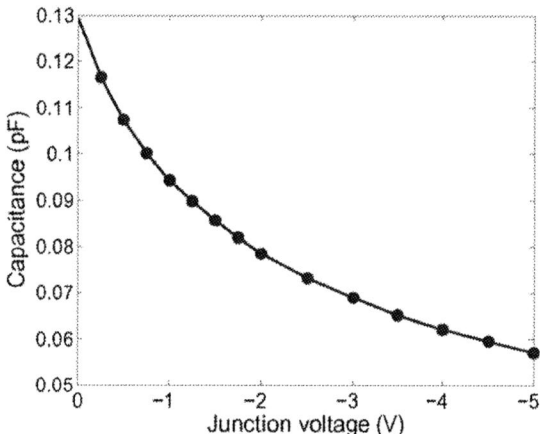

Fig. 7. Capacitance-voltage characteristics of the varactor ($A_E = 5 \times 1 \times 10\ \mu m^2$).

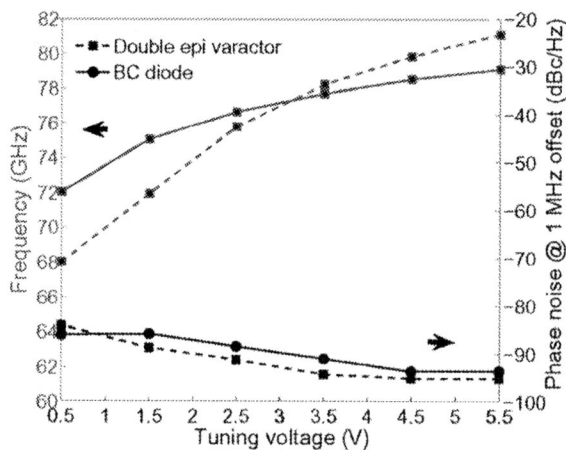

Fig. 9. Measured VCO tuning and phase noise characteristics. (Solid line: base-collector diode of HS transistor for frequency tuning; dashed line: new varactor (double epi concept) for frequency tuning.

REFERENCES

[1] J. -S Rieh et al., *IEDM Technical Digest*, 2004, pp. 247–250.
[2] B. Heinemann et al., *IEDM Technical Digest*, 2004, pp. 251–254.
[3] J. Böck et al., *IEDM Technical Digest*, 2004, pp. 255–258.
[4] Q. Z. Liu et al., *Compound Semiconductor Integrated Circuit Symposium (CSIC '05)*, 2005, pp. 117–120.
[5] K. K. O et al., *IEEE Electron Device Letters*, Vol. 19, No. 5, 1998, pp. 160–162.
[6] T. F. Meister et al., *BCTM Technical Digest*, 2003, pp. 103–106.
[7] H. Li et al., *Journal of Solid-State Circuits*, Vol. 39, No. 10, 2004, pp. 1650–1658.

TABLE I
DEVICE PARAMETERS

Parameter	High Speed npn	High Voltage npn
A_E	$0.14 \times 2.6\ \mu m^2$	$0.14 \times 2.6\ \mu m^2$
β	300	300
R_{BI}	$2.4\ k\Omega/\Box$	$2.4\ k\Omega/\Box$
BV_{CE0}	1.8 V	5.0 V
BV_{CB0}	6 V	17 V
C_{EB}	7.5 fF	7.5 fF
C_{BC}	5.8 fF	3.0 fF
C_{CS}	3.5 fF	5.5 fF
R_B	50 Ω	50 Ω
R_E	3.0 Ω	3.0 Ω
f_T	209 GHz ($V_{CB} = 0\ V$)	52 GHz ($V_{CB} = 2\ V$)
f_{Max}	237 GHz ($V_{CB} = 1\ V$)	174 GHz ($V_{CB} = 2\ V$)
τ_d	3.3 ps	–

RF, Analog and Mixed Signal Technologies for Communication ICs - An ITRS Perspective

W. Margaret Huang[1], Herbert S. Bennett[2], Julio Costa[3], Peter Cottrell[4], Anthony A. Immorlica Jr.[5], Jan-Erik Mueller[6], Marco Racanelli[7], Hisashi Shichijo[8], Charles E. Weitzel[1], and Bin Zhao[9]

[1]Freescale Semiconductor Inc., Tempe, Arizona, [2]National Institute of Standards and Technology, Gaithersburg, Maryland, [3]RF Micro Devices, Greensboro, North Carolina, [4]IBM, Essex Junction, Vermont, [5]BAE Systems, Nashua, New Hampshire, [6]Infineon Technologies, Munich, Germany, [7]Jazz Semiconductor, Newport Beach, California, [8]Texas Instruments, Dallas, Texas, [9]Skyworks Solutions, Irvine, California

Abstract — The International Technology Roadmap for Semiconductor (ITRS) Radio Frequency and Analog/Mixed-Signal (RF and AMS) Wireless Technology Working Group (TWG) addresses device technologies for wireless communications covering both silicon and III-V compound semiconductors. In this paper, we discuss the roadmap and the figures of merit (FoM) used to characterize both active and passive devices critical for typical radio front end designs. We review the trends, challenges and potential solutions and address the intersection of silicon and III-V compound semiconductors.

Introduction

RF and AMS technologies serve the rapidly growing wireless communications market and represent essential and critical technologies for the success of many semiconductor manufacturers. Recognizing wireless applications as a new system driver, ITRS formed the RF and AMS Wireless TWG in 2003.

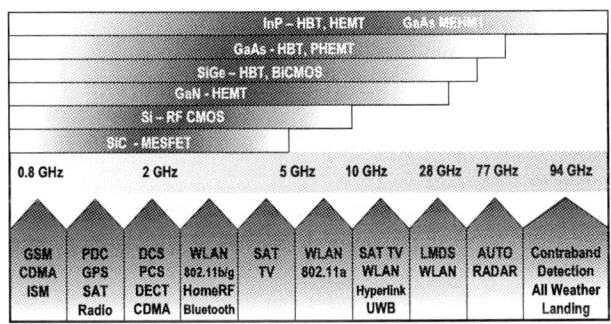

Fig. 1. Wireless communication application spectrum.

Figure 1 presents the interplay among commercial wireless communication applications, available spectrum, and the kinds of elemental and compound semiconductors likely to be used. The boundary between silicon and III-V semiconductors has been moving to higher frequencies with time. Today, Si and SiGe dominate below 10 GHz and III-V compound semiconductors dominate above 10 GHz products. In future years, it is expected that the frequency axis in Fig. 1 will lose its significance in defining the boundaries among technologies. Instead, the future boundaries will be dominated more by such parameters as noise figure, output power, power added efficiency, linearity and ultimately cost. Some of this is already true for power amplifier (PA) applications. The consumer portions of wireless communications markets are very sensitive to cost. With the different technologies capable of meeting the technical requirements, time-to-market and overall system cost will govern technology selection.

Figures of Merit

The wireless communication circuits considered as application drivers may be classified into AMS circuits including analog-to-digital and digital-to-analog converters, RF transceiver circuits including low noise amplifiers (LNAs), frequency synthesizers, voltage controlled oscillators (VCOs), driver amplifiers, PAs and filters. The device FoM tracked by the RF and AMS roadmap such as transit frequency at unity current gain f_T, maximum frequency of oscillation f_{MAX} at unit power gain, noise, voltage gain, mismatch, linearity, power, power added efficiency, breakdown, quality factor, and the like are chosen based on their impact to the performance of these key RF and AMS circuits. A more detailed description of the relation between circuit and device FoM can be found in [1].

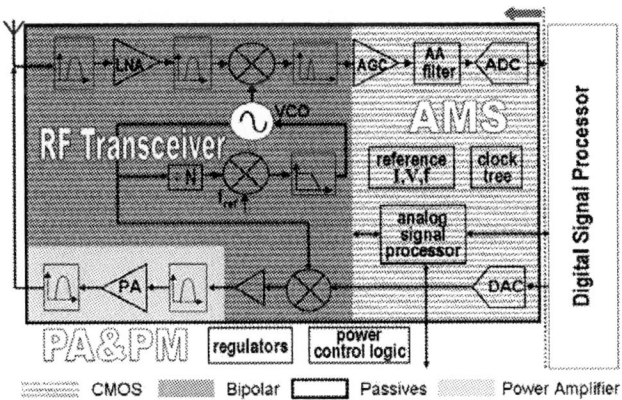

Fig. 2. Circuit functions of a typical wireless communication system and RF and AMS roadmap device partition.

The 2005 RF and AMS chapter was divided into five main sections [2]. Four sections focused on CMOS, Bipolar, Passives and PAs for the 0.8 to 10 GHz applications, while the last section, mm-Wave, covered the 10 to 100 GHz applications. The application frequencies in this context refer to the nominal carrier frequencies for the communications system and are not necessarily the clock or operating frequencies of the individual devices and circuits. Fig. 2 illustrates the typical radio circuit functions and how the roadmap partitions the different device sections to cover these functions.

CMOS

The trends of higher integration and the cost of technology development drove the need to align the RFCMOS roadmap with the logic platform (ITRS Process, Integration, Devices, and Structures roadmap). The Low-STandby Power (LSTP) CMOS roadmap which was targeted for portable applications was selected as the basis. The RF and AMS CMOS roadmap was delayed by one year compared to the logic LSTP roadmap to reflect the additional time needed to stabilize the technology for RF mixed signal products and allow time to introduce a mature RF analog design kit. Scaling digital CMOS performance (gate length and gate dielectric) results in an aggressive roadmap for device f_T, f_{MAX} (Figs. 3 and 4) and noise. A detailed description of device f_T, f_{MAX} and noise scaling can be found in [3]. The improved f_{MAX} compared to the 2004 roadmap is due to device layout optimization and improved data analysis obtained with complete de-embedding. The challenge for designers will be how to effectively use the increased f_{MAX} for higher frequency applications as supply voltage continues to scale down and the margin for model inaccuracies decreases [4].

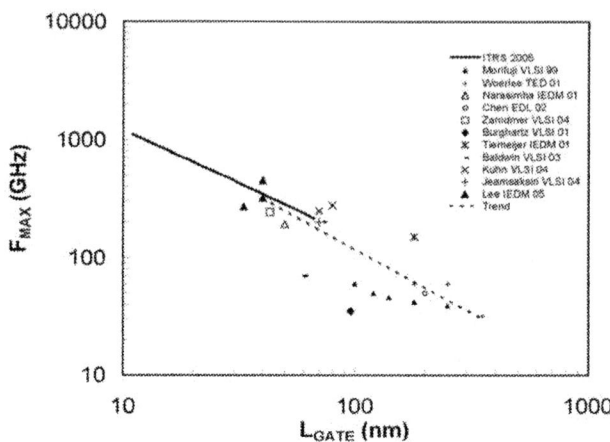

Fig. 4. NFET f_{MAX} versus L_{GATE}.

The requirement of low standby power limits the rate of gate oxide scaling relative to gate length scaling and, for conventional device structures, drives ever increasing doping concentration and non-uniform channel doping. These trends degrade voltage gain (Gm/G$_{ds}$) and increase the threshold mismatch. In the case of Gm/G$_{ds}$ scaling for the logic LSTP device, the roadmap saturates at 30, which is considered the minimum requirement for analog design (Fi.g 5). New gate electrode materials that enable threshold control via workfunction modulation, double gate structures or asymmetric doping profiles will mitigate this trend. However, it is possible that a unique device design for high gain may be required, supplementing the standard LSTP logic device.

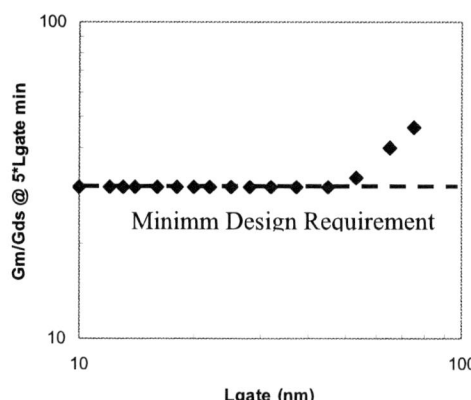

Fig. 5. ITRS 2005 Gm/G$_{ds}$ versus L_{GATE}.

Overall, the long-term prediction of device RF and noise performance becomes more uncertain with the introduction of metal gate electrodes (2009), high permittivity (high-K) gate dielectrics (2009), and new device structures such as fully depleted and/or double-gated SOI (2015) [5]. The introduction of metal gates may reduce threshold mismatch due to variations in gate doping and should increase

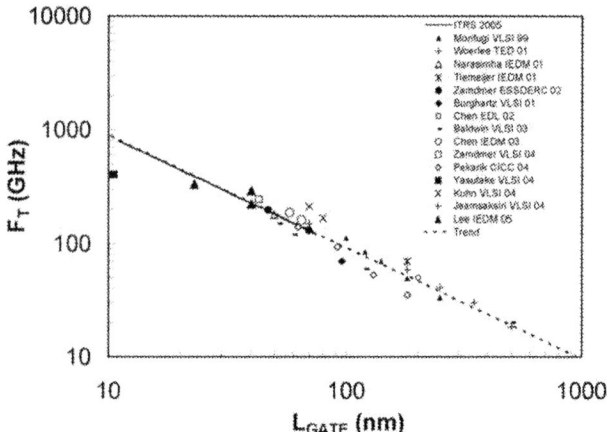

Fig. 3. NFET f_T versus L_{GATE}.

device f_{MAX} due to decreased gate resistance. The introduction of high-K gate dielectrics and channel strain increases uncertainty in device 1/f noise scaling. Fully-depleted, dual-gate SOI has low channel doping relative to conventional CMOS structures and so may have reduced mismatch. Finally, the introduction of multiple threshold voltages for digital power-delay optimization will also offer design options for mixed-signal and RF applications and possibly elevate some of the demands on individual devices as predicted by the roadmap.

In addition to the LSTP logic CMOS device, the RF and AMS roadmap added a thick-oxide device optimized for higher voltage precision analog designs and for driving RF signals off-chip. This device is often used as a second or a third input/output (I/O) transistor for logic applications to support interfaces to the outside world. The mixed-signal supply voltage scaling continues to lag that of digital by two or more generations driven by the need to maintain signal-to-noise ratios and low signal distortion in analog designs. The challenge in supporting this device comes as the LTSP logic CMOS device moves towards a new device structure such as fully depleted and/or double-gated SOI. The fabrication of the conventional precision analog device may require separate process steps resulting in higher wafer/die cost. The increased die cost may drive more interests in SiP/module solutions as opposed to the integrated SoC solutions. On the design research front, new architectures are being introduced that make use of the increased digital processing capabilities allowing more signal treatments to be done in the digital domain and enabling the potential replacement of most analog functions [6]. In addition, research in software defined radios to support low cost multi-mode multi-band applications will drive different technology requirements [7].

SiGe Bipolar

RF and AMS bipolar devices are most frequently used in RF transceiver blocks including LNAs, synthesizers and sometimes PA drivers or PAs. Technology requirements for bipolar devices used in wireless communications are driven by the need for lower power consumption, lower noise, and lower cost much in the same way as CMOS devices. Reduced power consumption and lower noise for bipolar devices are achieved through higher f_T and , f_{MAX} lower base resistance and base collector capacitance. Silicon bipolar devices have undergone strong performance gains since the introduction of SiGe [8]. The SiGe epitaxy process enabled use of bandgap engineering plus aggressive scaling of the base width and base doping. Bipolar scaling also benefits from lithography advances. By scaling the emitter width, device f_{MAX} increases, while noise and power consumption decreases.

In the near-term, the major challenge for scaling the BiCMOS technology is the cost of the technology and the increased difficulty with integrating bipolar devices in aggressively scaled CMOS with conflicting thermal budget requirements. In the 2005 roadmap, three separate bipolar technology requirements are monitored: 1) a high-speed bipolar devices where the device f_T and f_{MAX} continue to improve, challenging III-V technologies in mm_wave applications (Fig. 1); 2) a RF low-cost medium performance device suitable for <10Ghz mobile phone and connectivity solutions; 3) and finally, a high-voltage device optimized for PA drivers and PAs. All three bipolar devices may not be integrated in the state of the art CMOS technologies. The most cost effective solution may be to stay at a relative large geometry BiCMOS platform with the optimized bipolar device for increased frequency [9,10] or power [11] applications.

A more detailed discussion on the different SiGe bipolar device requirements for the emerging mm-wave and PA applications can be found in the following sections.

mm-Wave

The scope of this section includes both low-noise and power transistors that are based on several competing technologies: GaAs MESFET, GaAs PHEMT, InP HEMT, GaAs MHEMT, GaN HEMT, InP HBT and SiGe HBT. Today, compound semiconductors dominate the 10 to 100 GHz applications. The device types most commonly used for analog mm-wave applications are HEMT, PHEMT, and MHEMT while the device types most commonly used for mixed-signal and high-speed applications are predominately MESFET and HBT. There is great diversity in the nature and performance of these devices because device properties are critically dependent on the selection of materials, thickness and doping in the stack, which are proprietary to the manufacturer. Performance trends are driven primarily by a combination of "bandgap engineering" of the epitaxial layer stack in concert with shrinking lithography. Compound semiconductors do not enjoy the long-term heritage of silicon-based devices, nor do they follow Moore's Law. The biggest challenge facing compound semiconductors is production cost. Uniformity, reproducibility and yield metrics for compound semiconductors in general lag behind Si-based technologies. As production volume in a particular compound technology rises, unit costs decrease similar to that of silicon. In addition, the compound semiconductor wafer diameter needs to be within one or two generations of the silicon industry to take advantage of the advances in lithography and processing equipment. This is particularly challenging as silicon transitions from 200-mm to 300-mm wafers.

Recent advances in SiGe device performance enabled Si technology to venture into application frequencies that had been dominated by III-V compound semiconductors. Transceiver circuits for 60 GHz wireless personal area networks (WPAN) [12] and 77 Ghz automotive radar applications [13,14] using SiGe bipolar already show promising results. Figure 6 shows the projected roadmap for these high frequencies devices. The SiGe devices achieve comparable device f_T at a lower BV_{CEO}. However, the lower BV_{CEO} did not impose design limit since in actual design the base is never floating and a more relevant design parameter is device BV_{CBO} [15]. Collector-base breakdown for these fast SiGe devices is projected to be in 5.5 to 4.5 V range. f_{MAX} of these devices are projected to be between 250 to 450 GHz and associated noise figures (NFmin) at 77 Ghz between 5.5 and 3.5 dB. These projections are in line with recently reported performance for a 300 GHz f_T / f_{MAX} device with measured NFmin of 1.4dB at 24 GHz and extrapolated NFmin of ~ 3.5 dB at 60 GHz [9].

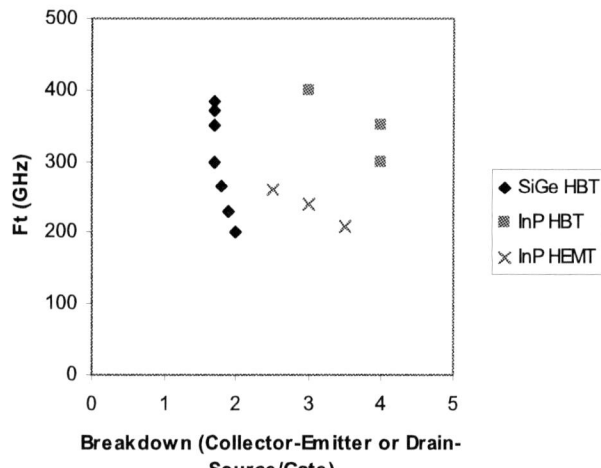

Fig. 6. ITRS 2005 mm_wave device breakdown versus f_T.

The SiGe bipolar processes offer relatively higher level of integration flexibility compared with III-V processes. The III-V device parameters depend heavily on the selection of materials, thickness and doping which makes it difficult to offer more than one device type on the same semiconductor wafer. On the other hand, more than one SiGe bipolar can be offered on the same Si wafer. Multiple SiGe bipolar devices optimized for higher f_T, higher f_{MAX} and lower noise or for higher breakdown [10] are commonly supported with one process flow. The multiple devices enable a higher level of design optimization. As reported in literature, comparable SiGe:C technology yielded a 77 GHz VCO with total output power of 18.5 dBm and phase noise of -97 dBc/Hz at 1-MHz offset [13] and a 77 GHz LNA with noise figure less than 5.5 dB and gain of 8.9 dB [14]. In addition, the CMOS devices in a BiCMOS process allow a high level of chip integration as reported recently for 60 GHz receiver and transmitter integrated circuits [12].

The challenge in designing mm-wave circuits using these high performance bipolar devices on Si is how to handle the lossy Si substrate. Careful layout optimization of the transmission line elements [16] and modeling of backend routing metal are critical at these frequencies. Extra attention in design and layout is also needed to maintain signal isolation. Integration of multi-channels receiver and transmitter will be more of a challenge in terms of ensuring signal integrity across the substrate and routing metal [17].

While SiGe devices have shown capability in mm-wave range, it is unlikely to replace III-V devices in applications where either ultra low noise or high power gain are required. For front end and output stages with challenging requirements, it is likely that III-V technologies such as PHEMT, MHEMT or even the evolving GaN HEMT technology will predominate. For example, InP HEMTS and MHEMTs consistently deliver less than 3 dB noise figure at 60 GHz [18]. GaN based HEMTs show similar excellent noise performance [19] with the added advantage of very high breakdown voltages on the order of 40 Volts. The high breakdown voltage of GaN will result in outstanding linearity and robustness along with low noise figure. Power results are also impressive, with single chip MHEMT MMICs on the order of 4 watts at 40 GHz [18] and for GaN, 8 watts at 30 GHz [20].

Power Amplifier

Wireless communications require both portable (handset) and fixed (base station) transmitters and receivers to form a connected network. Since the PA requirements are very different between the handset and base station, the ITRS RF and AMS roadmap has separate requirement tables for these applications. The PA section covers both III-V and Si-based technologies. The key driving forces for the PA applications are integration of components and cost. A cost parameter is included in the PA tables; Cost/mm^2 for handset PA and Cost/watt for base station PA.

Handset Power Amplifiers

In 2005, the bulk of the consumer market for PA technologies continues to be RFICs and modules for cellular subscriber handsets. Both cellular and connectivity applications typically have very strict performance specifications and are extremely sensitive to price/performance trade-off. This trade-off continuously drives the industry towards highly integrated low-cost system solutions. In the cellular subscriber area, the market for PAs is increasingly migrating away from packaged single die with RFIC to multi-band multi-mode integrated modules that deliver a complete amplifier solution. These RF modules typically integrate all or most of the matching and bypassing networks, and may also provide power detection, power management, filtering and RF switches for both transmit/receive and band selection. Contrary to 2004 predictions, we continue to see the proliferation of 3V-based systems and believe that the 3V systems will remain through the next two to three years. In terms of migrating towards a Si-based solution, the SiGe multi-band cellular PAs are being sampled but they are not yet present in any significant volumes. CMOS PAs are being discussed and sampled, but demonstrations of a viable and rugged PA is still not published.

The biggest advantage for Si-based PA applications is lower die cost (Fig. 7) and the potential for integration. The CMOS controller typically packaged in a front-end PA module today can be included in the same Si die, which reduces total system cost.

	LDMOS	SiGe HBT	GaAs HBT	GaAs FET
PA Die Size (mm²)	6	2.5	2.5	4
Cost / mm² ($$)	0.08	0.12	0.35	0.25
Product Cost ($$)	0.48	0.30	0.875	1.00

Fig. 7. ITRS projected PA die cost for 2006.

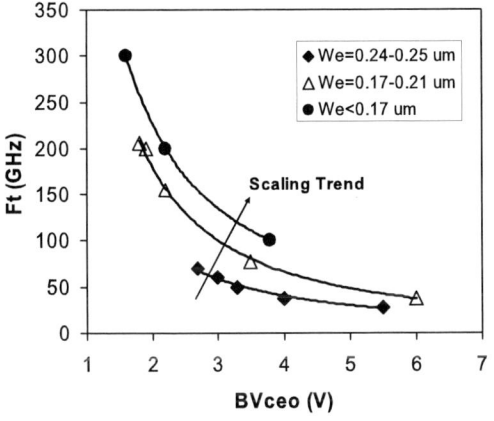

Fig 8. f_T of SiGe devices as a function of BVceo extracted from the literature for different ranges of emitter width [3]. Generally, more advanced geometry can result in improved f_T and breakdown.

SiGe devices are well suited for PA applications as they can support higher voltage levels for a given level of performance (f_T) vs. CMOS devices. Supporting high voltages is required in a PA despite the lower battery voltages because large antenna mismatch conditions can reflect RF power back into the device causing the collector voltage of a common emitter amplifier to swing to high levels. Figure 8 shows the evolution of breakdown and f_T for SiGe devices. It shows that each new generation of technology has

improved both the speed and breakdown voltage simultaneously. This differs from CMOS where the thinner gate oxides required in more advanced nodes make it difficult to improve speed and breakdown simultaneously.

Today, SiGe devices are used extensively in lower power standards such as WLAN and are now being commonly integrated with the rest of the transceiver in applications where output powers levels are in the range of 20dBm [3]. For higher power standards used for cellular applications that require 2 to 3W, SiGe has not yet significantly penetrated the market which is dominated today by GaAs HBTs and, to a lesser extent, discrete RF LDMOS devices. Some benchmark of SiGe and GaAs PA performance have been recently published [21] showing comparable performance levels at 900MHz with better performance for GaAs at 2GHz for a linear CDMA 28dBm PA application.

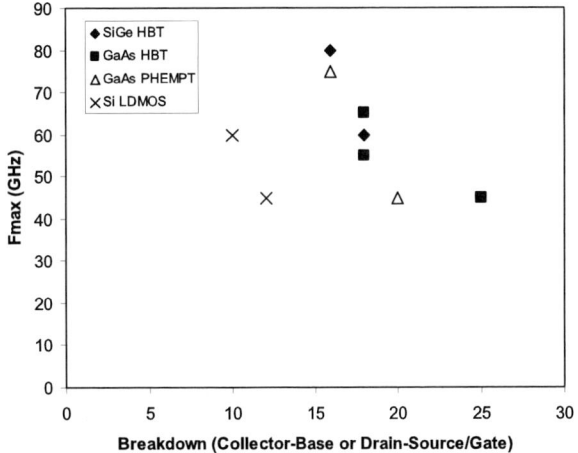

Fig. 9. ITRS 2005 PA device breakdown versus f_{MAX}

Figure 9 compares device f_{MAX} and breakdown voltage of the different PA technologies. However, as noted in earlier publication [3,21], PA performance cannot be extrapolated based purely on device parameters such as f_T, f_{MAX} and breakdown voltage. A PA design needs to be maximized for output power and power added efficiency (PAE) while passing ruggedness specifications from handset vendors. Recent advances in ruggedness protection circuitry and advances in SiGe PA technology is helping to close the performance gap making the reality of SiGe PAs in the cellular market more likely.

Base Station Power Amplifiers

The base station transmitter provides the outgoing data link to the cellular phone. Relatively high RF power (600 W) is required to achieve the desired cellular phone coverage. The major challenges for base station PAs include fabrication with a low amount of individual tuning. Silicon LDMOS transistors are now the technology of choice for cellular systems at 900 MHz and at 1900 MHz because of their technological maturity and low cost. Applications are moving from 2 GHz and below to higher frequencies, such as WiMAX at 3.5 GHz and from saturated power amplifiers to more linear power amplifiers for CDMA and WCDMA.

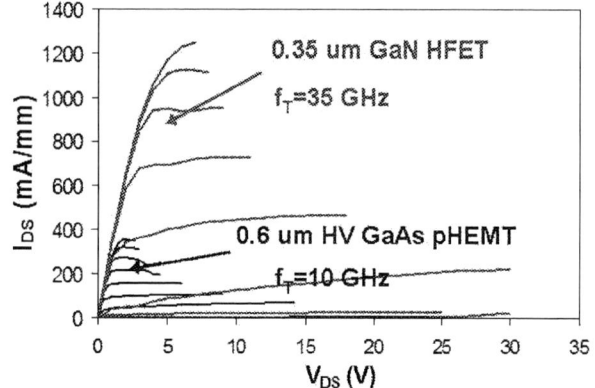

Fig. 10. Comparing GaN to GaAs. GaN device provides 3X higher in current density, 5X higher in breakdown voltage and 5-10X higher microwave power density than GaAs or Si devices.

As frequencies increase, LDMOS will experience challenges from GaAs FET and SiC MESFET. Gallium arsenide devices are more expensive, but offer higher efficiency than Si LDMOS. Gallium nitride is another technology coming over the technological horizon. Gallium nitride has power densities four times larger than silicon LDMOS or gallium arsenide due to the higher device breakdown voltage and current density (Fig. 10) [22]. The major challenge for GaN technology is demonstrating device reliability.

In spite of this move to higher frequencies, device cost as measured by dollars per RF Watt is still projected to steadily decrease from about $0.70/W today to less than $0.50/W by 2008.

Passives

Passive devices including capacitors, resistors, varactors, inductors, transformers and transmission lines are frequently used for impedance matching, resonance circuits, filters and bias circuits. The performance of these devices often plays a key role in determining RF and AMS circuits performance. For instance, the critical parameters of a VCO such as frequency tuning range, power consumption and phase noise are primarily determined by the tuning range of the varactor and the quality factor of the inductor and varactor. The biggest challenge for integrating passive elements into a digital CMOS process is the tradeoff between processing cost and device performance. When incorporating passives into a standard CMOS process, there are some additional processing steps and possibly new materials required. In addition, capacitors and inductors generally occupy more silicon area

than active devices which effectively increase total die cost. Both the SoC implementation where passive elements move from the board level to chip level and SiP implementation where passives are integrated into the package or module are being used and are dependent on the final system cost.

The long-term challenges for passive elements include the need to integrate new materials in a cost-effective manner to realize high-density, low leakage and high linearity metal-insulator-metal (MIM) capacitors, high density and high quality factor (Q) inductors, high temperature-linearity resistor and high tuning ratio varactor.

In order to achieve high capacitance density for RF MIM capacitors, various high-k dielectrics including Ta_2O_5 and HfO_2 are being explored. The key challenge is to keep the leakage current and voltage nonlinearity low as the film thickness is reduced. The 2005 roadmap is predicting a much slower scaling of the MIM capacitor density compared to the previous roadmap (Fig. 11) to account for the device need to meet all requirements including linearity and leakage and being integrated on a copper backend with no stacking.

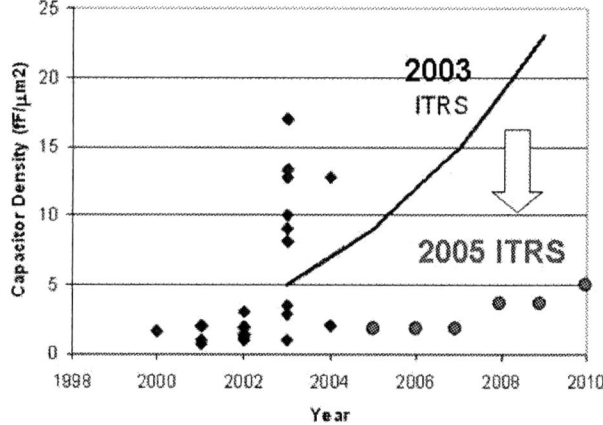

Fig. 11 MIM capacitor scaling.

One way to solve this tradeoff is to use multi-layered structures where the capacitance density and voltage linearity can be separately optimized. A low-cost alternative to the MIM capacitor is the inter-digitated lateral metal-oxide-metal capacitor using the standard multi-layer metal backend. As feature size scales, the unit capacitance of the inter-digitated, MOM capacitor can be close to or exceed that of the standard MIM capacitor. The MOM capacitor will have worse parasitic coupling with the Si substrate and proper layout and structure design is essential to achieve competitive mismatch performance. Nonetheless, the MOM capacitor requires no extra processing and is a good example of tradeoff between performance and cost.

Above-passivation inductors offer high Q-factors and resonant frequencies, but require special processing. On the other hand, improvements in inductance density are difficult to realize. Solutions of stacking inductors have been proposed, but add significant cost and jeopardize the resonant frequency. The use of magnetic materials has received some attention in recent publications. However, more research in this area is needed to render a practical solution.

In terms of resistor performance, increasing CMOS complexity makes it more difficult to produce stable, highly manufacturable front-end-of-line (FEOL) resistors. One solution to this problem is to provide high-resistance BEOL resistors comparable to the common p-type polysilicon resistors. However to address the concern with the thermal effect, new materials that provide good thermal control at high currents are needed.

Passives scaling is also directly affected by CMOS device scaling as in the case of MOS capacitors and varactors. The switch to high-k dielectrics in a CMOS device and subsequently the MOS varactors may affect the VCO phase noise. Research is needed to determine the effect on CMOS 1/f noise and MOS varactor Q.

Summary

An overview of the ITRS RF and AMS Wireless roadmap was discussed. Even though not included in the requirement tables today, the technology working group recognized the importance of signal isolation especially as demands for multi-mode, multi-in/multi-out integrated radio increase. Recent publication of a single chip radio [23] showed promising results that substrate crosstalk is manageable. The intention of the working group is to define a FoM for isolation in future roadmaps. Other elements under consideration for inclusion in the roadmap include above-IC integrated MEMs and BAW technologies [24] commonly used for filters and transmit receive (T/R) switch.

Acknowledgments

The authors gratefully acknowledge our many colleagues on the ITRS RF and AMS Wireless Technology Working Group for their excellent contributions to the roadmap.

References

[1] Ralf Brederlow, et al, "A mixed-signal roadmap for the international technology roadmap for semiconductors," *IEEE Design and Test for Computers*, vol.18, no. 6, pp. 34-46, 2001.

[2] 2005 Edition, Radio Frequency and Analog/Mixed-Signal Technologies for Wireless Communications, International Technology Roadmap for Semiconductors. Available: http://www.itrs.net/Common/2005ITRS/Home2005.htm.

[3] Herbert S. Bennett, et al., "Device and technology evolution for Si-based RF integrated circuits," *IEEE Trans. Electron Devices*, vol. ED-52, no. 7, pp. 1235-1258, July 2005.

[4] Chinh H. Doan, et al., "Design considerations for 60GHz CMOS radios," *IEEE Communications Magazine*, pp. 132-140, December 2004.

[5] Stefaan Decoutere, et al., "Process issues and ITRS relationship", *SEMI Europe Standards Conference*, October 2005.

[6] Chih-Ming Hung, et al., "A first RF digitally-controlled oscillator for SAW-less TX in cellular systems", *Symposium on VLSI Circuits Digest of Technical Papers*, pp. 402-405, June 2005.

[7] R. Bagheri, et al., "An 800MHz to 5GHz software-defined radio receiver in 90nm CMOS" *ISSCC Digest of Technical Papers*, pp. 480-481, February 2006.

[8] Marco Racanelli, et al., "SiGe BiCMOS technology for RF circuit applications", *IEEE Trans. Electron Devices*, vol. ED-52, no. 7, pp. 1259-1270, July 2005.

[9] J.-S. Riech, et al., "SiGe HBT's for millimeter-wave applications with simultaneously optimized f_T and f_{MAX} of 300GHz", *IEEE RFIC Symposium Digest of Papers*, pp. 395-398, June 2004.

[10] J. Bock, et al., "SiGe bipolar technology for automotive radar applications", *Proceedings BCTM*, pp. 84-87, September 2004.

[11] Jeffrey B. Johnson, et al., "Silicon-Germanium BiCMOS HBT technology for wireless power amplifier", *IEEE J. Solid-State Circuits*, vol. 39, no. 10, pp. 1605-1614, October 2004.

[12] Brian Flyod, et al., "A silicon 60GHz receiver and transmitter chipset for broadband communications", *ISSCC Digest of Technical Papers*, pp. 184-185, February 2006.

[13] Hao Li, et al., "Fully integrated SiGe VCOs with powerful output buffer for 77-GHz automotive radar systems and applicatons around 100GHz", *IEEE J. Solid-State Circuits*, vol. 39, no. 10, pp. 1650-1658, October 2004.

[14] Bernhard Dehlink, et al., "A low-noise amplifier at 77GHz in SiGe:C bipolar technology", *Compound Semiconductor Integrated Circuit Symposium* , pp. 287-290, October 2005.

[15] J. Kraft, et al., "Usage of HBTs beyond BVCEO", *Proceedings BCTM*, pp. 33-36, September 2005.

[16] P. Wennekers, et al., "SiGe Technology Requirements for Millimeter-Wave Applications", *Proceedings BCTM*, pp. 79-84, September 2004.

[17] A. Babakhani, et al., "A 77 GHz 4-element phased array receiver with on-ship dipole antennas in silicon", *ISSCC Digest of Technical Papers*, pp. 180-181, February 2006.

[18] P. M. Smith, et al., "Progress in GaAs Metamorphic HEMT Technology for Microwave Applications", *GaAs IC Symposium*, pp. 21-24, November 2003.

[19] J. S. Moon, et al., "Microwave Noise Performance of AlGaN-GaN HEMTs with small CD Power Dissipation", *IEEE Electron Device Letters*, vol. 23, pp. 637-639, November 2002.

[20] Y. F. Wu, et al., "8 Watt GaN HEMT's at Millimeter Wave Frequencies", *IEDM Technical Digest*, pp. 593-595, December 2005.

[21] K. Nellis and P. Zampardi, "A comparison of bipolar technologies for linear handset power amplifier applications," *Proceedings BCTM*, pp. 3-6, September 2003.

[22] Umeshk Mishara, et al., "AlGaN/GaN HEMTs—An overview of device operation and applications" *Proceedings of the IEEE*, vol. 90, no. 6, pp. 1022-1031, June 2002.

[23] Pierre-Henri Bonnaud, et al., "A fully integrated SoC for GSM/GPRS in 0.13um CMOS ", *ISSCC Digest of Technical Papers*, pp. 482-483, February 2006.

[24] *IEEE RFIC Symposium*, Workshop on Advanced Technologies, June 2005.

2006 Bipolar/BoCMOS Circuits and Technology Meeting Proceedings

A High-Slew Rate SiGe BiCMOS Operational Amplifier for Operation Down to Deep Cryogenic Temperatures

Ramkumar Krithivasan, Yuan Lu, Laleh Najafizadeh, Chendong Zhu, John D. Cressler,
Suheng Chen[1], Chandradevi Ulaganathan[1], and Benjamin J. Blalock[1]

School of Electrical and Computer Engineering
777 Atlantic Drive, N.W., Georgia Institute of Technology, Atlanta, GA, 30332-0250, USA
[1]University of Tennessee, Knoxville, TN 37996-2100, USA

Abstract— We investigate, for the first time, the design and implementation of a high-slew rate op-amp in SiGe BiCMOS technology capable of operation across very wide temperature ranges, and down to deep cryogenic temperatures. We achieve the first monolithic op-amp (for any material system) capable of operating reliably down to 4.3 K. Two variants of the SiGe BiCMOS op-amp were implemented using alternative biasing schemes, and the effects of temperature on these biasing schemes, and their impact on the overall op-amp performance, is investigated.

I. INTRODUCTION

Operational amplifiers (op-amps) represent a ubiquitous and essential analog building block that finds application in a wide variety of high-performance precision analog circuits such as switched-capacitor filters, analog-to-digital converters, and precision sensors. Recently, there has been a growing interest in using such high-performance analog circuits for niche applications such as "extreme environment" electronics, and in particular for electronics capable of operating down to deep-cryogenic temperatures, as might, for instance, be encountered on the Moon (+120 °C to -180 °C and even down to -230 °C) [1].

The key device parameters of bandgap-engineered silicon-germanium (SiGe) heterojunction bipolar transistors (HBTs), such the transconductance (g_m), current gain (β), and Early voltage (V_A), that are critical to op-amp performance are favorably impacted by cooling [2]. Thus, analog building blocks such as op-amps designed using a combination of scaled Si CMOS and SiGe HBTs could potentially offer an optimal solution for such cryogenic applications. Although the key device parameters of well-designed SiGe HBTs show remarkable improvement down to cryogenic temperatures, it remains to be demonstrated that this device-level performance in fact translates to superior circuit performance at low temperatures. NASA's upcoming lunar missions present a unique venue for practicing low temperature electronics in the SiGe material system. The envisioned lunar robotic electronics systems will be subjected to very large temperature variations (> 300 °C), and are cyclic in nature. Traditionally, the on-board electronics are housed in a centralized "warm box," which shields them from dramatic temperature changes, thus maintaining a narrow temperature range required for their reliable operation. Such warm boxes, apart from being power hungry, bulky, and heavy, compromise the ability to realize distributed system architectures [1].

Because of its attractive cryogenic performance and inherent robustness to ionizing radiation, SiGe-based mixed-signal electronic systems could potentially be deployed outside the spacecraft warm box in so-called remote electronics units (REUs), for needed sensing and data acquisition functions. Realization of key circuit blocks such as voltage and current references, op-amps, and ADCs in SiGe technology are a critical first step.

In this work we demonstrate a SiGe BiCMOS op-amp for such extreme temperature range lunar electronic systems. We achieve the first monolithic op-amp (for any material system) capable of operating reliably down to 4.3 K. The design methodology used in the present work is based on a circuit architecture described in [3]. The high-slew rate capability is achieved by using an auxiliary slew-buffer that enhances the output drive capability during slewing. Two variants of the SiGe BiCMOS op-amp were implemented using alternative biasing schemes and the effects of temperature on these schemes, and the overall performance of the op-amp, is carefully examined.

II. DEVICE TECHNOLOGY AND CRYOGENIC OPERATION

This SiGe BiCMOS op-amp was fabricated in a commercial, deep and shallow-trench isolated SiGe HBT BiCMOS technology, which integrates 0.5 μm, 3.3 V BV_{CEO}, 50 GHz f_T, 3.3 V BV_{CEO}, and 55 V V_A SiGe HBTs (at 300 K), together with 0.5 μm, 3.3 V Si CMOS devices [4]. It supports a full suite of passive elements including metal-insulator-metal (MIM) capacitors and low T_C poly resistors. This is a five metal layer process (all aluminum) with a top thick aluminum layer that enables high-Q spiral inductors.

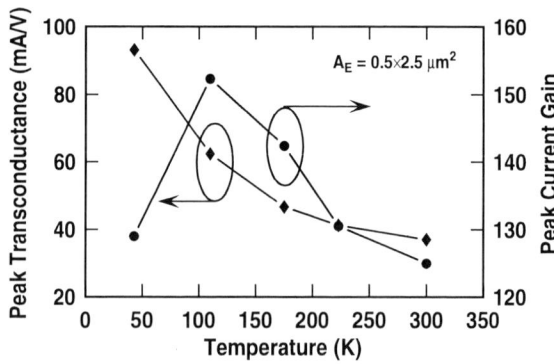

Fig. 1. Peak g_m and peak β versus temperature of the SiGe HBT with emitter area of 0.5×2.5 μm^2.

The SiGe HBTs from this technology generation show a strong monotonic increase in peak g_m with cooling (Fig. 1). In addition, with the base bandgap effects coupling strongly to the device equations in SiGe HBTs, β either increases or stays close

1-4244-0458-4/06/$20.00 ©2006 IEEE

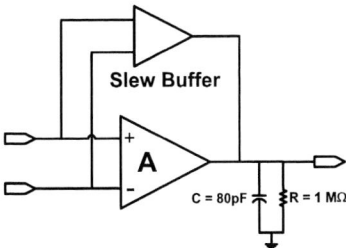

Fig. 2. The op-amp block diagram showing the core amplifier (A), the slew buffer, and a typical load condition.

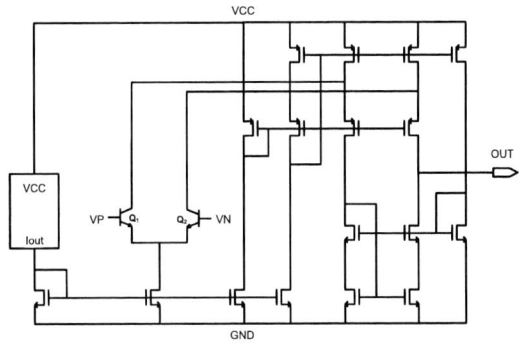

Fig. 3. Folded cascode amplifier core with bias block and SiGe HBT input stage.

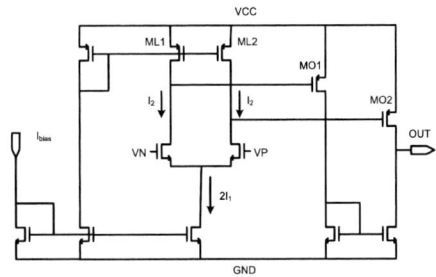

Fig. 4. Slew buffer with load transistors biased in the linear region.

to its room-temperature value down to deep cryogenic temperatures (Fig. 1). Although the increase in peak g_m should favorably translate into improved gain-bandwidth and slew rate performance in op-amps at low temperatures, this is highly dependent on the behavior of the bias current over temperature, as addressed below.

III. CIRCUIT DESIGN

The block diagram of the SiGe BiCMOS slew-enhanced op-amp is shown in Fig. 2. The overall amplifier consists of an amplifier core, which provides all the small-signal amplifier gain, and the auxiliary slew-buffer, which comes into action when the input is slewing, thus enabling the fast slewing of the output under very large capacitive loads. The amplifier is designed to work off of a single power supply V_{cc} of 3.3 V. The unity-gain buffer configuration of the amplifier is obtained by connecting the output of the op-amp to the negative input.

A. Amplifier Core Design

The amplifier core was implemented using the conventional folded cascode topology [5]. The folded cascode architecture achieves a wider input common mode range (ICMR), higher gain, more stable control over gain-bandwidth (due to self-compensation), higher output impedance, and higher power supply rejection, than a two stage op-amp with active loads. SiGe HBTs were used in the input differential pair to exploit their large g_m and inherent enhancement with cooling, and their superior matching properties. Due to the large output resistance of the SiGe HBT, the dominant pole of the op-amp is at the output, and hence the small-signal gain-bandwidth product (GB) of the self-compensated op-amp (including the slew buffer) is given by,

$$ GB = \frac{g_m}{C_L} \qquad (1) $$

where g_m is the transconductance of a SiGe HBT in the input differential pair, and C_L is the load capacitance.

B. Slew-Enhancement Technique

There are several techniques available for slew-enhancement of conventional op-amps, each with differing trade-offs. The technique described in [6] achieves slew-enhancement by internally switching the output buffer to high-drive mode when the inputs are slewing. This technique, however, can not be easily extended to the amplifier cores in a folded cascode architecture, and has been reported to have large variations in quiescent current with varying temperature [3], and thus not suitable for

wide temperature range operation needed here. Another slew-enhancement technique employs an adaptive biasing scheme [7] where the bias current to the amplifier core is increased by a factor of the difference in the branch currents of the input differential pair. When the input slews, there is a large mismatch in the branch current, resulting in large core bias current, and thus providing a higher drive at the output. This technique suffers from high power dissipation in the amplifier core during transients.

The slew-enhancement used in the current design is achieved by using a slew buffer (Fig. 4) whose load transistors ML1 and ML2 are biased in the triode region. This condition is achieved by sizing ML1 and ML2 such that $I_1 < I_2$, where I_1 is half the tail current and I_2 is the saturation current in ML1 and ML2. Thus, under small-signal conditions when the input terminals VP and VN are nearly equal, the gates of MO1 and MO2 (the output transistors) are pulled close to V_{cc}, turning them off. However, when the input is slewing, there is a large difference in the potentials of VP and VN, causing one of the load transistors to saturate, thereby pulling the gate terminal of the corresponding output transistor closer to ground potential. This turns on the output transistor very hard, causing it to deliver large currents to the output. Hence, during slewing the bias to the amplifier core is unaffected, which is clearly desirable. The threshold input differential voltage ($V_{in,th}$) required for triggering the slew buffer is given by [3],

$$ V_{in,th} = \alpha \sqrt{\frac{I_1}{\kappa}} \qquad (2) $$

where α is defined such that $I_2 = (1 + \alpha)I_1$, and κ is the MOS conductance parameter of the input transistors of the slew buffer.

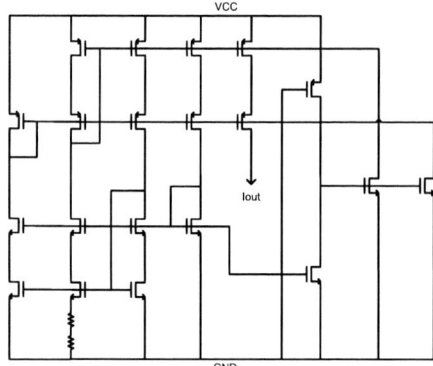

Fig. 5. Conventional wide-swing cascode current bias circuit with start-up.

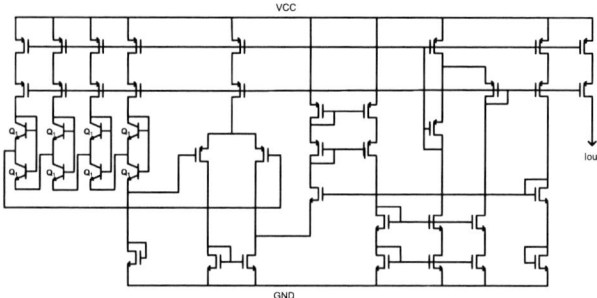

Fig. 6. Weak-T current bias circuit with SiGe HBT PTAT chain and a transconductor (start-up circuit not shown).

C. Biasing Schemes

The SiGe op-amp was implemented using two different bias schemes, each with a varying temperature response, for the output bias current. The first bias scheme utilizes a conventional wide-swing cascode bias circuit, as shown in Fig. 5 [8]. This bias scheme features a wide-swing bias current loop and a start-up circuit to avert the zero current bias condition. The efficacy of the start-up circuit was checked via simulations down to -55 °C, and worked well down to 4.3 K.

The second bias scheme, called "weak-T bias" here, is based on a resistor-free current reference that is designed to reduce overall analog performance variation over a wide temperature range (Fig. 6) [9]. The output bias current of this current source is given by, $2m\mu C_{ox}U_T^2$ where m is a temperature independent scaling factor, μ is the mobility ($\propto T^{-1.5}$ for PMOS), C_{ox} is the oxide capacitance, and U_T is the thermal voltage. It is therefore expected that the resultant g_m of the SiGe HBT biased using this current source be proportional to $\mu C_{ox}U_T$ ($\propto T^{-0.5}$) and the corresponding slew rate be proportional to $\mu C_{ox}U_T^2$ ($\propto T^{+0.5}$). Thus the residual temperature dependence of these key parameters are rendered weak by this biasing scheme, ensuring reduced small-signal and large-signal performance variation over temperature. The simulated SiGe HBT g_m behavior over temperature for weak-T bias is in qualitative agreement with above analysis (Fig. 7). In addition, observe the opposite trends in g_m for these two different bias schemes.

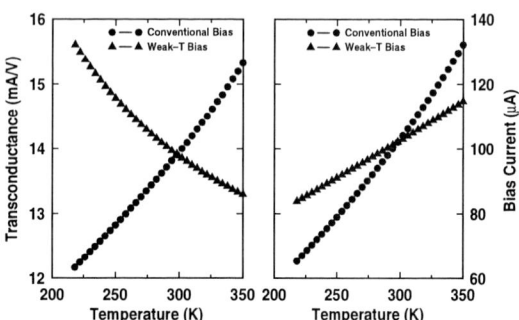

Fig. 7. Simulated input SiGe HBT g_m of the two circuits versus temperature for the corresponding bias tail current.

IV. MEASUREMENT RESULTS AND DISCUSSION

The *dc* and *ac* characterization was performed on the unity gain buffer configuration of the op-amp using a custom-designed cryogenic probe station capable of operating down to 4.3 K.

A. DC Characteristics

The op-amps with either of the current sources drew a total bias current of 4.3 mA nominally at room temperature. The total bias current dropped monotonically with cooling in the conventional bias case and to 143 K in the weak-T based circuit (Fig. 8). There was, however, an anomalous increase in bias current below 143 K in the weak-T op-amp. The input-referred offset voltage tracked the temperature behavior of the bias current. The input-referred offset power for the op-amp can be approximated as [10],

$$V_{os}^2 \;=\; V_{os,core}^2 + V_{os,slew}^2, \tag{3}$$

where,

$$V_{os,core}^2 \simeq U_T^2 \left[\frac{\triangle V_{TP}^2}{\left(\frac{V_{ov,p}}{2}\right)^2} + \frac{\triangle (W/L)_P^2}{(W/L)_P^2} + \frac{\triangle I_S^2}{I_S^2} \right], \tag{4}$$

and

$$V_{os,slew}^2 \simeq \triangle V_{TN}^2 + \left(\frac{V_{ov,N}}{2}\right)^2 \times$$
$$\left[\frac{\triangle V_{TP}^2}{\left(\frac{V_{ov,N}}{2}\right)^2} + \frac{\triangle (W/L)_P^2}{(W/L)_P^2} + \frac{\triangle (W/L)_N^2}{(W/L)_N^2} \right]. \tag{5}$$

Therefore, the offset contribution from the core amplifier is expected to drop naturally with cooling ($\propto U_T^2 \propto T^2$) and with the decrease in bias current ($\propto V_{ov}^2 \propto I_{bias}^2$), and for the same reason the contribution from the slew buffer is expected to also drop with bias current. Thus, the overall offset decreases with temperature for both bias schemes, clearly good news. The higher offset voltage in op-amp with weak-T bias is possibly due to the additional current mismatches introduced by PTAT chain and the transconductor, and is still under investigation.

The ICMR of both variants of the SiGe op-amp remained fairly constant across temperature, as evidenced by Fig. 9. Closer examination of the ICMR$_{min}$ does indicate a moderate increase attributable to the (expected) increase in the base-emitter turn-on voltage of the input differential pair with cooling.

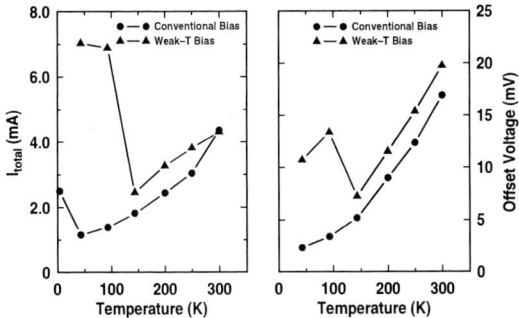

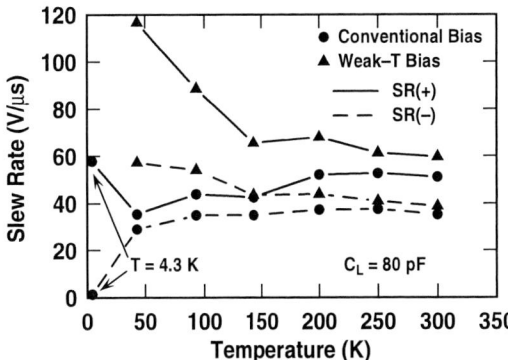

Fig. 8. Measured total bias current and input-referred offset voltage versus temperature for the two circuits.

Fig. 10. The positive and negative slew rates of the circuits across temperature.

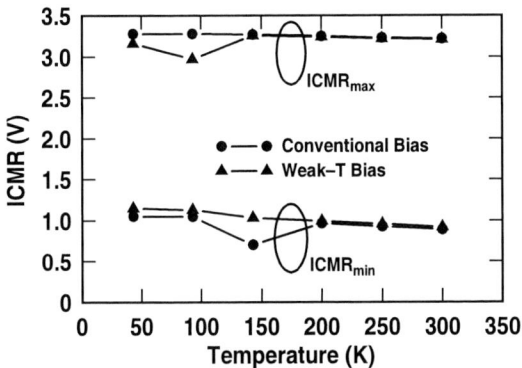

Fig. 9. Measured ICMR of the circuits as a function of temperature.

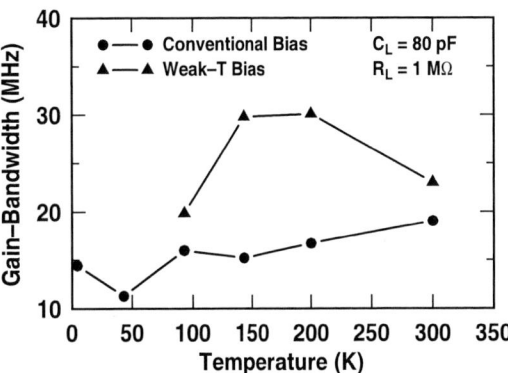

Fig. 11. Gain-bandwidth of the SiGe op-amps versus temperature.

B. Slew-Rate and Gain-Bandwidth

The op-amp with conventional bias displayed robust start-up down to 4.3 K, and had useful gain-bandwidth and appreciable slew rate at this temperature. To our knowledge, this is the first monolithic op-amp capable of operating at this low a temperature. The op-amp with weak-T bias showed a moderate increase in both positive and negative slew rates down to 143 K and further down in temperature a dramatic increase was observed in slew rate, which can be ascribed to the bias current behavior at these temperatures. The op-amp with conventional bias, on the other hand, showed only a moderate decrease in both the slew rates not readily explained using the bias current decrease. This slew rate behavior suggests that the slew-buffer effectively decouples the output slewing from the tail bias currents, and thus making it a weak function of temperature, a desirable attribute.

The small-signal gain-bandwidth is still tied intimately to the input transconductance of the op-amp. This observation was confirmed qualitatively by the comparing the simulated g_m of the input SiGe HBTs (Fig. 7) with the measured gain-bandwidth of the two circuits (Fig. 11). While the gain-bandwidth of the op-amp with weak-T bias increases with cooling down to 143 K, the gain-bandwidth of the conventional bias op-amp degrades steadily with cooling.

V. SUMMARY

We have designed and implemented a SiGe BiCMOS op-amp capable of operating across a very wide temperature range, and

down to temperatures as low as 4.3 K. We conclude that SiGe technology offers an ideal approach for developing a wide variety of analog and mixed-signal building blocks for emerging extreme environment electronics applications.

ACKNOWLEDGMENT

This work was supported by NASA, under grant NNL05AA7C, and the Georgia Electronic Design Center at Georgia Tech. We are grateful for the support of C. Moore, M. Watson, M. Beatty, L. Nadeau of NASA, E. Kolawa of JPL, and the IBM SiGe development group, as well the many contributions of the SiGe ETDP team, including: M. Mojarradi, B. Blalock, W. Johnson, R. Garbos, R. Berger, F. Dai, G. Niu, L. Peltz, A. Mantooth, P. McCluskey, M. Alles, R. Reed, A. Joseph, C. Eckert, and J.D. Cressler.

REFERENCES

[1] J.D. Cressler, *Proc. IEEE*, 93, p. 1559, 2005.
[2] J.D. Cressler *et al.*, *SiGe HBTs*, Norwood, MA: Artech House, 2003.
[3] K. Nagaraj, *Electron. Lett.*, 25, p. 1304, 1989.
[4] D.C. Ahlgren, *Tech. Dig. IEEE IEDM*, p. 859, 1996.
[5] P.E. Allen *et al.*, *CMOS Analog Circuit Design*, New York, NY: Oxford University Press, 2002.
[6] S.L. Wong *et al.*, *IEEE JSSC*, 21, p. 464, 1986.
[7] M.G. Degrauwe *et al.*, *IEEE JSSC*, 17, p. 464, 1982.
[8] D.A. Johns *et al.*, *Analog Integrated Circuit Design*, New York, NY: John Wiley & Sons, 1997.
[9] S. Chen *et al.*, *submitted to WOLTE-7*, 2006.
[10] P.R. Gray *et al.*, *Analysis and Design of Analog Integrated Circuits*, New York, NY: John Wiley & Sons, 2001.

2006 Bipolar/BoCMOS Circuits and Technology Meeting Proceedings

86dBΩ 10Gb/s SiGe Transimpedance Amplifier Using Photodiode Capacitance Neutralization and Vertical Threshold Adjustment

Adrian Maxim

Maxim Inc., 4201 Monterey Oaks, Austin TX 78749, Email: adrianmaxim@ieee.org

Abstract — **A high gain 10Gb/s transimpedance amplifier capable of directly driving a SERDES IC was realized in a 60GHz f_T 0.2µm SiGe HBT process. The shunt-feedback common-emitter input stage uses a bootstrap technique to neutralize the photodiode parasitic capacitance. Cascode configurations and cross-coupled Miller capacitance cancellation were used to minimize the input capacitance of the signal path stages. This reduces the number of inter-stage isolation emitter followers, allowing a low voltage operation. A signal amplitude dependent adjustable threshold was implemented in the back-end limiting stages by using inverse hyperbolic tangent circuits. The main TIA specifications include: 86dBΩ gain, 12µA input sensitivity, <8pA/√Hz input noise, 0.3W power dissipation from a 3.3V supply and 1.8x1.8mm² die area.**

Index Terms — **transimpedance amplifier, SONET.**

I. INTRODUCTION

Early 10Gb/s transimpedance amplifiers (TIA) were implemented in high cost and power hungry GaAs [1,2] and InP [3] processes. Si-bipolar TIAs were also developed [4-6] but they usually suffer of a lower sensitivity level. Recently, SiGe technology emerged as a cost effective alternative for 10Gb/s communications ICs [7,8]. The most popular TIA architectures are the common-base [8] and the common-emitter with shunt resistor feedback [1-7]. The first one offers a good stability, but suffers of poor noise performance. In contrast, the second one has excellent noise performance, but the stability needs to be carefully addressed. Cascode common-emitter TIAs were used to broadband the input stage by minimizing the input device Miller effect [2, 7]. Inductive peaking applied either in the base [2, 7] or in the collector [4] was used to increase the input stage bandwidth. Active inductive peaking [5] was used to generate an adjustable amount of peaking.

The second stage in the signal path provides additional voltage gain and usually performs the single-ended to differential conversion. The Cherry-Hooper (CH) stage [3,5], based on the impedance mismatching technique became a standard way of implementing wideband voltage amplifiers. Modified CH configurations [7,8] use additional emitter followers in the local shunt feedback loop to optimize the power-bandwidth compromise.

Most existing TIAs provide a low gain (0.5 to 1.5KΩ), requiring a following limiting amplifier IC. The power consumption of the optical transceiver can be significantly reduced by using a high gain TIA (20 to 50KΩ) that can drive directly a serializer-deserializer (SERDES) IC. The main challenge in

such high gain ICs is avoiding instability due to parasitic coupling between output and input stages.

This paper presents a high gain 10Gb/s limiting transimpedance amplifier that uses a bootstrap unity gain amplifier to increase the IC input bandwidth through a photodiode parasitic capacitance neutralization. Most existing designs use one or even two emitter followers between the adjacent gain stages, requiring a large value supply voltage. Present architecture reduces the number of inter-stage isolation emitter followers, being well suited for low supply voltage applications.

II. LIMITING TIA CIRCUIT DESCRIPTION

The top level diagram of the proposed 10Gb/s limiting transimpedance amplifier is presented in Fig.1. The signal path consists of a pseudo-differential input transimpedance stage having a main TIA and a dummy TIA that provides the common-mode voltage for the single-ended to differential conversion, three limiting stages using a modified Cherry-Hooper architecture (CH), and an output CML driver (DRV) that provides the required minimum 0.3V differential voltage swing on a 50Ω termination. To cancel the TIA offset voltage a low cut-off frequency DC feedback loop is closed around the front-end stages. It consists of a differential to single ended converter (A_{d_s}), a Miller gain amplifier (A_m) that magnifies the value of the on-chip C_m capacitance and a transconductance stage (g_{m_off}) that injects the feedback current back to the input of the main signal path TIA [7]. The DC offset cancellation current is proportional to the photodiode average current (I_{ave}). A second transconductance stage (g_{m_ave}) and a current mirror (MIR_{ave}) copy this current to the received signal strength indicator (RSSI), the loss of power detector (LOP) and to the

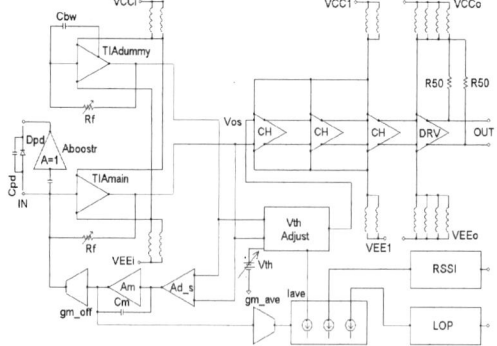

Fig.1 Limiting transimpedance amplifier IC top level

threshold adjustment block (Vth$_{ADJUST}$) that adds a signal dependent, adjustable vertical threshold voltage (V$_{os}$) at the input of the back-end limiting amplifier.

Separate power supply domains VCC$_i$/VEE$_i$, VCC$_l$/VEE$_l$, VCC$_o$/VEE$_o$ were used to isolate the sensitive front-end stages from the high current back-end stages. The high current stages use multiple supply bondwires to minimize the local supply bouncing due to L·di/dt voltage drops. Avoiding magnetic coupling between the different supply bondwires was ensured by a careful IC pin-out, which provides an orthogonal orientation between the critical coupling bondwires.

Fig.2 presents the schematic of the input transimpedance stage. In most designs the large parasitic capacitance of the PIN or avalanche photodiode limits the bandwidth at the input of the IC. Inductive peaking was used in the past to broadband the input stage [2, 4, 7, 8]. Its main drawback is the need to adjust the value of the peaking inductance based on the parasitic capacitance of the selected photodiode. This paper proposes an alternative way of broadbanding the input stage that does not need any adjustments. It is based on bootstrapping the photodiode parasitic capacitance with a high bandwidth emitter follower Q$_{boostr}$. The IC input voltage from the photodiode anode is applied through the C$_c$ isolation capacitance to the base of Q$_{bootstr}$ follower that keeps the cathode at a signal level virtually equal to the anode voltage. Having both diode terminals at the same AC potential results in a neutralization of its parasitic capacitance loading. A large value R$_b$ resistor was used to bias the base of Q$_{bootstr}$ in order to minimize its noise contribution. A supply filter R$_{filt}$, C$_{filt}$ was used to improve the PSRR and thus minimize the supply injected noise.

The input transimpedance stage uses a shunt resistor feedback (R$_{f1}$, R$_{f2}$) common-emitter (Q$_{10}$) architecture. A cascode device (Q$_{11}$) reduces the Miller gain of C$_\mu$(Q$_{10}$). Its base was biased with a local DC feedback loop built with Q$_{12}$, R$_{eb}$, R$_{cb}$ and C$_{casc}$. It ensures a lower impedance at the base of Q$_{11}$ in comparison with the stacked diode and diode multiplier bias networks [7]. The large value R$_c$ collector resistor is isolated from the IC input with the Q$_{13}$ emitter follower. The feedback resistors R$_{f1}$, R$_{f2}$ are trimmed in order to stabilize the gain and

bandwidth of the signal path. In overdrive conditions the D$_{10}$, D$_{11}$ Schotky diodes are shorting the feedback resistors, rolling-off the input stage gain. The input circuit is broadbanded with a feedback capacitive peaking realized by C$_f$ in conjunction with the parasitic capacitances of the D$_{10}$, D$_{11}$ diodes. The C$_{sh}$ shunt capacitance further improves the frequency response of the input stage by adding a second high frequency zero. The common-mode for the single-ended to differential conversion is provided by a dummy TIA that has its bandwidth rolled-off by C$_{bw}$ capacitance, avoiding thus input noise degradation. The differential output voltage is clamped by the D$_{14}$, D$_{15}$ Shottky diodes connected in parallel with R$_{out}$ resistor. The output current of the offset cancellation loop is injected back to the IC input by the low noise resistive degenerated current leg (Q$_{14}$, R$_{deg}$). A bi-directional offset cancellation was achieved by injecting an intentional I$_{off}$ offset current at the TIA input, having its noise rejected by the R$_n$, C$_n$ filter.

Fig.3 presents the detailed schematic of the first limiting stage (the other two stages have a similar architecture). It uses a differential-in, differential-out modified Cherry-Hooper architecture with 12dB gain. It consists of an input transadmitance stage (TAS) built with Q$_{20}$, Q$_{21}$ differential pair, having large emitter degeneration resistances R$_{e1}$ and an output transimpedance stage (TIS) realized with the Q$_{22}$, Q$_{23}$ differential pair, having the gain set by the R$_{f3}$ shunt feedback resistors. The overall voltage gain is equal to the ratio of R$_{f3}$ feedback and R$_{e1}$ degeneration resistors, having a low process and temperature variation. The Q$_{24}$, Q$_{25}$ emitter followers are added in the shunt feedback of the TIS stage to better control the gain and impedance levels. Despite the additional bias currents (I$_{ef}$) required by these emitter followers to keep them active at the peak overdrive current, the overall current consumption is reduced in comparison with standard CH due to a lower required g$_m$ gain for the TIS stage. Compensation capacitors (C$_C$) were connected in parallel with R$_{f3}$ to improve the stability of the local feedback loops (Q$_{22}$-Q$_{24}$, Q$_{23}$-Q$_{25}$).

The output of the input transimpedance stage is usually taken at the emitter of Q$_{13}$ follower [1-5]. This requires a large voltage headroom that is not compatible with a 3.3V supply. The alternative is to connect the second stage directly to the high

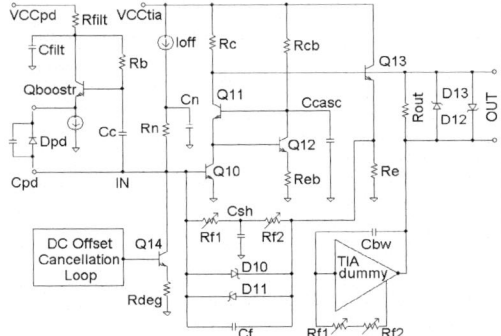

Fig.2 Pseudo-differential input transimpedance stage

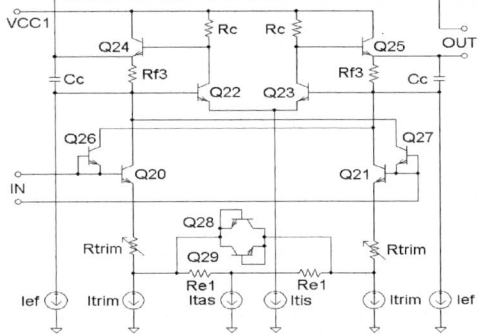

Fig.3 High bandwidth Cherry-Hooper limiting stage

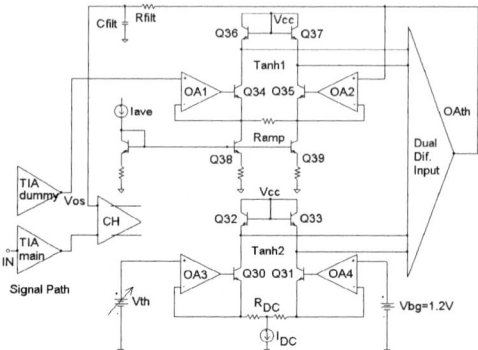

Fig.4 Signal dependent threshold adjustment circuit

impedance node of the TIA. To reduce the bandwidth degradation at the high impedance node of the TIA, the input capacitance of the second stage needs to be minimized. The $C_\pi(Q_{20},Q_{21})$ base-emitter capacitances are well degenerated by the R_{e1} resistors. On the other hand the $C_\mu(Q_{20},Q_{21})$ feedback capacitances may have a detrimental effect due to their Miller multiplication at the input. A cross-coupled C_μ neutralization technique was implemented with two shorted base-emitter transistors (Q_{26}, Q_{27}) that match the main amplifier devices (Q_{20} and Q_{21}). The bandwidth of the CH stage is increased using emitter capacitance peaking [7], implemented with Q_{28}, Q_{29} shorted base-emitter devices. All bias currents (I_{tas}, I_{tis}, I_{ef}) are PTAT type that ensures a constant g_m biasing and therefore stabilizes the gain of the local loops with temperature and minimizes residual peaking. The two feedback emitter followers Q_{24}, Q_{25} are used also to drive the following limiting amplifier with no additional isolation stages.

The threshold adjustment circuit converts an off-chip control voltage (V_{th}=0.2 to 2.2V) into a signal dependent offset injected in the signal path (20 to 80% of the signal amplitude). Avoiding high current consumption at input currents over the $600\mu A_{pp}$ level was realized by implementing a current clamp that limits the maximum I_{ave} average value that is passed to the threshold adjust circuit. At the maximum input level for threshold adjust the first differential Cherry-Hooper gain stage already operates in limiting regime. Therefore, the threshold adjustment needs to be placed in front of it. However, at the minimum signal level for threshold adjustment ($35\mu A_{pp}$) the signal amplitude is too low and the offset voltage of the C-H amplifier may prevent the stage from switching. Avoiding this failure mechanism was achieved by trimming-out the offset of the first CH stage with the R_{trim} resistors connected in series with the I_{trim} trimming bias currents (see Fig.3).

Fig.4 presents the detailed schematic of the threshold adjustment circuit implemented with inverse hyperbolic tangent differential circuits (tanh⁻¹). A first tanh⁻¹ realized with Q_{30}-Q_{33} that uses constant tail current I_{DC} converts the difference between the off-chip threshold adjust voltage V_{th} and the fixed 1.2V bandgap voltage into a control voltage:

$$V_{ctrl}=-2V_t \cdot \tanh^{-1}(V_{th}-1.2V)/(I_{DC} \cdot R_{DC}) \quad (1)$$

The second tanh⁻¹ circuit realized with Q_{34}-Q_{37} uses a variable tail current equal to the average photodiode current (I_{ave}) and has one input set by the desired offset voltage provided by the dummy TIA, while the second input is set by a negative feedback loop using the OA_{th} operational amplifier. The negative feedback keeps the differential outputs of the two tanh⁻¹ circuits equal. As a consequence the resulted offset voltage injected at the input of the first Cherry-Hooper limiting stage is given by:

$$V_{os}=(V_{th}-1.2V) \cdot I_{ave} \cdot R_{ave}/(I_{DC} \cdot R_{DC}) \quad (2)$$

being dependent both on the off-chip DC control voltage (V_{th}) and the average photodiode current I_{ave} (related to signal amplitude). Using the average current provided by the TIA offset cancellation loop results in a much better threshold adjustment accuracy ($\pm 10\%$) and stability ($\pm 3\%$) in comparison with true signal peak detector based implementations.

III. MEASUREMENT RESULTS

The proposed limiting transimpedance amplifier was realized in a 60GHz f_T 0.2μm SiGe technology. Fig.5 gives the output eye diagram for a $12\mu A_{pp}$ input current, obtained with a PRBS 2^{31}-1 NRZ sequence. It shows an 18ps combined deterministic jitter and pulse width distortion ($15ps_{pp}$ after subtracting the measurement setup contribution). Fig.6 presents the output eye diagram with a $600\mu A_{pp}$ input current, showing an 11.5ps output jitter. Fig.7 shows the dependence of the input equivalent noise current as a function of the photodiode input current level. A worst case input noise of 8pA/√Hz was achieved in the high overdrive regime ($3mA_{pp}$). Fig.8 illustrates the output eye diagram DJ variation with the photodiode current level. The worst case deterministic jitter $DJ_{max}=18ps_{pp}$ is obtained again for a $3mA_{pp}$ overdrive current. Fig.9 illustrates the IC die photo. The IC specifications are: $>20K\Omega$ transimpedance gain, $12\mu A_{pp}$ input sensitivity, >10GHz bandwidth, $3mA_{pp}$ input overdrive current, <15ps total jitter in non-overdrive regime, $>300mV_{pp}$ differential output voltage, 0.3W power consumption from a 3.3V+/-10%, supply voltage, and 1.8x1.8mm² die area. Table.1 presents a performance comparison with

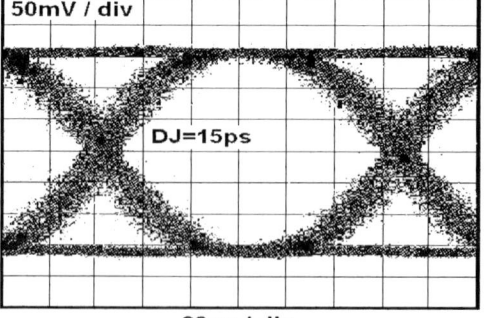

Fig.5 Eye diagrams for $12\mu A_{pp}$ & PRBS 2^{31} -1 sequence

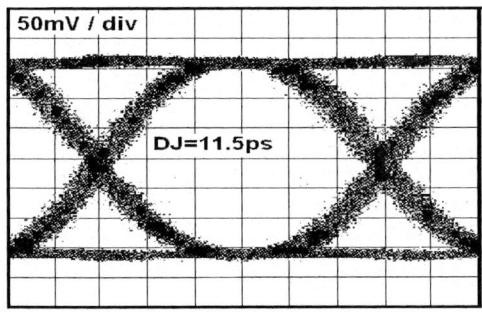

50mV / div

DJ=11.5ps

20ps / div

Fig.6 Eye diagrams for $600\mu A_{pp}$ & PRBS 2^{31}-1 sequence

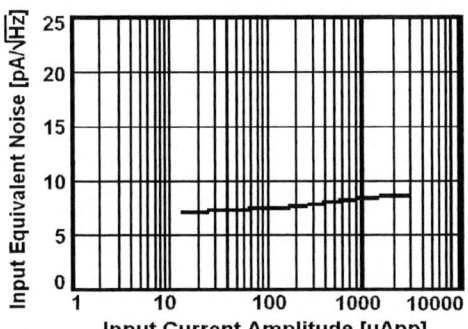

Fig.7 Input Noise versus photodiode current

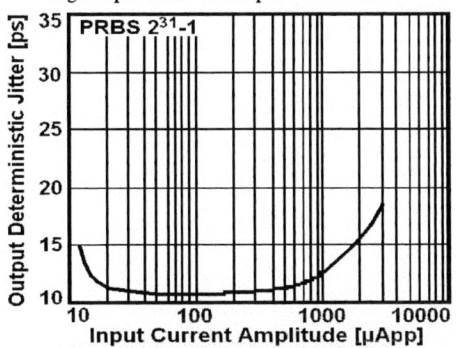

Fig.8 Deterministic jitter versus photodiode current

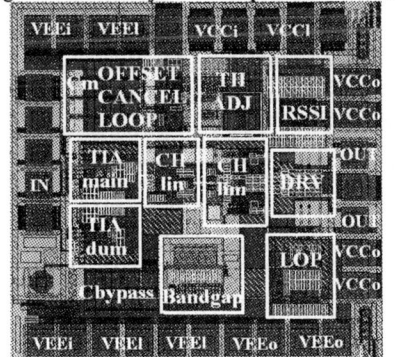

Fig.9 10Gb/s limiting TIA IC die-photo

Table. I. 10Gb/s TIA performance comparison

Ref.	[2]	[4]	[6]	This
Process	0.1μ HEMT	23GHz Si-bip	0.25μ Si-bip	0.2μ SiGe
Gain	63dBΩ	54dBΩ	55dBΩ	86dBΩ
Bandwidth	8GHz	9GHz	9GHz	10GHz
Input noise	6.5 pA/√Hz	10 pA/√Hz		8 pA/√Hz
Sensitivity	-21dBm optical	11μA electr.	-17dBm optical	12μA electr.
Overload	4.3dBm	0.9mA	1.4mA	3mA
DJ (no overload)	17ps	15ps	20ps	15ps
Supply V	5V	6.5V	5V	3.3V
Supply P	0.5W	0.25W	0.15W	0.3W
Die area	1.3x1.6	-	-	1.8x1.8

IV. CONCLUSION

A high gain limiting transimpedance amplifier for 10Gb/s optical communications was realized in a 60GHz f_T SiGe HBT technology. The input stage uses a unity gain bootstrap amplifier to neutralize the photodiode parasitic capacitance and thus increase the IC input bandwidth. Integrating together a transimpedance and a limiting amplifier resulted in a substantial power consumption reduction. Avoiding the use of on-chip inductors for the signal path broadbanding lead to a smaller die size. A modified Cherry-Hooper stage with Miller capacitance cancellation was used for the back-end limiting stages. Its low input capacitance allows the elimination of the inter-stage isolation emitter followers, allowing a low voltage operation (3.3V). The back-end limiting Cherry-Hooper stages use a signal level dependent threshold adjustment circuit built with $tanh^{-1}$ stages that provides a high accuracy and good stability for the threshold level.

REFERENCES

[1] Y. Suzuki, K. Honjo, "Wide band transimpedance amplifier using AlGaAs/InGaAs pseudomorphic 2-D EG FETs",IEEE JSSC,vol.33,pp.1559-1562,Oct. 1998.

[2] H. Ikeda, et al, "An auto-gain control TIA with low noise and wide input dynamic range for 10Gb/s optical systems",IEEE JSSC,vol.36, pp.1303-1308, Sep. 2001.

[3] J. Mullrich, et al, "High gain transimpedance amplifier in InP based HBT for the receiver in 40Gb/s optical fiber", IEEE JSSC, vol.35, pp.1260-1265,Sep. 2000.

[4] M. Neuhauser, H-M. Rein, "Low noise, high gain Si-bipolar preamplifiers for 10Gb/s optical fiber links", IEEE JSSC, vol.31, no.1, pp.24-29, Jan. 1996.

[5] K. Ohhata. Et al, "A wide dynamic range, high transimpedance Si-bipolar preamplifier IC for 10Gb/s optical links", IEEE JSSC, vol.34, pp.18-24, Jan. 1999.

[6] H. Kim, et al, "A Si BiCMOS transimpedance amplifier for 10Gb/s SONET receiver", IEEE JSSC, vol.36, no.5, pp.769-776, May 2001.

[7] A. Maxim, "A 10Gb/s SiGe TIA using a pseudo-differential iinput stage and a modified Cherry-Hooper amplifier", IEEE VLSI 2004 Symposium, pp.404-407.

[8] T. Masuda, et al, "40Gb/s analog IC chipset for optical receiver using SiGe HBTs", ISSCC Technical Digest, pp. 19.7.1-19.7.3, San Francisco, Feb. 1998.

several representative existing 10Gb/s TIAs. Using a limiting-transimpedance amplifier combination has lead to a 40% power consumption reduction in comparison with a two IC implementation.

2006 Bipolar/BoCMOS Circuits and Technology Meeting Proceedings

A 70 MHz – 4.1 GHz 5th-Order Elliptic g_m-C Low-Pass Filter in Complementary SiGe Technology

Yuan Lu, Ramkumar Krithivasan, Wei-Min Lance Kuo, Xiangtao Li,
John D. Cressler, Hans Gustat [1], and Bernd Heinemann [1]

School of Electrical and Computer Engineering
777 Atlantic Drive, N.W., Georgia Institute of Technology, Atlanta, GA 30332-0250, USA
Tel: (404) 894-5161 / Fax: (404) 894-4641 / E-mail: luyuan@ece.gatech.edu
[1] IHP, Im Technologiepark 25, 15236 Frankfurt (Oder), Germany

Abstract—We present the first demonstration of a continuous-time, fifth-order, elliptic, g_m-C low-pass active filter in 0.25 μm complementary (*npn* + *pnp*) silicon-germanium (C-SiGe) heterojunction bipolar transistor (HBT) technology. This C-SiGe technology features *npn* SiGe HBTs with peak f_T and f_{max} of 170 GHz and 170 GHz, respectively, as well as *pnp* SiGe HBTs having f_T and f_{max} of 90 GHz and 120 GHz, respectively. This C-SiGe active filter was implemented with Voorman transconductors [1] to fully exploit the complementary high-speed *npn* and *pnp* SiGe HBTs. The circuit occupies an area of 0.82 mm^2, and exhibits a filter cut-off frequency of 4.1 GHz. This C-SiGe active filter achieves a record continuous tuning range between 70 MHz and 4.1 GHz, attains an output noise power spectrum density (PSD) of -143 dBm/Hz, and operates off a 3.5 V supply, with a total power consumption of 100 mW at the maximum bandwidth of 4.1 GHz.

I. INTRODUCTION

As the data transfer rates of recording devices (e.g., DVDs and hard-disk drives) increases, very high-speed analog filters are required for the front-end electronics. In addition, ultra-wideband (UWB) wireless technology, which is capable of transmitting extremely low power signals at very high data rates, requires high-frequency anti-alias filters before the signals are digitized by the analog-to-digital converters (ADCs). The integrated filters for such applications can be implemented with on-chip passives, (i.e., capacitors and inductors), but given the frequency range of interest, from several hundred MHz to several GHz, excessively large chip area would be required by the inductors. Such passive filters, while attractively linear in their response, unfortunately attain very limited tuning capability, and often have significant pass-band loss due to low-Q inductors (especially in Si-based technology). For recording devices and communication channel selection applications, integrated active filters thus become an obvious alternative. Operational amplifier based filters (e.g., op amp RC filters) have higher linearity than transconductor based filters (i.e., g_m-C filters) [3]. However, it is very difficult to achieve very high-frequency operation (above 1 GHz) with op amp based filters, mainly because of insufficient loop gain at high frequency, which is limited by the trade-off between the gain-bandwidth and phase-margin for a multi-stage op amp [3]. Thus, to achieve maximum speed, g_m-C filters are preferred.

In this paper, the design and demonstration of an ultra-high-speed (4.1 GHz) g_m-C C-SiGe active filter using Voorman transconductors is presented. The 0.25 μm C-SiGe process technology used to implement this filter is discussed in Section II, details of the filter design are described in Section III, and measurement results are presented in Section IV, followed by a summary.

TABLE I
KEY PARAMETERS OF THE C-SIGE TECHNOLOGY.

	npn	*pnp*
β	200	100
$A_{E,min}$ (μm^2)	0.21×0.84	0.21×0.84
f_T (GHz)	170	90
f_{max} (GHz)	170	120
BV_{CEO} (V)	1.9	3.1
BV_{CBO} (V)	4.5	4.0
CMOS L_g (μm)	0.25	
Metal Layers	4	

II. C-SIGE HBT PROCESS TECHNOLOGY

Bandgap-engineered SiGe HBTs are receiving significant attention for communications IC applications because they enable a dramatic improvement in transistor-level performance while simultaneously maintaining strict compatibility with conventional low-cost, high-integration level, high-volume CMOS manufacturing [4]. SiGe technology is evolving rapidly, and has today reached a point where SiGe HBT technology is of comparable performance with best-of-breed III-V technologies. With the recent announcement of SiGe HBTs having peak cutoff frequency (f_T) above 300 GHz [5], and complementary (*npn* and *pnp*) SiGe HBTs having peak f_T above 180 GHz and 80 GHz, respectively [6], the application space for SiGe HBT technology now includes a wide variety of analog and RF through mm-wave applications. A key feature of the present work is the use of complementary (*npn* and *pnp*) SiGe HBTs. The high performance of the *pnp* SiGe HBTs is mainly the result of a highly tuned vertical doping profile, taking full advantage of the reduced phosphorus diffusion in the carbon-doped base, combined with the special collector construction of previously reported 200 GHz *npn* transistors [6]. In this C-SiGe technology, the formation of the whole SiGe HBT structure is made in one active area, without shallow trench isolation between the active emitter and the collector contact regions. This provides low-capacitance isolation from the substrate and low collector resistances [7]. Figure 1 shows a schematic cross-section of the *npn* and *pnp* transistors.

The current filter design makes use of six different geometrical variations of the SiGe HBTs, ranging from the minimum emit-

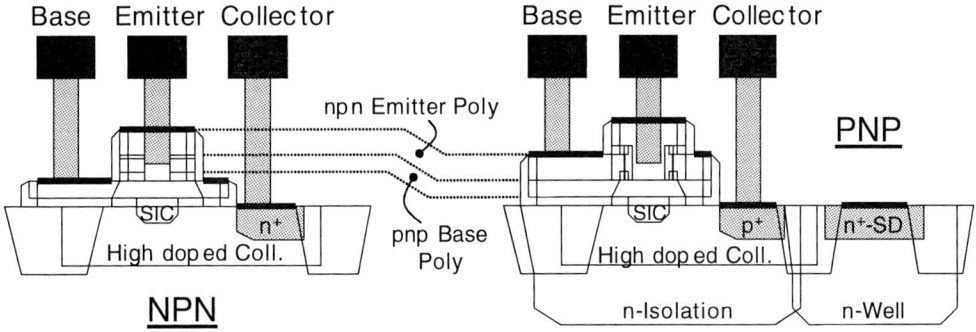

Fig. 1. Schematic cross-section of the *npn* and *pnp* SiGe HBTs.

ter size (0.21×0.84 μm^2) to eight times of the minimum emitter size. The following transistor parameters have been determined on an array of four minimum emitter size SiGe HBTs, from the same wafer the filter test chips are located on: peak f_T=170 GHz and peak f_{max}=170 GHz (with BV_{CEO}=1.9 V), for the *npn* SiGe HBTs; peak f_T=90 GHz, and peak f_{max}=120 GHz (with BV_{CEO}=3.1 V), for the *pnp* SiGe HBTs. The f_T and f_{max} were extrapolated from h_{21} and the unilateral gain (U), respectively, at 30 GHz using a -20 dB/decade slope at room temperature. In addition to the C-SiGe HBTs, ASIC compatible 2.5 V CMOS devices, and a full suite of passives (including metal-insulator-metal (MIM) capacitors, polysilicon resistors, and spiral inductors) are also available in this technology platform. Table I summarizes the key parameters of this C-SiGe technology.

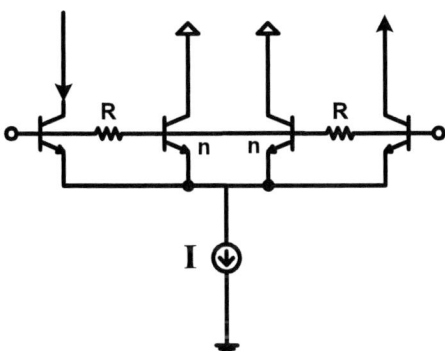

Fig. 2. Schematic of the Voorman transconductor.

III. FILTER DESIGN

The realized C-SiGe filter is based on Voorman transconductors (Figure 2 [1]). The emitter sizes of the central transistor pair are n times as large as the ones of the outer transistor pair. The large-signal output current is given by [1]

$$I_{out} = \frac{I}{2} \frac{\exp\left(\frac{V}{2V_T}\right) - \exp\left(-\frac{V}{2V_T}\right)}{2n + \exp\left(\frac{V}{2V_T}\right) + \exp\left(-\frac{V}{2V_T}\right)} \qquad (1)$$

where I is the tail current, V is the input voltage, and V_T is the thermal voltage (kT/q). Figure 3 shows the theoretical normalized transconductance ($g_m/g_{m,max}$) as a function of the input voltage. Curves 3, 4, and 5 correspond to $g_m - V$ curves when n

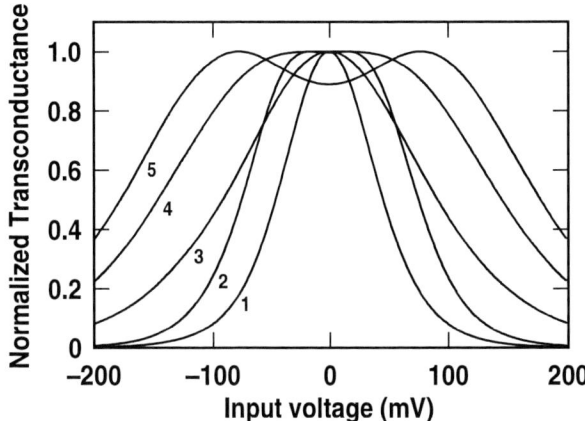

Fig. 3. Theoretical normalized transconductance ($g_m/g_{m,max}$) as a function of the input voltage.

= 1, 2, and 3, respectively. We can observe that when $n = 2$, the circuit has its maximum linear input range. Curve 1 corresponds to a simple differential pair, while curve 2 represents the Schmoock transconductor [2]. The linear input range of a Voorman transconductor ($n = 2$) is significantly larger compared to a simple differential pair or a Schmoock transconductor. The small-signal g_m of the optimum Voorman transconductor ($n = 2$) is [1],

$$g_m = \frac{I}{12V_T} \qquad (2)$$

Since the center node of the resistors is at virtual ground, the transistor's base can directly connect to its collector without any effect on the output current, and furthermore, the two center transistors can be combined together. Figure 4 shows the complementary Voorman transconductor used here [1]. The use of both *npn* and *pnp* transistors reduces the power consumption by effectively doubling the transconductance at the same current, or conversely, maintaining the same transconductance with only half of the current [1]. In addition, in the complementary implementation, the noise is reduced, because the noise generated by the bias current source is common-mode [1]. Another advantage of the complementary Voorman transconductor lies in its relatively low input capacitance, which is clearly important for very high-speed operation, due to the small input transistor sizes [1]. There-

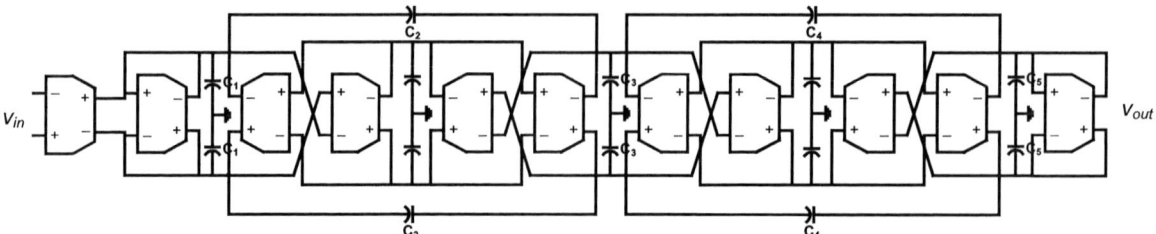

Fig. 6. Block diagram of the fifth-order low-pass filter.

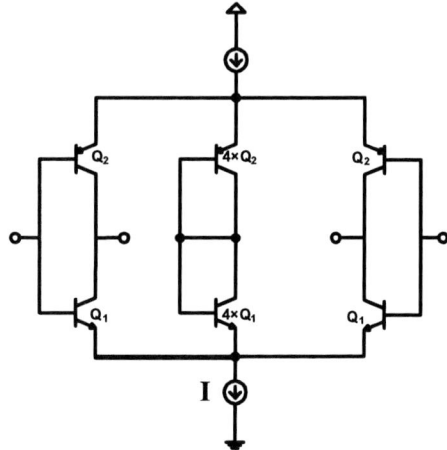

Fig. 4. Schematic of the complementary Voorman transconductor used here.

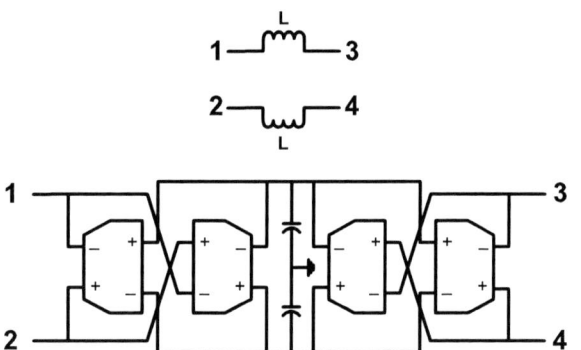

Fig. 5. Schematic of the differential tunable inductor (gyrator).

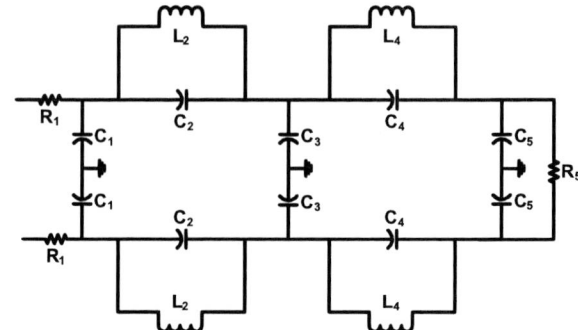

Fig. 7. Equivalent RLC circuit of the fifth-order low-pass filter.

current sources and the emitters of the *pnp* transistors are cross-connected in the gyrators in order to achieve common-mode stability [1]. In addition, all of the differential capacitors (i.e., C_1, C_3 and C_5) were connected to the real ground to improve the common-mode stability at high frequency [1]. Two 50 Ω shunt input resistors and a 50 Ω output buffer were added (not shown in Figure 6), in order to achieve broadband impedance matching to the test equipment.

IV. MEASUREMENT RESULTS

This C-SiGe filter was implemented in a commercially-available 0.25 μm 170/90 GHz C-SiGe HBT BiCMOS technology [6], and occupies a total area of 0.90×0.91 mm^2 including the probe pads. The chip micrograph is shown in Figure 8. The filter was tested on-wafer using 40 GHz probes and cables.

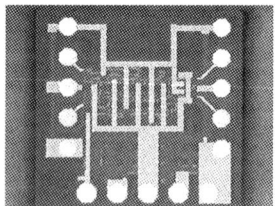

Fig. 8. Die micrograph of the C-SiGe g_m-C filter.

The filter operates off a 3.5 V power supply and has 28.5 mA of bias current flowing through the core circuit when achieving a maximum bandwidth of 4.1 GHz. Figure 9 shows a typical measured S_{21} from 0.05 GHz to 5 GHz. The 3-dB bandwidth is 430 MHz, and the attenuation at 670 MHz is about 42 dB.

fore, complementary Voorman transconductors are very suitable for high-speed, low-noise, and low-power applications. With the availability of very high speed C-SiGe HBTs, very attractive active filters can be achieved.

Figure 5 shows the tunable differential inductor (gyrator) used. The effective inductance is given by,

$$L = \frac{C}{g_m{}^2} \qquad (3)$$

The block diagram of the core fifth-order elliptic low-pass filter is shown in Figure 6. By changing the tail current of transconductor, the tuning function is realized. The RLC equivalent circuit of this filter is shown in Figure 7. The bases of upper

TABLE II

COMPARISON WITH PUBLISHED HIGH-FREQUENCY ACTIVE LOW-PASS FILTERS.

Reference	BW [MHz]	Power [mW]	VCC [V]	IIP3	Order	Topology	Technology
[3]	200-1000	90	1.8	13.5 dBV	5	op amp RC	0.18 μm CMOS
[8]	500	14	1.8	3.0 dBm	5	op amp RC	0.18 μm CMOS
[9]	80-200	270	3.0	-	7	g_m-C	0.25 μm CMOS
This work	70-4100	100	3.5	-5.5 dBm	5	g_m-C	0.25 μm C-SiGe

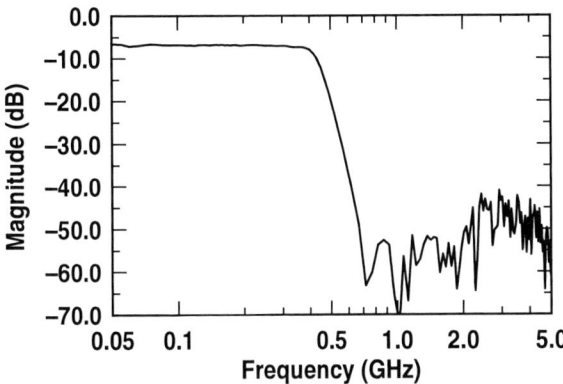

Fig. 9. Measured S_{21} of the C-SiGe g_m-C filter.

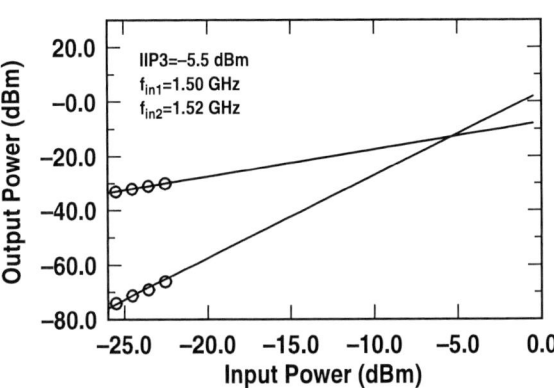

Fig. 11. Measured IIP3 of the C-SiGe g_m-C filter.

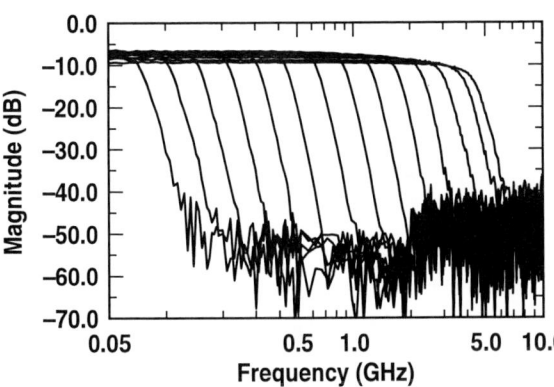

Fig. 10. Measured S_{21} of the C-SiGe g_m-C filter over its tuning range.

The pass-band shows a loss of 6.8 dB (compared to 0 dB simulated). Part of this loss is measurement-setup induced, originating from the lack of high-performance ultra-wideband (from 50 MHz to 10 GHz) differential signals for testing. We used a pair of back-to-back connected ultra-wideband baluns to achieve single-ended to differential and differential to single-ended conversions. However, large phase and amplitude mismatches were observed (and uncorrected for) at the outputs the balun. In addition, we can always increase the transconductance (current) of the input g_m cell to compensate any pass-band loss. Figure 10 shows the measured S_{21} over the tuning range. The bandwidth can be continuously tuned over a very wide range, from 70 MHz to 4.1 GHz. Figure 11 shows the linearity data of the filter. The third-order input intercept point (IIP3) is -5.5 dBm for a two-tone input signal containing 1.50 GHz and 1.52 GHz frequency components, and the output noise power spectrum density (PSD) is -143 dBm/Hz. Table II shows a comparison between the this C-SiGe filter and other published low-pass active filters. To the best of our knowledge, the present C-SiGe filter has the widest bandwidth and achieves a record continuous tuning range between 70 MHz and 4.1 GHz compared with the previous state-of-the-art.

V. SUMMARY

A high-frequency, continuous-time fifth-order elliptic g_m-C filter based on Voorman transconductors has been presented. The filter was fabricated in a high-performance complementary SiGe BiCMOS technology. The filter achieved a 3-dB bandwidth of 4.1 GHz and a tuning range of 70 MHz to 4.1 GHz. This C-SiGe g_m-C filter is well-suited for next-generation recording devices and UWB communications system applications.

VI. ACKNOWLEDGMENT

This work was supported by the Georgia Electronic Design Center at Georgia Tech and IHP. The authors are grateful to C. Zhu, as well as D. Knoll, B. Tillack, and the IHP SiGe team for their contributions.

REFERENCES

[1] H. Voorman *et al.*, *IEEE JSSC*, 35, pp. 1097-1108, 2000.
[2] J.C. Schmoock, *IEEE JSSC*, 10, pp. 407-411, 1975.
[3] J. Harrison *et al.*, *Tech. Dig. ISSCC*, pp. 481-483, 2003.
[4] J.D. Cressler and G. Niu, *SiGe HBTs*, Artech House, MA, 2003.
[5] B. Heinemann *et al.*, *Tech. Dig. IEDM*, pp. 251-254, 2004.
[6] B. Heinemann *et al.*, *Tech. Dig. IEDM*, pp. 117-120, 2003.
[7] B. Heinemann *et al.*, *Tech. Dig. IEDM*, pp. 775-778, 2002.
[8] S. Lida *et al.*, *Tech. Dig. ISSCC*, pp. 214-215, 2005.
[9] S. Dosho, *et al.*, *IEEE JSSC*, 37, pp. 559-565, 2002.

2006 Bipolar/BoCMOS Circuits and Technology Meeting Proceedings

A 3.5GHz Low Power Programmable Transversal Filter RFIC Implemented in 47GHz SiGe Technology

Vasanth Kakani, Xuefeng Yu, Foster F. Dai, Richard C. Jaeger

Department of Electrical and Computer Engineering

Auburn University, Auburn, AL 36849-5201, USA

Abstract — **This paper presents the design of a low power 3.5GHz analog programmable filter RFIC. The RF filter is a 7-tap transversal equalizer with cascaded Cherry-Hooper amplifiers for delay stages and Gilbert variable gain amplifier as tap weights. The delay stage using active devices greatly reduces the die area comparing to passive delay lines. The SiGe programmable filter RFIC consumes 250mW under 3.3V supply and occupies total 2.16mm^2 die size.**

Index Terms — **SiGe, transversal filter, analog filter, Cherry-Hooper amplifier.**

I. INTRODUCTION

Programmable RF filters find numerous applications in communication systems. In fiber communications, modal, chromatic and polarization mode dispersions are the major sources of transmission impairments. Electronic transversal filters can be used to compensate fiber dispersions by constructing an inverse transfer function of the dispersive channel [1]. For multi-band wireless transceiver designs, programmable RF notch filters are needed to selectively reject the bands based on various wireless standards. RF notch filters are critical for removing unwanted signals such as images and interferers. In [2], a fractionally spaced transversal filter is designed using passive transmission lines as delay elements. Passive delay elements occupy large die area and provide accurate delays only for a narrow frequency band. In this paper, we present a 3.5GHz programmable filter RFIC designed in 45GHz SiGe technology. In stead of using passive delay lines, we propose to use Cherry–Hooper amplifiers as the delay stages, which provide wide bandwidth and occupy small area. We used Gilbert variable gain amplifier for continuous tuning of the tap weights. Measured S21 of the programmable RF filter demonstrates various filter characteristics such low-pass, high pass and notch filters up to 3.5GHz frequency.

II. TRANSVERSAL FILTER DESIGN

The block diagram of the implemented transversal filter is shown in Fig. 1. The RF filter includes an analog tapped delay line with the feed forward taps forming an FIR filter. The transfer function of the integrated filter can be adaptively adjusted by changing its tap weights. Changing the tap weights affects only the locations of the zero's, while the poles of the programmable filter are fixed. Hence the filter is always stable.

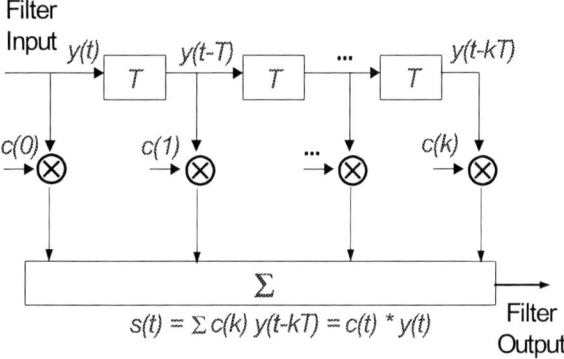

Fig. 1. Block diagram of the transversal filter. T denotes the delay amplifier and $\otimes$ denotes a variable gain amplifier.

Passive delay networks are either lossy (RC delay lines) or bulky (LC delay lines). Moreover, passive delay networks are always narrow band. In stead of using passive delay stages, we chose the series shunt cascaded Cherry-Hooper amplifier to implement the filter delay stages. As shown in Fig. 2, Cherry-Hooper amplifier [3] is a cascade of two feedback amplifiers, where the series feedback stage is a transconductance amplifier and shunt feedback stage is a trans-impedance amplifier. The amplifier with series feedback is driven from a low resistance voltage input obtained from the output of the shunt feedback stage. Conversely, the amplifier with shunt feedback is driven from a high resistance current input obtained from the output of the serial feedback stage. This arrangement is advantageous since the impedance requirement will be automatically

1-4244-0458-4/06/$20.00 ©2006 IEEE

satisfied at the input and output of the amplifier while cascading several such delay stages. Emitter followers are used in between the delay stages for level shifting and creating stronger impedance mismatch between succeeding stages to improve the bandwidth.

Referred to Fig.2, the transfer function of the Cherry–Hooper amplifier can be approximated as

$$\frac{V_{out}}{V_{in}} = \frac{g_{m1}(1+s \cdot C_E \cdot R_E)(1+s \cdot C_F \cdot R_F + g_{m3} \cdot R_F)}{g_{m3}(1+s \cdot C_F \cdot R_F)(1+s \cdot C_E \cdot R_E + g_{m1} \cdot R_E)} \quad (1)$$

where R_E and C_E are the degeneration resistance and capacitance, respectively; R_F and C_F are the shunt feedback resistance and capacitance, respectively. The degeneration and feedback capacitors C_E and C_F introduce zeros in the frequency response and thereby maximize the amplifier bandwidth. The first zero of the amplifier frequency response is created by the degeneration capacitor C_E, and the second zero is generated by the shunt feedback capacitor C_F. Fig. 3 gives the simulated magnitude response of the amplifier. As shown, the 3-dB cutoff frequency of the amplifier is about 9 GHz.

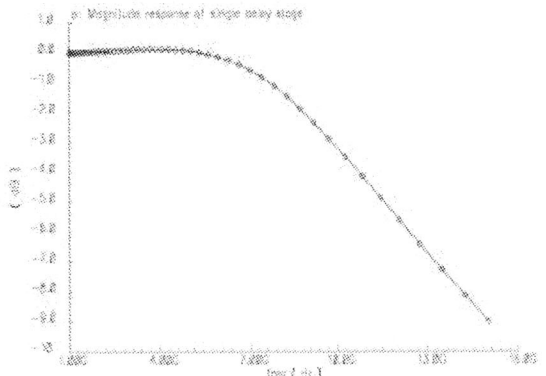

Fig. 3 Simulated magnitude response of the delay stage with 3-dB bandwidth at 9 GHz

Figure 5 shows the small-signal model of the gain control circuit in the Gilbert cell. The Gilbert cell is a current amplifier with the current gain transfer function given as

$$I(s) = \left\{ g_{m3}(1+sC_{\pi 4}r_b) - g_{m4}(1+sC_{\pi 3}r_b) \right\} \cdot$$
$$\left\{ \left(g_{m3}+sC_{\pi 3}+\frac{1}{r_{\pi 3}} \right)(1+sC_{\pi 4}r_b) + \left(g_{m4}+sC_{\pi 4}+\frac{1}{r_{\pi 4}} \right)(1+sC_{\pi 3}r_b) \right\}^{-1} \quad (2)$$

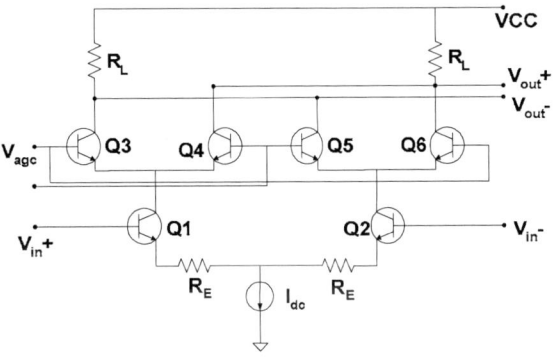

Fig. 4 Tap weights implemented using Gilbert variable gain amplifier.

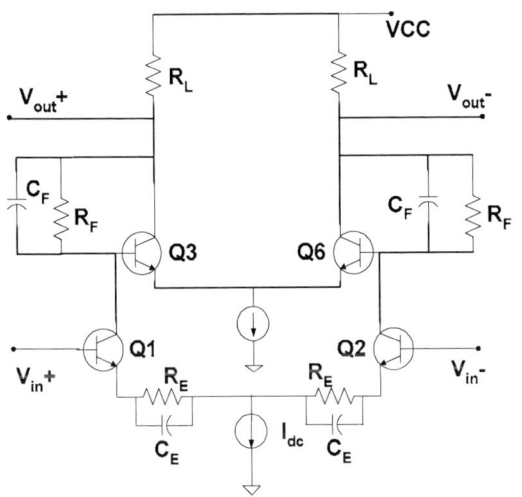

Fig. 2 Cherry–Hooper amplifier used to implement delay stages.

As shown in Fig. 4, the transversal filter tap with programmable gain is implemented using Gilbert variable gain amplifier. Tap weights are continuously adjustable between 0 and 1 of the CML logic level (about ± 100 mV differential). Thus the variable gain stage is infact a variable loss stage. Flipping the polarity of the gain control signal Vage provides a phase shift of 180^0 for negative tap coefficients.

Figure 6 shows the simulated magnitude response of the Gilbert variable gain amplifier. The 3-dB cutoff frequency is achieved at 14.5 GHz.

Finally the summation required in the transversal filter is performed in the current mode. The output current signals of all the taps are tied together to an external pull up resistor via a current buffer, which is a common-base amplifier.

85

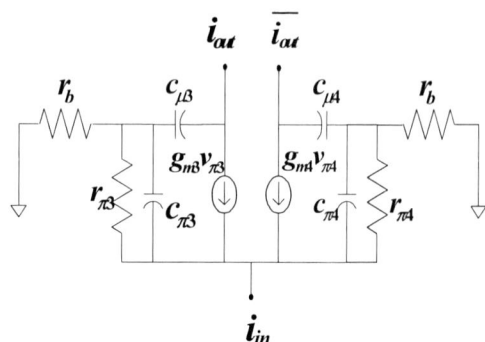

Fig. 5 Small-signal model of the gain control circuit.

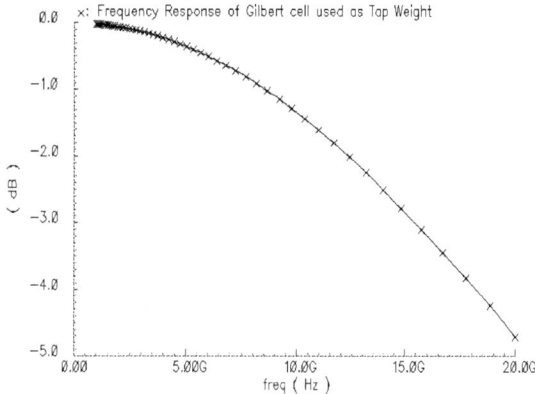

Fig.6 Simulated magnitude response of the Gilbert variable gain amplifier. The 3-dB frequency is 14.5 GHz.

IV. MEASURED RESULTS

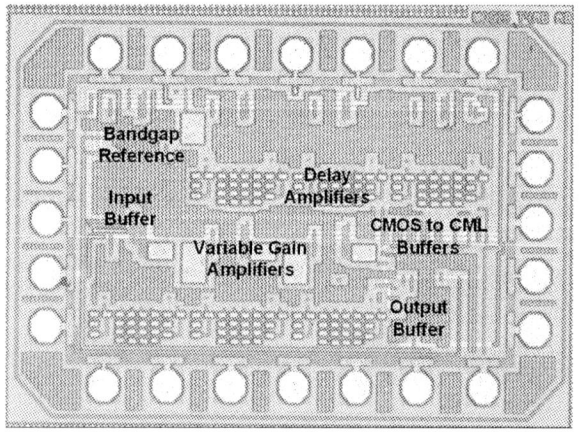

Fig. 7. Die photo of the integrated transversal filter RFIC.

The 3.5 GHz transversal filter is implemented in a 45GHz SiGe technology with a total $2.16mm^2$ die area including pads. The chip consumes 250mW power under a 3.3V supply voltage. The filter RFIC includes a bandgap reference to provide temperature independent

constant current sources for the amplifiers. The filter RFIC also includes an input buffer, output buffers and CMOS to CML buffers. Figure 7 shows the die photo of the transversal filter RFIC.

The frequency response of the integrated filter was measured using a vector network analyzer. The measured filter transfer functions under different tap weights are shown in Fig. 8 to Fig. 11.

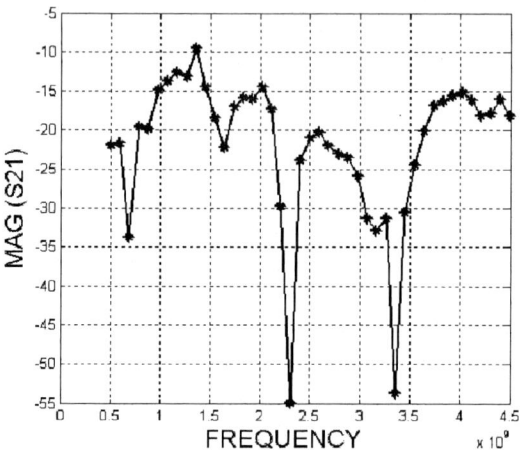

Fig. 8 Measured filter transfer function with double notches at 2.3GHz and 3.3 GHz. The tap coefficients are set as -40, 75, -40, 75, -40, 75, 90 [mV]. The magnitude of the notch is -55dB, which provides -37dB notch rejection compared to the pass band magnitude.

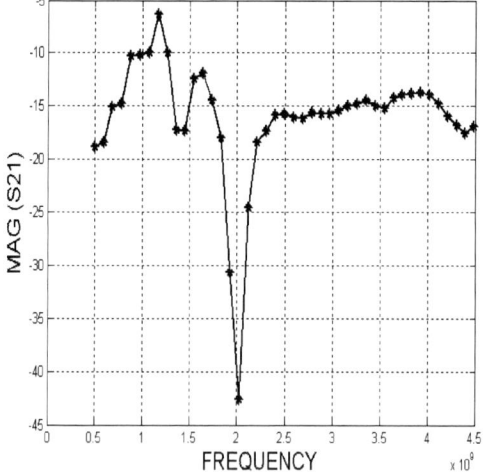

Fig. 9 Measured filter transfer function with notch at 2GHz. The filter coefficients are set as -85, 30, -20, 0, 30, 0, 0 [mV]. The magnitude of the notch is -43db, which provides -30db notch rejection compared to the pass band magnitude.

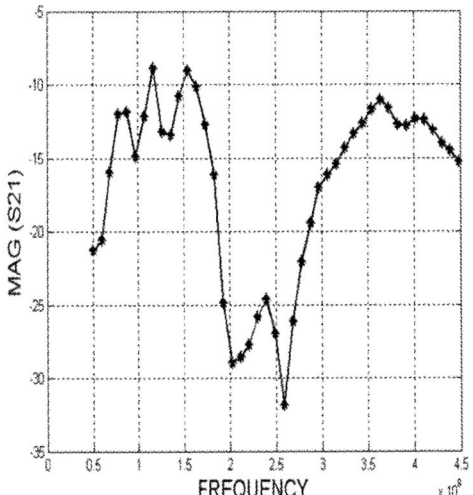

Fig. 10 Measured filter transfer function with band-rejection from 2GHz to 2.7GHz. The coefficients are set as 0, 60, 0, 25, 0, 100, 60 [mV]. The filter achieves a band-rejection of -20dB from 2GHz to 2.7GHz.

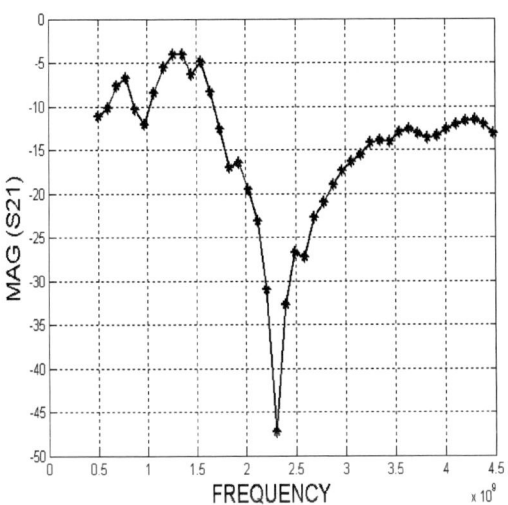

Fig. 11 Measured filter transfer function with notch at 2.3GHz. The filter coefficients are set as 100m, -10m, -10m, 0, 55m, 0, 20m. The magnitude of the notch is -47db, which provides -33db notch rejection compared to the passband magnitude.

Figure 8 demonstrates the filter characteristic with double notches at 2.3 GHz and 3.3 GHz. The magnitude of the notch is -55dB, which provides -37dB notch rejection compared to the pass band magnitude. Figure 9 shows the single notch characteristic of the tunable filter. Figure 10 illustrates the band–rejection characteristic of the filter, in which the filter achieves a band-rejection of -20dB from 2GHz to 2.7GHz. The magnitude of the notch is -43db, which provides -30db

notch rejection compared to the pass band magnitude. Thus, the integrated programmable RF filter is able to tune zeros to any frequency up to 3.5 GHz.

V. CONCLUSIONS

We have implemented a low power 3.5GHz analog transversal filter in a 47GHz SiGe technology. The RF filter utilizes cascaded Cherry-Hooper amplifiers for delay stages and Gilbert variable gain amplifier as for continuous gain tuning. The delay stage using active devices greatly reduces the die area comparing to passive delay lines. Measured results show that by adjusting the tap coefficients, the integrated programmable filter IC is capable to adapt zeros at various frequencies up to 3.5GHz with various filter characteristics. Thus, the integrated transversal filter can be used to mimic the inverse transfer function of dispersive communication channels for dispersion compensation. It can also be used a programmable notch filters in wireless transceiver designs.

ACKNOWLEDGEMENTS

We acknowledge MOSIS for fabrication support under the MEP program.

REFERENCES

[1] J. H. Winters, R. D. Gitlin and K. Sanjay, "Reducing the effects of Transmission Impairments in Digital Fiber Optic Systems," *IEEE Communications Magazine*, June 1993.

[2] H. Wu, J. A. Tierno, P. Pepeljugoski, J. Schwab, S. Gowda, J.A. Kash, A. Hajimiri, "Integrated Transversal Equalizers in High-Speed Fiber-Optic Systems," *IEEE J. Solid State Circuits*, vol.38, pp. 2131-2137, 2003.

[3] H. M. Rein and M. Moller, "Design Considerations for very high speed Si-bipolar IC's operating up to 50Gb/s", *IEEE J. Solid State Circuits*, vol.31, pp. 1076-1090, August 1996.

2006 Bipolar/BoCMOS Circuits and Technology Meeting Proceedings

SiGe BiCMOS Precision Voltage References for Extreme Temperature Range Electronics

Laleh Najafizadeh, Chendong Zhu, Ramkumar Krithivasan, John D. Cressler, Yan Cui [1], Guofu Niu [1], Suheng Chen [2], Chandradevi Ulaganathan [2], Benjamin J. Blalock [2], and Alvin J. Joseph [3]

School of Electrical and Computer Engineering, 777 Atlantic Drive, N.W.
Georgia Institute of Technology, Atlanta, GA, 30332-0250 USA
[1] Auburn University, Auburn, AL 36830 USA
[2] University of Tennessee, Knoxville, TN 37996 USA
[3] IBM Microelectronics, Essex Junction, VT 05452 USA

Abstract — We present the first investigation of the optimal implementation of SiGe BiCMOS precision voltage references for extreme temperature range applications (+120 °C to -180 °C and below). We have developed and fabricated two unique Ge profiles optimized specifically for cryogenic operation, and for the first time compare the impact of Ge profile shape on precision voltage reference performance down to -180 °C. Our best case reference achieves a 28.1 ppm/ °C temperature coefficient over +27 °C to -180 °C, more than adequate for the intended lunar electronics applications.

Index Terms — analog circuits, SiGe HBT, cryogenic temperature, voltage reference, device physics.

I. INTRODUCTION

For many space exploration applications, it is highly desirable to have electronic components capable of operating robustly over extreme temperature ranges, since it can profoundly improve robotic system architecture and performance, reduce system power drain, dramatically reduce launch weight, and improve overall mission reliability. On the surface of the Moon (NASA's next mandated exploration venue), for example, ambient temperatures range from +120 °C (lunar day) to -180 °C (lunar night), and even down to -230 °C (shadowed polar craters). Designing robust electronic systems for > 300 °C (cyclic) temperature variations has never been attempted, until now. Bandgap-engineered SiGe heterojunction bipolar transistors (SiGe HBTs) are known to have superior performance down to deep cryogenic temperatures [1] and therefore have recently emerged as a leading contender for such extreme temperature range electronics applications. SiGe IC design platforms offer both high speed SiGe HBTs and Si CMOS, and a host of passive elements for developing highly-integrated system-on-a-chip and system-in-a-package components for use in emerging lunar electronic systems. SiGe HBTs have a desirable side benefit of possessing a natural hardness to ionizing radiation (a key concern for most space applications).

Key to the success of this vision of developing extreme temperature range electronics is to realize

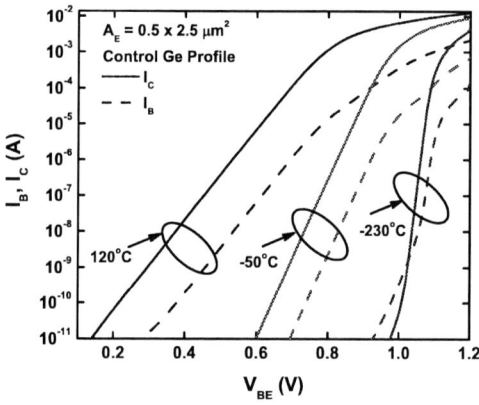

Fig.1. Gummel characteristics at +120°C, -50°C, and -230°C for a 0.5×2.5 μm^2 SiGe HBT.

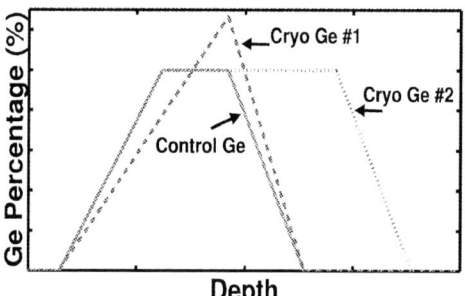

Fig. 2. Shapes of the three Ge profiles used.

a robust precision voltage reference. Voltage references (e.g., bandgap voltage references - BGR) are used extensively in space electronic systems for A/D and D/A converters, voltage regulators, and many other mixed-signal circuits, and can be viewed as a necessary (and inherently difficult) analog primitive for the successful realization of this vision. Simply stated, if robust voltage references cannot be achieved, all bets are off for developing extreme temperature range systems.

We present here the first investigation of the optimal implementation of SiGe BiCMOS precision voltage references for such extreme temperature range applications, and demonstrate that SiGe BiCMOS can indeed achieve the necessary performance for

1-4244-0458-4/06/$20.00 ©2006 IEEE

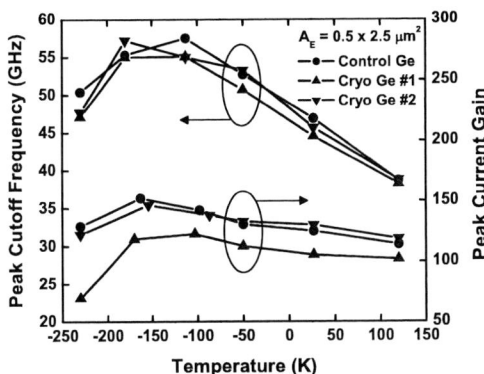

Fig. 3. Measurement results for the maximum current gain and peak f_T as a function of temperature for three Ge profiles.

lunar electronics systems. In addition, we have developed and fabricated two unique Ge profiles optimized specifically for cryogenic operation, and for the first time compare the impact of Ge profile shape on precision voltage reference performance over temperature.

II. SiGe HBT BiCMOS Technology

The SiGe technology platform used here is the commercially-available IBM 5AM SiGe BiCMOS technology, featuring 0.5 μm, 50 GHz (27 °C) SiGe HBTs. Fig. 1 shows the Gummel characteristics of a 0.5×2.5 μm^2 SiGe HBT at 120 °C, -50 °C, and -230 °C (the lowest temperature for lunar applications). Due to the exponential decrease of the intrinsic carrier concentration with cooling, the base-emitter turn-on voltage increases as the temperature decreases, as expected. At -230 °C, this device has a maximum current derive in excess of 4mA/μm^2. We call the Ge profile used in this commercial technology the "Control Ge" profile. A comparison of the three Ge profiles used in this investigation is shown in Fig. 2. "Cryo Ge #1" and "Cryo Ge #2" are the optimal cryogenic profiles that were designed based upon calibrated 2-D simulations over temperature, with constant stability and deeper retrograding, respectively (the latter to improve immunity to heterojunction barrier effect). The shape of the Ge profile at the base-emitter side for both the conventional Cryo Ge #2 and the Control Ge profile is similar. Measurement results for the peak current gain and peak cutoff frequency as a function of temperature for the three Ge profiles are shown in Fig. 3. As can be seen, all three Ge profiles exhibit excellent characteristics down to -230 °C, while the two optimized Ge profiles give significantly better DC and AC performance at peak f_T and into high injection, as intended.

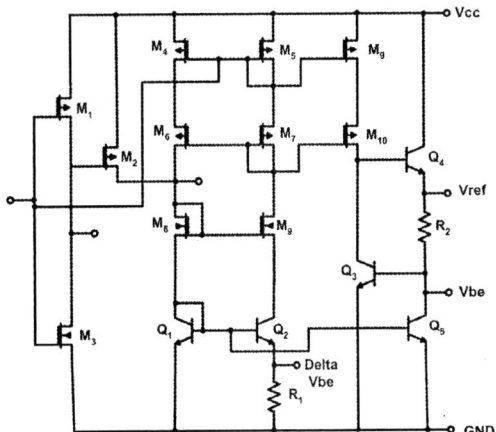

Fig. 4. Schematic of the SiGe voltage reference.

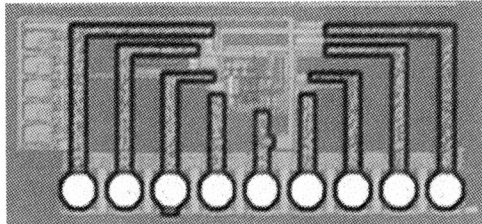

Fig. 5. Die micrograph of the SiGe voltage reference.

III. Circuit Design

The BGR implemented for this investigation is based on an exponential curvature compensation technique [2] and exploits the temperature characteristics of the current gain of the SiGe HBT for curvature compensation. To enable meaningful circuit design, the calibrated VBIC models within the design kit were first modified and fit to low temperature transistor data. The schematic of the BGR circuit is shown in Fig. 4. Each of the bipolar transistors shown in this figure, except for Q_2, consists of four parallel copies of the 0.5×2.5 μm^2 SiGe HBT. The area of transistor Q_2 is eight times larger than that of the other transistors. Transistors M_1-M_3 comprise the startup circuit, and transistors M_4-M_9 and Q_1-Q_2, along with the resistor R_1, generate the proportional to the absolute temperature (PTAT) bias current for the other stages. Several pads have been inserted to monitor the nodes of some of the transistors inside the circuit as the temperature was varied. The circuit was designed for a power supply voltage of 3.3 V. With the pads, the circuit occupies an area of 1.3×0.4 mm^2 (Fig. 5). Careful layout techniques were employed to reduce mismatch effects.

IV. EXPERIMENTAL RESULTS & DISCUSSION

The circuits were mounted in a 28 pin ceramic package, wire-bonded, and placed inside a closed-cycle helium cryostat for characterization.

TABLE I
SUMMARY

	Control Ge		Cryo Ge #1		Cryo Ge #2	
Vref (V)	1.1885	(27 °C)	1.1723	(27 °C)	1.1762	(27 °C)
@	1.1778	(-180 °C)	1.1662	(-180 °C)	1.1625	(-180 °C)
Vcc=3.3 V	1.1629	(-230 °C)	1.1519	(-230 °C)	1.1408	(-230 °C)
Icc (μ A)	139	(27 °C)	126	(27 °C)	130	(27 °C)
TC (ppm/°C)	17.1	(-50:27) °C	10.6	(-50:27) °C	7.8	(-50:27) °C
@	49.8	(-180:27) °C	57.6	(-180:27) °C	28.1	(-180:27) °C
Vcc=3.3 V	89.1	(-230:27) °C	118.4	(-230:27) °C	69.9	(-230:27) °C

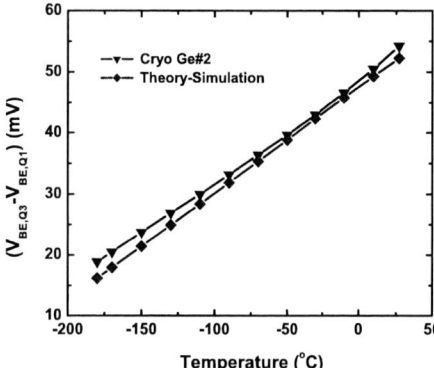

Fig. 6. Base-emitter voltage difference as a function of temperature.

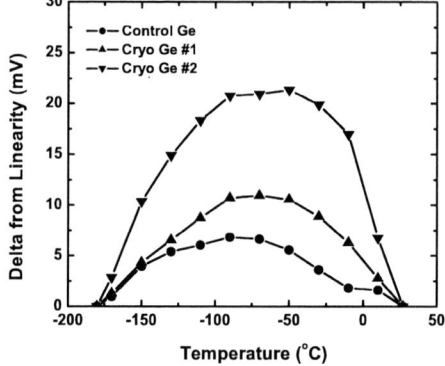

Fig. 7. Measured deviation from linearity for a SiGe HBT biased with a PTAT current for the three Ge profiles.

Measurements were performed using an Agilent 4155 Semiconductor Parameter Analyzer over a temperature range of 27 °C to –230 °C (the current limits of the cryostat system). The results are shown in Figs. 6-9.

The difference between the base-emitter voltages of transistors Q_2 and Q_1 was measured across temperature and is shown in Fig. 6. Ideally, this curve should follow equation (1) and should be linear over temperature. However, as reported in [3], the deviation from V_{BE} linearity for the two SiGe HBTs with different areas is different, resulting in a finite mismatch. As a result, the difference between the base-emitter voltages is not perfectly linear.

$$V_{BE,Q_2} - V_{BE,Q_1} = \frac{kT}{q} \ln(8). \qquad (1)$$

The deviation from V_{BE} linearity for transistor Q_3 as a function of temperature for the three Ge profiles is shown in Fig. 7. Note that the transistor is biased with a PTAT current. This deviation from linearity has usually been reported for the case when the transistor is biased at fixed collector current [3]. However, in many BGR topologies, the collector current for the transistors inside the circuit is instead the PTAT current. Calibrated 2-D simulations of the deviation from linearity for a single transistor with a size of Q_3, at fixed biased current and comparable collector-base voltage, are different from the measured results at the PTAT current for each of the three Ge profiles.

Measurement results show that the Cryo Ge #2

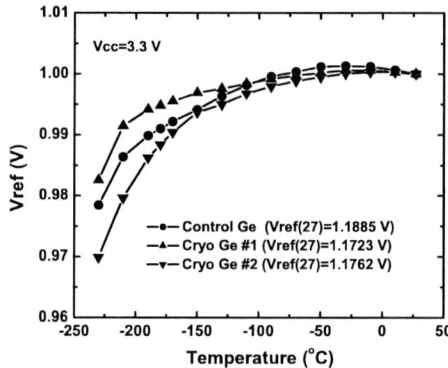

Fig. 8. Measured output voltage of the voltage reference as a function of temperature for three Ge profiles.

profile has the maximum deviation from linearity (at the PTAT current), while the Control Ge profile results in the minimum deviation from linearity over temperature, as expected.

Fig. 8 shows the output voltage of the reference as a function of temperature for the three Ge profiles. The output voltage for each circuit was normalized by dividing it by its nominal voltage at 27 °C, for ease of comparison. It can be seen that circuits with Cryo Ge #2 and Cryo Ge #1 profiles have the best and the worst temperature stability among the three profiles, respectively, compared to the Control Ge profile. The temperature coefficients (TC in ppm/°C) for the 3 profiles are summarized in Table I. Our best case result of 28.1 ppm/°C for the cryo Ge profile #2 from

TABLE II
PERFORMANCE COMPARISON

	This Work	[7]	[8]	[2]	[9]	[10]
Technology	SiGe 5AM	SiGe 8HP	SiGe5HP	1.5 μm BiCMOS	6 μm CMOS	0.6 μm CMOS
Vcc (V)	3.3	2.5	2.5	5	-	2
Vref (V) @ 27 °C	1.172	1.026	1.328	1.264	1.192	1.142
TC (ppm/°C)	7.8 (-50:27) / 28.1 (-180:27) / 69.9 (-230:27)	8.4 (-50:27) / 20.2(-180:27)	36.5(-50:150) / 25.1 (0:75)	3.5 (0:70) / 8.9 (-55:125)	13.1 (0:70) / 25.6(-55:125)	5.3 (0:100)

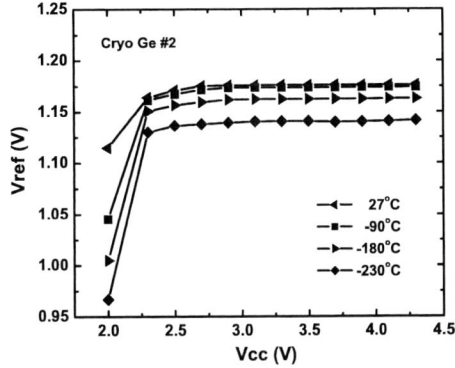

Fig. 9. Measured output voltage as function of power supply at 120 °C, 27 °C, -60 °C, -180 °C, and -230 °C for the Cryo Ge #2 profile.

27 °C to -180 °C, more than satisfies the required specifications for lunar systems.

The dependence of TC on Ge profile shape can be explained by reviewing the technique used here for curvature compensation. Referring to Fig. 4, the output voltage of the BGR is approximately by [2]

$$V_{ref}(T) \approx V_{BE,Q_3}(T) + K_1 R_2 T + K_1 R_2 T / \beta(T) \quad (2)$$

where T is the absolute temperature, R_2 is the resistor in the last circuit branch, and K_1 and K_2 are coefficients corresponding to the PTAT currents. Note that both $V_{BE,Q3}$ and the current gain (β) depend on the shape of the Ge profile in the base, thus affecting the overall output temperature dependence of the reference output [3]-[6]. The temperature dependence of the current gain, in particular, is used to minimize the temperature drift of the output voltage. Detailed expressions for $V_{BE,Q3}(T)$ and $\beta(T)$ for SiGe HBTs as a function of Ge profile shape can be found in [4], and when combined with (2) are generally consistent with our measurements.

Fig. 9 shows the change in the output voltage as the power supply changes, at five different temperatures, for the Cryo Ge #2 version of the reference circuit. As can be seen, the line regulation remains well-behaved across the extreme temperature range.

A comparison of the BGRs presented in this work with the existing state-of-the-art ([2],[7]-[10]) across a more standard operating temperature range is given in Table II. It is expected that the circuits in [2],[9] and [10] will not function down to -180 °C.

V. SUMMARY

We have presented results on SiGe BiCMOS precision voltage references designed for operation across extreme temperature ranges (+120 °C to -180 °C). Our best case voltage reference achieves a 28.1 ppm/°C temperature coefficient over +27 °C to -180 °C, more than adequate for the intended lunar electronics applications, suggesting that SiGe technology is an ideal candidate for such space exploration applications.

VI. ACKNOWLEDGEMENT

This work was supported by NASA, under contract NL06AA29C, and the Georgia Electronic Design Center at Georgia Tech. We are grateful for the support of M. Watson, C. Moore, D. Hope, D. Frazier, M. Beatty, and L. Nadeau of NASA, E. Kolawa of JPL, and the IBM SiGe development team, as well the many contributions of the SiGe ETDP team, including: M. Mojarradi, W. Johnson, R. Garbos, R. Berger, F. Dai, L. Peltz, A. Mantooth, P. McCluskey, M. Alles, R. Reed, A. Joseph, and C. Eckert.

REFERENCES

[1] J.D. Cressler, *Proc. IEEE,* vol. 93, p. 1559, 2005.
[2] I. Lee *et al., IEEE JSSC,* vol. 29, p. 1396, 1994.
[3] S. Salmon *et al., IEEE TED,* vol. 47, p. 292, 2000.
[4] J.D. Cressler and G. Niu, *Silicon-Germanium Heterojunction Bipolar Transistors,* Boston, MA: Artech House, 2003.
[5] J.D. Cressler *et al., IEEE EDL,* vol. 15, p. 472, 1994.
[6] A. J. Joseph *et al., IEEE EDL,* vol. 16, p. 268, 1995.
[7] L. Najafizadeh *et al.,* to appear in *Proc. WOLTE-7.*
[8] H. A. Ainspan *et al., Elect. Lett.,* vol. 34, p. 1441, 1998.
[9] B. S. Song *et al., IEEE JSSC,* vol. 18, p. 634, 1983.
[10] K. Leung *et al., IEEE JSSC,* vol. 38, p. 561, 2003.

2006 Bipolar/BiCMOS Circuits and Technology Meeting Proceedings

Quantitative Analysis of Errors in On-Wafer S-Parameter De-embedding Techniques for High Frequency Device Modeling

Rob Groves, Jing Wang, Lawrence Wagner and Ava Wan[1]

IBM Semiconductor Research and Development Center
2070 Route 52, Hopewell Junction, NY 12533 USA
(845) 892-3993 (office), (845) 892-6042 (fax), grovesr@us.ibm.com
(1) summer student (6/05-8/05) from Cornell University, Ithaca, NY 14850 USA

Abstract — The de-embedding of intrinsic device parameters from on-wafer measurements is a central problem in high frequency device measurement and modeling. The first *quantitative* analysis of the errors associated with de-embedding on-wafer s-parameter measurements of 90nm bulk FETs and 130nm SiGe HBTs taking into account the effects of non-ideal standards is presented. Four different on-wafer de-embedding techniques are examined. Electromagnetic (E-M) simulations accounting for these non-idealities are used to compare different methods. It is demonstrated that unwanted parasitics in standards can significantly affect parameters extracted using more complex de-embedding techniques. The most sensitive standard is identified and an optimized design is presented.

Index Terms — De-embedding, calibration, s-parameter, on-wafer, parameter extraction, high frequency.

I. Introduction

Model parameter extraction from on-wafer s-parameter measurements is essential to the creation of device models useful in high frequency circuit design. FET and HBT cut-off frequencies of 300GHz+ [1-2] require accurate determination of device characteristics that may be as much as two orders of magnitude smaller than raw measurements. Furthermore, as device speed increases, accurate models of device behavior at frequencies up to and beyond 100GHz are essential. Accurate determination of device terminal characteristics at these frequencies requires an effective de-embedding technique. Industry standard on-wafer de-embedding techniques are known to be inadequate to completely remove the parasitic effects of on-wafer measurement pads [3-6]. While much has been published on techniques designed to improve de-embedding accuracy, results have only been evaluated on a *qualitative* basis or with the assumption that all de-embedding standards are ideal [6]. What is desired is a technique for *quantitatively* evaluating the effectiveness of the various proposed de-embedding techniques, taking into account the non-ideal nature of the de-embedding standards used. E-M simulation, carefully done, provides a means of achieving this quantitative comparison.

The on-wafer measurement pad environment is simulated within a planar E-M simulator (Sonnet®) as a five-port network and the *intrinsic* Device-Under-Test (DUT) network characteristics are derived from a compact model simulation utilizing an industry standard device model. The resulting s-parameters of the measurement pad environment and the device are combined to create a virtual measurement environment, emulating the actual results achieved in on-wafer s-parameter measurements (see Fig. 1).

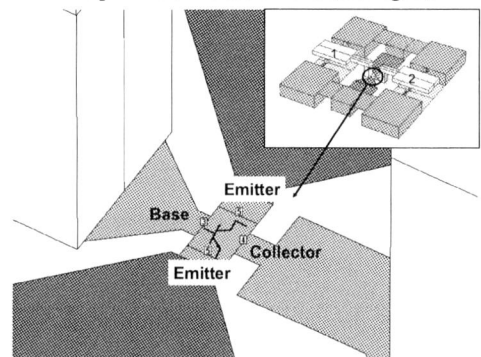

Fig. 1 E-M simulator (Sonnet®) model of on-wafer pad environment showing NPN model terminal connections.

Furthermore, various de-embedding standards (i.e. Open, Short, Thru, …) can be created within the simulated measurement environment and the s-parameters of these non-ideal standards can be used to carry out *virtual de-embedding* of the combined E-M/compact model simulated s-parameters. The results of this *virtual de-embedding* can be used to evaluate the advantages and disadvantages of each technique. The unique advantage of *virtual de-embedding* is that it provides a means of optimizing a de-embedding methodology, taking into account all unwanted parasitic effects, without requiring expensive wafer fabrication.

II. Error Network Models and De-embedding

The goal, in the de-embedding of high frequency device network parameters, is to characterize the parasitics (error network) that exist between the

1-4244-0458-4/06/$20.00 ©2006 IEEE

measurement instrument reference plane and the DUT reference plane.

Often, the determination of the error network is carried out in a two-step process. First, the measurement reference plane is moved from the instrument to the microwave probe tips in a "calibration" step. The second step is designed to determine the on-wafer error network and to move the reference plane to the device terminals in a step generally called "de-embedding". The de-embedding step is intended to correct for any residual error left over from the calibration step and to account for parasitic coupling mechanisms present in the on-wafer measurement environment. The remainder of this paper is concerned with the de-embedding step.

In order to characterize the on-wafer error network, a model topology must be constructed to describe its behavior. The four de-embedding techniques [3-6] make various assumptions regarding the configuration of the error network.

Once an assumption of the configuration of the error network is made, a series of standards with known characteristics can be connected in place of the DUT. From measurements of these standards, connected to the measurement pads, the various elements of the assumed error network can be determined. These measured standards must have ideal characteristics (i.e. open has perfect reflection, thru has zero phase shift and loss, etc.) in order to satisfy the assumptions made in the de-embedding algorithms. Any deviation from these ideal characteristics will manifest itself as an inaccuracy in the resulting de-embedded DUT s-parameters. Quantifying the magnitude of the resulting error in extracted DUT parameters is a prime motivator behind this work.

Fig. 2 shows the four de-embedding models used in this comparison: Open-Short (OS) [3], Pad-Open-Short (POS) [4], 3-STEP [5], and COMPLETE [6-7]. The number of standards required for de-embedding is in parenthesis above each model.

The simplest technique is Open-Short (OS) and only requires two standards to determine the six elements of the error network. The most complex technique is COMPLETE de-embedding and requires a minimum of four standards to determine the 15 elements of the error network. The only technique that completely accounts for all parasitics is COMPLETE de-embedding [6-7], which may also utilize more standards containing various load configurations to improve dynamic range.

As can be seen in Fig.2, the most popular de-embedding methodologies (top three in figure) ignore

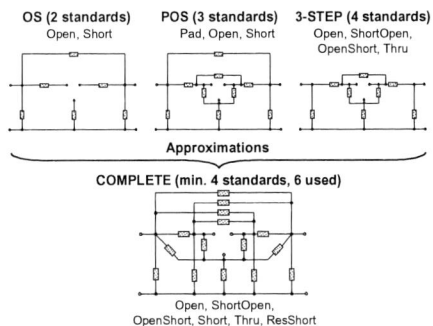

Fig. 2 De-embedding models, simple to complex

many coupling and loss mechanisms inherent in the complete pad error network. The approximations made in the OS, POS, and 3-STEP methods can lead to undesired frequency-dependent behavior in parameters extracted from de-embedded measurements, even when idealized standards are used.

III. *Virtual De-embedding* Methodology

Through E-M simulation, an N-port network describing the on-wafer pad parasitic network can be synthesized. The number of ports in the simulation is dictated by the number of external ports (probes) and the number of internal connections to the DUT. For a transistor (HBT or FET) in a typical on-wafer measurement environment, a five port network is appropriate, consisting of two external ports, representing the wafer probes, and three internal ports representing the Base (Gate), Collector (Drain), and Emitter (Source) connections to the HBT (FET) (see Fig. 3).

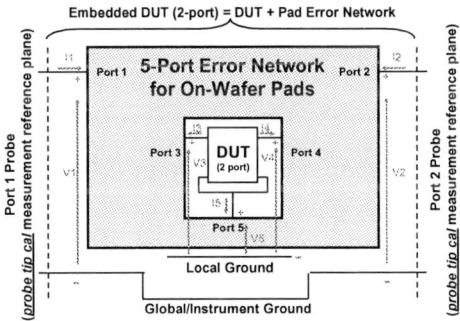

Fig. 2 5-Port error network cascaded with 2-port DUT

Fig. 3 5-Port error network cascaded with 2-port DUT to yield 2-port embedded network parameters (PAD+DUT).

The resulting 5-port network (PAD) can be cascaded with the simulated *intrinsic* device 2-port s-parameters (DUT). The DUT s-parameters are derived from an AC analysis of a compact device model under appropriate bias conditions. These two networks are cascaded to yield 2-port s-parameters (PAD+DUT) that are analogous to "embedded" on-wafer s-parameter measurements of the device. The interconnection of the PAD and DUT networks is shown in Fig. 3, where it can be seen that the 2-port DUT ground return is connected to port 5 of the PAD error network. Cascading of the PAD network with the DUT network is accomplished in a linear simulator.

IV. Ideal vs. Physical Standards

The first step in evaluating each de-embedding technique is to compare the resulting DUT parameters when de-embedding is done with both "ideal" and "physical" standards.

Once the PAD network has been simulated, it can be inserted into the linear simulator and the three internal ports connected to "ideal" parasitic-free standards (i.e. ideal Open = no connection to three internal ports, ideal Short = connect all three internal ports together, etc.). The resulting simulations represent the characteristics of a parasitic-free standard embedded inside the measurement error network. These "ideal" standards are used to de-embed the PAD+DUT simulation results. This result is unachievable with physical measurements and allows us to carry out the de-embedding techniques under the idealized conditions assumed by the de-embedding algorithms. Next, physically realizable standards are created within the E-M measurement pad environment and the resulting structure simulated as a 2-port network, with the two-ports representing the probe pads. These results represent the characteristics of actual measured de-embedding standards that would be realizable on-wafer. The de-embedding algorithms are then applied to the PAD+DUT simulation results using the "physical" standards.

This procedure was carried out for a 0.12μmx3μm single stripe SiGe NPN model biased at it's peak Ft current (Vbe=0.9V,Vcb=0.5V) and a 100nmx3μm single stripe bulk nFET model biased in saturation (v_g=1.2V, v_d=1.2V). The FET results were used for the "ideal" vs. "physical" comparison due to the fact that the intrinsic parasitics of the FET were between one and two orders of magnitude smaller than the HBT, providing a more sensitive measure of the de-embedding results. The three "approximate" techniques, OS, POS, and 3-STEP, demonstrated little difference between the results with "ideal" and "physical" standards. This can be attributed to the large residual errors that result from the approximations themselves. COMPLETE de-embedding, on the other hand, was sensitive to the parasitics of the standards and was found to be most sensitive to the characteristics of the various resistor standards. It was determined that coupling from the

resistor to the substrate was the primary driver of this sensitivity. To quantify this effect, two different Resistor-Short standards were created (see Fig. 4), one with a large physical area, as would be required to satisfy design groundrules for a polysilicon resistor, and one with a smaller, non-groundrule compliant area (LARGE=24μm², SMALL=6μm²).

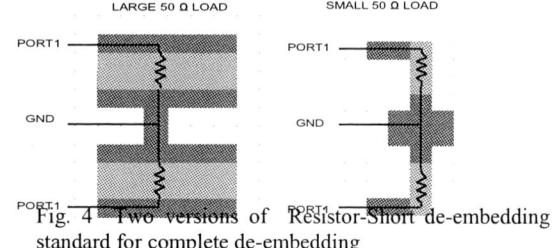

Fig. 4 Two versions of Resistor-Short de-embedding standard for complete de-embedding

Figure 5 shows the "ideal" and "physical" de-embedded results using OS and COMPLETE de-embedding for two key FET parameters: , $\sqrt{U} * F$, where U denotes the unilateral gain and F is the frequency, and $Re\{H_{11}\}$, which is related to gate resistance.

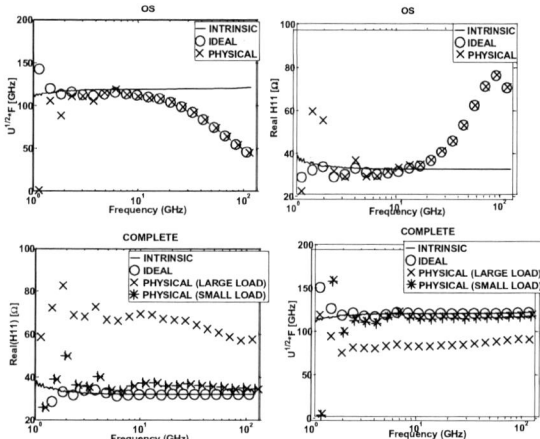

Fig. 5 "Ideal" vs. "Physical" de-embedding for COMPLETE and OS de-embedding techniques (FET). COMPLETE de-embedding results with two size loads

It can be seen from Fig. 5 that the errors from OS de-embedding are indeed large enough to hide any inaccuracy due to the non-ideal physical standards. COMPLETE de-embedding, with less residual error, shows a greater sensitivity to the quality of the standards used. With an optimized load standard, excellent results were achieved with COMPLETE de-embedding.

V. HBT Results (0.12μmx3μm emitter)

Fig. 6 show the simulated intrinsic and de-embedded data (using OS, COMPLETE with small load, POS and 3-STEP) for the two previously described parameters: $\sqrt{U} * F$ and $Re\{H_{11}\}$, which is related to base resistance in the NPN.

It can be seen from Fig. 6 that COMPLETE de-embedding outperforms the other three techniques up to 110GHz. In general, POS is closest to the COMPLETE results followed by 3-STEP and OS, with deviations occurring above 30GHz. It should be noted that the base resistance (Rbb) extracted using the s-parameter circle fit method from OS yields a value that is quite different from the intrinsic result and the result using COMPLETE de-embedding (OS = 34.5±5Ω, COMPLETE = 44.0±0.4Ω, intrinsic = 43.5±0.3Ω). This difference is attributed to a very non-ideal circle realized from the OS technique, while the COMPLETE technique generates a circle essentially identical to the intrinsic model.

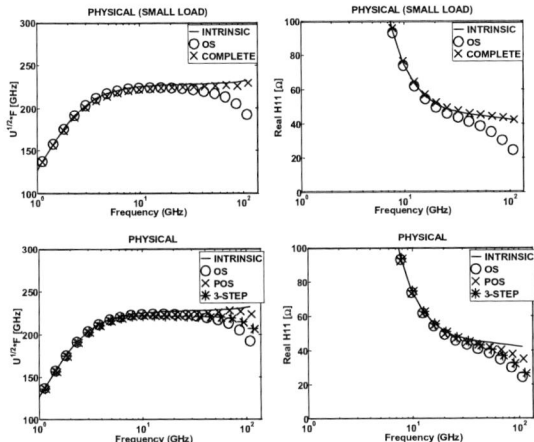

Fig. 6 Top two figures OS and COMPLETE de-embedding. Bottom two figures OS, POS, 3-STEP.

VI. FET Results (0.10μm x3μm gate)

Fig. 7 shows the simulated intrinsic and de-embedded data (using OS, COMPLETE with small load, POS and 3-STEP) for the same two parameters. It is clear from Fig. 7 that COMPLETE de-embedding, again, accurately reproduces the intrinsic characteristics up to a very high frequency (110GHz). For the FET, 3-STEP and OS yield essentially identical results with significant errors above 20GHz. POS, while improving extraction of $\sqrt{U} * F$, yields non-physical results for $Re\{H_{11}\}$, where the extracted parameter is negative at low frequencies. Examination of other extracted FET parameters not shown here, indicates that POS performs better than OS and 3-STEP in all other cases. The errors in OS and 3-STEP and the lack of consistent results for POS indicate that caution must be observed when interpreting de-embedded results from techniques other than COMPLETE de-embedding.

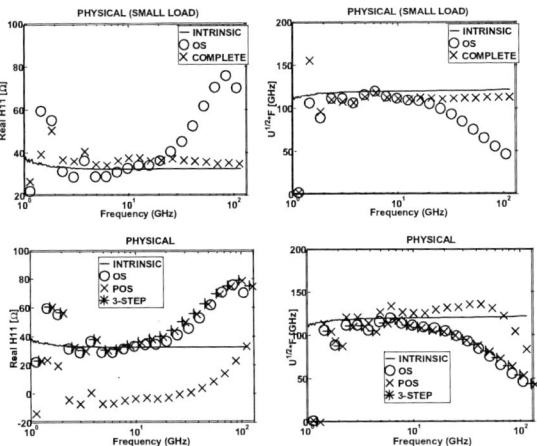

Fig. 7 Top two figures OS and COMPLETE de-embedding. Bottom two figures OS, POS, 3-STEP.

VII. Conclusion

Use of *virtual de-embedding* has shown, quantitatively, the errors introduced when using non-ideal standards and de-embedding techniques that use a simplified model. From these results, it is now clear that, above 20 GHz, parameters extracted using approximate de-embedding techniques are suspect and should not be used to optimize the high frequency device model. COMPLETE de-embedding overcomes the limitations of the approximate methods, yielding data without high frequency anomalies as long as the load standard has minimized parasitics. Creating the required optimal load standards for COMPLETE de-embedding may necessitate violation of design groundrules. The inconsistent results from the POS technique dictate caution when evaluating the effectiveness of any particular de-embedding technique. A variety of extracted parameters should be examined before reaching any conclusions on the accuracy of the technique.

References

[1]. B. Jagannathan, et al., Digest of SiRF Conf. 2006, pp. 259-264, Jan. 2006.

[2]. J.-S. Rieh et al., Tech. Dig. IEEE IEDM, pp. 771-774, Dec. 2002.

[3]. M. Koolen, J. Geelen, M. Versleijen, BCTM, pp.188-191, 9-10 Sep 1991.

[4]. T. Kolding, IEEE Trans. Electron Devices, vol. 47, no. 4, Apr. 2000.

[5]. E. Vandamme, D. Schreurs, and C. van Dinther, IEEE Trans. Elect. Dev., vol. 48, no. 4, pp. 737–742, Apr. 2001.

[6]. Qingqing Liang, et al., RFIC Symposium, 2003 IEEE , pp. 357- 360, 8-10 June 2003

[7]. L. Wagner, unpublished.

2006 Bipolar/BoCMOS Circuits and Technology Meeting Proceedings

BJT Base and Emitter Resistance Extraction from DC Data

Colin C. McAndrew

Freescale Semiconductor, Tempe, AZ, Colin.McAndrew@freescale.com

ABSTRACT

This paper presents a new technique to determine the base and emitter resistances of BJTs. The method is based on analysis of intrinsic and extrinsic conductances, which can be calculated from forward Gummel DC data at different V_{cb}, and explicitly accounts for the difference between DC and AC values of base resistance.

1. INTRODUCTION

Bipolar transistors (BJTs) inherently include parasitic resistances, which affect device behavior and are important parts of BJT models. The main parasitic resistances to be characterized are the (DC) base resistance $r_b = 1/g_b$, which has both bias dependent (intrinsic, r_{bi}) and bias independent (extrinsic, r_{bx}) components, and the emitter resistance r_e, which is assumed to be independent of bias.

BJT base and emitter resistance extraction is a topic of perennial interest. Too many approaches have been proposed to be referenced here, but several general approaches have proven fruitful. The Ning-Tang method [1] is now classic, but assumes r_{bi} to be proportional to the DC current gain $\beta = I_c/I_b$, requires an independent measurement of the pinched base sheet resistance, and so is not based solely on device specific measurements; the analysis it provides is still important though. The open collector method [2] and later extensions [3] are useful, but they bias BJTs under conditions significantly different from those used in real circuits; models are approximations, and extraction of parameters based on such bias conditions does not guarantee model accuracy in the bias range of interest for circuit design. AC techniques have been proposed [4][5], but details of AC behavior depend on details of the small-signal model used, such as the partitioning of r_b and of base-collector capacitance, and the derivatives of r_b. All of these details are not accounted for in the analyses; the results are thus inconsistent with circuit simulation models.

Special test structures and biasing procedures have been proposed for BJT resistance extraction [6][7]. However, it is still desirable to be able to extract BJT resistances from standard devices under normal DC bias conditions, rather than requiring special structures or biases.

Parasitic series resistances cause DC de-biasing and affect the small-signal characteristics of a device. In particular, the extrinsic conductances, seen from the terminals of a device, differ from the intrinsic conductances. This paper shows how the low frequency intrinsic and extrinsic conductances can be used to calculate BJT resistances. Specifically, a new method to calculate r_e is derived that, unlike previous work, does not require a model for I_b and accounts for the difference between the small-signal base resistance r_b^{ac} and its DC value.

Accounting for the bias dependence of r_b explains the differences seen between results of previously proposed AC and DC methods to characterize base resistance.

In the analyses, the extrinsic emitter is used as a voltage reference, upper and lower case distinguishes extrinsic and intrinsic voltages, $V_b = V(b)$, $v_b = V(bi) - V(ei)$, $V_c = V(c)$, $v_c = V(ci) - V(ei)$, and a tilde denotes small-signal quantities. Upper and lower case are also used to distinguish extrinsic and intrinsic 2-port conductances.

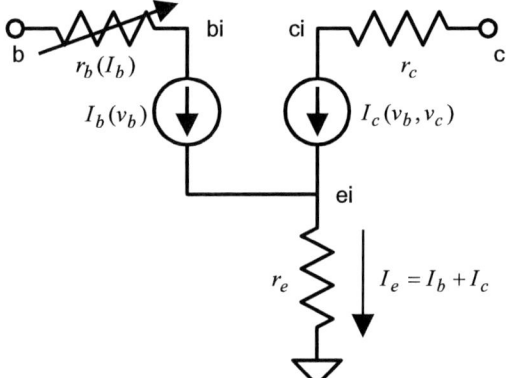

Fig. 1   BJT Equivalent Network

2. BASIC FORMULATION

Fig. 1 shows the BJT model equivalent network used here. The analysis is based on low frequency conductances, so capacitance components are not included. The reason r_b is considered to be a function of I_b rather than v_b (as it is in many models; the dependence on v_c is small and will be neglected here) will become apparent. In practice it does not matter which is used as application of the chain rule can convert derivatives from one dependency to the other.

Consider a BJT biased in forward active operation, and assume that self-heating, base-collector impact ionization and neutral base recombination are negligible. The small-signal admittance representation of the intrinsic device is

$$(1) \quad \begin{bmatrix} \widetilde{i_b} \\ \widetilde{i_c} \end{bmatrix} = \begin{bmatrix} g_\pi & 0 \\ g_m & g_o \end{bmatrix} \begin{bmatrix} \widetilde{v_b} \\ \widetilde{v_c} \end{bmatrix}$$

1-4244-0458-4/06/$20.00 ©2006 IEEE   96

where $g_\pi = \partial I_b/\partial v_b$, $g_o = \partial I_c/\partial v_c$, and $g_m = \partial I_c/\partial v_b$. Note that these small-signal conductances cannot be measured directly, only extrinsic conductances with respect to terminal voltages can be measured. The extrinsic small-signal admittance representation is

$$(2) \quad \begin{bmatrix} \tilde{i}_b \\ \tilde{i}_c \end{bmatrix} = \begin{bmatrix} G_\pi & G_r \\ G_m & G_o \end{bmatrix} \begin{bmatrix} \tilde{V}_b \\ \tilde{V}_c \end{bmatrix}$$

where (because of the assumptions noted above) the intrinsic and extrinsic currents are the same, and conductances are calculated with respect to variations in terminal, not intrinsic, biases. Although there is no variation of I_b with v_c, it may depend on the extrinsic bias V_c because of resistive debiasing, so $G_r = \partial I_b/\partial V_c$ must be included in the analysis. The conductances in (2) can be directly calculated from DC data by numerical differencing. Specifically, from forward Gummel (FG) data at different V_{cb} (noting that in (2) $\tilde{V}_c = \tilde{V}_{ce}$).

It would at first appear that the small-signal intrinsic biases could be easily calculated from the extrinsic biases, by subtracting drops across the parasitic resistances (e.g. $\tilde{V}(bi) = \tilde{V}(b) - r_b \tilde{i}_b$); however, the bias dependence of r_b complicates this (which is one reason it has not been included in previous analyses). For the base resistance,

$$(3) \quad I_b = g_b(V(b) - V(bi)) = g_b(I_b)V_{rb}$$

therefore for small-signal excitation

$$(4) \quad \tilde{i}_b = g_b\tilde{V}_{rb} + d_b\tilde{i}_b, \quad d_b = \partial \log(g_b)/\partial \log(I_b).$$

The term that involves the derivative of g_b is not taken into account in previous analyses of BJT resistances, but is included here because it affects the small-signal characteristics of a device, and is included in circuit simulation models. The AC bias on the intrinsic base node is therefore

$$(5) \quad \tilde{V}(bi) = \tilde{V}_b - \tilde{V}_{rb} = \tilde{V}_b - r_b^{ac}\tilde{i}_b, \quad r_b^{ac} = r_b(1 - d_b)$$

which shows that by considering the base resistance to be base current (rather than voltage) dependent, the bias dependence enters the AC analysis as a correction to the DC base resistance.

The resistive debiasing is therefore

$$(6) \quad \begin{bmatrix} \tilde{v}_b \\ \tilde{v}_c \end{bmatrix} = \begin{bmatrix} \tilde{V}_b \\ \tilde{V}_c \end{bmatrix} - \begin{bmatrix} r_b^{ac} + r_e & r_e \\ r_e & r_c + r_e \end{bmatrix} \begin{bmatrix} \tilde{i}_b \\ \tilde{i}_c \end{bmatrix}.$$

This is the obvious direct resistive debiasing between extrinsic and intrinsic voltages, with a correction because of the bias dependence of the base resistance.

Substituting (6) into (1) and comparison to (2) gives

$$(7) \quad G_\pi = g_\pi \frac{1 + g_o(r_c + r_e)}{\Delta}$$

$$(8) \quad G_r = -\frac{g_o g_\pi r_e}{\Delta}$$

$$(9) \quad G_m = g_m \frac{1 - g_o g_\pi r_e/g_m}{\Delta}, \quad G_m - G_r = \frac{g_m}{\Delta}$$

$$(10) \quad G_o = g_o \frac{1 + g_\pi(r_b^{ac} + r_e)}{\Delta}$$

where

$$(11) \quad \begin{aligned} \Delta &= \begin{aligned} &1 + g_\pi(r_b^{ac} + r_e) + g_m r_e + g_o(r_c + r_e) \\ &+ g_o g_\pi(r_b^{ac} r_e + r_c r_b^{ac} + r_c r_e) \end{aligned} \\ &\approx 1 + g_\pi(r_b^{ac} + r_e) + g_m r_e + g_o r_c(1 + g_\pi(r_b^{ac} + r_e)) \end{aligned}$$

Clearly, the parasitic resistances cause the extrinsic and extrinsic conductances to differ. Fig. 2 shows measured extrinsic conductances, along with intrinsic conductances calculated from an I_b model.

Note that (7) through (11) include the effect of the intrinsic output conductance g_o and the collector resistance r_c. The $g_o r_c$ product is assumed small in the analyses below. A result similar to (9) was reported in [5]; however, it was assumed that $r_b^{ac} = r_b$; the significance of this inaccuracy will become apparent below.

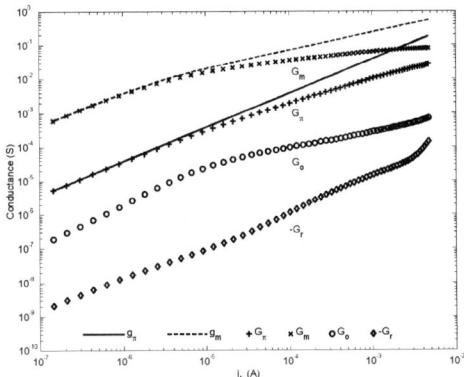

Fig. 2 Extrinsic and Intrinsic Conductances from Data

3. AUXILIARY INFORMATION

As in the Ning-Tang method it is assumed that the base current can be extrapolated from low bias values, so $I_b(v_b)$ is known (from a model). From this g_π follows directly, and from the FG data

$$(12) \quad g_m = \partial \beta I_b/\partial v_b = \beta g_\pi(1 + \partial \log(\beta)/\partial \log(I_b))$$

gives the intrinsic transconductance (where β is taken to be a function of I_b). Note that previously a separate I_c model was required to calculate g_m; (12) shows that this is not necessary.

The Ning-Tang method also gives

$$(13) \quad r_b + (1 + \beta)r_e = (V_b - v_b)/I_b$$

97

where $v_b(I_b)$ follows from the base current model extracted from low bias data.

So measured data gives the four extrinsic conductances in (2), V_b, I_c, I_b, and β, and a fitted model for $I_b(v_b)$ gives v_b and the intrinsic conductances g_π and g_m. Rather than considering certain quantities to be functions of v_b they are taken to be functions of I_b; this facilitates determination of these quantities (and their derivatives) from DC data.

g_o is not known from direct measurement or from analysis of the base current variation of device characteristics, and as it is a function of depletion pinching, base push-out and high-level injection, varies with both base and collector bias. The analyses below either cancel the effect of g_o, or assume its effect on the quantities analyzed (except for G_r/G_o) is small.

4. COMPARISON OF AC AND DC MODELS

If the DC and AC currents gains (β and g_m/g_π) are different, it would appear that combining the DC result (13) and AC conductance information would allow r_e and r_b to be determined. E. g., ignoring the bias dependence of r_b,

$$(14) \quad r_b + \left(1 + \frac{g_m}{g_\pi}\right)r_e \approx \frac{1}{G_\pi} - \frac{1}{g_\pi} \approx \frac{g_m}{g_\pi}\left(\frac{1}{G_m} - \frac{1}{g_m}\right)$$

which is clearly independent of (13) for $\beta \neq g_m/g_\pi$ and so combining (14) with (13) should enable r_e and r_b to be computed as they represent two independent equations.

However, r_b is bias dependent, and its small-signal AC value differs from its DC value. Assume that $r_b \propto \beta$, which is reasonable under the bias conditions where the effects of r_e and r_b are observable. Then

$$(15) \quad d_b = \left(1 - g_m/\beta g_\pi\right)$$

and therefore

$$(16) \quad r_b^{ac} = r_b g_m/\beta g_\pi < r_b$$

so substituting this in (14) gives

$$(17) \quad r_b + \left(\frac{\beta g_\pi}{g_m} + \beta\right)r_e \approx \frac{\beta g_\pi}{g_m}\left(\frac{1}{G_\pi} - \frac{1}{g_\pi}\right).$$

As $\beta g_\pi/g_m \sim 1 \ll \beta$ it is apparent that (17), derived from the difference between extrinsic and intrinsic conductances, is essentially equivalent to the DC result (13). It is not initially obvious that this should be the case.

The conclusion is clear: although it appears that r_e and r_b (and its bias dependence) could be determined from a combination of DC and AC conductance data, the obvious approach of looking at the degradation of g_π or g_m does not work. Further, previous approaches using AC data, for example impedance circle analysis, are inaccurate because the base resistance they determine is actually r_b^{ac}, and they do not account for the fact that this is different from the DC base resistance r_b.

5. RESISTANCE CALCULATION

Combining extrinsic conductance expressions gives

$$(18) \quad -\frac{G_r}{G_o G_\pi} = r_e \frac{\Delta}{\left(1 + g_o(r_c + r_e)\right)\left(1 + g_\pi(r_b^{ac} + r_e)\right)} \geq r_e$$

or using g_π in place of G_π gives

$$(19) \quad -\frac{G_r}{G_o g_\pi} = r_e \frac{1}{\left(1 + g_\pi(r_b^{ac} + r_e)\right)} \leq r_e$$

(there is a cancellation of g_o in forming (18) and (19); for these to be reasonable the device must have a finite output resistance). These can provide (unfortunately, fairly wide) upper and lower bounds for r_e.

From (7) and (19) (dropping terms in g_o and r_c)

$$(20) \quad r_b^{ac} + \left(1 + \frac{g_m}{g_\pi}\right)r_e = \frac{1}{G_\pi} - \frac{1}{g_\pi}$$

$$(21) \quad r_b^{ac} + \left(1 + \frac{G_o}{G_r}\right)r_e = -\frac{1}{g_\pi}$$

and combining these, and introducing (12), gives

$$(22) \quad r_e = \frac{1/G_\pi}{\beta\left(1 + \partial \log(\beta)/\partial \log(I_b)\right) - G_o/G_r}.$$

The AC base resistance is then

$$(23) \quad r_b^{ac} = -\frac{1}{g_\pi} - \left(1 + \frac{G_o}{G_r}\right)r_e,$$

and the DC base resistance follows by substituting (22) in (13). Note that although an $I_b(v_b)$ model has been assumed in some of the analyses, the expression (22) involves only measured data.

Certain assumptions underlie the derivation of (22). First, the g_o terms in the numerator of the right hand side of (7) and in the expression (11) for r_e are taken to be small, and so are dropped. Second, g_o is taken to be sufficiently large to make determination of G_o and G_r from measured data reliable. These are not overly stringent requirements in practice.

Fig. 3 and Fig. 4 show results from the above analysis. The data are from an NPN BJT in a 0.25μm SmartPower silicon BiCMOS process. Some bias dependence is apparent for r_e (this was also the case in [5]). At low I_b the effects of resistive de-biasing are small so the precision of the extraction is reduced compared to high bias. At high bias some variation of r_e is apparent; this has not been investigated in detail. In the "flat" portion of the curve, the value of r_e from (22) agrees well with the 5.02Ω determined from the open collector method. The difference between DC and AC base resistance in Fig. 4 is clear, and there is good agreement between r_b^{ac} from (23) and $r_b(1 - d_b)$ from (13).

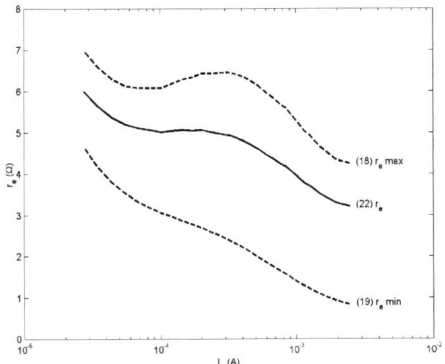

Fig. 3 Emitter Resistance from FG Data

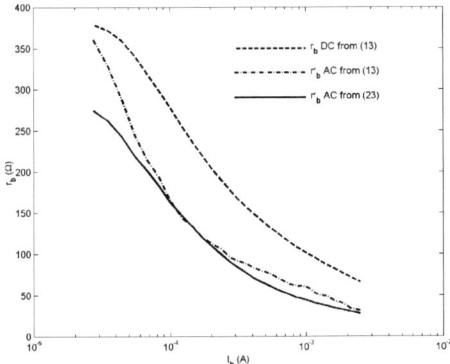

Fig. 4 DC and AC Base Resistance from FG Data

6. ADDITIONAL COMMENTS

Most models assume that r_e is independent of bias. If this is not so, the bias dependence of r_e can be included in the analysis in a similar manner to that used for r_b. If

$$(24) \quad I_e = g_e(I_e)V(ei) = g_e V_{re}$$

then under small signal-excitation

$$(25) \quad \tilde{i}_e = g_e \tilde{V}_{re} + d_e \tilde{i}_e, \quad d_e = \partial \log(g_e)/\partial \log(I_e).$$

This can be accounted for in (6) by substituting $r_e^{ac} = (1 - d_e)r_e$ for r_e (assuming it can be modeled as a function of I_e).

Although g_o is not used explicitly here (apart from its contribution to Δ and G_π being assumed to be negligible), it can be calculated from (8) or (10). This would allow evaluation of the accuracy of some of the assumptions that underlie the procedures defined above (although r_c needs to be known to completely evaluate terms that are omitted from the analyses).

7. CONCLUSIONS

This paper presents analyses of the difference between intrinsic and extrinsic BJT small-signal conductances, for the purpose of determining base and emitter resistance.

That the difference between the intrinsic and extrinsic transconductance can be used to calculate r_e has been noted previously [5], however the analyses provided here are based on a more-accurate small-signal model.

Importantly, the bias dependence of r_b is taken into account in the analysis of the small-signal conductances; this is done by introducing the effective AC base resistance $r_b^{ac} = r_b(1 - \partial \log(1/r_b)/\partial \log(I_b))$. This has a significant effect on AC modeling; previous small-signal analyses for BJT resistance extraction have omitted this component. The simplicity of the analyses presented is in part due to considering I_b to be the quantity other parameters depend on.

When the proper AC base resistance is included in small-signal analysis, it is shown that this gives essentially the same results as a DC analysis. *A priori* it would seem that as long as $\beta \neq g_m/g_\pi$ there would be additional information in the conductance data; the analysis shows this is incorrect.

A simple estimate of r_e is presented, based on FG data and not biasing schemes completely different from those used in real circuits. This expression (22) for r_e is based purely on measured data, it does not require fitting of an $I_b(v_b)$ model.

ACKNOWLEDGEMENTS

I would like to thank Tamara Bettinger and Zhong Lu for the measurement data and discussions on this work.

REFERENCES

[1] T. H. Ning and D. D. Tang, "Method for Determining the Emitter and Base Series Resistances of Bipolar Transistors," *IEEE Trans. Electron Dev.*, vol. ED-31, no. 4, pp. 409-412, Apr. 1984.

[2] L. J. Giacoletto, "Measurement of Emitter and Collector Series Resistances," *IEEE Trans. Electron Devices*, vol. ED-19, pp. 692-693, 1972.

[3] R. Gabl, M. Reisch, and M. Pohl, "Improved Extraction Method for the Emitter Resistance of Bipolar Transistors," *Proc. IEEE BCTM*, pp. 211-214, 1998.

[4] W. M. C. Sansen and R. G. Meyer, "Characterization and Measurement of the Base and Emitter Resistances of Bipolar Transistors," *IEEE J. Solid-State Circuits*, vol. SC-7, no. 6, pp. 492-498, Dec. 1972.

[5] W. J. Kloosterman, J. C. J. Paasschens, and D. B. M. Klaassen, "Improved Extraction of Base and Emitter Resistance from Small Signal High Frequency Admittance Measurements," *Proc. IEEE BCTM*, pp. 93-96, 1999.

[6] M. Linder, F. Ingvarson, K. O. Jeppson, J. V. Grahn, S.-L. Zhang, and M. Ostling, "Extraction of Emitter and Base Series Resistances of Bipolar Transistors from a Single DC Measurement," *IEEE Trans. Sem. Manufacturing*, vol. 13, no. 2, pp. 119-126, May 2000.

[7] W. Z. Cai, G. H. Loechelt, and S. Shastri, "A New Method for Extracting Base Resistance in Bipolar Transistors," *IEEE Trans. Electron Dev.*, vol. 52, no. 4, pp. 626-629, Apr. 2005.

2006 Bipolar/BoCMOS Circuits and Technology Meeting Proceedings

Simultaneous Extraction of Collector and Substrate Series Resistance by Simple DC Measurement

Tianbing Chen, Tracey L. Krakowski, Andy Strachan, Yun Liu, Alexei Sadovnikov, and Jeff Babcock

National Semiconductor Corporation
2900 Semiconductor Drive, Santa Clara, CA 95052-8090, USA
Tel: (408) 721-5671 / Fax: (408) 721-5100 / E-mail: tianbing.chen@nsc.com

ABSTRACT—**A new forced emitter current method is proposed for the simultaneous measurement of collector and substrate series resistance in bipolar transistors. Compared with conventional series resistance extraction method, this new method does not need any prior knowledge of certain device parameters, or any pre-selected bias condition. It can be used for any bulk bipolar technology.**

I. INTRODUCTION

Parasitic series resistances are important modeling parameters for bipolar transistors. For example, the collector series resistance affects not only the static parameters such as the DC current gain at high currents, but also limits the frequency response of the transistors. Accurate extraction of substrate resistance is critical for modeling the impact of substrate coupling on the frequency dependent output conductance of bipolar transistors [1]. The extrinsic collector series resistance is typically modeled as a lumped resistance R_{CX}, which includes both the vertical portion from collector contact to the n^+ buried layer and the lateral part of resistance of the n^+ buried layer. It can be measured through transistor's DC characteristics [2], high-frequency Z-parameters [3], or a special test structure [4]. The DC measurement is generally the preferred method for collector resistance extraction, since the high-frequency measurements are time-consuming and special test structures incur design cost.

Two main methods have been proposed for the extraction of collector resistance, the force-beta technique [2], [5] and the substrate current method [6], [7]. However, the major problem with the force-β method is that the substrate current is controlled by $V_{BC} - I_C R_{CX}$ [6] rather than by V_{BC} [6], where V_{BC} is the base-collector voltage and $I_C R_{CX}$ is the collector series resistance induced voltage drop. The optimized force-β technique requires pre-determination of β and R_E [5], and the substrate current method needs to pre-select a bias condition [6], [7]. A new optimized substrate current method is proposed in this work, which provides an efficient way of R_{CX} extraction. This new method can be incorporated easily into an automated test program, and is addressed in detail in the next section.

II. PROPOSED METHOD

The simplified equivalent circuit schematic of an n-p-n transistor including all series resistances is shown in Figure 1.

When an n-p-n transistor is biased in a high current condition, the collector current I_C passing through R_{CX} can forward-bias the bottom substrate-collector junction, thus turning on the parasitic p-n-p transistor. The collector resistance can be measured

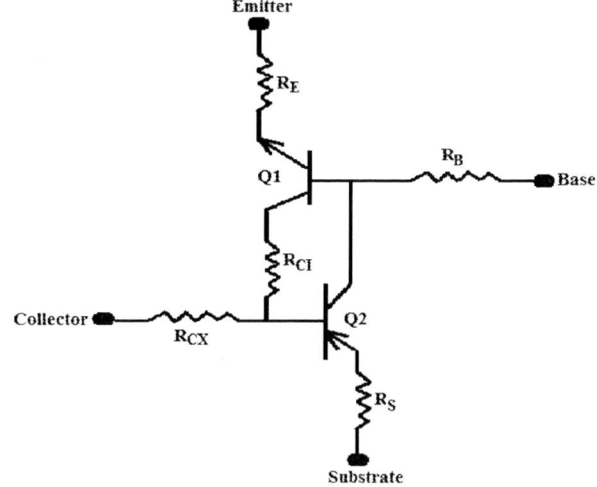

Fig. 1. Equivalent circuit schematic of an n-p-n transistor. Q1 is the main n-p-n transistor and Q2 is the parasitic p-n-p transistor.

from the bias dependence of substrate current, where the internal collector is inaccessible and the internal V_C has to be measured indirectly. In the VBIC manual [8], substrate current I_S is recorded during the forward Gummel measurement. By applying simple p-n junction theory, R_{CX} can be calculated as

$$R_{CX} = \frac{V_T ln\left(\frac{I_{S1}}{I_{S2}}\right)}{I_{C1} - I_{C2}} \quad (1)$$

where V_T is the thermal voltage, I_{S1}, I_{S2}, I_{C1}, and I_{C2} are the substrate and collector currents at two different bias points, respectively. Observe that I_S is an exponential function of the forward bias of collector-substrate junction, which changes linearly with $I_C R_{CX}$. During the Gummel-Poon measurement, I_C varies exponentially when V_{BE} is swept linearly. This double exponential function makes I_S increase enormously with V_{BE} and saturate in a very narrow range of V_{BE}. A very fine sweeping of V_{BE} in pre-selected range is preferred for the accurate calculation of R_{CX}. It is, therefore, desirable to use a forced current method such that we have better control of the forward bias of collector-substrate junction. When the I_E of an n-p-n transistor is swept with the collector, base, and substrate at constant bias, R_{CX} can be calculated from I_C and I_S. By assuming that R_{CX} is independent of I_C due to the heavily doped buried layer, the effective

1-4244-0458-4/06/$20.00 ©2006 IEEE   100

forward bias voltage can be calculated as

$$V_{bias} = V_S - I_S R_S + I_C R_{CX} - V_C \qquad (2)$$

Note that R_S is fairly large for many modern bipolar process, and the second term on the right of Equation 2 may not be negligible even for relatively small I_S. Then I_S can be expressed as

$$I_S = I_{SS} \exp \left(\frac{V_S - V_C + I_C R_{CX} - I_S R_S}{nVT} \right) \qquad (3)$$

where I_{SS} is the saturation current of collector substrate junction, and the ideal factor n is assumed to be its default value 1. With $I_{SX} = I_{SS} \exp \left(\frac{V_S - V_C}{nVT} \right)$, Equation 3 can be rearranged as

$$I_C = \frac{V_T}{R_{CX}} \ln \left(\frac{I_S}{I_{SX}} \right) + I_S \frac{R_S}{R_{CX}} \qquad (4)$$

Taking the first-order derivative of Equation 4 we obtain

$$\frac{\partial I_C}{\partial I_S} = \frac{V_T}{R_{CX}} \frac{1}{I_S} + \frac{R_S}{R_{CX}} \qquad (5)$$

It can be seen from Equation 5 that by plotting the first derivative of I_C over I_S versus $1/I_S$, a straight line, is obtained. First R_{CX} is calculated from the slope of the line, then R_S can be calculated from the intercept.

III. DEVICE TECHNOLOGY

The n-p-n bipolar transistors used in this investigation are from a high-performance 24 V Power BiCMOS/DMOS process [9]. The emitter widths for the transistors used in this work are 0.5 μm and 2 μm, and emitter lengths vary from 2 μm to 64 μm. Fig. 2 shows the typical cross section of a vertical n-p-n transistor.

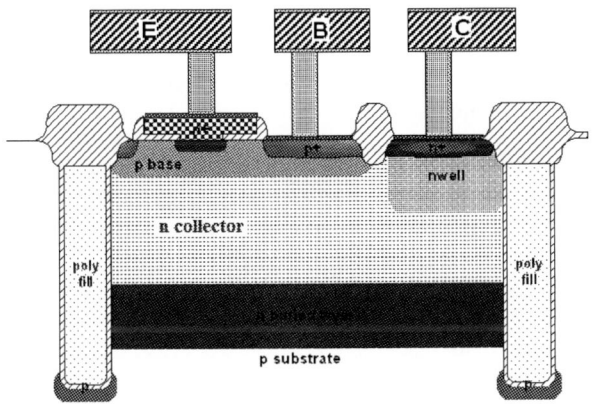

Fig. 2. Cross-sectional schematic of an n-p-n bipolar transistor.

In this BiCMOS/DMOS process, the CMOS n- and p-well implants are reused to contact the n^+ and p^+ buried layers. A deep poly filled trench with a drawn width of 1.0 μm is used to isolate individual devices. The high performance n-p-n transistor features 30V BV_{CEO}, 1.8 GHz peak f_T, and 30,000 βV_A product.

IV. RESULTS AND DISCUSSION

Firstly, a forward Gummel measurement was performed on an n-p-n transistor with an emitter width of 0.5 μm and 64 μm

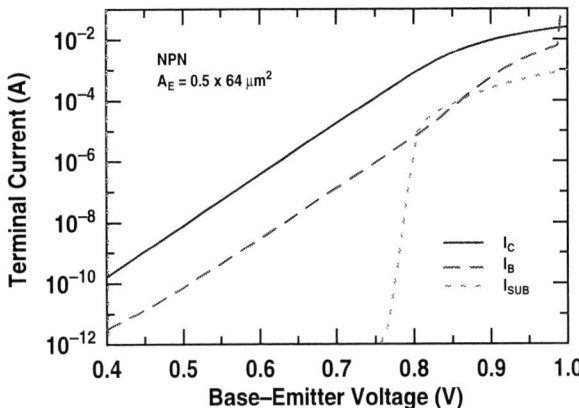

Fig. 3. Forward Gummel characteristics of a n-p-n transistor. Collector, base, and substrate were grounded during the measurement.

emitter length. The collector, base, and substrate were grounded during the measurement, and V_E was swept from 0 to -1 V. The measurement results are plotted in Figure 3.

It can be seen from Figure 3 that as V_{BE} increases to 0.75 V, $I_C R_{CX}$ product (the forward voltage drop across collector-substrate junction) starts to turn on this pn junction and I_S increases exponentially with I_C, which increases exponentially with V_{BE}. Thus, within less than 50 mV increase of V_{BE}, I_S increases sharply by more than seven orders of magnitude and saturates at about 10^{-4}A due to large R_S induced voltage drop. According to Equation 1, a careful examination of the data for V_{BE} between 0.75 and 0.80 V gives the R_{CX} value of 674 Ω.

The forced I_E measurement was then performed on the same transistor. The collector, base, and substrate were grounded during the I_E sweep. The I_C and I_S as a function of I_E are plotted in Figure 4.

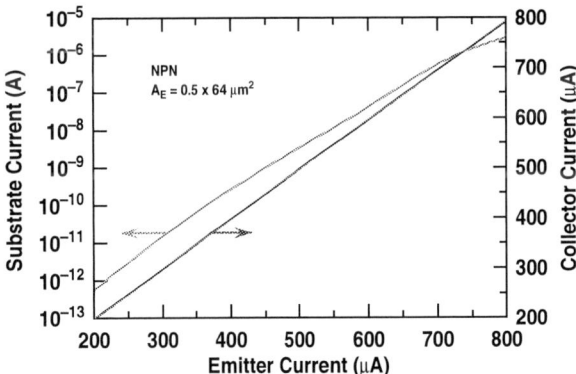

Fig. 4. I_C and I_S versus I_E for a n-p-n transistor during the forced I_E sweeping.

Since the n-p-n transistor was biased in the forward-active region, I_B is fairly small when compared with I_C and I_E. Figure 4 confirms that I_C increases linearly with the increasing I_E, thus the forward bias on the collector-substrate junction is being linearly swept during this forced I_E measurement. As a consequence, an exponential increase of I_S is observed in Figure 4, similar to I_C and I_B in a typical Gummel plot when V_{BE} is linearly swept. The first-order derivative of I_C over I_S versus $1/I_S$ during the forced I_E measurement is plotted in Figure 5.

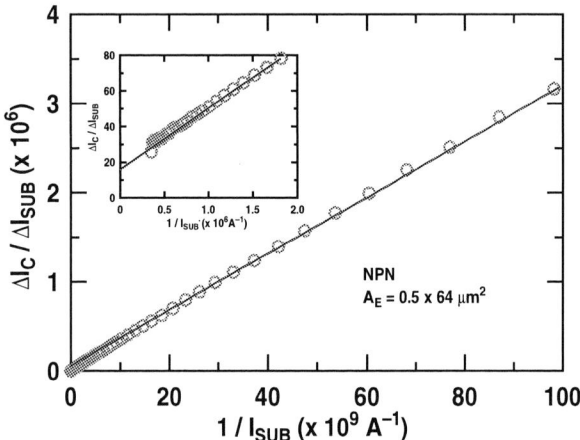

Fig. 5. The first-order derivative of I_C over I_S versus $1/I_S$ for a n-p-n transistor during forced I_E sweeping.

It can be seen from Figure 5 that $\partial I_C / \partial I_S$ is a linear function of $1/I_S$. According to Equation 5, R_{CX} can be calculated from the slope of the line, which is extracted by linear fitting. The calculated R_{CX} is 695 Ω, in excellent agreement with the R_{CX} value extracted from forward Gummel characteristics. The inset is the zoom-in plot of $\partial I_C / \partial I_S$ at large I_S, and the linear fitting gives an intercept value of about 16. R_S is calculated as 10 KΩ based on Equation 5. R_{CX} can also be calculated if we take another derivative over $1/I_S$ from Equation 5

$$\frac{\partial \left(\frac{\partial I_C}{\partial I_S} \right)}{\partial (1/I_S)} = \frac{V_T}{R_{CX}} \qquad (6)$$

This calculation can be easily included in measurement software such as ICCAP, and Figure 6 shows the ICCAP output of R_{CX} from the forced I_E measurement.

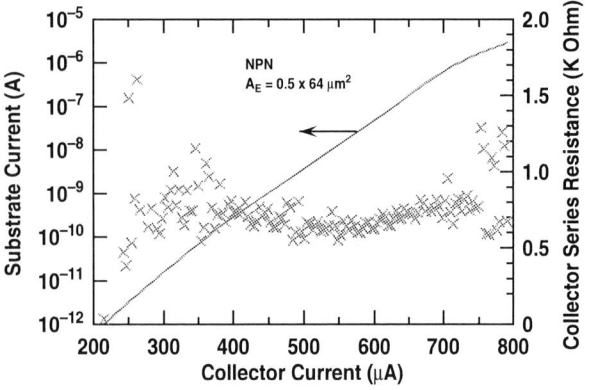

Fig. 6. I_C and I_S versus I_E for a n-p-n transistor during forced I_E sweeping.

The median of the R_{CX} values extracted based on Equation 6

is 700 Ω, consistent with both the R_{CX} calculated from I_S in Figure 3 and that from linear extraction in Figure 5. Furthermore, this allows an alternative way to check the bias-dependence of R_{CX}. It can be seen from Figure 2 that the collector series resistance is dominated by the resistance in the collector epilayer since both nwell and n^+ buried layer are much more heavily doped than collector epi layer. By checking the layout design rules, we know that R_{CX} is inversely proportional to the collector length. R_{CX} for n-p-n transistors with various device geometries were measured, and their normalized values were plotted in Figure 7.

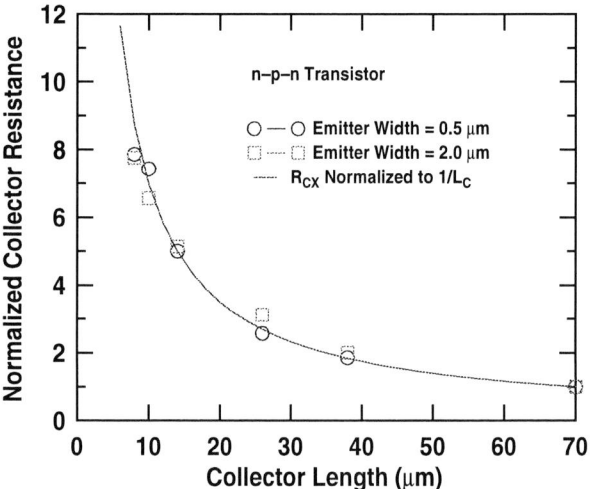

Fig. 7. Normalized R_{CX} for n-p-n transistors with various emitter widths and emitter lengths.

It can be seen from Figure 7 that the normalized R_{CX} for n-p-n transistors with two different emitter widths and seven different emitter lengths align extremely well with $1/L_C$ line as predicted by the layout analysis. A similar geometry dependence of R_{CX} has been observed elsewhere [4]. The lack of an efficient measurement method to extract R_{CX} has been an limiting factor for bipolar lateral scaling [4]. The proposed forced I_E method can potentially be used to fit that need. The measured R_S for n-p-n transistors with multiple geometries were plotted in Figure 8.

The measured substrate resistance is actually the series resistance for a two-dimensional distributed structure. For a collector-substrate junction with collector width of W, we assume that the I_C during the forced I_E measurement is uniformly distributed in the n^+ buried layer. Consider the extreme case that the forward bias on collector-substrate junction increases from 0 at X=0 to $I_C R_{CX}$ at X=W. If the forward bias voltage is approximated as a linear function of X

$$V_{bias} = I_C R_{CX}(X/W) \qquad (7)$$

By proper consideration of the distribution effect of the bias voltage, Equation 3 can be rearranged as

$$I_S \exp \left(\frac{I_S R_S}{n V_T} \right) = \frac{1}{W} \int_0^W I_{SX} \exp \left\{ I_C R_{CX} \frac{X}{W} \right\} dx \qquad (8)$$

102

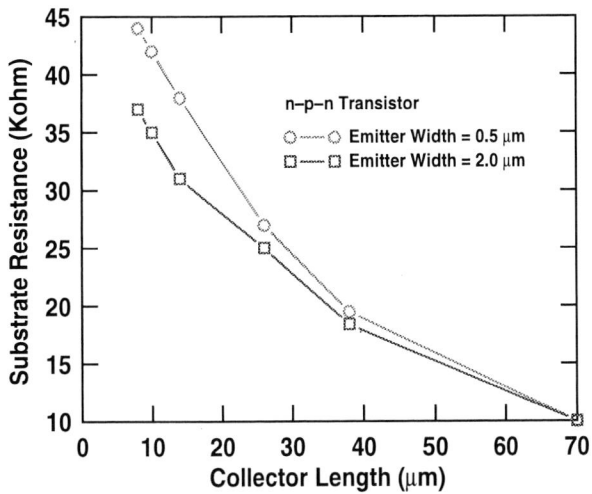

Fig. 8. Measured R_S for n-p-n transistors with various emitter widths and emitter lengths.

By solving the integration Equation 8 becomes

$$I_S \exp\left(\frac{I_S R_S}{nV_T}\right) = I_{SX}\frac{I_C R_{CX}}{nV_T}\exp\left(\frac{I_C R_{CX}}{nV_T}\right) \quad (9)$$

Taking the first derivative over I_S from Equation 9 we obtain

$$\left(1+\frac{nV_T}{I_C R_{CX}}\right)\frac{\partial I_C}{\partial I_S} = \frac{nV_T}{R_{CX}}\frac{1}{I_S}+\frac{R_S}{R_{CX}} \quad (10)$$

Note that $I_C R_{CX}$ needs to be more than 0.5 V in order to turn on the collector-substrate junction, while V_T is about 26 mV at room temperature. Neglecting the two-dimensional effect in the calculation based on Equation 5 will cause only about 5% error at room temperature when compared with Equation 10. From a physical point of view, the junction current is exponentially related to the bias voltage, the current will be lumped across the points that are the most forward biased. Neglecting distributed effects for the R_S calculation in this work did not cause any significant error.

V. SUMMARY

We present a new forced current method for the simultaneous extraction of both collector and substrate resistance. This optimized substrate current method allows better control of the forward bias of collector-substrate junction, and provides an easy yet efficient way to extract R_{CX} and R_S. It can be easily incorporated into automated test, and fit the need for accurate R_{CX} extraction for bipolar lateral scaling models. The devices used in this work are from a power technology, but the method can be extended in general to any bulk bipolar technology.

VI. ACKNOWLEDGMENT

The authors would like to thank all the members of Advanced Process Technology Development in Santa Clara, CA, especially Yen Nguyen and Kevin Li for their technical support. The continuous support of Linda Smith and Dr. Reda Razouk is gratefully acknowledged.

REFERENCES

[1] T-Y. Lee et al., Proc. IEEE BCTM, p. 101-104, 1999.
[2] J. Choma et al., IEEE JSSC, p. 318-322, 1976.
[3] S. Lee et al., IEEE Trans. Elec. Dev., 41, p. 233-238, 1994.
[4] N. Kauffmann et al., 5th European HICUM Workshop., 2005.
[5] H-C. Wu et al., Proc. SAFE., p. 25-26, 2004.
[6] W.D. Mack, et al., IEEE JSSC, 17, p. 767-773, 1982.
[7] J-S. Park et al., IEEE Trans. Elec. Dev., p. 365-372, 1991.
[8] VBIC manual, Chapter 6, 2004.
[9] A. Strachan et al., IEEE BCTM, p. 41-44, 2002.

2006 Bipolar/BiCMOS Circuits and Technology Meeting Proceedings

Joint Extraction of the Base and Collector Resistances with the Base-Collector Capacitance Split of HBT/BJT Transistors

Z. Huszka, E. Seebacher and W. Pflanzl

austriamicrosystems AG, Schloss Premstätten A8141 Unterpremstätten, Austria

e_mail: zoltan.huszka@axelero.hu

Abstract — A novel approach is described for the simultaneous extraction of the parasitic base and collector resistances along with the internal collector capacitance *cjc* of HBT/BJT devices. A strong impact of the external access resistances *rbx* and *rcx* on the accuracy of *cjc* extraction is revealed. The interaction is strictly observed at the determination of these and the internal base and collector resistance parameters.

Index Terms — Bipolar transistor modeling, base resistance, collector resistance, collector capacitance split.

I. INTRODUCTION

The well known equivalent network elements *re, rbx, rbi, rci, rcx* and *cjc* are all shared by Hicum, Mextram and VBIC which are the three most powerful bipolar transistor models today. Though the physical meaning of these components is very clear it is by far not straightforward to extract them in a unique way from electrical measurements. It only contributes to the difficulties that *rbi, rci* and *cjc* are bias dependent hence it is necessary to obtain their values at several external voltage/current forces. Then additional model parameters and physical equations are requested to arrive at their 'zero-bias' values what are included in the model cards. That is why the overall procedure is quite complex. Special test structures have been suggested for finding e.g. *rcx*, and *rbi* [1]-[2], or 2D and 3D simulations were used for the same [3]-[4]. A detailed summary of the DC methods for extracting parasitic resistances can be found in [5]. Extraction of *cjc* from high frequency cold and active *s*-parameter measurements were suggested in [6] and [7] respectively. Active mode *s*-parameters are used in combination with the solution of a quadratic equation in [8] for the determination of the base-collector capacitance splitting.

In former methods the referred components were extracted individually disregarding important physical interdependence or correlation with other parameters. Recently an explicit effect of *cjc* and *rci* was revealed in [9] on the extraction of *rbi*. Additional correlations will be shown in this paper with reference to the external access resistances and a new approach will be presented handling and utilizing these dependencies in parameter extraction.

II. FUNDAMENTAL EQUATIONS

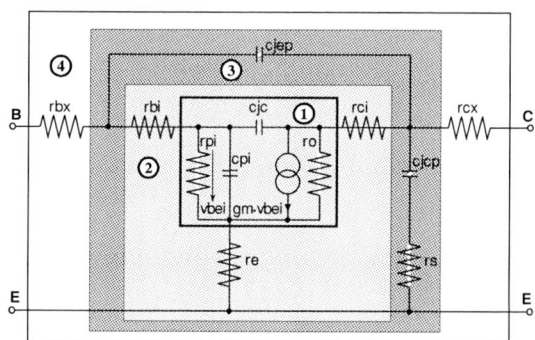

Fig. 1. VBIC equivalent in common emitter (CE) rotation

The intrinsic transistor block#1 on Fig. 1 can be described by its common emitter (CE) or common base (CB) admittance matrices

$$\mathbf{Y}_e^{(1)} = \begin{bmatrix} gpi + j\omega(cpi + cjc) & -j\omega \cdot cjc \\ gm - j\omega \cdot cjc & go + j\omega \cdot cjc \end{bmatrix} \quad (1)$$

$$\mathbf{Y}_b^{(1)} = \begin{bmatrix} gm + go + gpi + j\omega \cdot cpi & -go \\ -gm - go & go + j\omega \cdot cjc \end{bmatrix} \quad (2)$$

Denote the 2-port parameter determinants of the kth and $(k\text{-}1)^{th}$ blocks by $\Delta \mathbf{Z}^{(k)}$ and $\Delta \mathbf{Y}^{(k-1)}$. A series embedding by an impedance 2-port of generic T elements (Z_1, Z_T, Z_2) transforms the y parameters as

$$y_{ii}^{(k)} = \frac{y_{ii}^{(k-1)} + \Delta \mathbf{Y}^{(k-1)} \cdot (Z_j + Z_T)}{\delta^{(k)}};$$

$$y_{ij}^{(k)} = \frac{y_{ij}^{(k-1)} - \Delta \mathbf{Y}^{(k-1)} \cdot Z_T}{\delta^{(k)}}; \quad \delta^{(k)} = \Delta \mathbf{Z}^{(k)} \cdot \Delta \mathbf{Y}^{(k-1)} \quad (3)$$

Parallel embedding by Π admittances (Y_1, Y_T, Y_2):

$$y_{ii}^{(k)} = y_{ii}^{(k-1)} + Y_i + Y_T; \quad y_{ij}^{(k)} = y_{ij}^{(k-1)} - Y_T \quad (4)$$

The T or Π transfer immittance (of index T) cancels out of the "unilateralized" (*UL*) parameters below

$$\tilde{y}_{ii} = y_{ii} + y_{12}; \quad \tilde{y}_{21} = y_{21} - y_{12}; \quad \tilde{h}_{11} = 1/\tilde{y}_{11}$$

$$\tilde{z}_{ii} = z_{ii} - z_{12}; \quad \tilde{z}_{21} = z_{21} - z_{12}; \quad \tilde{h}_{21} = \tilde{y}_{21}/\tilde{y}_{11} \quad (5)$$

1-4244-0458-4/06/$20.00 ©2006 IEEE

$$\tilde{y}_{ii}^{(k)} = \frac{\tilde{y}_{ii}^{(k-1)} + \Delta \mathbf{Y}^{(k-1)} \cdot Z_j}{\delta^{(k)}}; \quad \tilde{y}_{21}^{(k)} = \frac{\tilde{y}_{21}^{(k-1)}}{\delta^{(k)}} \qquad (6)$$

$$\tilde{y}_{ii}^{(k)} = \tilde{y}_{ii}^{(k-1)} + Y_i; \quad \tilde{y}_{21}^{(k)} = \tilde{y}_{21}^{(k-1)} \qquad (7)$$

With $p>q> \dots >r$ series embedding steps in block k the following "gm-transparency" condition results

$$\tilde{y}_{21}^{(k)} = \frac{\tilde{y}_{21}^{(1)}}{\delta^{(p)} \cdot \delta^{(q)} \cdot \dots \cdot \delta^{(r)}}; \quad k \geq p \qquad (8)$$

(6) and (7) can be re-formulated to either (9) or (10)

$$\frac{\tilde{y}_{ii}^{(k)}}{\tilde{y}_{21}^{(k)}} = \frac{\tilde{y}_{ii}^{(k-1)} + \Delta \mathbf{Y}^{(k-1)} \cdot Z_j}{\tilde{y}_{21}^{(k-1)}}$$
$$\frac{y_{12}^{(k)}}{\tilde{y}_{21}^{(k)}} = \frac{y_{12}^{(k-1)} - \Delta \mathbf{Y}^{(k-1)} \cdot Z_T}{\tilde{y}_{21}^{(k-1)}} \qquad (9)$$

$$\frac{\tilde{y}_{ii}^{(k)}}{\tilde{y}_{21}^{(k)}} = \frac{\tilde{y}_{ii}^{(k-1)} + \Delta \mathbf{Y}^{(k-1)} \cdot Z_j}{\tilde{y}_{21}^{(1)}} \cdot \frac{\delta^{(p)} \cdot \delta^{(q)} \cdot \dots \cdot \delta^{(r)}}{\delta^{(k)}}$$
$$\frac{y_{12}^{(k)}}{\tilde{y}_{21}^{(k)}} = \frac{y_{12}^{(k-1)} - \Delta \mathbf{Y}^{(k-1)} \cdot Z_T}{\tilde{y}_{21}^{(1)}} \cdot \frac{\delta^{(p)} \cdot \delta^{(q)} \cdot \dots \cdot \delta^{(r)}}{\delta^{(k)}} \qquad (10)$$

For the model of Fig. 1 with $go = y_{12b}^{(1)} \approx 0$ one has

$$\Delta \mathbf{Y}^{(1)} = y_{11b}^{(1)} \cdot y_{22}^{(1)}; \quad \tilde{y}_{21e}^{(1)} = gm; \quad \tilde{y}_{21b}^{(1)} = -gm;$$
$$\delta^{(2)} = 1 + re \cdot y_{11b}^{(1)} + rbi \cdot y_{11e}^{(1)} + rci \cdot y_{22}^{(1)} + \Delta r \cdot \Delta \mathbf{Y}^{(1)}$$
$$\Delta r = re \cdot rbi + re \cdot rci + rci \cdot rbi \qquad (11)$$
$$\frac{1}{\delta^{(4)}} = 1 - rbx \cdot y_{11e}^{(4)} - rcx \cdot y_{22e}^{(4)} + rbx \cdot rcx \cdot \Delta \mathbf{Y}^{(4)}$$

The following results were derived in [9]

$$rbce = rbi + re + cf \cdot (rbi + rci) = \frac{\Im(\tilde{h}_{11}^{(3)} conj(\tilde{h}_{21}^{(3)}))}{-\Im(\tilde{h}_{21}^{(3)})} \qquad (12)$$

$$regm = re + \frac{1}{gm} = \frac{\Im(\tilde{h}_{11e}^{(3)})}{\Im(\tilde{h}_{21e}^{(3)})} \qquad (13)$$

$$cf = \frac{\tilde{\omega}_T^{(3)} \cdot cjc}{gm} = \tilde{\omega}_T^{(3)} \cdot cjc \cdot (regm - re) \qquad (14)$$

Missing e, b subscripts mean CE-CB invariance.

Existing methods were advised in [9] for the determination of cjc and for obtaining rbx and rcx to generate block#3 data. In the frame of a complete extraction scheme for all elements of the equivalent on Fig. 1 here, an improved rcx extraction procedure is presented first. Extraction of rbx will be based on the invariance of the SPICE parameter $xcjc$ from the cold to the active state until V_{cbi} is kept constant.

III. EXTRACTION OF *RCX*

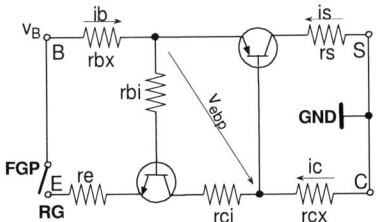

Fig.2 Measurement scheme for rcx extraction

The forward Gummel plot (*FGP*) of the parasitic *pnp* and the reverse Gummel plot (*RG*) of the main *npn* transistor are measured in the open and closed positions of the switch of Fig. 2 respectively.

Assume that the currents of Fig. 2 have been cascaded in vectors **ib**, **ic** and **is** including the merged bias set of the *FPG* and *RG* configurations. The emitter-base voltage of the *pnp* can be expressed by the VBIC model parameters *NFP*, *ISP* and *IKP* as

$$\mathbf{v}_{ebp} = NFP \cdot V_T \cdot \ln \left[-\frac{\mathbf{is}}{ISP} \left(1 - \frac{\mathbf{is}}{IKP} \right) \right] \qquad (15)$$

Introducing

$$\mathbf{x} = \begin{bmatrix} rbx \\ rcx \end{bmatrix} \quad \text{and} \quad \mathbf{A} = [\mathbf{ib}, -\mathbf{ic}] \qquad (16)$$

the total voltage drop across rbx and rcx reads as

$$\mathbf{v}_{bc} - \mathbf{v}_{ebp} = \mathbf{A} \cdot \mathbf{x} \qquad (17)$$

The best fit vector **x** in the least square sense

$$\mathbf{x} = \mathbf{P} \cdot (\mathbf{v}_{bc} - \mathbf{v}_{ebp}); \quad \mathbf{P} = (\mathbf{A}^T \mathbf{A})^{-1} \cdot \mathbf{A}^T \qquad (18)$$

is producing a simulated base-collector voltage

$$\mathbf{u}_{bc} = \mathbf{A} \cdot \mathbf{x} + \mathbf{v}_{ebp} = (\mathbf{A} \cdot \mathbf{P}) \cdot (\mathbf{v}_{bc} - \mathbf{v}_{ebp}) + \mathbf{v}_{ebp} \qquad (19)$$

with an error of

$$\mathbf{\Delta} = \mathbf{v}_{bc} - \mathbf{u}_{bc} = (\mathbf{E} - \mathbf{A} \cdot \mathbf{P}) \cdot (\mathbf{v}_{bc} - \mathbf{v}_{ebp}) \qquad (20)$$

to the physically applied base-collector bias. The dimension of unit matrix **E** is the number of bias points. The matrix derived from the measured **ib**, **ic**

$$\mathbf{Q} = (\mathbf{E} - \mathbf{A} \cdot \mathbf{P})^T \cdot (\mathbf{E} - \mathbf{A} \cdot \mathbf{P}) \qquad (21)$$

provides the total error square

$$S(NFP, ISP, IKP) = (\mathbf{v}_{bc} - \mathbf{v}_{ebp})^T \cdot \mathbf{Q} \cdot (\mathbf{v}_{bc} - \mathbf{v}_{ebp}) \quad (22)$$

S can be minimized as a function of the model parameters in (15). Generally *NFP=1* can be fixed at Si based BJT/HBTs.

Error form (22) allows a one-piece optimization over the complete bias range. The addition of *RG* to the conventionally used single *FGP* data [10] is greatly increasing robustness. Even so rbx is weakly defined by this and other [12] DC methods.

IV. THE COLLECTOR TIME-CONSTANT

Applying the 1st equation of (10) to network#2 then considering (7) between network#3 and #2

$$\frac{\tilde{y}_{11b}^{(1)} + \Delta \mathbf{Y}^{(1)} \cdot rci}{\tilde{y}_{21b}^{(1)}} = \frac{\tilde{y}_{11b}^{(2)}}{\tilde{y}_{21b}^{(2)}} = \frac{\tilde{y}_{11b}^{(3)}}{\tilde{y}_{21b}^{(3)}} = \frac{1}{\tilde{h}_{21b}^{(3)}} \qquad (23)$$

With $\Delta \mathbf{Y}^{(1)}$ from (11)

$$\frac{1}{\tilde{h}_{21b}^{(3)}} = \frac{1}{h_{21b}^{(1)}} (1 + j\omega \cdot rci \cdot cjc) \qquad (24)$$

The core network#1 is a single pole model with transit radian frequency $\omega_T^{(1)}$. Making use of the general condition $(h_{21e})^{-1} + (h_{21b})^{-1} = -1$ one gets

$$\frac{1}{\omega} \Im \left(\frac{1}{\tilde{h}_{21e}^{(3)}} \right) = \frac{1}{\tilde{\omega}_T^{(3)}} = \frac{1}{\omega_T^{(1)}} + rci \cdot cjc \qquad (25)$$

$$-\Re\left(\frac{1}{\tilde{h}_{21e}^{(3)}}\right) = -\frac{1}{\tilde{h}_{21e0}^{(1)}} + \omega^2\frac{rci \cdot cjc}{\omega_T^{(1)}} \qquad (26)$$

A quadratic is made up by $(\tilde{\omega}_T^{(3)})^{-1}$ and the regression coefficient of ω^2 for the intrinsic transit time $(\omega_T^{(1)})^{-1}$ (of high importance for Hicum modeling) and the collector time-constant $\tau_{ci}=rci*cjc$. The latter will be used to obtain rci once cjc has been extracted.

V. CAPACITANCES

A. Cold mode

The - iteratively accountable - access resistances will be omitted from the discussion of the cold state. At such biases $gm=gpi=0$ and the network becomes reciprocal. In terms of (11) $\delta^{(4)}=1$ hence

$$-y_{12e}^{(4)} = -y_{12e}^{(3)} = \frac{-y_{12e}^{(1)} + re\cdot\Delta\mathbf{Y}^{(1)}}{\delta^{(2)}} + j\omega\cdot cjep \qquad (27)$$

Neglecting the term with $\Delta\mathbf{Y}^{(1)} = -\omega^2\cdot cpi\cdot cjc$ the denominator simplifies to

$$\delta^{(2)} \approx 1 + j\omega\cdot\tau_i; \quad \tau_i = (re+rbi)\cdot cpi + (rci+rbi)\cdot cjc \quad (28)$$

With the same neglection in (27)

$$-\frac{1}{\omega}\Im(y_{12e}^{(4)}) = \frac{cjc}{1+\omega^2\cdot\tau_i^2} + cjep \qquad (29)$$

As it is widely known the total collector capacitance

$$cjc_T = cjep + cjc \qquad (30)$$

can be obtained from the LHS of (29) by averaging it over the low-frequency range. Then $xcjc$ in SPICE is found by regression from the reciprocal of (31) or by directly optimizing it at higher frequencies:

$$\frac{\Im(y_{12e}^{(4)})}{\omega\cdot cjc_T} + 1 = xcjc\cdot\frac{\omega^2\cdot\tau_i^2}{1+\omega^2\cdot\tau_i^2} \qquad (31)$$

This $xcjc$ will be used as a reference in active mode for the determination of rbx.

The ratio of the two equations of (9) leads to the following regression for the emitter capacitance cpi:

$$\frac{|ry|^2}{\Re(ry)} = 1 - xcjc + \frac{cpi}{cjc_T} + b\cdot\frac{\omega\cdot\Im(ry)}{\Re(ry)}; \quad ry = -\frac{\tilde{y}_{11e}^{(4)}}{y_{12e}^{(4)}} \quad (32)$$

The term with the constant b is usually negligible.

B. Active mode

(3) and (8) gives for the complete transistor

$$\frac{y_{12e}^{(4)}}{\tilde{y}_{21e}^{(4)}} = \frac{y_{12e}^{(1)} - re\cdot\Delta\mathbf{Y}^{(1)}}{gm} - j\omega\cdot cjep\cdot\frac{\delta^{(2)}}{gm} \quad (33)$$

With the results of paragraph IV and substituting

$$\frac{\delta^{(2)}}{gm} = regm + \frac{rbi+re}{\tilde{h}_{21e}^{(3)}} + (rbi+rci)\cdot\frac{j\omega\cdot cjc}{gm} - re\cdot\frac{j\omega\cdot rbi\cdot cjc}{h_{21b}^{(1)}}$$

one arrives at

$$-\frac{1}{\omega\cdot regm}\Im\left(\frac{y_{12e}^{(4)}}{\tilde{y}_{21e}^{(4)}}\right) = cjc_T - a\cdot\omega^2 \qquad (34)$$

where a is a constant. The active-mode total collector capacitance yields by regression against ω^2.

The 2nd equation in (10) for block#3

$$y_{12e}^{(3)} = \frac{\tilde{y}_{21e}^{(3)}}{gm}(y_{12e}^{(1)} - re\cdot\Delta\mathbf{Y}^{(1)}) - j\omega\cdot cjep \qquad (35)$$

provides the following result for the active-mode internal collector capacitance

$$ktc = 1 - \tilde{\omega}_T^{(3)}\cdot rci\cdot cjc$$

$$cjc = \frac{1}{\omega}\frac{\Re(y_{12e}^{(3)})}{\Im(\tilde{y}_{21e}^{(3)})\cdot regm + re\cdot\Re(\tilde{y}_{21e}^{(3)})\cdot\Im\left(\frac{1}{\tilde{h}_{21e}^{(3)}}\right)\cdot ktc} \quad (36)$$

The second term in the denominator contributes less than a few percent to the result and can be omitted.

(7) for block#3 provides the substrate admittance for extracting the model-dependent substrate network

$$Y_{cs} = -y_{12b}^{(3)} + \omega\cdot rbi\cdot cjc\cdot\tilde{y}_{21b}^{(3)}\cdot\left[j - \Im\left(\frac{1}{\tilde{h}_{21e}^{(3)}}\right)\cdot ktc\right] \quad (37)$$

Finally gm results from condition (8)

$$gm = \frac{1 + j\omega\cdot cjc\cdot(rbi+rci)}{\frac{1}{\tilde{y}_{21e}^{(4)}\cdot\varepsilon^{(4)}} + \frac{Z}{\tilde{h}_{21b}^{(3)}} + rbi}$$

$$Z = \frac{re+rbi+j\omega\cdot cjc\cdot\Delta r}{1+j\omega\cdot cjc\cdot rci} \qquad (38)$$

The emitter resistance can be upgraded by (38) as

$$re = regm - \Re(gm^{-1}) \qquad (39)$$

VI. EXPERIMENTS AND EXTRACTION RESULTS

The samples have been selected from a 0.35um SiGe HBT technology [11] of fmax=70GHz.

The extraction was started with an estimated re. Only rcx has been retained from the optimization result of Chapter III. Filled symbols on Fig. 3 show rbx obtained by the method of [12]. Next, $xcjc$ was determined from cold y-parameters by (30) and (31). Declining $xcjc$ vs. Vcb as on Fig.4 is typical for SIC collectors: the cjc part is a hyperabrupt diode. The reference value marked on Fig. 4 was used to determine rbx from the active mode $xcjc$ contour plot over the rcx-rbx plane by (34), (36) as shown on Fig. 5. With the correct (rbx,rcx) pair known, the block#3 parameters could be easily obtained by deembedding. The collector time constant was extracted by (25)-(26) and the computation of cjc with (36) allowed for the determination of rci as shown on Fig. 6. All data was available that point for calculating the net rbi from (14) and (12), and to upgrade re using (38) and (39) as shown on Fig. 7 and 8. The first cycle usually provides the correct re value.

VII. CONCLUSION

Novel methods have been suggested for the determination of the complete AC equivalent of Fig. 1

for HBT/BJTs. The analysis and extraction steps were described in details.

REFERENCES

[1] M. Schroeter, "Test Structures for Bipolar Transistor Compact Modeling and Parameter Extraction," *University of Technology Dresden,* June 2000.

[2] J. Weng, J. Holz and T. F. Meister, "New Method to Determine the Base Resistance of Bipolar Transistors," *IEEE Electron Device Lett.* vol. 13, No3, pp. 158-160, March 1992

[3] D. Roulston, "Three dimensional Effects in Bipolar Transistors Using Fast Combined 1D and 2D Numerical Simulation," *Delhi University Conference*, December, 1997.

[4] M. Schröter, "Simulation and Modeling of the Low-Frequency Base Resistance of Bipolar Transistors and its Dependence on Current and Geometry," *IEEE Trans. on Electron Devices*, vol. 38. No. 3, pp. 538-544, March 1991.

[5] J. Berkner, „Kompaktmodelle für Bipolar-transistoren," expert verlag, 2002, D-71272 Renningen, Germany.

[6] B. Ardouin, T. Zimmer, H. Mnif, P. Fouillat, „Direct Method for Bipolar Base-Emitter and Base-Collector Capacitance

Splitting Using High Frequency Measurements," *IEEE BCTM*, pp. 114-117, September 2001.

[7] S. Lee, "A Simple Method to Extract Intrinsic and Extrinsic Base-Collector Capacitances of Bipolar Transistors," *IEEE Trans. on Electron Devices*, vol. 51. No. 4, pp. 538-544, pp. 647-650, April 2004.

[8] D. Berger, N. Gambetta, D. Celi, C. Dufaza, „Extraction of the Base-Collector Capacitance Splitting Along the Base Resistance," *IEEE BCTM*, pp. 180-183, September 2000.

[9] Z. Huszka, E. Seebacher and W. Pflanzl, "An Extended Two-Port Method for the Determination of the Base and Emitter Resistances," *IEEE BCTM*, pp. 188-191, October 2005, Santa Barbara, CA.

[10] J.C.J. Paasschens, W.J. Kloosterman, and R.J. Havens, "Parameter Extraction for Mextam, Level 504, http://www.semiconductors.philips.com/Philips_Models/bipolar/mextram/

[11] www.austriamicrosystems.com

[12] T. Zimmer, A. Meresse, Ph. Cazenave and J. P. Dom, „Simple Determination of BJT Extrinsic Base Resistance," *Electronic Letters*, vol. 27, No. 21, pp. 1895-1896, 10th Oct. 1991

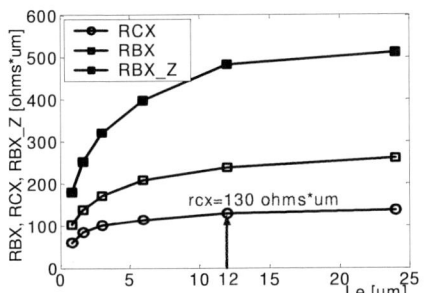

Fig. 3. Access resistances by (22), RBX_Z from [12]

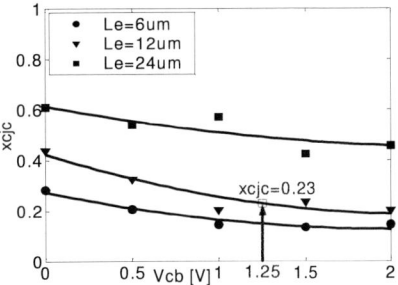

Fig. 4. *xcjc* from cold measurements by (31). Vbe=0V.

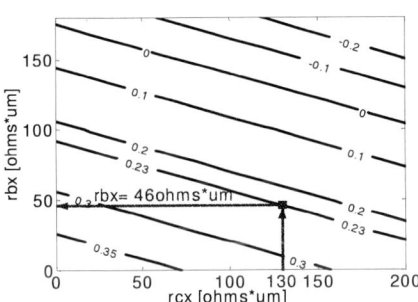

Fig. 5. *xcjc* from (36) and (34) over the *rcx-rbx* plane. Le=12um, Vce=2V, Vbe=0.75V

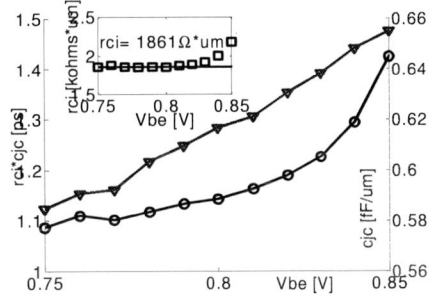

Fig. 6. Determination of *rci*. Le=12um, Vce=1V

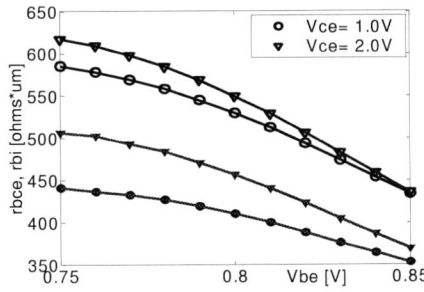

Fig. 7. *rbce* (open) and net *rbi* (filled) Le=12um

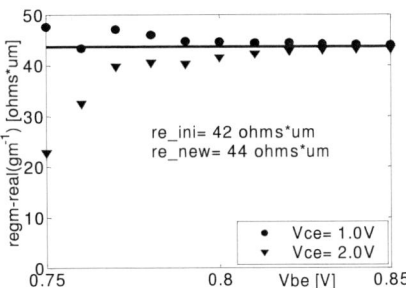

Fig. 8. Upgrading *re* by (39), Le=12um

2006 Bipolar/BoCMOS Circuits and Technology Meeting Proceedings

Compact Modeling of GaAs Heterojunction Bipolar Transistors using the new Mextram 3500 model

R. van der Toorn[1], J.C.J. Paasschens[2], J.J. Dohmen[3], R.M.T. Pijper[1], B.N. Balm[4]

[1]Philips Research, High Tech Campus 5, 5656 AE Eindhoven, The Netherlands
[2]Philips Semiconductors, Process & Library Technology, Gerstweg 2, 6534 AE Nijmegen, The Netherlands
[3]Philips Research, Electronic Design&Tools, High Tech Campus 5, 5656AE Eindhoven
[4]Philips Semiconductors, Business Line Cellular Systems, Gerstweg 2, 6534AE Nijmegen, The Netherlands

Abstract — We adapt the Mextram 504 compact model for Si and SiGe bipolar transistors to make it suitable for compact modeling of GaAs HBT's. We discuss the different physics included in the new model, Mextram 3500, and demonstrate the capabilities of the new model on GaAs HBT characteristic simulations. We also show an example of advanced GaAs PA-circuit simulations that have been performed with our model.

I Introduction

In modern power amplifier (PA) modules for RF applications, such as GPRS and EDGE for mobile communication, GaAs heterojunction bipolar transistors (HBT's) are applied in high-power amplifier stages. To meet the tight specifications posed by such applications on linearity and power efficiency for example, sophisticated circuit design schemes are applied. In order to reduce development costs and time-to-market such design efforts rely on circuit simulations. Therefore the need for an industrially supported, advanced compact bipolar transistor model capable of describing the characteristics of GaAs HBT's in the relevant regimes of operation is actual. Typically, in this field of application, advanced simulation techniques are used, such as harmonic balance (HB) analysis in the frequency domain, to study non-linear large signal behaviour, in a regime of significant power dissipation.

In this paper we introduce a variant of the Mextram compact model, Mextram level 3500, that is capable of modeling GaAs HBT's. We shall demonstrate its ability to model modern GaAs HBT characteristics as well as its ability to support advanced large signal circuit simulations relevant to modern RF PA design.

Based on device physics, in the Mextram models, both for Si/SiGe and GaAs, transit times and capacitances are derived from internal charges [1], as derivatives of a relevant charge with respect to current or voltage, respectively. The interdependence of transit times and capacitances that results from the common origin of these quantities in charges was discussed recently in detail for GaAs HBT's [2]; it was stressed that this interdependence forms a criterion for model consistency. In Mextram 3500 we have included a *physics-based* model [3, 4] for the collector-base depletion charge Q_{t_C} for GaAs HBT's, which makes our model unique among the currently available GaAs HBT mod-

els [2, 5, 6]. This model for Q_{t_C} takes into account the dependence of electron drift velocity v_d on electric field E, in particular the *decrease* of v_d with increasing $|E|$ that is observed in GaAs for $|E| > 3\mathrm{kV/cm}$. (In silicon v_d saturates at a constant value for large $|E|$.) The relevance of this for the modeling of nonlinear behaviour of GaAs HBT amplifier stages was experimentally confirmed by [7]. Because our Q_{t_C} model is physics-based, the value of most of its parameters can be estimated from material and technological data of the HBT to be modeled, which is an advantage for parameter extraction.

II Model description

Mextram 3500 is based on the Compact Model Council world standard compact model Mextram [8] for Si/SiGe- based bipolar transistors. Mextram 3500 has inherited many of the advanced aspects of the Mextram model. Examples of these are extensive modeling of the extrinsic regions of the transistor, advanced formulations for temperature scaling and geometric scalability. Derivatives with respect to device temperature are completely implemented in the model's source code in order to achieve quadratic convergence of simulations that include self-heating. Given the relatively high thermal resistance of GaAs HBT's, modeling of self-heating is essential for the regime of application of the model.

To account for differences between Si/SiGe-based bipolar transistors on the one hand and GaAs-based heterojunction bipolar transistors on the other hand, several parts of the Mextram model for Si/SiGe based transistors have been changed to derive a model suitable for GaAs based HBT's.

The heterojunction at the base-emitter junction in GaAs HBT's often causes deviations from ideal $\exp[q_e V/(k_B T)]$-behaviour of the collector current of the transistor and even stronger deviations from ideal behaviour in the base current. To account for these non-idealities, the dc-current model of Mextram 3500 was generalized with respect to the Si/SiGe Mextram model, by adding non-ideality parameters.

The steep doping profiles that are commonly applied in GaAs HBT's are reflected by pronounced reach-through behaviour of capacitance versus voltage ($C(V)$) characteristics; by reach-through we denote the

1-4244-0458-4/06/$20.00 ©2006 IEEE

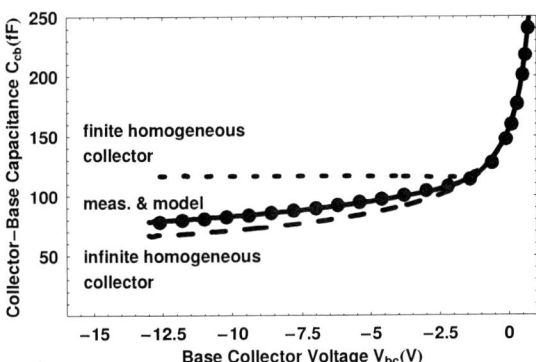

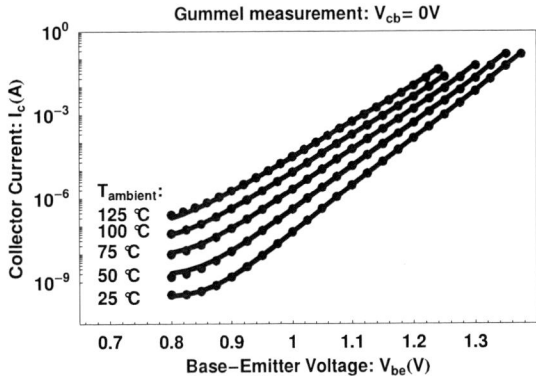

Figure 1: Collector-base capacitance C_{cb} as a function of base-collector voltage (V_{bc}). Shown are measured values (dots) for an industrially applied GaAs HBT, modeled values (curve) and model curves for a finite homogeneous collector that shows pronounced reach-through (short dashes) and an infinite collector (long dashes), that does not show reach-through. Our model interpolates between the finite and infinite homogeneous collector cases.

Figure 2: Measured (dots) and modeled (curves) collector current (I_c) of a typical modern GaAs HBT for RF PA applications as a function of base-emitter voltage (V_{be} (V)) for a constant collector-base voltage of 0V, and for various ambient temperatures T.

situation in which the depletion layer in the collector just reaches all the way to the sub-collector. To account for this, the model of the depletion charge of the collector-base junction was modified. In order to be able to model also the $C(V)$ characteristics of HBT's that have non-homogeneous doping profiles in the collector, we implemented a collector depletion charge model that interpolates between the case of an infinitely long collector, for which reach-through does not occur, and a collector of finite length and homogeneous doping profile. It turns out that this model captures the $C(V)$ characteristics of actual GaAs HBT's very well. An example of this is shown in figure 1, where we show the collector-base capacitance C_{cb} as a function of collector base voltage V_{cb}, at zero collector current, for a GaAs HBT that has a stepped doping profile in its collector. The measured $C_{cb}(V_{cb})$ characteristic can neither be modeled as an infinitely long collector nor as a finite homogeneous collector showing sharp reach-through. The observed characteristic is modeled very well by interpolation of the two cases however.

Furthermore, to this model for collector-base charge, we have added a physically based correction that is induced by the anomalous dependence of the electron drift velocity v_d on the electric field E [3, 4]. This anomalous $v_d(E)$ is typical for III-V materials, and causes strong V_{cb} dependence of collector transit time, clearly observable in measured $f_T(I_c, V_{ce})$ characteristics of GaAs HBT's. It also induces anomalous (as compared to silicon based bipolar transistors) dependence on I_c of C_{cb}.

In Mextram, collector-base depletion charge Q_{tc} depends on both the internal collector-base voltage and on the collector current, the latter dependence describing the modulation of the electric field in the collector, most notably in the depletion layer, by the charge of

the electrons that carry the collector current. In Mextram collector-base capacitance C_{cb} and transit time τ_c are derived quantities: they are derivatives with respect to voltage and current respectively, of Q_{tc}. Taking the anomalous $v_d(E)$ characteristic of GaAs into account in the model of Q_{tc} in Mextram 3500, the consequences of this material property of GaAs for C_{cb} and τ_c are modeled in a unified way.

Mextram 3500 has inherited the full self-heating implementation from the Si/SiGe Mextram model. We have added an optional model for non-linear self-heating [9].

III   Characteristics

Figures 2 to 5 show measured dc-characteristics of a typical modern GaAs HBT for RF PA applications for various bias conditions and temperatures, together with corresponding characteristics simulated using Mextram 3500. The device has two emitter fingers and a single InGaP/GaAs base-emitter heterojunction.

In the Gummel measurements, at base-emitter voltages below 0.9V, ohmic leakage currents are observed, between base and emitter and between collector and emitter. These are currently not modeled by the Mextram model itself but are modeled by addition of external parasitic resistors in the GΩ range. The resistances were observed to be exponentially dependent on temperature, which indicates that they are associated with thermally activated conduction, possibly hopping conduction [10]. Figure 3 shows simulated base-current characteristics with (full curves) and without (dashed curves) externally added resistances.

GaAs HBT's commonly show strong and complicated dependence of cut-off frequency on collector voltage, as is shown in figure 6. In the case of the particular GaAs HBT's considered here, this dependence is induced by modulation of the thickness of the collector depletion layer and by anomalous dependence of electron drift velocity on electric field. This is demon-

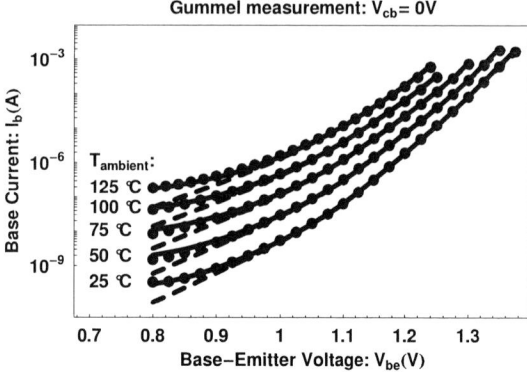

Figure 3: Measured (dots) and modeled (curves) base current (I_b) corresponding to the HBT of figure 2 as a function of base-emitter voltage (V_{be} (V)) for a constant collector-base voltage of 0V, and for various ambient temperatures T. Dashed curves show the simulated characteristics without the externally added leakage resistances.

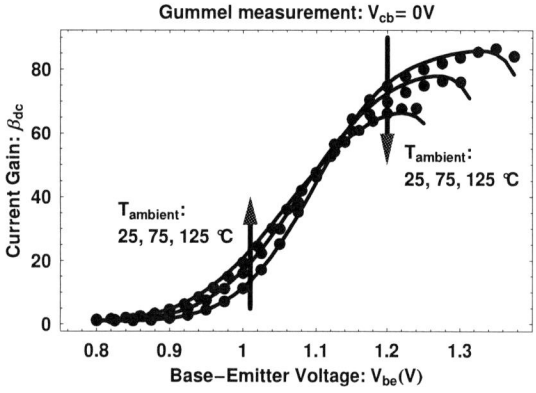

Figure 4: Measured (dots) and modeled (curves) dc current gain β_{dc} corresponding to the HBT of figure 2 as a function of base-emitter voltage (V_{be} (V)) for a constant collector-base voltage of 0V, and for several fixed ambient temperatures $T_{ambient}$.

strated in figure 7, in which simulated characteristics are shown while both or only the second of the two effects are suppressed in the model. The simulations with suppressed modulations of the width of the collector-base depletion layer correspond to the $C_{cb}(V_{bc})$ characteristic of a finite homogeneous collector in figure 1.

The model for quasi-saturation of Mextram 3500 is similar to the model used in Mextram 504, for Si/SiGe-based devices. As can be seen in the f_T characteristics, at high V_{ce} this model is not accurate for GaAs HBT's and in Mextram 3500 there is room for improvement here. For $V_{ce} \leq 1$V, the model accurately describes measured $f_T(I_c, V_{ce})$ characteristics in the high current, quasi-saturation regime, as is shown in the inset of figure 6. Figure 8 shows measured and simulated collector base capacitance as a function of bias conditions; dashed curves show the simulated results without the $v_d(E)$ correction on Q_{tC}.

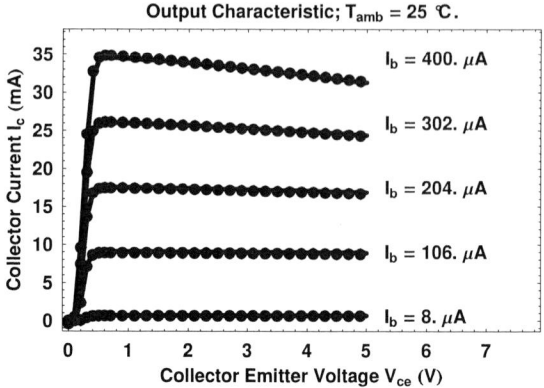

Figure 5: Measured (dots) and modeled (curves) dc current output characteristics, corresponding to the HBT of figure 2: collector current (I_c (mA)) as a function of collector-emitter voltage (V_{ce} (V)), for several fixed values of the base-current I_b and fixed ambient temperature of 25°C. GaAs HBT's commonly show substantial offset voltage; the decrease of I_c with increasing V_{ce} is due to self-heating.

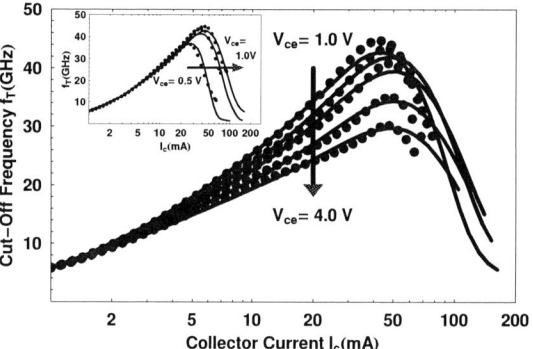

Figure 6: Measured (dots) and modeled cut-off frequency f_T as a function of collector current (I_c), for several fixed collector-emitter voltages: $V_{ce} = 1.0, 1.5, 2.0, 3.0, 4.0$V, at an ambient temperature of 25° C. The inserted plot shows characteristics for $V_{ce} = 0.5, 0.75, 1.0$V, in the saturation regime.

Mextram 3500 was applied in ADS Harmonic Balance (HB) simulations of a Philips BGY502 front-end module [8] for GSM/GPRS applications. Figure 9 shows measured and simulated antenna output power of a BGY502 as a function of an internal control voltage for power control, both for open and closed power control loop. This plot shows that Mextram 3500 is ready for use in advanced industrial circuit simulations.

IV Conclusion

We have introduced a variant of the CMC world standard model Mextram for Si/SiGe- based bipolar transistors. Our new variant of the Mextram model, Mextram 3500, is tailored to model GaAs heterojunction bipolar transistors. The dc-current model is generalized to capture the deviations from $\exp(q_e V/(k_B T))$ and

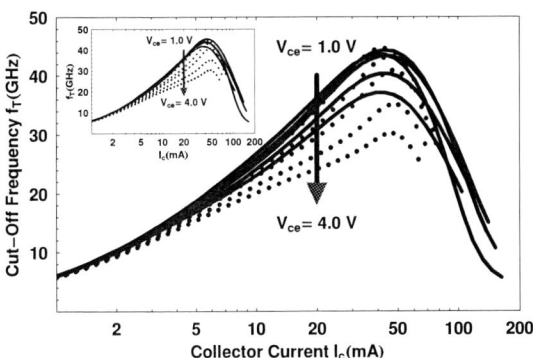

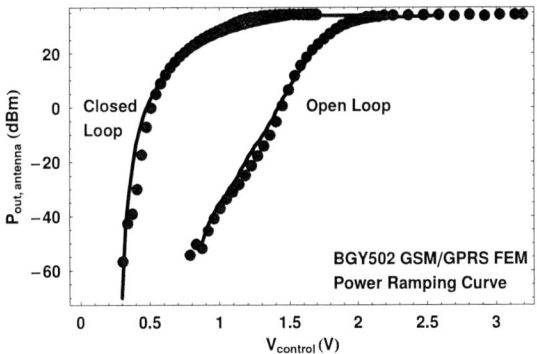

Figure 7: This figure, to be compared with figure 6, shows how Mextram 3500 physically models transit time. Dots are as in figure 6. The inserted plot shows corresponding simulated characteristics with strongly restricted modulation of the thickness of the depletion layer in the collector and without correction on the collector-base depletion charge due to the drift-velocity versus electric field ($v_d(E)$) characteristic of GaAs. The main plot shows simulations with modulated thickness but still without the $v_d(E)$ effect.

Figure 9: Antenna output power of a full Philips BGY502 PA front-end module, at 849MHz and 1mW input power, measured (dots) and simulated (curves) using the Mextram 3500 model for GaAs HBT's in the output stage. Inaccuracies of the simulation results are most probably due to inaccuracies in modeling of passives.

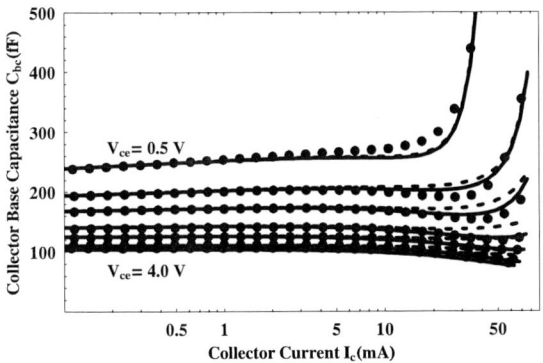

Figure 8: Measured (dots) and modeled collector base capacitance C_{cb} as a function of collector current (I_c), for several fixed collector-emitter voltages: $V_{ce} = 0.5, 0.75, 1.0, 1.5, 2.0, 3.0, 4.0$V, at an ambient temperature of $25°$ C. Capacitance $C_{bc} = -Im(Y_{12})/\omega$, measured at $f = 3.7$ GHz, as a function of collector current. Dashed curves show simulated characteristics without the collector-base depletion charge correction due to the drift-velocity versus electric field characteristic of GaAs.

$\exp(q_e V/(2k_B T))$ behaviour in the collector and base currents that is commonly observed in GaAs HBT's. The collector depletion charge model was changed in order to model commonly observed $C_{cb}(V_{cb})$ characteristics in GaAs HBT's. The consequences of anomalous electron-drift velocity dependence of electric field for collector-base capacitance and collector transit-time are modelled by a physics-based correction [3, 4] of the collector-base depletion charge.

We have shown that Mextram 3500 is capable of modeling observed characteristics of modern GaAs HBT's and that it can be used to perform advanced

simulation based large signal analysis of industrial RF-power circuits, such as large signal harmonic balance simulations of PA-circuits for mobile communication.

References

[1] H. C. de Graaff and F. M. Klaassen, *Compact transistor modelling for circuit design*. Springer-Verlag, 1990.

[2] M. Rudolph and R. Doerner, "Consistent modeling of capacitances and transit times of GaAs-based HBT's," *IEEE Trans. Electron Devices*, vol. 52, no. 9, pp. 1969–1975, September 2005.

[3] R. van der Toorn, J. C. J. Paasschens, and R. J. Havens, "A physically based analytical model of the collector charge of III-V heterojunction bipolar transistors," in *IEEE GaAs Digest*. IEEE, 2003, pp. 111–114.

[4] ——, "Physically based analytical modeling of base-collector charge, capacitance and transit time of III-V HBT's," in *IEEE CSICS Digest*. IEEE, 2004, pp. 283–286.

[5] M. Iwamoto, D. E. Root, J. B. Scott, A. Cognata, P. M. Asbeck, B. Hughes, and D. C. D'Avanzo, "Large-signal HBT model with improved collector transit time formulation for GaAs and InP technologies," in *IEEE MTT-S Digest*. IEEE, 2003, pp. 635–638.

[6] D. E. Root, M. Iwamoto, and J. Wood, "Device modeling for III-V semiconductors – an overview," in *Proc. IEEE Compound Semicond. Integrated Circ. Symp.*, 2004, pp. 279–282.

[7] M. Rudolph and R. Doerner, "Large-signal hbt model requirements to predict nonlinear behaviour," in *Microwave Symposium Digest*, ser. MTT-S International, vol. 1. IEEE, 2004, pp. 43–46.

[8] www.semiconductors.philips.com.

[9] J. C. J. Paasschens, S. Harmsma, and R. van der Toorn, "Dependence of thermal resistance on ambient and actual temperature," in *Proc. BCTM*, 2004, pp. 96–99.

[10] N. F. Mott, *Metal-Insulator Transitions*. Taylor & Francis LTD, 1974.

Substrate Current Injection and Latchup in Complementary Vertical Bipolar Processes

Andy Strachan

Advanced Process Technology Development,
National Semiconductor Corporation, Santa Clara, CA
Tel: (408) 721 2223   Fax: (408) 721 5100
Email: andy.strachan@nsc.com

Abstract — **Challenges and solutions for substrate current injection and latchup phenomena in complementary BiCMOS technologies are described. The vertical PNP presents specific challenges that differ from CMOS or NPN-only BiCMOS cases.**

Background

An important advantage for BCD processes is the ability to operate over a wide voltage range, particularly at low voltages for battery driven applications. To enable effective low voltage operation, it is ideal to have both NPN and PNP devices with high cutoff frequency (f_T) and Early Voltage (V_A) [1]. High voltage processes with truly complementary bipolar devices meet this demand. However, the addition of vertical PNP devices to a typical BCD process presents additional issues with respect to latchup and substrate current crosstalk.

A 24 V BCD technology with complementary vertical bipolar devices has been used for this work. A description of the process technology used was presented in [2] and illustrated in Figure 1. Device isolation is made by a deep trench between adjacent devices. The effectiveness of deep trench isolation

in preventing latchup and improving substrate current effects has been described in [3], [4] and [5]. In these papers, a heavily doped n-buried layer was available to isolate all devices from p-substrate, which is not the case with a complementary vertical PNP.

The PNP collector is isolated from the p-substrate by a deep n-isolation layer. The n-isolation region must be doped to withstand the 24 V operating voltage of the PNP collector and so is much more lightly doped than a typical n+ buried layer. This presents unique challenges in latchup prevention not seen in lower voltage technologies.

This paper will cover the latchup mechanism in complementary bipolar along with a description of the test structure and experimental method for characterization. Results from these experiments will be presented on the effect of epitaxial substrates in suppressing latchup, on the effect of guard rings on NPN to PNP spacing design rules and on limits to the size of PNP devices driven by latchup risks. Finally, there is some discussion on the effect of pre-latchup leakage on analog circuit operation.

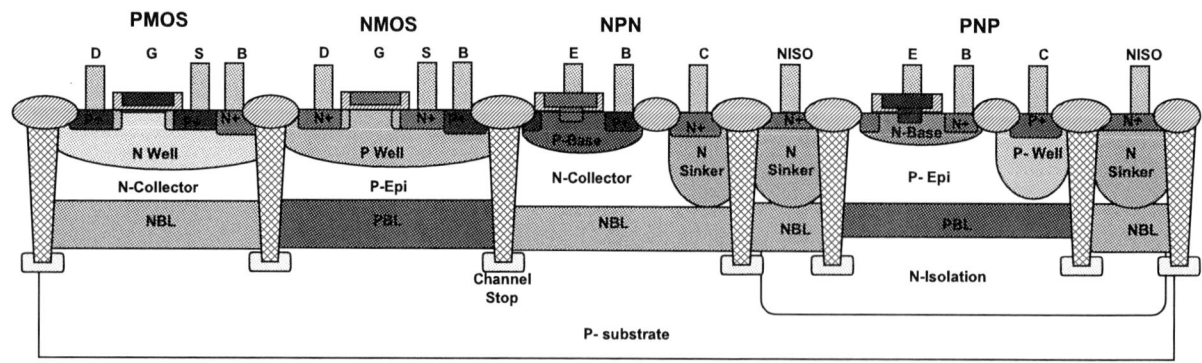

Figure 1: Cross section of low voltage CMOS and complementary vertical NPN and PNP for a 24 V trench isolated smartpower technology.

Latchup Mechanisms

Latchup is most commonly considered as an npnp parasitic effect within a CMOS environment. Table 1 compares the equivalent terminals between the CMOS and the complementary bipolar cases. To trigger positive latchup the PNP collector is forward biased to the PNP n-isolation layer. To trigger negative latchup the NPN collector is required to inject substrate current. Typically, this might be when the NPN collector is below ground, but can also occur when the NPN is in heavy saturation. For CMOS latchup there are usually just two parasitic bipolar transistors present. The goal is then to set the sum of the gains of these two parasitics to less than unity. For complementary bipolar processes the situation is more complex with six bipolar devices to consider. Figure 2 shows that for neighboring NPN and PNP transistors, there are four parasitic devices in addition to the two main bipolars.

The combination of trench isolation and heavily doped n+ and p+ buried layers suppress the parasitic gains from bipolar bases to p-substrate to less than unity. However, there are two key parasitic bipolars to consider. The first is the vertical PNP formed by PNP collector, n-isolation and p-substrate. The gain of this PNP is high since the n-isolation doping is required to be relatively low to support a high breakdown voltage. This gain is fixed in the process technology and cannot be adjusted by layout. The other key parasitic bipolar is the lateral NPN in the substrate between the NPN collector and the PNP n-isolation region. The gain of this NPN depends on the separation between the NPN and PNP and the presence of guard rings between the devices. Layout design rules must be determined to limit the gain of this lateral substrate NPN.

CMOS Latchup	cBipolar Latchup	Bias
NMOS Source/Drain	NPN Collector	Below ground for negative injection
Pwell Contact	P-substrate Contact	Ground
PMOS Source/Drain	PNP Collector	Above ground for positive injection
Nwell Contact	PNP N-Isolation	Supply

Table 1: Equivalent terminals between CMOS latchup and latchup in complementary bipolar technologies.

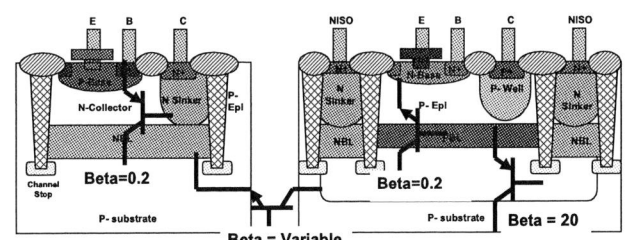

Figure 2: Parasitic devices created within a complementary bipolar structure. The two key parasitics are the PNP collector, N-isolation and P-substrate high gain vertical PNP and the variable gain NPN with p-substrate as base.

Experimental Test Structures and Methodology

To determine the gain of the lateral substrate NPN transistor, dedicated test structures were created. These are shown in cross section in Figure 3. They consist of an NPN device surrounded by one or more guard rings. The double guard ring configuration is a grounded p-type guard ring and a false collector (or n-type) guard ring connected to a separate terminal. The single guard ring configuration has only the grounded p-type guard ring with no n-type false collector. A PNP transistor is positioned as a detector of substrate current at various distances shown in Figure 3 as "Z".

The test procedure is to inject current into the substrate by forward biasing the NPN collector to p-substrate and monitor the currents in all the other terminals of the test structure. A device simulation of the current flow is shown in Figure 4. Device simulations can be used to estimate the effectiveness of various guard ring options with quantitative results depending on accurate tuning of the minority carrier lifetime in the substrate.

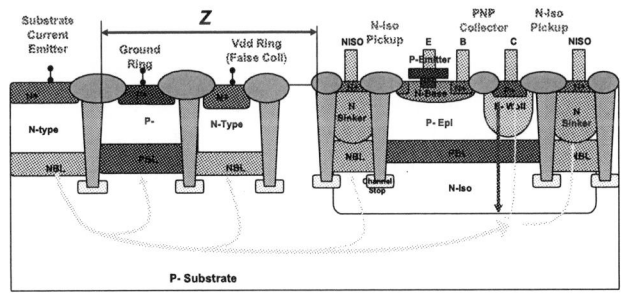

Figure 3: Substrate injection by NPN to an adjacent PNP transistor. Experiments were performed to determine amount of substrate current capture by PNP as a function of device separation (Z).

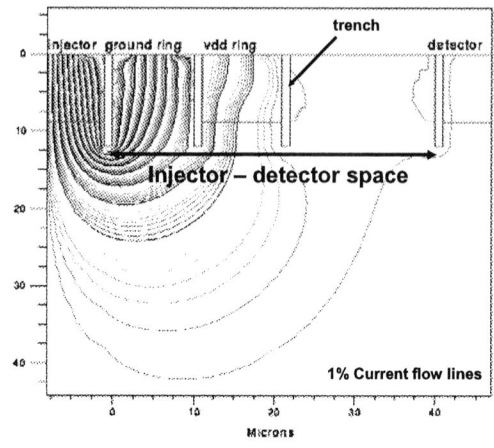

Figure 4: TCAD Simulation of current flow from NPN to PNP as NPN collector is biased below ground.

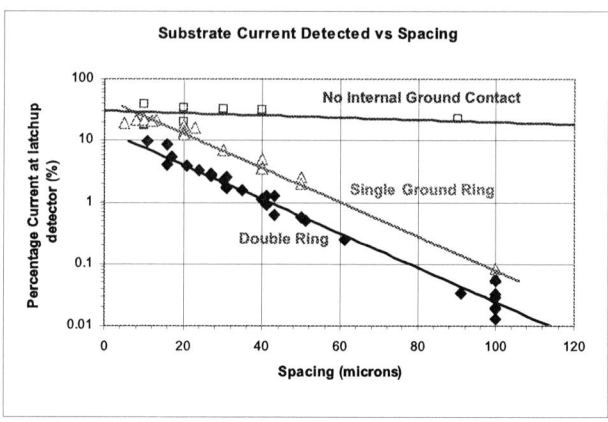

Figure 5: Percentage of injected substrate current measured at the detector as a function of spacing (Z) and number of guard rings. Multiple data points at the same spacing represent different guard ring widths.

Substrate Injection Test Results

Results of the measured substrate injection experiments are shown in Figure 5. The effectiveness of the double guard ring approach over a single guard ring is an approximately 5x reduction in substrate current at a given distance. Multiple data points at the same separation values are from multiple guard ring widths. These show that guard ring width is a secondary factor. At 100 μm separation there is an 8x difference between two 50 μm guard rings (the maximum possible) and two minimum width guard rings. Also shown in Figure 5 is the case where there are no substrate contacts between the NPN and the PNP. In this case the gain of the substrate lateral NPN remains above 0.1 for over 150 μm of separation.

From the data in Figure 5, it is possible to select a separation that reduces the gain of the parasitic substrate NPN to a value such that the gain of this NPN multiplied by the high gain PNP from p-collector to p-substrate is less than unity. However, this is not a guarantee of latchup immunity since these parasitics operate in conjunction with the high gain of the main bipolars.

Epitaxial Substrates

Epitaxial substrates are commonly used in CMOS processes to suppress latchup and have also been used in smartpower technologies [4]. In the case of complementary bipolar technologies, there are additional constraints that limit the usefulness of epitaxial substrates for latchup prevention.

The test structures described earlier were used to evaluate various thicknesses of p-/p+ epitaxial substrates as a function of p- layer thickness on the starting material. The results in Figure 6 show that, as expected, the substrate parasitic gain is strongly suppressed and that the effectiveness depends on the p- layer thickness. This is an effective solution for low voltage complementary bipolar processes.

For higher voltages the thickness of the p- layer on the epitaxial substrate has to be set to maintain the breakdown voltage of the PNP device. The n-isolation layer beneath the PNP is deep to allow a high punchthough voltage from PNP collector to p-substrate. This deep layer is formed by a high temperature diffusion which will also diffuse the boron epitaxial substrate doping towards the surface of the wafer. The interaction of the n-isolation with the out-diffusing p+ substrate will limit the voltage of the process.

Figure 7 shows the breakdown voltage for high side operation of a PNP transistor as a function of substrate epitaxial thickness. It shows that 30 microns of p- over p+ is required for a 24 V process. This reduces the effectiveness of the substrate current protection towards the same values as the p-starting material. For high voltage complementary bipolar processes, the ability to reduce NPN to PNP spacing by a small amount may not justify the higher initial wafer cost.

114

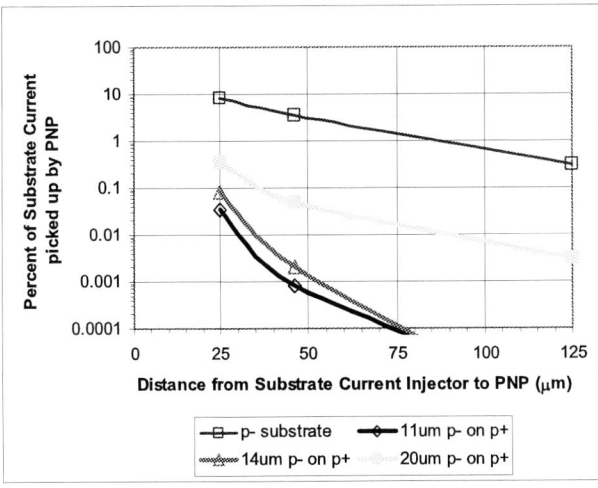

Figure 6: Effect of p-/p+ epi starting material on current collected by PNPs at various distances from an emitter of substrate current.

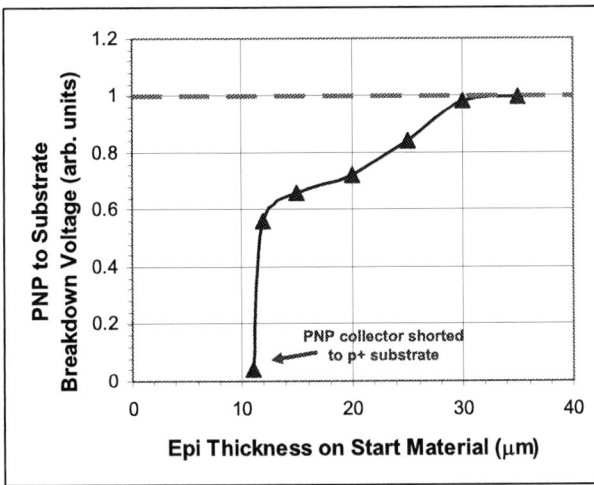

Figure 7: Breakdown from PNP collector through the n-isolation to the p+ substrate as a function of p- starting material thickness. The dashed line represents the breakdown on p- starting material.

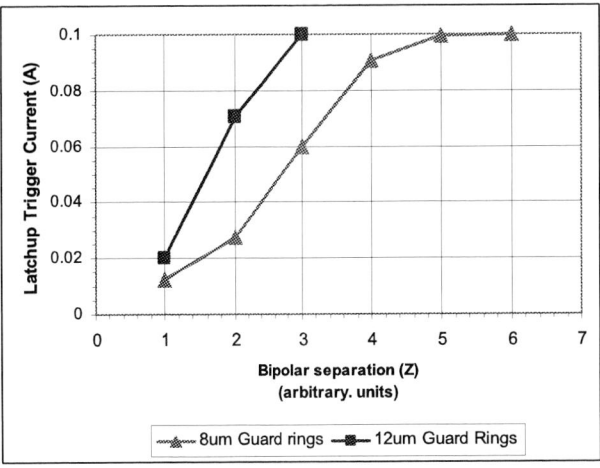

Figure 8: Latchup trigger current as a function of NPN to PNP separation (Z) using double guard rings. The values of Z are scaled relative to the minimum possible separation of the bipolars. This data is taken at 125 degrees C.

Latchup Test Results

The test structure shown in Figure 3 can also be used to evaluate the latchup trigger current. A dedicated automated test routine has been developed in conjunction with this test structure to detect and report latchup triggering current and associated parameters. The test routine ramps the amount of current injected from the NPN into the p-substrate and monitors the voltage required to do so. A collapse in this voltage indicates that the trigger current has been reached. Current and voltage data at the point prior to the trigger current is reported. The test routine is implemented on a fully automated electrical tester with thermal chuck capability allowing statistical quantities of data to be collected at wafer level.

Using this automated test routine the latchup trigger current was measured for various spacings (Z in Figure 3) and guard ring widths. These automated latchup tests are run at hot temperatures to determine the worst case characteristics.

Figure 8 shows the results of latchup trigger current for two different guard ring widths. In both cases the structures have double guard rings of ground and false collector as shown in Figure 3. For protection against 100 mA of injected substrate current, a NPN surrounded by 12 micron guard rings can be positioned at a distance of 3 on the arbitrary scale of Figure 8 from an PNP. If 8 micron guard rings are used the separation goes up by 60% to 5 on the arbitrary scale. These results are used to define layout design rules to meet product latchup criteria.

Size limits for PNP devices

One key component of latchup robustness for the vertical PNP is the resistance of the n-isolation layer shown in Figure 1. The sheet resistance of this layer is set by process voltage requirements. However, the n-isolation resistance of large PNP devices is also determined by the distance between n-isolation contacts areas using the n+ buried layer and n+ sinker. This distance sets the maximum PNP collector size. The test structures and automated test routine described in the previous sections were used to measure latchup trigger current on various sizes of PNP at a two different separations (Z).

Figure 9 shows the results of these tests performed at an elevated temperature of 125 degrees C. The PNP sizes are relative to a 1x emitter size transistor for the technology. For the larger separation it is clear there is a sharp drop-off in trigger current beyond a 20x device. This is used to set design rules for the maximum distance between n-isolation pick-up regions.

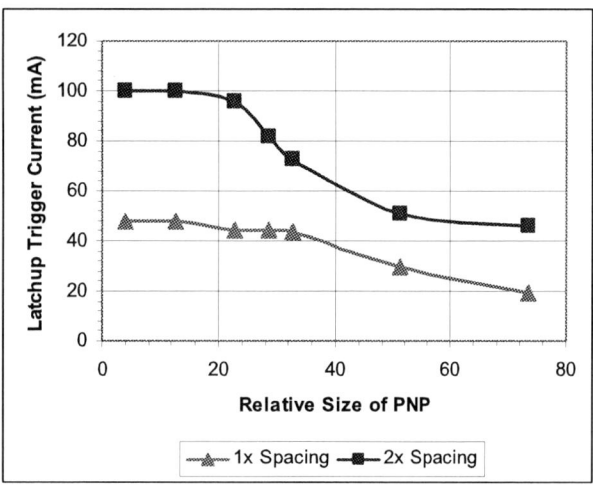

Figure 9: Latchup trigger current at 125 degrees C as a function of PNP size. The larger the PNP the more series resistance in the n-isolation region leading to a reduction in trigger current.

Analog Circuit Effects

The previous results described the conditions for triggering latchup between NPN and PNP devices. However, for sensitive analog circuits it is not necessary to trigger latchup for problems with circuit operation to occur. The operation of circuits can be disturbed at much lower substrate injection levels than cause latchup. Figure 10 shows a sweep of substrate current injection into a PNP structure. Latchup occurs when the injected current reaches 0.12 A. At this point, the supply current jumps to the compliance limit. Also shown is the current measured at the PNP collector terminal. This current has passed though the n-isolation region. The PNP collector current received from the substrate is in the order of microamps prior to latchup.

An example of how small parasitic currents cause problems in complementary bipolar circuits is shown in Figure 11. A typical bipolar output stage is a large PNP (QP1) and large NPN (QN1). Latchup between QP1 and QN1 is not usual as the collectors of these transistors are tied together. However as the output node goes below ground a parasitic NPN transistor forms between the collector of QN1 and portions of the base drive circuitry for QP1. In this example, the parasitic forms to adjacent NPN, QN2. The parasitic substrate current collected by QN2 is small. However, this small current will be magnified by the forward gains of QP2 and QP1 and appear as collector current in QP1. For PNP gains of 100 this becomes 10000x the original leakage. If QN1 is conducting current as the output goes below ground, then the amplification of the substrate parasitic can cause high currents in both QP1 and QN1 leading to circuit failure. Therefore, it is necessary to use the test structures and methods described above to develop layout design rules for spacing of complementary bipolar devices that preserve the integrity of the circuit operation over and above providing protection against latchup.

Conclusion

Providing protection against latchup in complementary bipolar processes is more complex than CMOS or NPN-only BiCMOS processes with more parasitic bipolar devices available. Dedicated test structures and automated test routines are needed to determine key design rules for NPN to PNP spacing, guard ring widths and limits to PNP size. A method for extraction of these rules has been demonstrated for a 24 V technology.

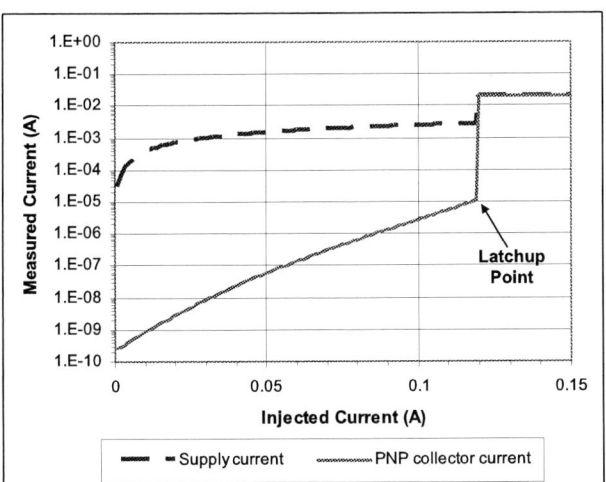

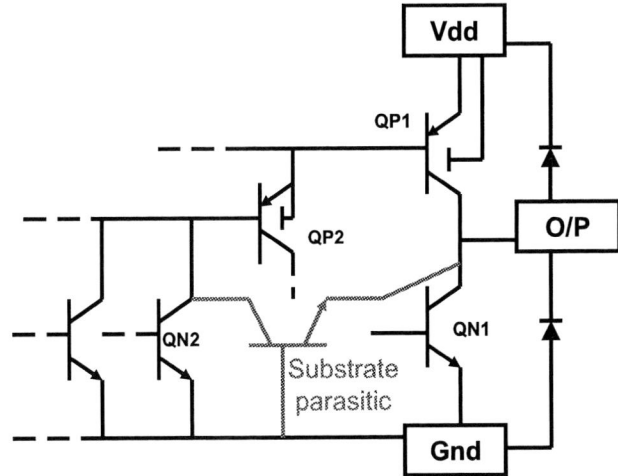

Figure 10: Example of the increase in PNP collector current leakage prior to the actual latchup trigger. The sharp rise in supply current at 0.12A is the latchup trigger point. However significant current is seen in the PNP collector at lower levels of substrate injection. This data is taken at 125 degrees C.

Figure 11: Partial schematic of a bipolar output stage. As the output node goes below ground, a parasitic NPN formed in the substrate between QN1 and QN2. This parasitic NPN may only conduct a small current. However, amplification of this current through the base drive circuit can turn on the large output PNP, QP1.

Acknowledgements:

The author would like to acknowledge all members of the Advanced Process Technology Development group in both Santa Clara, CA and Arlington, TX who contributed to this work. Particular thanks to Lisa Rozario for development of the automated test program to detect latchup trigger current.

References:

[1] D. Monticelli, "The Future of Complementary Bipolar", *BCTM 2004*.

[2] A. Strachan et al., "A Trench-Isolated Power BiCMOS Process with Complementary High performance Bipolars", *BCTM 2002*.

[3] A. Watson et al., "The Effect of Deep Trench and Sub-collector on the Latchup Robustness in BiCMOS Silicon Germanium Technology", *BCTM 2004*.

[4] V. Parthasarathy et al., "A Multi Trench Analog + Logic Protection for Substrate Crosstalk Prevention in a 0.25μm SmartPower Platform with 100V High-side Capability". *ISPSD 2004*.

[5] S. Voldman, "The Influence of a Novel Contacted Polysilicon Filled Deep Trench Biased Structure and its Voltage Bias State on CMOS Latchup". *IRPS 2005*.

10A 12V 1 chip DC/DC converter IC using bump technology

Kazutoshi Nakamura, Kenichi Matsushita, Norio Yasuhara
Koichi Endo, Fumito Suzuki, Morio Takahashi and Akio Nakagawa

Discrete Semiconductor Division, Toshiba Corporation Semiconductor Company
1,Komukai Toshiba-cho,Saiwai-ku,Kawasaki,212-8583,Japan
Phone:+81-44-549-2602,facsimile:+81-44-549-2883,
E-mail:kazutoshi.nakamura@toshiba.co.jp

Abstract — **In this paper, we demonstrate high frequency and 10A operation of 1 chip DC/DC converter. The chip adopted low impedance metal bump technology and a high speed gate driving technique for large LDMOS, what we call ▯distributed driver circuit▯ The fabricated chip achieves the switching time 3ns at load current 10A and high efficiency 88.9% for the input voltage, the output voltage and switching frequency of 12V, 1.3V and 780kHz, respectively.**

Index Terms ▯ **Power device, DC/DC converter, LDMOS.**

I. INTRODUCTION

With recent increase in clock speed of microprocessors, high efficiency, high power density, high current slew rate di/dt is strongly demanded for DC/DC converters. In order to improve the conversion efficiency, Multi Chip Module has been proposed to reduce parasitic inductances[1,2].MCM consists of a driver IC and two power MOSFETs(Low Side and High side) in one package. Not only small size but also high frequency operation is simultaneously realized.

In the case of 1chip DC/DC converters, although the parasitic inductances are negligible and package size can be even smaller, the output current of 1 chip converter has been limited to several amperes because of the large on-resistance of output lateral devices. In this paper, we demonstrate 10A operation and high speed switching in one-chip converter by applying a metal interconnect with bump technology and adopting distributed driver circuit layout.

II. DEVICE STRUCTURES

Figure 1 shows a cross-sectional view of 20V output LDMOS devices based on the low cost 0.6um BiCD process. The device is fabricated in the p-well so that the gate drain capacitance is minimized. The buried N+ layer is electrically connected to the source electrode to reduce the coupling between the drain and the substrate. Three metal layers with a 3um thick top metal are utilized. The drift region is self-aligned to the gate electrode in order to reduce the parasitic

gate-drain capacitance which affects the switching power loss. In case of N-ch LDMOS in figure 1(a), the breakdown voltage degrades as the drain current increases, and the I-V curves show snapback characteristics. We have applied 2-step shallow n-implant structure (Adaptive Resurf) to the N-ch LDMOS as shown in figure 1(b)[3]. Figure 2 shows the measured I-V characteristics of the fabricated N-ch LDMOS. The applied gate-source voltage is 5V and the channel width is 157um. The 2-step shallow n-implant structure improves the on-state breakdown voltage of 25V, as shown in figure 2. The measured values of the device characteristics are listed in Table I. The optimized breakdown voltage, on-state voltage, threshold voltage and specific on-resistance is 24.4V, 25.0V, 0.85V and 23.6mΩmm^2, respectively.

III. POWER IC USING BUMP TECHNOLOGY

The on-resistance of lateral MOSFETs with bonded wires deteriorates considerably with increasing device size due to the parasitic resistances. The interconnect resistance, including that of bonding wires, not only increases overall device resistance but also causes a debiasing effect in active cells. In order to reduce the interconnect resistance, we have adopted bump technology.

Figure 3 shows the assembled image, the layout of the top metal in IC and the Cu pattern on a printed circuit board (PCB). The chip is attached to the intermediate PCB through bump balls. We have adopted P-ch LDMOS for high side switching device in DC/DC converter. The source and drain metals are alternately formed and the drain metals of P-ch LDMOS and N-ch LDMOS are electrically connected. The drain and the source bumps are electrically connected by parallel running thick Cu metals in the PCB. The resistance between two adjacent bumps in the top metal is made as small as possible.

Figure 4 shows the output characteristics of a large area N-ch LDMOS (the effective area 3.6mm^2). The on resistance is 9.7mΩ when the drain current and the gate voltage is 5A and 5V, respectively.

IV. DISTRIBUTED DRIVER CIRCUIT LAYOUT

When the area of LDMOS is large, the gate current between the gate driver and LDMOS becomes increasingly large. The whole LDMOS device doesn't uniformly turn-on or -off because the gate drive delay may occur within the large LDMOS. The parasitic resistances and capacitances of the gate interconnects induce the gate signal delay. Especially in the turn-off period, the gate delay may cause significant non-uniform switching because of the small threshold voltage (0.85V). Thus, we propose "distributed driver circuit layout."

Figure 5 compares the distributed driver circuit layout with the conventional concentrated driver circuit layout. In the concentrated driver circuit layout (a), a large gate charging or discharging current flows in one large gate driver signal bus line and the parasitic resistance of the signal bus line causes non-uniform gate voltage distribution. In the distributed driver circuit layout (b), a number of gate driver circuits are located beside the segmented LDMOS and the length of the gate charging or discharging current flow path is made as short as possible. The current magnitude of the signal bus line is substantially small, and does not cause the gate signal delay.

Table II shows the simulated switching loss of the distributed driver circuit and the concentrated driver circuit under resistive switching when the input voltage, the load resistance and switching frequency is 12V, 1.2ohm and 780 kHz, respectively. The simulated condition is that the channel width of the gate driving MOSFET in the concentrated driver circuit is equal to the total channel width of the gate driving MOSFETs in the distributed driver circuit. As shown in table II, the turn off loss of distributed driver circuit is reduced to 54% of that of concentrated driver circuit.

Figure 6 shows the micrograph of fabricated chip based on the low cost 0.6um process. The chip size is 20.3mm². A number of driver circuits are placed between N-ch and P-ch LDMOS.

Figure 7 shows the equivalent circuit of the fabricated chip. This circuit is consisted of driver circuits for N-ch and P-ch LDMOS, level shift circuit and a regulator circuit which generates 5V and VIN1-5V (VIN1: input voltage) for driver circuits.

Figure 8 shows the switching characteristics of the fabricated chip at the condition of an inductance, 2uH. The rise time of switching node is 3ns. The fabricated devices in the one-chip DC/DC converter indicates high speed and high current switching capability of 10A.

Figure 9 shows the measured dependence of efficiency on output current, when the input voltage Vin, the output voltage Vout and switching frequency fsw is 12V, 1.3V and 780kHz, respectively. The maximum efficiency is 88.9% at output current 5A. The fabricated chip has accomplished the high efficiency.

V. CONCLUSION

In this paper, we demonstrated high speed 10A DCDC converter by applying a metal interconnect with bump technology and adopting distributed driver circuit layout.

ACKNOWLEDGEMENT

The authors would like to thank Mr. M. Yamaguchi for providing the opportunity of this study and Prof. D.Maksimovic from Colorado University for providing the RTL code of FPGA, which was used to drive our chip in 780kHz.

REFERENCES

[1] Y.Kawaguchi et al., "Multi Chip Module with Minimum Parasitic Inductance for New Generation Voltage Regulator,"ISPSD'15, pp.371-374

[2] M.Shiraishi et al.,"Low Loss and Small SiP for DC-DC Converters,"ISPSD'15, pp.175-178

[3] K. Kinoshita et al.,"A New Adaptive Resurf Concept for 20V LDMOS Without Breakdown Voltage Degradation at High Current,"ISPSD'18, pp. 65-68.

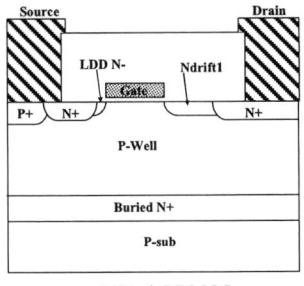

(a)N-ch LDMOS
(1-step n-implant)

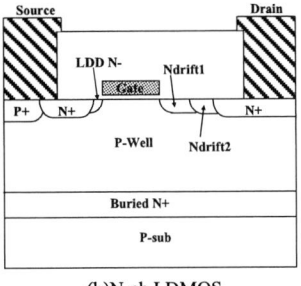

(b)N-ch LDMOS
(2-step n-implant)

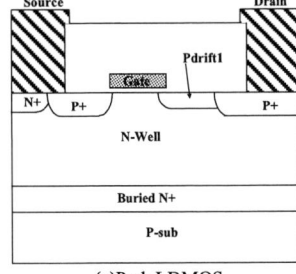

(c)P-ch LDMOS

Figure.1 Cross-sectional view of 20V output power devices based on the low cost 0.6um BiCD process. In case of N-ch LDMOS, we have adopted 2-step shallow n-implant structure (Ndrift1 & Ndrift2) which improves the on-state breakdown voltage.

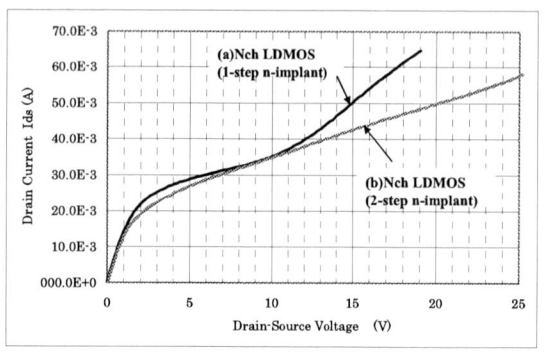

Figure.2 Measured I-V characteristics for fabricated N-ch LDMOS (@Vgs=5V, Channel width=157um). The 2-step n-implant structure improves the on-state breakdown voltage of 25V.

Table I Device characteristics

	Nch LDMOS (1-step n-implant)	Nch LDMOS (2-step n-implant)	Pch LDMOS
Static Breakdown Voltage (V)	25.0	24.4	23.4
On-state Breakdown Voltage (V)	19.0	25.0	25.0
Threshold Voltage (V)	0.85	0.85	-0.85
On-resistance (mohm · mm^2)	23.1	23.6	42.1

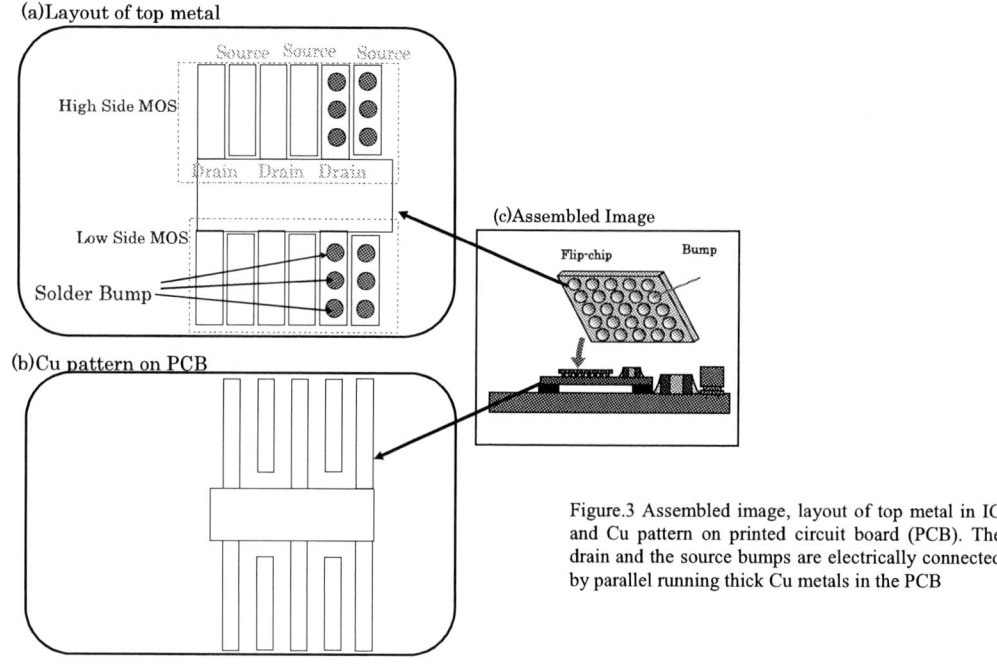

Figure.3 Assembled image, layout of top metal in IC and Cu pattern on printed circuit board (PCB). The drain and the source bumps are electrically connected by parallel running thick Cu metals in the PCB

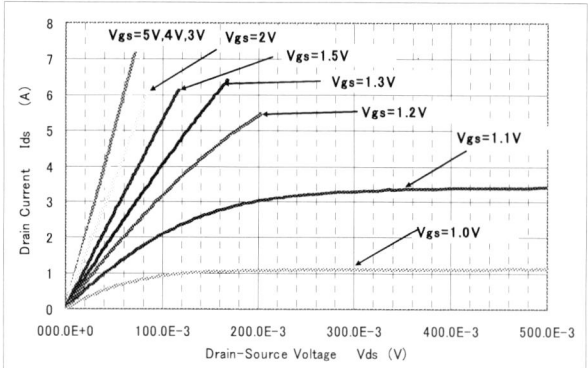

Figure.4 Output characteristics of a large area device (@effective area =3.6mm^2). Ron is 9.7m Ω (@Vgs=5V, Ids=5A)

120

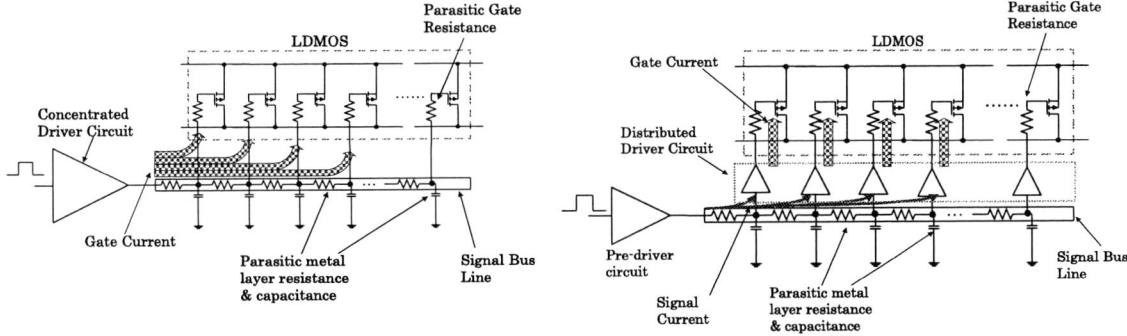

(a) Concentrated driver circuit layout (b) Distributed driver circuit layout

Figure.5 Comparison of the distributed driver circuit layout with the conventional concentrated driver circuit layout

Table II. Simulated Device characteristics

	Concentrated driver circuit layout	Distributed driver circuit layout
Turn-on loss (mW)	12.1	5.48
Turn-off loss (mW)	48.6	26.6

@Vin=12V, Resistance =1.2ohm, Switching frequency=780kHz

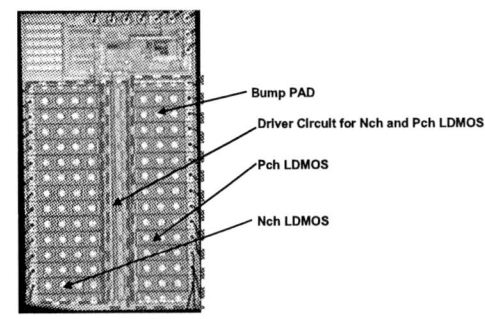

Figure.6 Micrograph of fabricated chip based on low cost 0.6um process. The chip size is 20.3mm^2

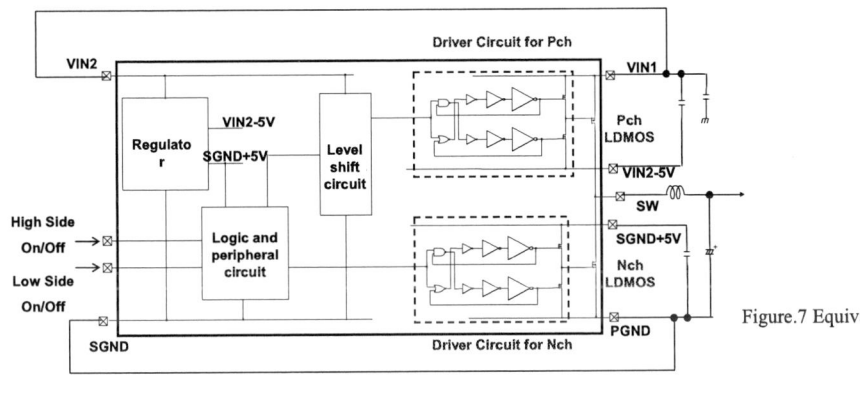

Figure.7 Equivalent circuit of fabricated chip

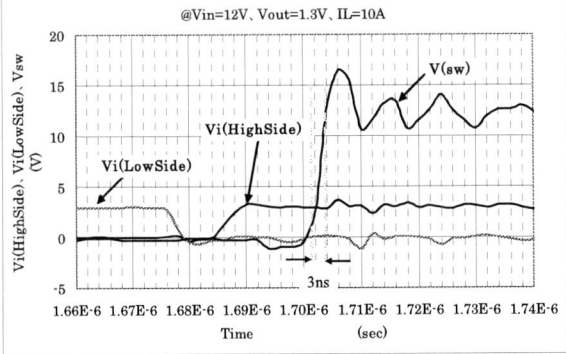

Figure.8 The switching characteristics of the fabricated chip at the condition of an inductance 2uH. The rise time of switching node is 3ns.

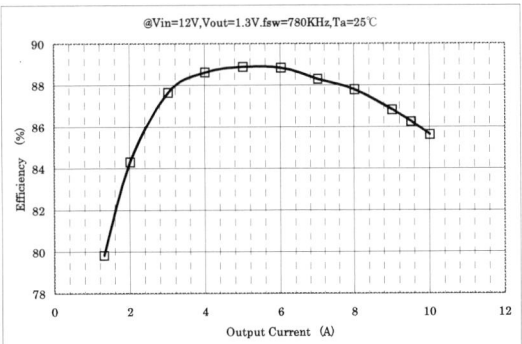

Figure.9 The measured dependence of efficiency on output current, (Vin=12V, Vout=1.3V, fsw=780kHz)

2006 Bipolar/BoCMOS Circuits and Technology Meeting Proceedings

High RF performances asymmetric spacer NLDMOS integration in a 0.25µm SiGe:C BiCMOS Technology

Bertrand Szelag, Dorothée Muller, Jocelyne Mourier, Caroline Arnaud, Halim Bilgen, Fabienne Judong, Alexandre Giry, Denis Pache, Agustin Monroy

STMicroelectronics, Centre Commun de Microélectronique de Crolles,
BP16, F-38921 Crolles, France

Abstract — An asymmetrical spacer LDMOSFET integrated in a 0.25µm BiCMOS technology is presented. Improved RF performances are obtained with this new architecture: f_T close to 35GHz with BVds larger than 15V. Process integration strategy is discussed. Impact on the other devices is described.

Index Terms — Power MOSFETs , Power Amplifier, RF optimization.

I. INTRODUCTION

Integration of high voltage devices in advanced technology is mandatory to satisfy modern wireless circuit needs [1,2]. Drain extension MOS transistors are widely used to address this challenge. These devices fit power management requirements where low voltage circuits interface with other circuits operating at higher voltage or with battery. In the case of RF power applications these transistors are in competition with hetero-junction bipolar transistors which traditionally present better dynamic performances. Specific RF optimization of lateral drain extension DMOS transistors has been presented in [3]: reduction of parasitic capacitances and achievement of a fully salicided gate have been identified as the main architecture improvement axes to get better RF performances.

In this paper, the architecture of an advanced NLDMOS FET is described. Achievement of high dynamic performances is reached thanks to an asymmetric spacer structure. Motivations for such architecture are given; its process and its integration in STM advanced RF BiCMOS process are described. Electrical performances, including large signal characteristics, are presented. Finally the impact of this NLDMOS FET integration on the other devices is discussed.

II. RF-NLDMOS ARCHITECTURE AND PROCESS DESCRIPTION

Two approaches can be used to achieve unsalicided drain extension: either by using the STI/LOCOS or by using a photolithographic process to define a salicide protection layer. These two architectures are presented on figure 1. The second solution (right side cross-section) leads to better dynamic performances and is preferred to RF Power Amplifier designs. Meanwhile, this architecture can be more optimized. As shown on the cross-section, the salicide protection, which defines the drain extension, overlaps the poly-silicon gate in order to take into account potential misalignments of the photolithographic process. It leads to a poly-silicon gate length larger than what would be required to control the P-body region (channel) and then to a large gate to drain capacitance.

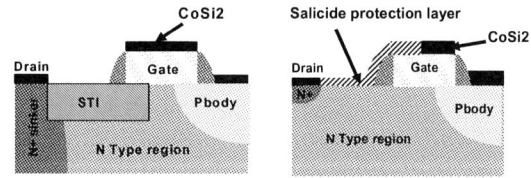

Fig. 1: Drain extension DMOSFET. Left side: extension under STI; right side: extension using salicide protection photo-lithographic process.

Our new architecture is based on a fully salicided gate. The salicide protection overlap on the gate is suppressed thanks to the definition of a large spacer on drain side, which can 'absorb' the potential misalignments of the photolithographic process as illustrated in figure 2. Then, the gate length can be reduced to its minimum value, slightly larger than the P-body diffusion length, which will minimize the gate to drain capacitance.

1-4244-0458-4/06/$20.00 ©2006 IEEE   122

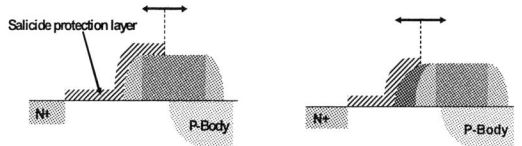

Fig. 2: Salicide protection photo misalignment (arrow). Left: conventional architecture. Right: asymmetric spacer.

The process integration scheme is illustrated in figure 3. Our 0.25µm SiGe:C Bicmos technology has been extensively described in [4]. The asymmetric NLDMOS is realized in 3 steps. First, N-type drift region is defined during CMOS well formation by a set of specific implantations. The P-body implantation which defines the transistor channel is realized after the SiGe:C HBT module to limit the boron diffusion under the gate due to the HBT thermal budget and just before the CMOS spacer formation to be self-aligned with the gate. The resulting channel length is slightly larger than 0.15µm. The last step, specific to this device architecture, consists in the formation of an additional spacer on drain side. A thick nitride is used in order to achieve a sufficiently wide spacer. As this spacer is needed only on NLDMOS drain side, a dedicated mask is used to protect it while all the 'parasitic' spacers are removed on the other devices and on the source side. Two devices are proposed: one high voltage (HV) transistor featuring a source-drain breakdown voltage (Bvds) around 15V; the other one, low voltage (LV), with a BVds around 9V. These two transistors are using the same P-Body implantation but are realized in different N-drift regions.

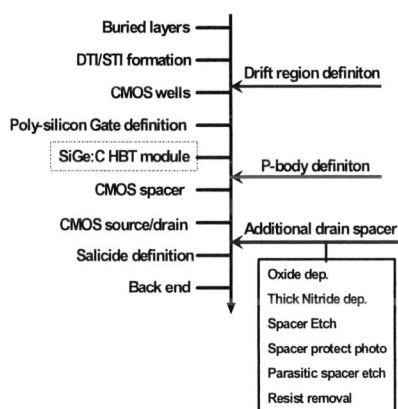

Fig. 3: Process integration scheme

Integration of such a device in our already qualified BiCMOS process is quite challenging since the overall thermal budget applied on other devices is modified. Special care must be taken in the choice of the material used to form the additional spacer. Another major

integration issue is the parasitic spacer removing. A TEM picture showing the device is given in figure 4. It is clearly illustrated that the gate is completely salicided. The additional spacer is located only on drain side and totally removed on source side. The salicide protection on the drain extension is well defined on top the drain spacer.

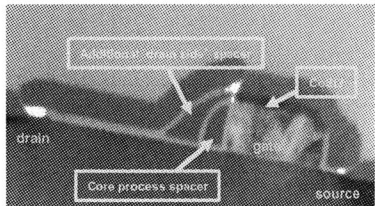

Fig. 4: Asymmetric LDMOS TEM picture.

III. RF-NLDMOS ELECTRICAL PERFORMANCES AND NON LINEAR CHARACTERIZATION

In this section, the electrical performances of HV and LV transistors are presented. Table 1 summarizes the main parameters. These results can be compared to our first generation NLDMOS presented in [5]. Since the optimization of this new device is mainly focused on dynamic performances through a reduction of the parasitic capacitances, the major improvement is observed on f_T which is enhanced of more than 10 GHz while keeping almost the same DC characteristics. The f_TxBVds product of the new device is 505 V.GHz compare to 300 V.GHz in the previous published results.

	HV	LV	Conditions
W.Ron (Ω.mm)	2.80	2.05	Vd=0.1V; Vg=2.5V
Sat. Current (µA/µm)	440	505	Vd=3.6V; Vg=2.5V
Leak. Current (pA/µm)	5	300	Vd=3.6V; Vg=0V
Breakdown Voltage (V)	15.8	9.5	-
Peak fT (GHz)	32	40	Vd=3.6V

Table 1: HV and LV NLDMOS electrical performances

Figure 5 presents the output characteristics of HV and LV LDMOS FETs at Vg=1V and 2.5V. Saturation is well observed for both devices. Lower Ron and larger saturation current for LV transistor are clearly highlighted. Figure 6 presents the breakdown characteristics. As explained in the previous section, HV and LV devices are made in different N-drift regions. A higher doping level is used for LV device; it leads to a smaller channel length since the Pbody compensation occurs sooner. It explains the higher leakage current observed for the LV transistor. Breakdown voltage is defined by the doping levels at the N-drift/Pbody junction. Finally, figure 7 shows the transit frequency variations with the drain current for 1.5mm gate width HV and LV transistors operating at Vds=3.6V.

Lower N-drift resistance mainly explains the higher f_T observed for the LV transistor.

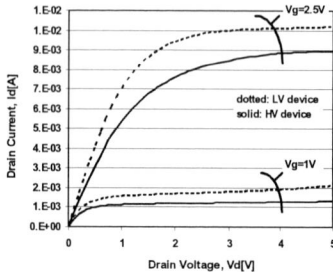

Fig. 5: Output characteristics (20µm gate width HV & LV devices).

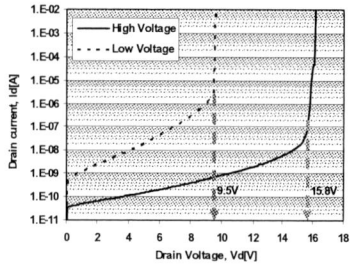

Fig. 6: Breakdown characteristics (20µm gate width HV & LV devices).

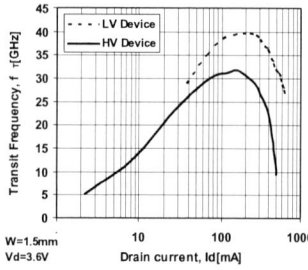

Fig. 7: Transit frequencies (HV & LV devices)

To complete this study, power characterizations have been done with a Focus Microwaves multi-harmonic load-pull bench based on automated tuners. This bench enables to control the source and load impedances presented to the device at the fundamental frequency, the 2nd and the 3rd harmonic frequencies, independently one from the others. Frequency and drain/collector voltage have been set respectively to 1.8GHz and 3.6V. Figure 8 presents the optimal source impedance, Zs, of 1.5mm gate width HV NLDMOS transistors. Higher input impedance is observed for the new architecture is well correlated to a reduced gate to drain capacitance compared to the old device. The power characteristics of a 1.5mm gate width asymmetric HV NLDMOS are shown in figure 9. For these curves, drain to source voltage is 3.6V, quiescent current is 15mA. The saturated output power is close to 25dBm with almost

85% efficiency. Complementary ruggedness measurements are as important as electrical characteristics. A 10:1 VSWR is sustained at Vds=5V for the asymmetric NLDMOS which is comparable to the old device showing that robustness of the device has not be sacrificed to gain in performance.

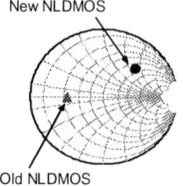

Fig. 8: Optimal source impedance

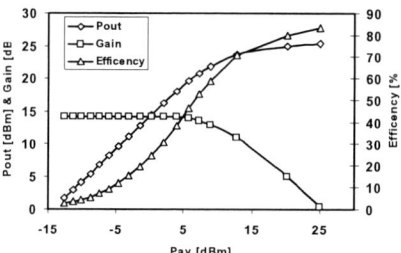

Fig. 9: Pav-Pout characteristics in the load pull measurement setup (W=1.5mm) at 1.8GHz and Vds=3.6V.

IV. RF-NLDMOS INTEGRATION IMPACTS ON TECHNOLOGY

In this section, LDMOS transistor integration constraints will be reviewed and discussed. Since the new device is implemented in an already qualified technology, it must not impact the other devices to keep the compatibility with the core process. Knowing the process integration scheme, two different aspects must be considered:

    1. The impact of the thermal budget modification
    2. The removing of the different layers added on the other devices.

First, the modification of the thermal budget is related to the addition in the core process of the deposition of an oxide/nitride stack to form the additional spacer. It adds thermal budget while all devices have been finished. Diffusion and activation processes can then be affected, leading to device electrical characteristic shifts. Since one can not avoid the addition of these materials to form the additional spacer, they must be chosen with the lowest thermal budget to minimize the impact. The selected oxide is a TEOS with a sufficient thickness of few hundreds of angstroms. The deposition temperature is slightly larger than 500°C which is low enough. The nitride conformity must be good to form the spacer. Several types of nitride have been evaluated to satisfy this need, taking also into account the thermal aspect. Finally, CVD nitride has been found to be the best candidate. It is processed during few minutes at a temperature lower than 500°C to get a thick enough layer for the achievement of the additional spacer width.

Secondly the etch process, which is necessary to remove the parasitic spacers on the source side and on all the other devices is very critical. The dry-nitride etching process must satisfy the following characteristics: good selectivity with respect to the oxide and low resist consumption to be sure that the drain-side-spacer will be protected. The first aspect will guarantee the other device integrity, the oxide layer being removed by a wet etching process done just before salicide material deposition. Unfortunately, good oxide selectivity is often obtained to the detriment of the resist consumption. Etching chemistry has been optimized to find out the best trade-off between these two aspects. Design rules of the drain-spacer protection mask is also a degree of freedom that can help for this etching process.

The impact of the NLDMOS integration is illustrated in table 2, 3 and 4 which compare the main electrical parameters of the technology without ('std' column) and with ('new' column) the integration of the NLDMOS. The CMOS parameters (table 2) are equivalent which shows that the limitation of the additional thermal budget has been efficient for these devices. Table 3 presents results on the resistance values. Here also, no impact is seen for any type of resistance. P- resistance, boron doped, is not affected by thermal budget modification. Salicided resistance remains the same, which indicates that the oxide layer used in the additional spacer has been well removed and does not affect the salicide formation. Finally, table 4 presents the electrical parameters of the Low voltage (LV) and high voltage (HV) NPN SiGe:C HBT. It is clearly shown that the devices are not impacted. The use of carbon in the p-type base makes the integration of the devices easier since it blocks the boron diffusion and then reduces the impact of the thermal budget modification.

Parameters	Std	New	Units
Vt NMOS	645	640	mV
Vt PMOS	546	548	mV
Ion NMOS	5.82	5.90	mA
Ion PMOS	2.43	2.46	mA
Ioff NMOS	12	16	pA
Ioff PMOS	18	22	pA

Table 2: CMOS transistor parameters with and wo integration of NLDMOS.

Parameters	Std	New	Units
P+ Poly	89	89	Ω/sq
N+ Poly	181	182	Ω/sq
P- Poly	994	1005	Ω/sq
N+ Active	72	72	Ω/sq
N+ Active sal.	6.14	6.37	Ω/sq

Table 3: Sheet resistances with and wo integration of NLDMOS.

Parameters	Std	New	Units
Beta	183	180	-
BVceo (LV)	3.41	3.50	V
BVceo (HV)	6.46	6.35	V
BVcbo (LV)	13.0	13.1	V
BVcbo (HV)	19.0	18.8	V

Table 4: : NPN HBT parameters with and wo integration of NLDMOS.

V. CONCLUSION

In this paper, the architecture and the process of a new high RF performance lateral drain extension DMOS have been described. This device features a fully salicided gate achieved by the use of an asymmetrical spacer system. The specific process modules have been described and the main electrical characteristics of this new device are given. High voltage transistor exhibits a f_T around 32GHz for Vds=3.6V while BVds is larger than 15V. For the low voltage device, f_T reaches 40GHz with BVds=9.5V. Finally, the integration constraints in our already qualified 0.25µm RF BiCMOS process are discussed. No variations of the electrical characteristics of the other devices have been noticed showing the successful integration of this asymmetric spacer NLDMOS transistor.

REFERENCES

[1] D. Harame et al, "Imagine the Future in Communications Technology", ESSDERC 2002 proceedings, pp 53-60

[2] L. Larson, "RFIC Requirements for RF Devices", IEDM 2002 short course

[3] B. Szelag et al, "NLDMOS RF Optimization Guidelines For Wireless Power Amplifier Applications", BCTM 2005 proceedings, pp.280-283

[4] H. Baudry et al., "BiCMOS7RF: a highly-manufacturable 0.25µm BiCMOS7RF- application-dedicated technology using non selective SiGe:C epitaxy", BCTM 2003, pp207-210.

[5] B. Szelag et al., "Integration and Optimisation of a high performance RF Lateral DMOS in an advanced BiCMOS technology", ESSDERC 2003, p39-42.

2006 Bipolar/BoCMOS Circuits and Technology Meeting Proceedings

A Modular 0.18 um Analog / RFCMOS Technology Comprising 32 GHz F_T RF-LDMOS and 40V Complementary MOSFET Devices

Zachary Lee, Robert Zwingman, Jie Zheng, Will Cai, Paul Hurwitz, and Marco Racanelli

Jazz Semiconductor, Inc., 4321 Jamboree Road, Mailstop H01-109, P.O. Box 7720,
Newport Beach, CA 92658, USA
Phone: 949-435-8576 Fax: 949-435-8187 Email: zachary.lee@jazzsemi.com

Abstract — **High-performance RF-LDMOS, medium voltage (8-12V), and high voltage (20-40V) NLDMOS and PLDMOS devices, integrated in a modular 0.18 um analog CMOS / RFCMOS technology platform, are described. Device design, process integration, manufacturability, and reliability issues are discussed. These devices are among the highest performance in their class. The successful integration of these devices has enabled us to address our roadmap in the growing wireless, analog, mixed-signal, and power management markets.**

Index Terms — **Silicon RF-LDMOS technology, LDMOS, power devices, CMOS analog integrated circuits.**

I. INTRODUCTION

The ability to integrate wireless, analog, digital, and power management functionalities on silicon in an advanced technology node requires a modular technology platform that is high-performance, manufacturable, reliable, expandable, and flexible with respect to integration and design. In order to address these needs, high-performance RF-LDMOS devices as well as complementary MOSFET devices addressing the medium voltage range (8-12V) and the high-voltage range (20-40V) applications, integrated in Jazz's modular 0.18 um technology platform, have been developed. Markets being addressed include cellular phone power amplifiers, power management, MEMs driver, display driver, class D audio amplifier, as well as other analog and high voltage applications.

Jazz's 0.18 um process platform is an analog / RFCMOS process consisting of core 1.8V CMOS, with either 3.3V or 5V FETs in a dual gate architecture. Other features include 2 fF/um^2 (unstacked) or 4 fF/um^2 (stacked) MIM capacitors, low Vt native NFETs, triple well isolation, varactor, salicided and unsalicided poly resistors, NWELL resistors, vertical and lateral PNP transistors, metal fuses, 6 layers aluminum metal, 2.8 um top metal, as well as high Q inductors. The RF-LDMOS, medium voltage, and high voltage devices that are introduced in this paper are additional features developed for this family.

Two versions of RF-LDMOS devices exist, differing primarily by gate oxide thickness (1.8V gate oxide or 5V gate oxide) as well as poly gate length. Cutoff frequency (Ft) and off-state drain breakdown voltage (BV$_{DSS}$) of 32 GHz and 14V, respectively, have been achieved in the 1.8V gate oxide devices, while the 5V gate oxide devices exhibited Ft = 19 GHz with BV$_{DSS}$

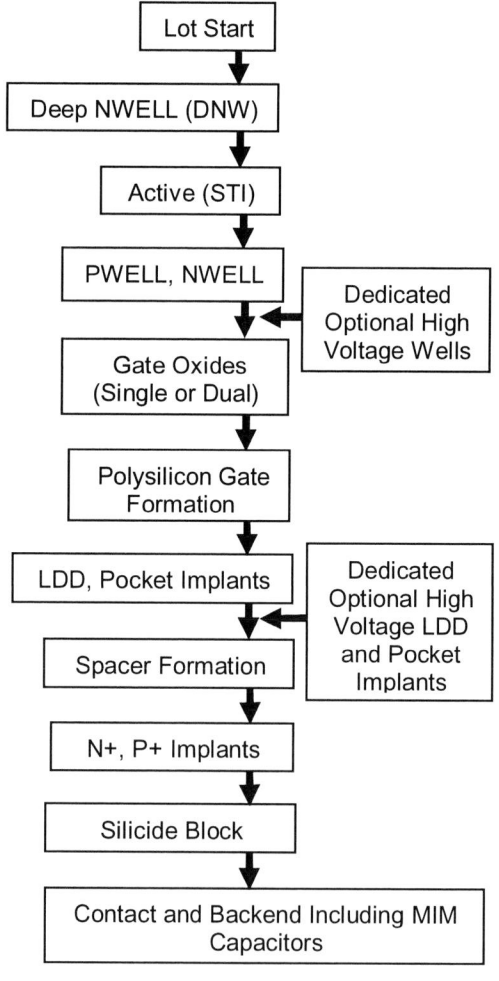

Figure 1. Simplified process flow of Jazz's 0.18 um process platform.

1-4244-0458-4/06/$20.00 ©2006 IEEE

TABLE I

SPECIFICATIONS OF VARIOUS DEVICE AND DEVICE OPTIONS AVAILABLE IN JAZZ'S 0.18 um ANALOG AND RFCMOS PLATFORM

Feature	VGS (V)	VDS (V)	BV$_{DSS}$ (V)	Vt (V)	Rdson (mohm-mm^2)	Idsat (uA/um)	Notes
LV Core NFET / PFET	1.8	1.8		0.52 / -0.44		600 / -255	
Dual Gate NFET / PFET	3.3	3.3		0.62 / -0.74		600 / -290	
Dual Gate NFET / PFET	5	5	> 8	0.65 / -0.77	2.3 / 7.9	520 / -260	
Medium Voltage NLDMOS / PLDMOS	3.3	8	12 / -10	0.63 / -0.83	16 / 35	420 / -108	
Medium Voltage NLDMOS / PLDMOS	5	12	17 / -14	0.72 / -0.81	16 / 36	428 / -203	
High Voltage NLDMOS	5		38	0.63	22.5	350	
High Voltage NLDMOS	5		47	0.63	40	340	
High Voltage NLDMOS	5		52	0.63	59	280	
High Voltage Isolated NLDMOS	5		25	0.63	22.5	350	
High Voltage PLDMOS	5		-42	-0.84	80	-140	
High Voltage PLDMOS	5		-52	-0.84	120	-140	
High Voltage PLDMOS	5		-55	-0.84	170	-140	
RF-LDMOS NFET	1.8		14-18	0.36	3.9 ohm-mm	480	Ft > 30 GHz
RF-LDMOS NFET	5		20-25	1.2	2.9 ohm-mm	530	Ft > 18 GHz

= 21V. Other high-breakdown process variants of the same device exhibited Ft > 30 GHz with BV$_{DSS}$ = 18V (1.8V gate oxide) and Ft = 18 GHz with BV$_{DSS}$ = 25V (5V gate oxide).

Medium and high voltage complementary LDMOS devices are also built. The 8V devices (with 12V / -10V NLDMOS / PLDMOS BV$_{DSS}$) are built with a 3.3V gate oxide variant without additional processing steps, while the 12V devices (with 17V / -14V NLDMOS / PLDMOS BV$_{DSS}$) are built with a 5V gate oxide variant, also without additional processing steps. The 20-40V devices (with BV$_{DSS}$ to 55V) are available with the addition of two masking and implant steps. A summary of specifications of various devices and process variants is provided in Table I.

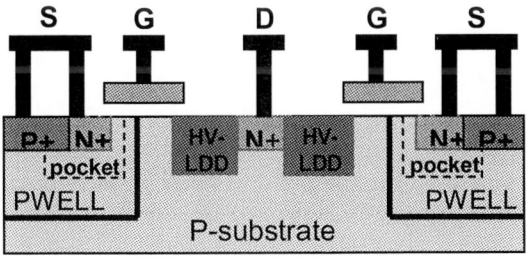

Figure 2. Schematic cross-section of an RF-LDMOS device.

II. PROCESS FLOW

The simplified process flow is shown in Figure 1. Device isolation from the p-substrate is accomplished by a multi-MeV deep NWELL (DNW) implant near the beginning of the flow, followed by shallow trench isolation (STI) based active area formation. NWELL and PWELL implants, as well as dedicated wells for the specific process modules (depending on process options) are then performed, followed by gate oxidation in which single or dual gate oxides are formed. Polysilicon is then deposited and patterned, followed by various LDD and pocket implants (depending on process options), spacer formation, N+ and P+ implants, silicide block formation, contact module, and then a six layer aluminum metal backend, which also includes MIM capacitors.

III. RF-LDMOS DEVICES

The basic device structure of the RF-LDMOS device is shown in Figure 2. The POLY, N+, and P+ layers are shared with the base CMOS devices, while the HV-LDD is formed with the implantation of a light n-type dopant after polysilicon gate formation. A silicide-blocking layer, part of the standard CMOS process to form unsalicided poly-resistors, is needed to prevent silicidation of the region above the HV-LDD to achieve high BV$_{DSS}$. Since the PWELL is not self-aligned to the polysilicon gate, and is subject to manufacturing variations, a source-side pocket implant is needed to suppress leakage, snapback, and device parameter roll-off in order to account for the worst-case alignment condition. A comparison of devices with and without pocket implant is shown in Figure 3. It can be seen that the pocket implant plays a crucial role in suppressing the off-state leakage (among other parameters) caused by PWELL alignment variation. In designing these devices, the alignment of the PWELL with respect to the polysilicon gate is chosen such that the device performance is maximized without device parameter roll-off in the worst-case alignment condition. Measured Ft and Vt of an RF-LDMOS device having a 1.8V gate oxide (with pocket implant) as a function of PWELL misalignment is shown in Figure 4. Typical Ft performance as a function of VGS and VDS for 1.8V and 5V RF-LDMOS devices is shown in Figure 5. A comparison of this work with recent published results is given in Figure 6. The devices shown in this work are among the highest performance in their class today.

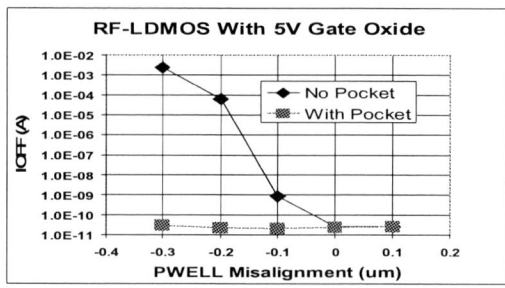

Figure 3. Comparison of off-state leakage of a 100 um wide RF-LDMOS device with and without pocket implant as a function of PWELL misalignment.

For the 1.8V gate oxide device, the source-side pocket implant is a dedicated masking layer, while for the 5V gate oxide device, this implant is shared with the CMOS devices in a single masking step. In a dual gate oxide process with 1.8V / 3.3V gate oxides, the 1.8V RF-LDMOS devices are fabricated with two additional masking layers (source side LDD and pocket, plus drain side HV-LDD), while in a 1.8V / 5V process, the 5V RF-LDMOS devices are fabricated with either one or two additional masking layers (drain side HV-LDD with optional dedicated PWELL), depending on the particular specifications.

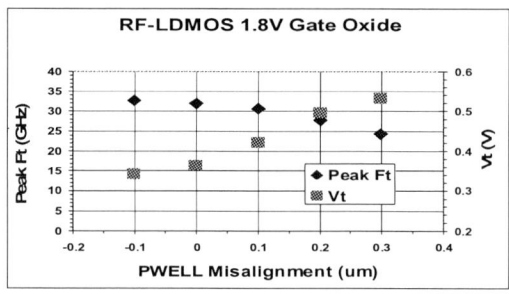

Figure 4. Measured Ft and Vt as a function of PWELL misalignment of a 1.8V gate oxide RF-LDMOS device.

The demand for very high performance of these devices makes them vulnerable to degradation due to

hot carrier (HCI) effects. For this reason, a significant amount of effort has been put in the optimization of these devices to improve their HCI reliability. It was found that the overlap region of the HV-LDD by the polysilicon gate plays a very important role in the device degradation. Optimization of the PWELL alignment and HV-LDD implant has allowed us to significantly increase the maximum operating VDD while achieving very low device degradation. The Rdson lifetime as a function of 1/VDS stress of a 5V RF-LDMOS device under worst-case bias condition is given in Figure 7.

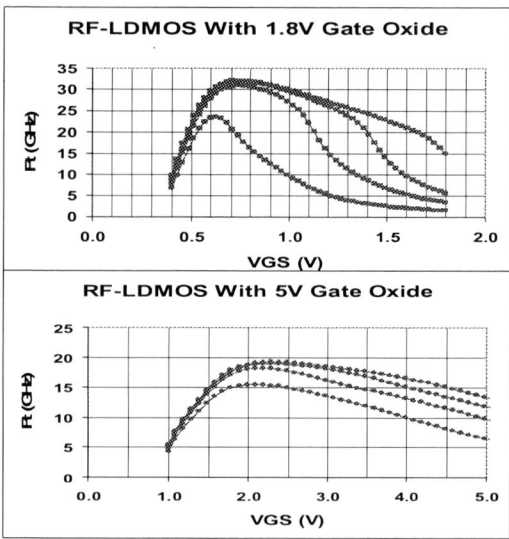

Figure 5. Measured Ft as a function of bias. For the 1.8V device, Ft was measured at VDS = 1, 2, 3, and 5V. For the 5V device, Ft was measured at VDS = 2, 3, 4, and 5V.

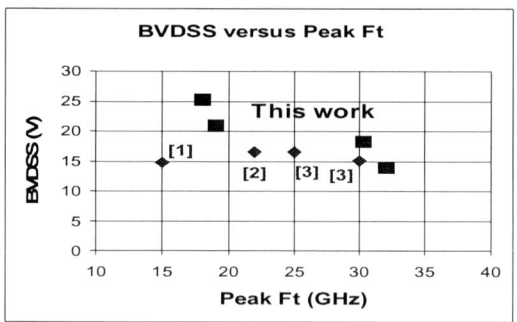

Figure 6. Comparison of various recent published results with this work [1-3].

IV. MEDIUM AND HIGH VOLTAGE LDMOS DEVICES

Power management, display driver, MEMs driver, class D audio amplifier, as well as other advanced applications can also be integrated onto the same silicon using the available medium and high voltage LDMOS devices. The 8V drain-extended devices (with 12V / -10V NLDMOS / PLDMOS BVDSS) are

128

built with a 3.3V gate oxide variant, while the 12V devices (with 17V / -14V NLDMOS / PLDMOS BVDSS) and 20-40V devices (with BVDSS to 55V) are built with a 5V gate oxide variant.

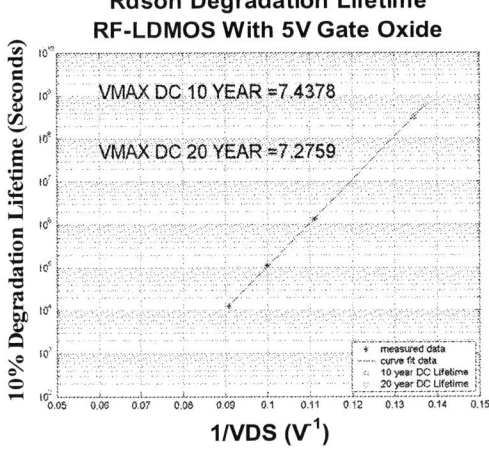

Figure 7. Rdson degradation of an RF-LDMOS device having 5V gate oxide as a function of 1/VDS under worst-case VGS bias = 1.9V.

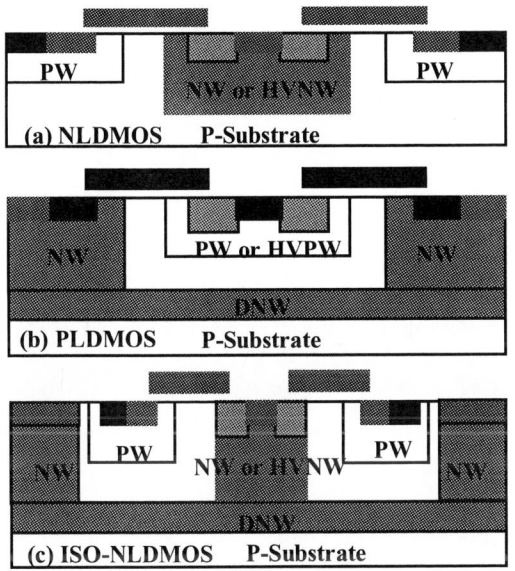

Figure 8. Schematic cross-sections of various types of medium and high-voltage LDMOS devices.

Cross-sections of NLDMOS, PLDMOS, as well as isolated NLDMOS devices are shown in Figure 8. The 8V or 12V devices make use of the PWELL or NWELL used in the low voltage devices as channels or drain extensions, and do not require additional processing steps. Isolation of the NLDMOS body as well as PLDMOS drift regions is provided by a high-energy DNW implant, which is shared with the triple well isolated low voltage devices. The high voltage devices (20V-40V) require two additional masking

(and implant) steps to form the dedicated drain extensions. Along with layout optimization, the addition of dedicated layers has allowed us to optimize the extended drain implants to achieve the best-in-class RdsonA-BVDSS tradeoff (Figure 9), while achieving good HCI reliability. In order to maximize flexibility in their applications, these devices feature scalable drain extensions to achieve different ratings and SOA requirements versus RdsonA tradeoff. The drain extension of the isolated NLDMOS is not scalable, however, since it is limited by PWELL to DNW breakdown, with BVDSS = 25V, RdsonA = 22.5 mOhm-mm^2. Optimization of the PWELL should allow a higher BVDSS to be realized without penalty on RdsonA. DNW to substrate breakdown in this process is 55V. The re-use of various low voltage and high voltage wells has also allowed us to construct high-voltage diodes and bipolar transistors without additional processing steps.

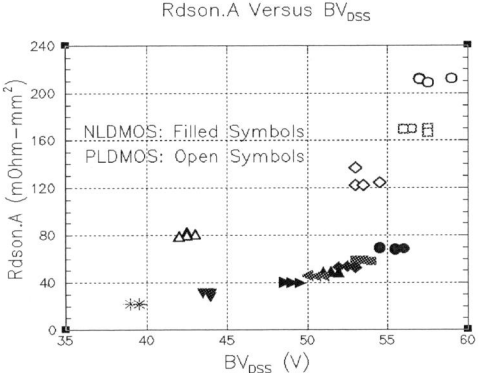

Figure 9. Rdson.A versus BVDSS tradeoff of NLDMOS and PLDMOS.

V. CONCLUSION

We have successfully demonstrated very high performance RF-LDMOS as well as medium and high voltage NLDMOS and PLDMOS devices in a modular 0.18 um analog / RFCMOS process. These devices further enhance our ability to integrate wireless, power management, as well as other advanced functionalities in the growing analog and mixed-signal markets.

REFERENCES

[1] S. Matsumoto and M. Mino, "Impact of the pattern layout on RF performance of TF-SOI power MOSFETs,", *Proc. IEEE ISPSD*, pp. 245-248, 2004

[2] D. Muller, A. Giry, D. Pache, J. Mourier, B. Szelag, A. Monroy, "Architecture optimization of an N-channel LDMOS device dedicated to RF-power application," *Proc. IEEE ISPSD,"* pp. 159-162, 2005.

[3] B. Szelag, D. Muller, J. Mourier, F. Judong, A. Giry, D. Pache, A. Monroy, M. Roche, "NLDMOS RF optimization for wireless power amplifier applications," *Proc. IEEE BCTM*, pp.280-283, 2005.

2006 Bipolar/BoCMOS Circuits and Technology Meeting Proceedings

A Concurrent Fully-Integrated BiFET LNA
for W-CDMA/IEEE 802.11a Applications

C. P. Moreira[1], E. Kerherve[1], P. Jarry[1], D. Belot[2]

[1] IXL Laboratory UMR CNRS 5818– University of Bordeaux – 33405 – Talence – France.
[2] ST Microelectronics – R&D – 38926 – Crolles – France.

Abstract— **This paper presents a new BiFET concurrent fully-integrated low noise amplifier (LNA) targeted to W-CDMA/WLAN IEEE 802.11a standards. The use of a concurrent topology enables saving important die area and power consumption compared to the parallel solution that employs two separated LNAs. A design methodology that helps selecting input/output matching network element values is also presented.**

I. INTRODUCTION

The success of GSM 2G system, essentially voice-based, has increased the demand for multimedia wireless services. In order to meet this market requirement, higher data rates are currently being deployed with recent installed 3G networks. In parallel with cellular developments, wireless data-based systems (e. g., IEEE 802.11b/g/a, HiperLAN2) experiment an exponential growth. Instead of competitive, these technologies are complementary, which requires flexible and optimized devices.

The low noise amplifier is the first active block of an RF receiver front-end. Commercially, several implementations propose dual-band LNAs, mainly covering the whole W-LAN b/g/a standards. However, they mostly use two parallel LNAs, which is obviously not a cost-efficient solution. A concurrent LNA is most suited because it enables saving die area (~25%) and power consumption (~30%) compared to the parallel LNAs approach. A reduced number of CMOS concurrent LNAs implementations have been proposed for W-LAN systems applications [1], [2]. However, in all these contributions, any information concerning the selection of dual-band input/output network elements values is presented. In this paper, we explore the concurrent LNA topology for W-CDMA/W-LAN 802.11a applications. In order to improve LNA linearity, mainly in low-band, we have used a mixed cascode topology, the BiFET, which combines a low-noise input stage (HBT) with a linear MOSFET cascode stage [3], [4]. Also, we propose an original design methodology that allows choosing input/output matching network element values. All concurrent LNA elements are on-chip and it is intended to on-chip probe measurements.

II. CONCURRENT BIFET LNA DESIGN

A. Topology

The concurrent LNA is built based on a mixed BiFET cascode topology. The LNA input impedance matching at two different frequencies (ω_{rm1}, ω_{rm2}) is possible by connecting a parallel LC network (*Lr, Cr*) in series with a traditional single standard LNA input matching network, Fig.1. This LC tank resonates with the single standard LNA matching network (*Lb, Cbe, Le*) at both frequencies. The output matching is performed by connecting a series LC branch (*Ls, Cs*) in parallel with a single standard LNA LC resonator load. Two notches are introduced in S_{11} and S_{21} functions at ω_{no1} and ω_{no2} frequencies, respectively.

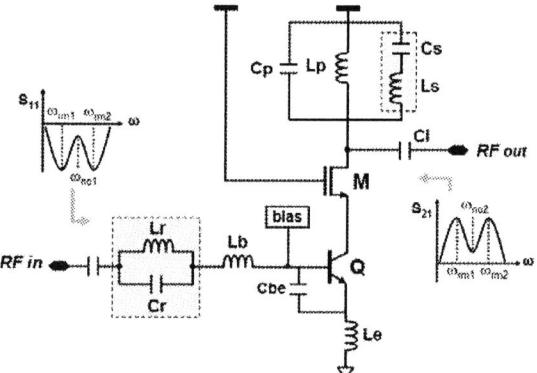

Fig. 1. Concurrent BiFET dual-band LNA topology.

B. Design Methodology
1) HBT Input Stage and Input Matching Network

If we assume a simplified transistor model (*Cbc* is neglected), an expression for the input impedance (*Zin*) of the concurrent input stage can be derivate, Fig. 2.

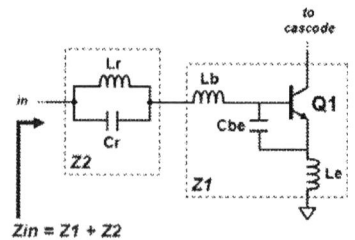

Fig. 2. Concurrent input matching network.

Then, by assuming that the matching elements are lossless, and under the mutual resonance condition ($Im[Z1]=-Im[Z2]$), we can calculate the two resonance frequencies (ω_{rm1}, ω_{rm2}) that enable the concurrent dual-band operation:

$$\omega_{rm1,2} = \sqrt{\frac{\frac{Lt}{Lr}+\frac{Cr}{Cbe}+1}{2CrLt} \pm \frac{\sqrt{\left(\frac{Lt}{Lr}+\frac{Cr}{Cbe}+1\right)^2 - 4\left(\frac{CrLt}{CbeLr}\right)}}{2CrLt}} \quad (1)$$

where $Lt=Lb+Le$. In practice, correctly setting the two desired frequencies with (1) by varying element values can be very time consuming. Then, in order to reduce design and simulation time, a reverse synthesis methodology is here developed.

In this case, the two desired resonant frequencies are first selected. After that, by assuming a constant value for Cbe (it comes from input transistor dimensions [4]), it is possible to choose among a large range of Cr and Lt values that fulfill the mutual resonance condition, for a given range of Lr values. Based on expression (1), we define two variables K^2 and ΔK^2:

$$K^2 = \frac{\frac{Lt}{Lr}+\frac{Cr}{Cbe}+1}{2CrLt} \quad (2)$$

$$\Delta K^2 = \frac{\sqrt{\left(\frac{Lt}{Lr}+\frac{Cr}{Cbe}+1\right)^2 - 4\left(\frac{CrLt}{CbeLr}\right)}}{2CrLt} \quad (3)$$

Then, (1) becomes:

$$\omega_{rm1,2} = \sqrt{K^2 \pm \Delta K^2} \quad (4)$$

with:

$$\omega_{rm1}^2 = K^2 + \Delta K^2 \quad (5)$$

$$\omega_{rm2}^2 = K^2 - \Delta K^2 \quad (6)$$

From (5) and (6), we can write:

$$K^2 = \frac{\omega_{rm1}^2 + \omega_{rm2}^2}{2} \quad (7)$$

$$\Delta K^2 = \frac{\omega_{rm1}^2 - \omega_{rm2}^2}{2} \quad (8)$$

Then, once selected the two desired resonance frequencies, ω_{rm1} (for higher frequency) and ω_{rm2} (for lower frequency), expressions (7) and (8) will acquire numerical values. These values will be further used in expressions (2) and (3). In order to obtain the related plots that enable selecting matching network element values, we first isolate Cr in (2):

$$Cr = \frac{Cbe(Lt+Lr)}{Lr(2K^2LtCbe-1)} \quad (9)$$

where Cbe and K are known. After that, if we replace (9) in (3), we obtain:

$$\left(A^2+B^2\right)Lt^2 + \left[\left(A^2+B^2\right)Lr - \frac{2A}{Cbe}\right]Lt + \frac{1}{Cbe^2} = 0 \quad (10)$$

where $A=K^2$ and $B=\Delta K^2$. Expression (10) is now a function of only two unknown parameters, Lt and Lr. This expression can be solved for Lt, resulting in two possible solutions. Thus, based on these two roots, a first plot of Lt as a function of Lr [$Lt(Lr)$] can be built. Afterward, using the data of the first plot and expression (9), a second plot showing Cr as a function of Lr can be constructed.

This methodology is then applied to assist the design of a concurrent dual-band LNA. The two resonant frequencies ω_{rm1} and ω_{rm2} are set to 5.25GHz and 2.14GHz, respectively. Cbe is 0.9pF. Then, we can now build four plots, two for $Lt(Lr)$ and two for $Cr(Lr)$, Fig. 3 and Fig. 4, respectively.

We can note from these figures that expression (10) has two real solutions (Lt_1, Lt_2) up to an Lr value. Besides this Lr value, a forbidden region is established. Then, Lr values inside this region can not be selected because it would result in resonant frequencies different from the specified. Similarly, this forbidden region is present in Cr plot, as shown in Fig. 4. Also, the two possible Lt_1, Lt_2 values will determine, respectively, Cr_1 and Cr_2 values.

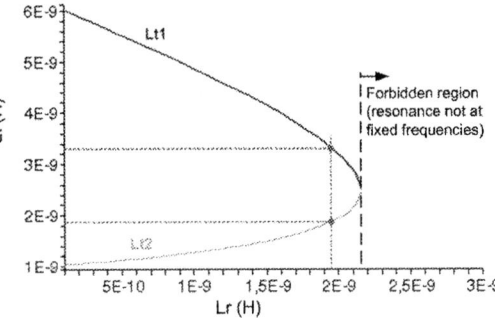

Fig. 3. Variation of Lt as a function of Lr (Cbe fixed).

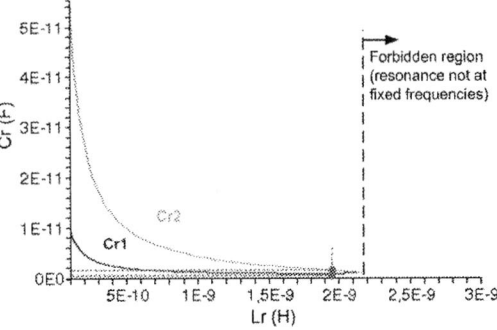

Fig. 4. Variation of Cr as a function of Lr (Cbe fixed).

For Lr=1.95nH, two couple of (Lt, Cr) values are obtained:

$$Lt_1 = 3.335\text{nH}, \; Cr_1 = 0.868\text{pF} \quad (1^{st} \text{ solution}) \quad (11)$$
$$Lt_2 = 1.881\text{nH}, \; Cr_2 = 1.540\text{pF} \quad (2^{nd} \text{ solution})$$

If we consider the first solution, for example, the final values for the passive lossless network components are: Cbe=0.9pF (fixed by input transistor dimensions), Lr=1.95nH (selected), Lt=3.335nH (calculated), Cr=0.868pF (calculated).

These synthesized values are used as initial values when designing the concurrent input impedance matching network. After that, the frequency shifts introduced by inductances (and other components) losses are compensated through an error correction procedure. This procedure feeds back error quantities (frequency shifts) in the lossless analysis described above, and new element values (Lt', Cr') are calculated. This sequence is repeated as long as the minimum error condition, set by the designer, is not fulfilled.

2) MOSFET Cascode Stage

In order to provide higher linearity than a bipolar cascode stage, for the same current consumption, we may choose appropriated dimensions for the MOSFET transistor ($M1$). The optimal gate width (W_{opt}) value, for a fixed gate length (L), is set using equation (12) [3]:

$$W_{opt} = \left(\frac{1}{2\pi} \sqrt{\frac{9 I_C \mu_n}{8 C_{ox} L^3 f_o^2}} + \frac{1}{2\pi} \sqrt{\frac{9 I_C \mu_n}{8 C_{ox} L^3 f_o^2} - \frac{3 C_{bc}}{2 C_{ox} L}} \right)^2 \quad (12)$$

The maximum operating frequency f_o of the BiFET LNA must satisfy (13):

$$f_o \le \frac{1}{2\pi} \sqrt{\frac{3 I_C \mu_n}{4 C_{bc} L^2}} \quad (13)$$

Expression (12) gives an insight for setting the initial value of the optimum gate width (W_{opt}) for the cascode MOSFET transistor, for a given gate length L. Since the LNA operates in two distinctive frequencies, two different gatewidths are calculated. For a bias current of 6mA, we have fixed a gate width of 400μm (16μm and 25 fingers). C_{bc} capacitance is fixed by the input stage optimization (60fF) and a gate length of 0.5μm is selected. C_{ox} and μ_n values come from technologic process. Using expression (13), the maximum LNA operating frequency is found to be 14GHz, which is largely greater than the LNA higher operating frequency (f_{o_high}=5.25GHz). From [4], the minimum required supply voltage (Vdd) is 2V. We have set the voltage supply to 2.2V to have some implementation margin.

3) Output Matching Network

A similar analysis as in the input impedance case can be carried out to assist selecting output concurrent resonator element values. However, the additional challenge when using a concurrent resonator load is to adjust the impedances levels at the resonance frequencies to appropriate levels (14) by using on-chip inductors:

$$Z\left(\omega_{rm1,rm2}\right) = \cfrac{1}{\cfrac{R_S}{R_S^2 + \left(\omega_{rm1,rm2} L_S - \cfrac{1}{\omega_{rm1,rm2} C_S}\right)^2} + \cfrac{R_P}{R_P^2 + \omega_{rm1,rm2} L_P^2}}$$

$$(14)$$

where R_S and R_P represent the series resistances of output resonator inductances. Expression (14) does not take into account the series capacitance (Cl is a bypass capacitor).

III. LAYOUT ISSUES

The microphotograph of the proposed BiFET concurrent LNA is shown in Fig.5. All components are on-chip, differing from the solution proposed in [1]. It takes 1.0x0.95 mm², including pads.

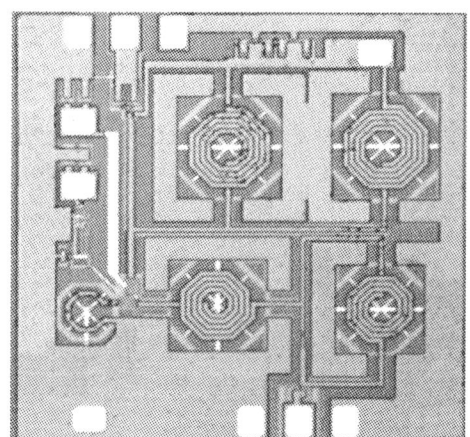

Fig. 5. Concurrent BiFET LNA (microphotograph).

IV. MEASUREMENT RESULTS

Measurements have been undertaken using the PA200 on-chip probe and an HP-based test-set, for both S-parameters and noise figure. S-parameter test-set is commanded by the ICCAP software. Some concurrent BiFET LNA measurement results are presented in Fig.6 and Fig.7. From these figures, we can note that for the higher band, process variation has shifted the centre frequency from 5.25GHz to 4.9GHz. In Fig.7, this frequency shift reduces power gain at 5.25GHz to 13.8dB, while a maximum gain of 14.8dB is achieved at around 4.9GHz.

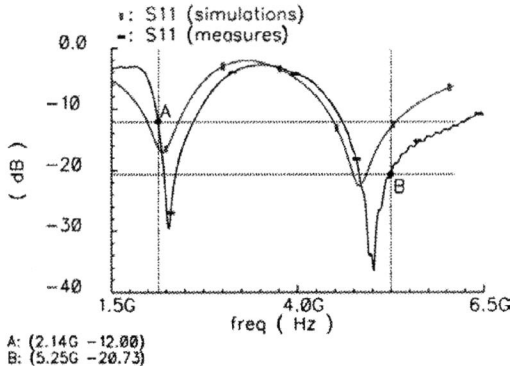

A: (2.14G −12.00)
B: (5.25G −20.73)

Fig. 6. Concurrent BiFET LNA return loss (S_{11}).

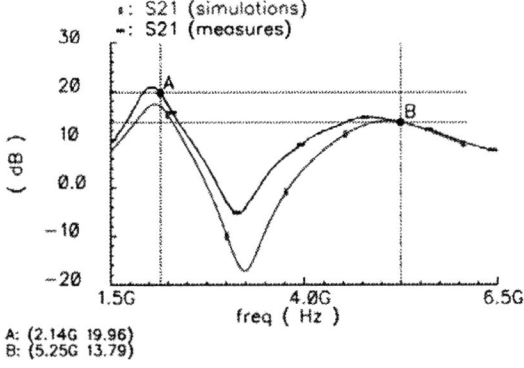

A: (2.14G 19.96)
B: (5.25G 13.79)

Fig. 7. Concurrent BiFET LNA power gain (S_{21}).

Table I compares measurement results of existing dual-band concurrent LNAs. It should be noted that the BiFET LNA proposed in this paper is fully-integrated (FI), with no external matching circuitry. In such a way, the low quality factor of integrated inductors (if compared to external inductors) contributes to additional noise and gain reduction. Process variation can also changes output resonator impedance level, which in consequence reduces the gain of the concurrent LNA. Measured IIP3 is -12.8dBm (low-band) and -10.3dBm (high-band).

TABLE I
DUAL-BAND CONCURRENT LNAs COMPARISON.

	Technology	FI	Freq. (GHz)	NF (dB)	S_{21} (dB)	S_{11} (dB)	Power Cons.
this work	0.25µm BiCMOS SiGe:C (HBT/MOS)	yes	2.14	2.8	20	-12	6mA/2.2V (13.2mW)
			5.25	3.7	13.8	-20.7	
[1]	0.35µm BiCMOS (only MOS)	no	2.45	2.3	14	-25	4mA/2.5V (10mW)
			5.25	4.5	15.5	-15	
[2]	0.18µm CMOS	yes	2.44	5.7	7.6	-9.5	6mA/1.8V (10.8mW)
			5.76	6.8	8.6	-6.2	

V. CONCLUSION

This paper discusses guidelines to design a dual-band concurrent fully-integrated BiFET low noise amplifier (LNA). It is dedicated to W-CDMA/IEEE 802.11a (European band) applications. For circuit manufacturing, we have used a 0.25µm BiCMOS SiGe:C process from STMicroelectronics.

An original synthesis methodology that helps selecting input/output matching network element values is also presented. This method is quite useful since it allows reducing design/simulation time, while achieving good input/output impedance matching at two selected frequencies.

The proposed concurrent dual-band BiFET LNA is an original solution that integrates, for the first time, two devices (HBT and a MOS) in a dual-band concurrent context. Also, it is a fully-integrated solution with no external elements. Circuit measurement results, in conformity with simulation results, enable validating the proposed concurrent design methodology. Finally, this methodology could also be applied to other communications systems, for a dual-band application.

REFERENCES

[1] H. Hashemi, A. Hajimiri, "Concurrent Multi-band Low-Noise Amplifiers: Theory, Design, and Applications", *IEEE TMTT*, Vol. 50, No. 1, pp. 288-301, Jan. 2002.

[2] Q.-H. Huang, D.-H. Huang, H.-R. Chuang, "A Fully Integrated 2.4/5.7 GHz Concurrent Dual-Band 0.18µm CMOS LNA for an 802.11 WLAN Direct Conversion Receiver", *Microwave Journal*, pp. 76-88, Feb. 2004.

[3] P. Ma, M. Racanelli, J. Zheng, M. Knight, "A Novel Bipolar-MOSFET Low-Noise Amplifier (BiFET LNA), Circuit Configuration, Design Methodology, and Chip Implementation". *IEEE TMTT*, Vol. 51, No.11, pp. 2175-2180, November 2003.

[4] C. P. Moreira, E. Kerherve, P. Jarry, D. Belot, "Fully-Integrated Low-Consumption BiFET Low-Noise Amplifier for W-CDMA Applications", *IEEE ICECS'2005*, Gammarth, Tunisia, pp.53-56.

2006 Bipolar/BiCMOS Circuits and Technology Meeting Proceedings

A 77 GHz (W-band) SiGe LNA with a 6.2 dB Noise Figure and Gain Adjustable to 33 dB

Ralf Reuter, Yi Yin

Freescale Halbleiter GmbH – TSO Technology Solutions Organization.

RF/IF Innovation Center Munich
Schatzbogen 7, 81829 Munich, Germany
Phone: +49-(0)89-92103803, FAX: +49(0)89-92103809
Email: Ralf.Reuter@freescale.com

Abstract - **This paper presents a state-off the art low-noise SiGe-amplifier (LNA) for the frequency range from 75 up to 85 GHz integrated in a 0.18µm BiCMOS technology [1, 2]. The LNA shows noise figures of about 6.2 dB at 77 GHz and simultaneously extremely high gain adjustable from nearly 0 dB up to 33 dB at 77 GHz. The possibility to fully disable the LNA completes the functionality of the presented circuit. To the knowledge of the author this combination demonstrates the best performance in noise and gain ever reported for a commercial BiCMOS process [3, 4]. Microstrip transmission lines in combination with integrated MIM capacitors are used as matching elements. The SiGe(C)-HBT demonstrates a typical maximum f_T and f_{max} performance of approx. 200 GHz, respectively.**

Index Terms – **Silicon bipolar/BiCMOS, RF circuits, amplifier noise**

I. INTRODUCTION

Radar applications in the W-band are one of key customer markets in the near future. Up to now, this application is dominated by III/V semiconductor circuits. Recently, SiGe-HBT's have demonstrated excellent microwave and even millimeter-wave performance. Especially the demands for modern rf-systems at higher frequencies makes it necessary to optimize passive components, such as matching elements, to obtain full potential of the active devices. Furthermore, one key issue to minimize costs and to enable a rapid production is to realize as many components (Si-MMIC) and functions on one chip (SoC) [4].

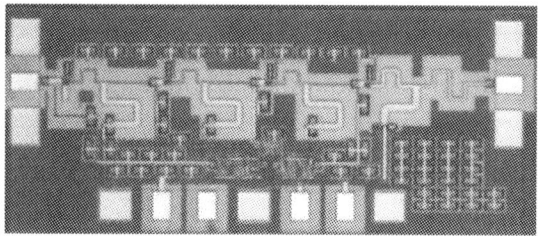

Fig. 1: Chip photo of the 4 stage low-noise amplifier, incorporating a variable gain stage, complete biasing circuitry with enable/disable switching logic; chip size: approx. 1.8×0.7 mm^2.

The presented low-noise amplifier (s. fig. 1) is designed as key element in a receiver architecture to simultaneously obtain high gain and lowest noise figure to minimize the overall system noise figure.

II. TECHNOLOGY

0.18 µm-SiGe(C)-HBTs of Freescale's commercially available BiCMOS process are used as active devices [1, 2]. Recent progress in the development of reliable high-speed SiGe(C)-HBT, results in a typical performance of the maximum cut-off frequencies f_T and f_{max} of approx. 200 GHz, respectively (Fig. 2). The high-speed version of the HBT also includes a collector enhancement. A low resistivity sub-isolation buried layer ("SIBL") under selected regions of the shallow trench isolation, gives the device significantly lower collector resistance.

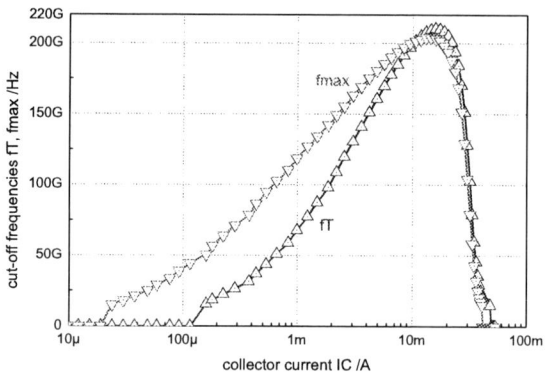

Fig. 2: Current-gain cut-off frequency f_T and maximum oscillation frequency f_{max} of a 0.18µm × 10µm in dependence on the collector current I_C.

The collector-emitter breakdown voltage BV_{CE0} is ≥ 2.0 V and the base-collector breakdown voltage BV_{BC0} is ≥ 6.2 V, respectively. The BiCMOS process backend provides a stack of five Cu-metal and dielectric layers. Matching circuits are designed with microstrip transmission lines. The backplane has been realized using metal 1 incorporated in a SiO$_2$ layer, to

1-4244-0458-4/06/$20.00 ©2006 IEEE

minimize high frequency losses, which occur in resistive Si-substrates. Thick top metallization layer forms the active signal line. With this combination a typical 50 Ω transmission line shows a loss of approx. 1.25 dB/mm at 77 GHz (s. fig. 3). Transmission lines with various centre line width (W = 3 μm up to 20 μm) were used to realize different characteristic wave impedances, which enables a variety of matching elements.

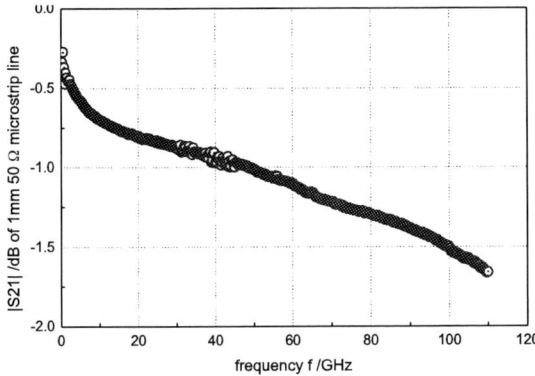

Fig. 3: Measured transmission coefficient $|S_{21}|$ of a 50 Ω microstrip transmission line in dependence on the frequency.

For fast and accurate simulations of the high frequency behavior, VBIC models for the SiGe(C)-HBTs, scalable models for the MIM capacitors and for the microstrip transmissions have been implemented into the Advanced Design System (ADS) from Agilent. All models have been extracted on the basis of on-wafer s-parameter measurements in the frequency range from 0.2 GHz up to 110 GHz.

III. AMPLIFIER CIRCUIT

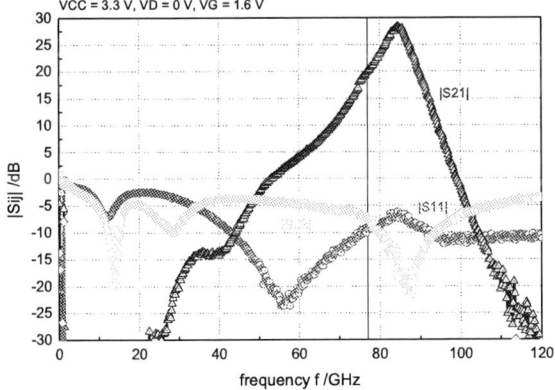

Fig. 4: Typical measured s-parameters of the low-noise amplifier with an applied gain control voltage of $V_G = 1.65$ V.

A chip photo of the fabricated W-band low-noise is shown in figure 1. The design was guided by the goal, to achieve lowest noise figures at 77 GHz in combi-

nation with an adjustable gain from nearly 0 dB up to maximum achievable gain of an integrated SiGe-amplifier. The amplifier is composed of 4 cascode-HBT stages, which provide high gain with reasonable noise performance in the desired frequency range. The amplifier is designed for a single supply voltage V_{CC} of 3.3 V. The bias conditions of all cascode-HBT stages are internally set by switchable current mirrors. The maximum switching speed is mainly limited by internal RC-combinations to get quiescent bias conditions. In the presented design the maximum switching speed is limited to about 100 MHz. In the enable condition the supply current I_{CC} is approx. 31 mA for variable gain control voltage of $V_G = 1.65$ V (medium gain condition) and 37 mA for a variable gain control voltage of $V_G = 2.2$ V (maximum gain condition), respectively. In the disable condition the total supply current decreases to 3 mA. The first two cascade stages mainly dominate the overall noise figure of the LNA. Both stages are optimized by the noise measure method (a compromise between gain and noise). In the first stage an inductive degeneration realized by a microstrip transmission line is used to get an optimum compromise between gain and noise matching (s. Fig. 5).

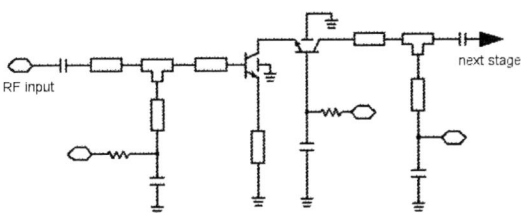

Fig. 5: First cascode-HBT stage with the inductive degeneration realized by a microstrip transmission line.

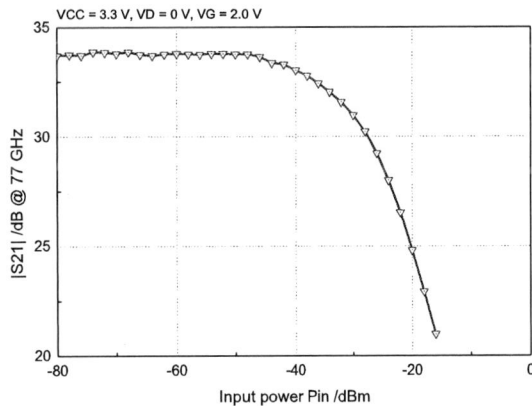

Fig. 6: Forward gain $|S_{21}|$ at 77 GHz in dependence on the leveled port power at constant variable gain control voltage V_G of 2.0 V.

The third stage is biased at high gain conditions and the last stage is biased via an adjustable current mirror to realize a variable gain stage. All stages a DC decoupled by integrated MIM capacitors. The compromise between optimum noise and gain matching of the low-noise amplifier, results in maximum gain, which is not exactly centered at 77 GHz. The input and output of the circuit are matched to 50 Ω by a set of microstrip transmission lines. Also the inter-stage matching is realized with transmission lines. The main three s-parameters $|S_{11}|$, $|S_{22}|$ and $|S_{21}|$ of the LNA at a typical bias condition are shown in fig. 4.

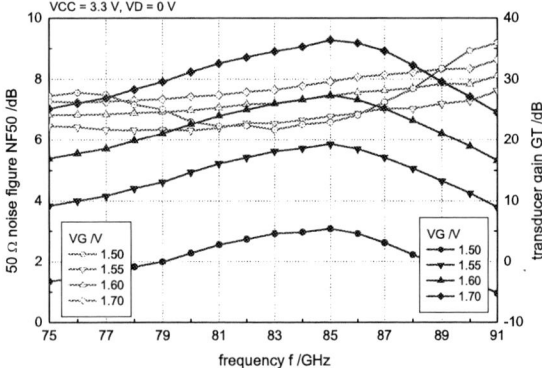

Fig. 7: 50 Ω noise figure and corresponding transducer gain of the LNA at various variable gain control voltages V_G.

The input reflection coefficient $|S_{11}|$ at 77 GHz is close to -10 dB and the output reflection coefficient $|S_{22}|$ is about -6 dB at 77 GHz. Due to small deviation of output stage cascode-HBT s-parameters in combination parasitic inductances from the DC-blocking MIM in the output path, the output matching is shifted to slightly higher frequencies. The reverse transmission coefficient $|S_{12}|$ of the complete circuit is in the range of -45 dB (not shown in the graph). The on chip achievable maximum isolation from the input to the output pad (45 dB) is the main limiting factor for the maximum available gain of the 4 stage LNA. Further improvements on the inter-stage isolation and rf-pad arrangements will give potential to enhance the rf-isolation at 77 GHz. Especially the combination of high gain and high isolation requires an extreme carefully set-up of the measurement parameters, more than 80 dB dynamic range at low input power levels are needed to fully characterize the LNA within one frequency sweep. A LRRM and/or LRM+ calibration with a Cascade W-Band ISS on an absorber material have been applied for all measurements. The leveled port power of the used HP8510XF system was set to -60 dBm to avoid gain compression of the LNA. In fig. 5 the behavior of the gain at 77 GHz (with constant variable gain control voltage V_G) is shown in dependence on the leveled port power. Taking into account additional losses from the coaxial 1 mm cable and the probe-tip, the input power compression point (-1 dB) of the LNA in the maximum gain condition is

about -40 dBm at 77 GHz. The typical noise and gain characteristics of the realized low-noise amplifier are depicted in fig. 7. Precisely calibrated ELVA-1 noise sources in combination with a Millitech mixer and a HP down-converter are used in the noise set-up. At a frequency of 77 GHz the LNA shows lowest noise figure of 6.2 dB with associated high gain of 12 dB. Up to a maximum peak gain of about 40 dB, the amplifier is unconditional stable in the entire frequency range. The only limiting factor in this case, is the typical isolation from the output to the input of about -45 dB caused by the silicon substrate. The maximum gain of the amplifier is obtained at a frequency of 85 GHz.

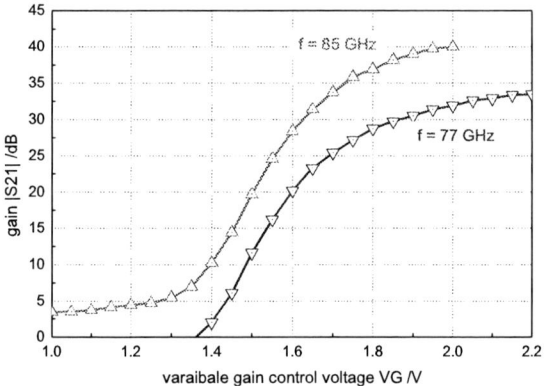

Fig. 8: Measured forward gain $|S_{21}|$ at 77 GHz and 85 GHz in dependence on the variable gain control voltage V_G. Input power was kept constant at -60 dBm.

The functionality of the gain control is shown in fig. 8. At 77 GHz the gain can be adjusted in the range from 0 dB up to 33 dB without influencing the input and output matching.

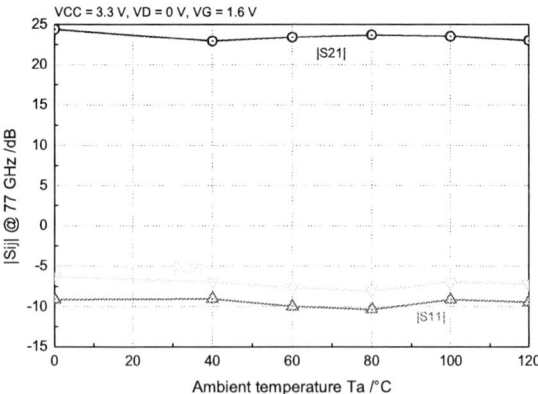

Fig. 9: Measured forward gain $|S_{21}|$, input and output reflection coefficients $|S_{11}|$, $|S_{22}|$ at 77 GHz, respectively, in dependence on the ambient temperature T_a.

To keep the LNA always unconditional stable in the whole frequency range the maximum gain at 85 GHz is limited to approx. 40 dB. Due to the natural biasing of all stages and the fact that the maximum gain of the LNA is not centered at 77 GHz, the temperature dependence of the gain, input- and output reflection coefficients at 77 GHz is extremely low (s. fig. 9).

IV. OUTLOOK

The on-going SiGe-HBT development with respect to improved rf- and especially noise behavior [1], results in a BiCMOS technology with competitive rf-performance and high level CMOS circuitry [2, 5]. The combination of the here presented low-noise amplifier with a down-converter stage, forms one of the basis elements in a receiver circuitry. Multiple channel architecture, each channel equipped with high speed digital enable and disable functionality can be envisioned, which has the capability to enhance the complexity of integrated receiver modules.

V. CONCLUSIONS

The presented low-noise amplifier demonstrates the potential of the Freescale production SiGe-BiCMOS technology for W-band applications. In contrast to III/V-semiconductor MMICs, the combination of high-speed SiGe(C)-HBTs with lumped and distributed passive elements enables low-cost Si-MMICs. Without process modification, transmission lines (microstrip and also coplanar) can be realized by the standard metal/dielectric stack of the process backend. SiGe-BiCMOS will lead to higher integration densities of MMICs and an additional use of CMOS circuits will drastically raise the functionality.

ACKNOWLEDGMENTS

The authors thankfully acknowledge, that Jim Kirchgessner of Freescale Semiconductor Tempe, USA, supported this work by making the SiGe(C)-HBT wafers available for measurements. Finally we wish to acknowledge the continuous support by the work of Margaret Huang, Freescale Semiconductor, Tempe, USA.

REFERENCES

[1] Jay P. John, Francis Chai, Dave Morgan, Theresa Keller, Jim Kirchgessner, Ralf Reuter, Hernan Rueda, Jim Teplik, Jan White, Sandy Wipf, Dragan Zupac, "Optimization of a SiGe:C HBT in a BiCMOS Technology for Low Power Wireless Applications", IEEE BCTM 11.2, Sept. 2002, pp. 193-196

[2] Jay P. John, Jim Kirchgessner, Matt Menner, Hernan Rueda, Francis Chai, Dave Morgan, Jill Hildreth,, Ralf Reuter, and Hao Li, "A 180nm, Cost-Effective, Selective-Epi, SiGe:C Bipolar Technology for 77GHz Automotive Radar",BCTM 2006

[3] Brian A. Floyd, "V-Band and W-Band SiGe Bipolar Low-Noise Amplifiers and Voltage-Controlled Oscillators", IEEE Radio Frequency Integrated Circuits Symposium, 2004, pp. 295-298

[4] Brian A. Floyd, "SiGe Bipolar Transceiver Circuits Operating at 60 GHz", IEEE Journal of Solid-State Circuits, vol. 40, no. 1, 2005, pp. 156-167

[5] A. Natarajan, A. Komijani, X. Guan, A. Babakhani, Y. Wang, A. Hajimiri,, "A 77GHz Phased-Array Transmitter with Local LO- Path Phase-Shifting in Silicon", Proceeding of the ISSCC 2006, pp. 182-183

A 5GHz Series Coupled BiCMOS Quadrature VCO with Wide Tuning Range

Vasanth Kakani, Foster F. Dai and Richard C. Jaeger

Dept. of Electrical and Computer Engineering
Auburn University, Auburn, AL 36849-5201, USA

Abstract — This paper presents a novel quadrature VCO (QVCO) implemented in a 47GHz SiGe technology. The QVCO is a serially coupled LC VCO that utilizes SiGe HBTs for oscillation and MOSFETs for coupling. The SiGe BiCMOS QVCO prototype achieves about 14.6% tuning range from 4.3 to 5 GHz. The phase noise of the QVCO is measured as -115dBc/Hz @ 2MHz offset. The 5GHz QVCO core consumes 6mA current under 3.3V power supply and occupies 0.88mm2 area.

Index Terms — quadrature VCO, oscillator, phase noise, BiCMOS, SiGe.

I. INTRODUCTION

For image-reject-mixing carrier's and modulated signals, quadrature local oscillation signals are needed. There are various ways to generate quadrature signals: *(i)* A divide-by-two frequency divider following the VCO running at the double the LO frequency. This approach generally shows poor phase noise and quadrature accuracy, as it requires 50% duty cycle VCO. *(ii)* A VCO followed by a passive poly-phase RC complex filter. An integrated poly-phase network is narrow band with poor quadrature accuracy. It suffers from process variation on the RC time constant's that lead to amplitude imbalance between the quadrature signals. A typical RC phase filter also loads the VCO and has large loss such that a power hungry buffer is needed after the filter. *(iii)* Two VCOs are forced to run in quadrature using transistor or transformer coupling. This technique provides wide-band quadrature accuracy and superior phase noise performance with tradeoff of increased power and silicon area.

By coupling two symmetric *LC*-tank VCO's to each other, a quadrature VCO (QVCO) generates wideband quadrature signals at high frequency. There are various ways to couple the two VCOs and inject-lock their oscillation frequency. The most common QVCO topology shown in Fig.1 is the parallel coupling proposed by Rofougaran *et al.* [1], where each

oscillator consists of a cross-coupled feedback circuit and each oscillator output is connected to another oscillator using transistors in parallel to the cross-coupled transistors. The parallel QVCO (P-QVCO) delivers quadrature signals with low phase and amplitude errors, yet it consumes large current to bias both the oscillation and coupling transistors. QVCOs can also be serially coupled by placing the coupling transistors in series with the oscillation transistors [2,4]. By connecting the coupling transistors in series in a cascode current reuse topology, the serial QVCO (S-QVCO) reduces the noise from the cascode devices and provides better isolation between the VCO outputs and its current sources. The two QVCO topologies have been compared in [2]. For the pure MOSFET S-QVCO, the coupling transistors should be about five times larger than the oscillation transistors for good phase noise. For optimal coupling, the P-QVCO appears to have better quadrature amplitude and phase matching. However, under optimal coupling, the S-QVCO achieves better phase noise. In the P-QVCO and S-QVCO reported so far, the same type of transistors has been used for oscillation and coupling.

II. PROPOSED S-QVCO TOPOLOGY

This paper presents a novel QVCO implemented in a 47GHz SiGe technology. The proposed QVCO is a serially coupled LC VCO that utilizes SiGe HBTs for oscillation and MOSFETs for coupling. The oscillation NPN transistors achieve high oscillation frequency and low phase noise, while the NMOS coupling transistors provide more headroom, better isolation and increased tuning range.

The proposed S-QVCO circuit is illustrated in Fig. 2, in which the NPN transistors Q1 and Q2 form a cross-coupled negative transconductance LC VCO and Q3 and Q4 form another identical LC VCO. The coupling between the two VCOs is realized using four NMOS transistors M1, M2, M3 and M4. Thus, the proposed S-QVCO utilizes different types of transistors for oscillation and coupling. The advantages of this technique will be discussed in the coming paragraphs.

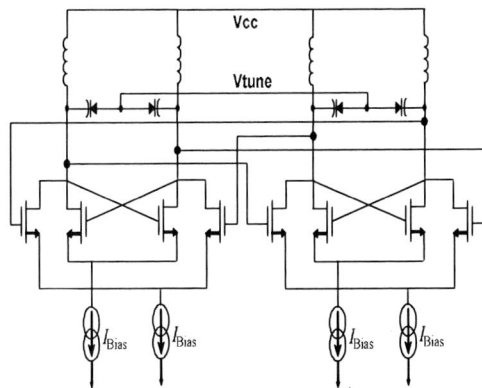

Fig. 1 P-QVCO circuit schematic

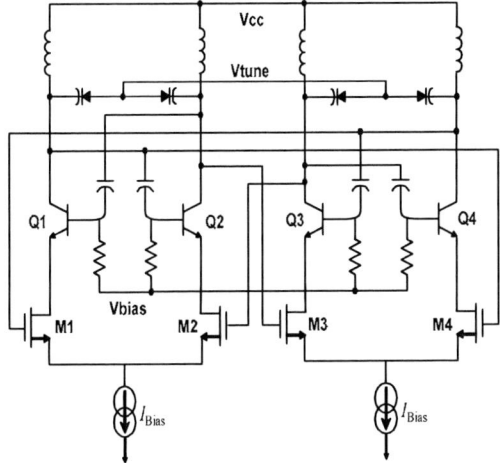

Fig. 2 Proposed S-QVCO using NPN for oscillation and NMOS for coupling.

Compared to a P-QVCO, the S-QVCO achieves lower current consumption, since the coupling and oscillation transistors share the same bias current. In a P-QVCO, the coupling pair of transistors usually consumes an additional 25-30% of the core oscillator current for reasonable compromise between phase noise and phase error. In an S-QVCO, the coupling transistors are in series with the $-g_m$ transistors. Additional current sources are not required for coupling transistors, resulting in considerable power saving. Also, the coupling and oscillation transistors are connected in a cascode manner such that the noise coming from the coupling transistors and current sources are isolated from the tank. The phase noise and phase error are relatively independent of each other in this topology. However, under the same voltage supply, the S-QVCO has less voltage headroom for output swing due to the coupling transistors.

Previously, the same types of transistors were used in the S-QVCO topology. As known, NPN transistors can achieve higher oscillation frequency due to enhanced f_{max}. On the other hand NMOS transistors

have higher output impedance, reduced voltage headroom, and much relaxed bias requirements compared to their BJT counterparts. Since both types of transistors are available in a BiCMOS technology, we can combine the advantages of both types and achieve better QVCO performance. By using NPN transistors for oscillation, high oscillation frequency can be achieved and by using MOS transistors for coupling, higher voltage headroom can be provided for output swing due to the reduced headroom required by the MOS transistors. Moreover, better isolation between the current source and oscillation tank can be achieved due to the large output impedance of the MOS transistors. In the proposed S-QVCO, the MOS coupling transistors are directly connected to the VCO output nodes, providing a much easier biasing scheme.

In the S-QVCO design, the NPN transistor size is chosen for maximum speed. Their emitter length is selected to operate the transistors at peak f_T and the bias current is selected to provide the transconductance of

$$g_{m,osc} \geq \frac{\chi}{R_{\tan k}} \qquad (1)$$

Where $R_{\tan k}$ is the effective resistance (loss) of the tank and χ is an empirical parameter with value from three to five to provide design margin for reliable VCO startup. The oscillation frequency is given by $f_{osc} = \left(2\pi\sqrt{LC}\right)^{-1}$, where C is the total capacitance seen across the inductors which includes the varactor capacitance C_v .and the fixed parasitic capacitance C_p. The frequency tuning range can be found as

$$FTR = \frac{C_p + C_{v\max}}{C_p + C_{v\min}} \qquad (2)$$

This fixed capacitance is made up of the parasitic capacitance of the inductor, varactor, capacitance contributed by the coupling transistors and oscillating transistors. At high frequencies this fixed capacitance can contribute a large portion to the overall capacitance and limit the tuning range and maximum attainable oscillation frequency. To first order the fixed capacitance provided by the oscillating transistor for the P-QVCO is $\frac{C_\pi}{2} + 2C_\mu$ for the bipolar version and $\frac{C_{gs}}{2} + 2C_{gd}$ for the MOS version.

The total fixed capacitance seen across the LC tank for S-QVCO and P-QVCO shown in Fig 2 and Fig.1 are given by (3)and (4) respectively

139

$$C_p = 2C_\mu + \frac{C_{gs}}{2} + \frac{C_\pi C_{gd} C_{gs}}{2(C_\pi C_{gd} + C_\pi C_{gs} + C_{gs} C_{gd})} \qquad (3)$$

$$C_p = 2C_{gd} + C_{gs} + \frac{1}{2}\left(\frac{C_{gd} C_{gs}}{C_{gd} + C_{gs}}\right) \qquad (4)$$

As can be seen from the above equations the use of MOS transistors for coupling in serial fashion considerably reduces the fixed parasitic capacitance seen across the tank due to the increased capacitive degeneration (C_{gs} and C_{gd}) of the serially coupled MOS transistors at the emitter terminals of the $-g_m$ pairs. This enables more tuning range and maximum attainable oscillation frequency. At high frequencies where the influence of parasitics becomes dominant the S-QVCO should be a better topology for quadrature signal generation compared to the P-QVCO.

The NMOS coupling transistors are sized such that they provide about the same g_m as that of the NPN transistors. This balances the load impedance of the tanks, which leads to smaller quadrature phase/amplitude errors and improves the phase noise, as well. While the above argument would also be valid if we use bipolar transistors for coupling instead of MOS, the fully bipolar version would require current consuming level shifters and the power saving advantage of this current reuse S-QVCO topology compared to P- QVCO would be lost.

III. Measured Results

The 5 GHz serially coupled quadrature oscillator was implemented in a 47GHz SiGe BiCMOS technology with four metal layers. The circuit consumes a die area of 0.88mm^2 as shown in Fig. 3. The layout is suboptimal since the long interconnect between the two corner inductors introduces harmful parasitic resistance. The tank uses a pair of on-chip inductors with an inductance of 971pH and a Q of approximately 12 at 5GHz. The two varactors are implemented as a collector base diode and exhibit a Q of around 50 at 5GHz. The VCO core consumes 6mA of current from 3.3V supply. As shown in Fig. 4, the serially coupled quadrature oscillator achieves phase noise of -115dbc/Hz @ 2MHz offset from 4.67GHz. The measured oscillation frequency versus the reverse bias voltage across the varactor is given in Fig. 5. As shown, the tuning range of the oscillator is from 4.32 to 5 GHz with tuning voltage covering the entire possible range from 0V to supply voltage. This high tuning range of 15% is due to the coupling MOS transistors in series with $-g_m$ pairs. With the wide tuning range, the oscillator also achieves fairly linear tuning gain (k_{vco}), which is important to avoid VCO

chirping and pulling in PLL synthesizer designs. The measured VCO gain is 206 MHz/V. Figure 6 shows the measured oscillator phase noise versus the tuning voltage. The maximum variation in phase noise at 2MHz offset over the entire tuning range is 2.3dBc/Hz. The VCO achieved better than -113 dBc/Hz phase noise @ 2MHz offset over the entire tuning range. Table 1 summarizes the measured oscillator performance. Table 2 compares the performance of the quadrature oscillator to prior woks.

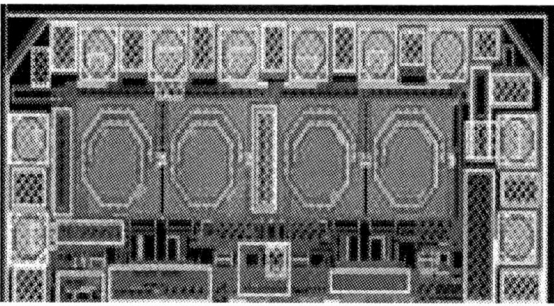

Fig. 3 QVCO chip layout diagram.

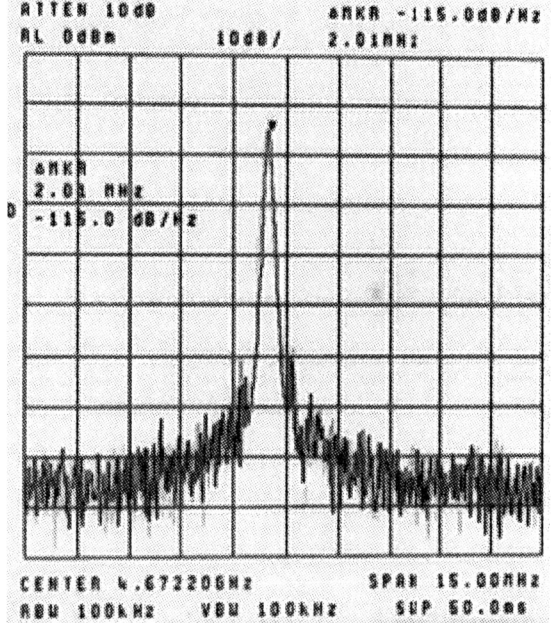

Fig. 4 Measured QVCO output spectrum. The phase noise is
- 115dBc/Hz @2MHz offset from the carrier frequency.

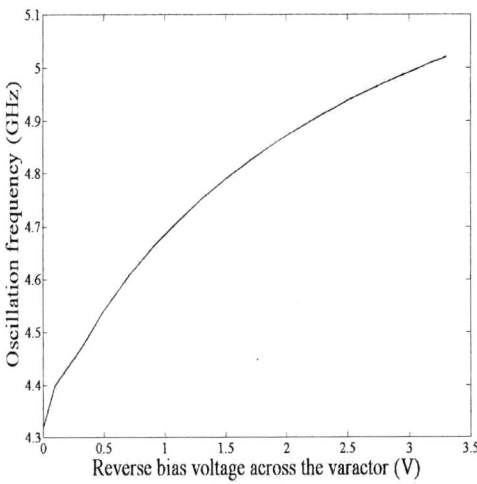

Fig. 5 The above plot shows the Oscillation frequency versus the reverse bias voltage across the varactor. The tuning range is from 4.3-5GHz.

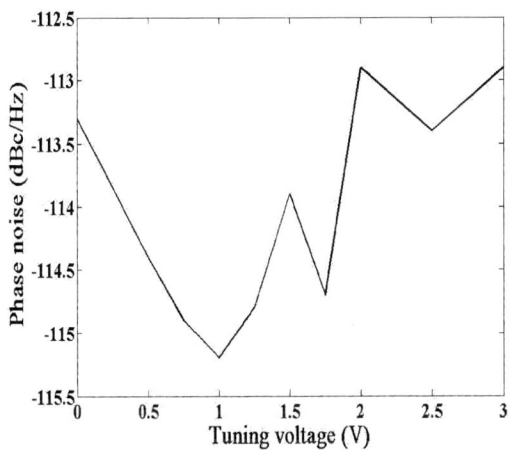

Fig. 6 The above plot shows the Phase noise versus applied tuning voltage.

Table 1. Measured performance of the LC-QVCO.

Supply Voltage	3.3 V
Core Current	6 mA
Oscillation Frequency	4.32 – 5 GHz
Tuning Range	14.6 %
Phase noise @ 2MHz	-115 dBc/Hz

Table 2. Quadrature VCO performance comparison.

Ref.	f_{osc} (GHz)	Tuning range	Phase Noise	Core power (mW)
1	0.9	17%	-110@1MHz	30
2	1.8	18%	-140@3MHz	50
3	5	6.4%	-115@2MHz	21.2
5	13	37%	112.3@10MHz	39
This work	5	14.6%	-115@2MHz	19.8

IV. CONCLUSIONS

We presented a novel series coupled quadrature oscillator implemented in a 47GHz BiCMOS SiGe technology. The BiCMOS S-QVCO uses NPN for oscillation and NMOS devices for coupling. The oscillator prototype achieves greater than 14% tuning range from 4.3 to 5 GHz the phase noise of the oscillator is measured as -115dBc/Hz @ 2MHz offset from the carrier. The power consumption of the 5GHz QVCO core is 3mA per tank from a 3.3V supply.

ACKNOWLEDGEMENTS

We acknowledge MOSIS for fabrication support under the MEP program.

REFERENCES

[1] A. Rofougaran, J. Rael, M. Rofougaran, and A. Abidi, "A 900 MHz CMOS LC-oscillator with quadrature outputs," *Digest ISSCC* 1996, Feb. 1996, pp. 392–393.

[2] P. Andreani, A. Bonfanti, L. Romano, and C. Samori, "Analysis and design of a 1.8 GHz CMOS LC quadrature VCO," *IEEE J. Solid-State Circuits*, vol. 37, no. 12, pp. 1737-1747, 2002.

[3] P. Van de Ven et al., "An optimally coupled 5 GHz quadrature LC oscillator," Digest of the *Symposium on VLSI Circuits*, pp. 55-56, 2001.

[4] P. Andreani, A. Bonfanti, L. Romano, and C. Samori, "On phase noise and phase error performances of quadrature VCO," *IEEE J. Solid-State Circuits*, vol. 39, no. 11, Nov.2004, pp. 1883-1893.

[5] A.L. Coban *et al.,* "A Highly-Tunable 12 GHz Quadrature LC-VCO in SiGe BiCMOS process," Digest of the *Symposium on VLSI Circuits,* 2001, pp.119-120.

2006 Bipolar/BoCMOS Circuits and Technology Meeting Proceedings

Design and Scaling of SiGe BiCMOS VCOs Above 100GHz

S. T. Nicolson[1], K.H.K Yau[1], K.A. Tang[1], P. Chevalier[2], A. Chantre[2], B. Sautreuil[2], and S. P. Voinigescu[1]

1) Edward S. Rogers Sr. Dept. of Elec. & Comp. Eng., Univ. of Toronto, Toronto, ON M5S 3G4, Canada

2) STMicroelectronics, 850 rue Jean Monnet, F-38926 Crolles, France

Abstract — This paper presents a comparison of 100 GHz Colpitts VCOs fabricated in two generations of SiGe BiCMOS technology, with MOS and HBT varactors and with integrated inductors. Based on a study of the optimal biasing conditions for minimum phase noise, it is shown that VCOs can be used to monitor the noise performance of the SiGe HBTs at mm-waves. Measurements and simulations show a 104 GHz VCO operating from 2.5 V with phase noise of -101.3 dBc/Hz at 1 MHz offset, which delivers +2.7 dBm of differential output power at 25°C, with operation up to 125°C.

Index terms — Millimeter-wave circuits, phase-noise, SiGe BiCMOS technology, voltage-controlled oscillators, W-band.

I. INTRODUCTION

The latest SiGe processes with f_T/f_{MAX} above 150 GHz [1-6] allow them to compete directly with III-V technologies for applications in the W-band (75-110 GHz). Many W-band radar transceivers and communications circuits require a phase-locked loop, in which the VCO [7] and the frequency divider are critical components. Fig. 1 compares the phase noises of state-of-the-art W-band SiGe and CMOS VCOs [8]-[13] along with those presented in this paper.

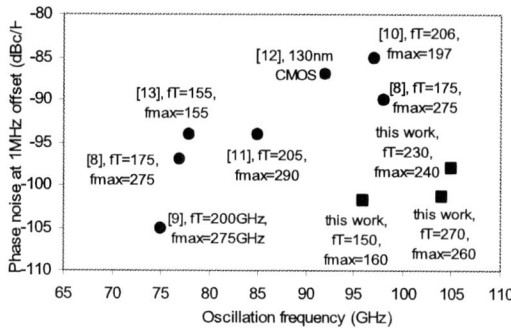

Fig. 1. Phase noise of state-of-the-art SiGe HBT and CMOS W-band VCOs (inc. process f_T/f_{MAX}).

As the level of integration in W-band circuits increases, high yield processes must be developed, and circuit scaling between successive SiGe technology generations must be understood. Key process monitor circuits are required to investigate both issues. Given the complexity and variability of noise parameter measurements above 50 GHz, Colpitts oscillators can be employed to monitor the phase noise at W-band in much the same way as ring oscillators are used as process speed monitors.

This paper presents a Colpitts oscillator topology with compact layout, suitable for low-voltage, low-phase

noise oscillators in the W-band, and explores the impact of technology scaling on VCO performance.

II. VCO TOPOLOGY

The differential Colpitts topology [7] is a common choice for low-phase noise, mm-wave VCOs [8]-[11], [13], [14]. Fig. 2 shows the schematic of our SiGe BiCMOS implementation of this topology.

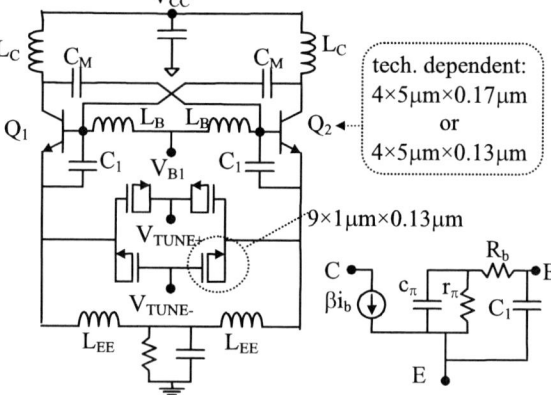

Fig. 2. Colpitts oscillator schematic.

In an effort to reduce phase noise and to simplify layout, which is critical at 100 GHz, the cascode is replaced by a single-transistor topology. A MiM capacitor (C_1) in parallel with c_π improves negative resistance and reduces phase noise by shunting the base resistance (R_B). Negative Miller capacitors (C_M) are placed at the BC junctions to cancel the effect of the C_{BC}. Finally, fully differential tuning with MOS varactors is used for better supply noise rejection. This also reduces the modulation of the varactor capacitance (C_{VAR}) by the tank voltage, which helps to suppress the maximum in phase noise at the center of the tuning range seen in other VCOs [13]. The varactor layouts are optimized for high-Q [13]. Wherever possible, spiral inductors are used in place of transmission lines to achieve a compact layout.

III. VCO DESIGN METHODOLOGY

The VCO shown in Fig. 2 was designed and fabricated in a SiGe BiCMOS process with f_T/f_{MAX} of 150/160 GHz (referred to as "BiCMOS9" [1]), and two SiGe HBT processes with f_T/f_{MAX} of 230/300 GHz and 270/260GHz (referred to as "BipX" and "BipXF" respectively [2]). The technologies have identical back-ends and HBT layouts (except for the emitter window, see Fig. 2), which allows investigation of the effects of

1-4244-0458-4/06/$20.00 ©2006 IEEE

SiGe technology scaling on mm-wave VCO performance.

Because MOSFETs are not yet available in BipX and BipXF, two versions of the VCO were designed, one with the differential MOS varactor tuning illustrated in Fig. 2, and another with single ended HBT varactors. Apart from varactors and emitter width, *the VCOs are otherwise identical in design and layout,* which allows measurement results from both VCOs to be compared directly.

The oscillation frequency (f_{osc}) of the VCO is given by (1), and was designed to be 100 GHz. Note that C_M is chosen to remove the $C\mu$ contribution to C_{EFF}.

$$ f_{osc} = \frac{1}{2\pi\sqrt{L_B C_{EFF}}} \quad C_{EFF} = C_\mu + \frac{C_{VAR}(C_1 + C_\pi)}{C_{VAR} + C_1 + C_\pi} \quad (1) $$

The negative resistance provided by Q_1, given by (2), must be large enough to overcome losses in the tank.

$$ R_{NEG} = R_B + \frac{\omega_{osc}L_B}{Q_{L_B}} + \frac{1}{\omega_{osc}Q_{C_{VAR}}C_{VAR}} - \frac{g_m}{\omega_{osc}^2 C_\pi C_{VAR}} \quad (2) $$

At 100 GHz, the finite Q of the varactor (C_{VAR}) and base inductor (L_B) add substantial losses to the tank. Therefore, as illustrated in Fig. 3, C_1 should be chosen to maximize the negative resistance at the desired f_{osc}.

Low phase noise is an important specification in radar applications [9]. In [13] and [14], it was shown that to minimize phase noise in a mm-wave VCO, Q_1 should be biased at the optimum NF_{MIN} current density. However, in the W-band, the correlation between base and collector noise currents pushes the optimum NF_{MIN} current density closer to the peak f_T/f_{MAX} current density [15]. Therefore, the VCOs were designed with Q_1/Q_2 biased at peak f_T/f_{MAX} current density.

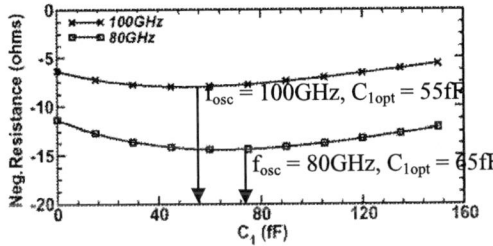

Fig. 3.   Simulation to find optimum value of C_1.

IV. EXPERIMENTAL RESULTS

A die microphotograph of the VCO is shown in Fig. 4. The VCO occupies 300μm × 400 μm including all DC pads (not shown). The die area is smaller than other W-band SiGe VCOs [8], [9], [11] because inductors are used in place of transmission lines. The VCO consumes 140 mW from a 2.5 V supply, which to the authors' knowledge is the lowest supply voltage published for a W-band VCO in SiGe BiCMOS technology.

Fig. 5 illustrates a spectral plot of phase noise at 1MHz offset for the 104 GHz BipX VCO, and Fig. 6 shows the output power and phase noise of all three VCOs over the tuning range. The measured phase noises of -101.6 dBc/Hz at 1MHz offset for the 96 GHz

BiCMOS9 VCO with MOS varactor, and -101.3 dBc/Hz for the 104 GHz BipX VCO with HBT varactor are records for SiGe VCOs above 80 GHz. Table 1 summarizes the performance of all fabricated VCOs, with probe, adapter, DC-blocking capacitor, and cable losses de-embedded.

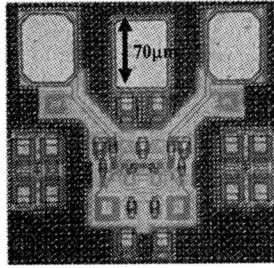

Fig. 4.   104 GHz VCO microphotograph.

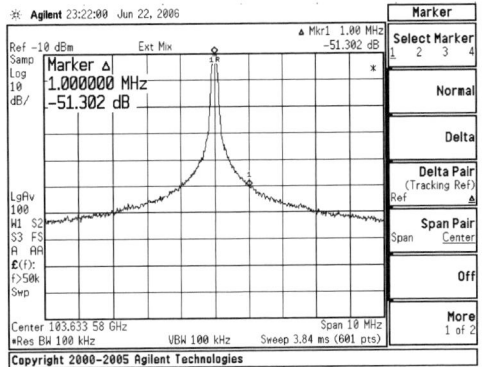

Fig. 5.   Averaged spectral plot of phase noise in 104 GHz BipXF VCO with HBT varactor.

TABLE 1: SUMMARY OF VCO PERFORMANCE

	BiCMOS9, MOS var.	BiCMOS9, HBT var.	BipX, HBT var.	BipXF, HBT var.
Differential Pout (dBm)	+0.7 sim. +5.5	-1.3 sim. -1.15	+2.7 +6.5	+2.5
SSB PN @ 1MHz (dBc/Hz)	-101.6	-80	-98	-101.3
Osc. Freq. (GHz)	96 sim. 96	100 sim. 100	106 sim. 108	104

Measurement results indicate that, in the 150/160 GHz BiCMOS technology, the VCO with MOS varactor achieves 14 dB lower phase noise than the one with junction varactor, demonstrating that MOFETs, not just HBTs, are required to optimize the phase noise of W-band SiGe VCOs. Interestingly, The VCO center frequency is practically immune to the change in technology. Although the transistor f_T and f_{MAX} improve by 40%, the VCO center frequency changes by only 6%. Clearly, accurate design and modeling of passive components has a greater effect on the center frequency than the transistor itself. So long as back-end-of-line geometry remains unchanged, mm-wave circuits can be ported quickly between successive generations of SiGe technology, mitigating some of the cost of moving to the next technology node.

143

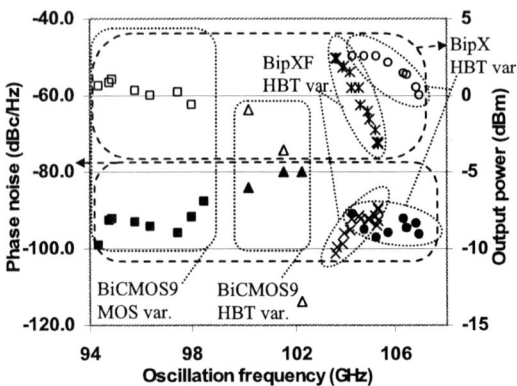

Fig. 6. Phase noise versus oscillation frequency.

The VCO phase noise was also measured as a function of HBT bias current density. Shown in Fig. 7 are the averaged phase noise and output power measurements for four BipX VCOs. The minimum in phase noise corresponds to the peak f_T/f_{MAX} current density. It confirms the shift to higher current densities of the NF_{MIN} current density as frequency increases. The transistor noise figure was obtained from Y parameter measurements, as in [15], with a noise transit time of 0.3 ps. Note that NF_{MINn} @ 65 GHz is only 1.7 dB, the lowest reported for a SiGe HBT at this frequency.

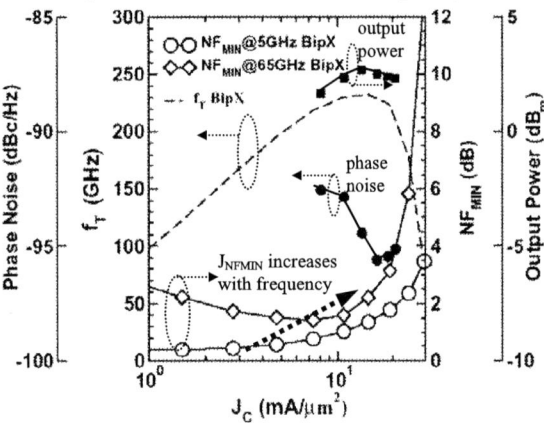

Fig. 7. Measured minimum phase noise is at measured peak f_T/f_{MAX} current density.

Shown in Fig. 8 are simulated and measured tuning characteristics of the VCOs with HBT varactors. The BipX VCO displays 2.5 GHz of tuning range up to 125°C. The tuning is linear enough to allow frequency modulation of the VCO output by applying a triangular wave to the tuning input – a modulation technique commonly employed in FMCW radar systems. The resulting spectrum obtained at 100°C using the BipX HBT varactor VCO, is shown in Fig. 9.

Fig. 10 compiles the measured output power across the tuning range at 25°C, 70°C, and 125°C. The BiCMOS9 VCO with MOS varactor oscillates up to 70°C and the BiCMOS9 VCO with HBT varactor

operates up to 50°C. In contrast, the BipX oscillator functions up to at least 125°C.

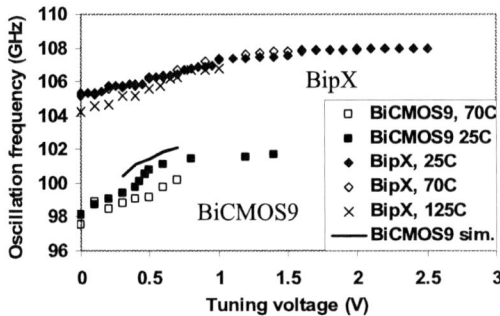

Fig. 8. Tuning characteristics across temperature for HBT varactor VCOs in BipX and BiCMOS9.

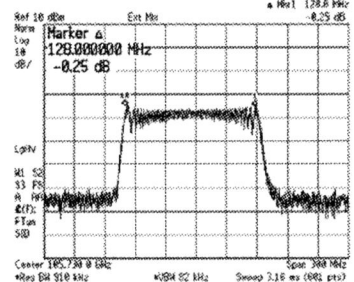

Fig. 9. Frequency modulation of BipX VCO output.

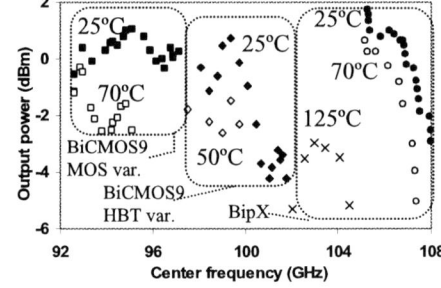

Fig. 10. P_{out} versus f_{osc} at 25°C, 70°C, and 125°C.

To gauge the impact of process variations on VCO performance, the mm-wave and DC characteristics of both BiCMOS9 VCOs were collected from 60 dice from 4 different wafers. Tables 2 and 3 summarize the results for the MOS varactor and HBT varactor VCOs, respectively. Of the 120 VCOs tested, 4 had significantly below average performance, and another 2 VCOs failed to oscillate. Those VCOs are not included in the averages given.

To further characterize the VCOs over process variations, in Fig. 11 we have plotted measured output power versus oscillation frequency for 1 die on each of the 4 wafers, beside simulation results. A 2 dB variation in output power between BiCMOS9 VCOs on different

144

wafers is illustrated. Note however that the tuning range and center frequency remain constant over the wafers.

Fig. 12 reproduces wafer maps of oscillation frequency and phase noise as functions of location for BipX VCOs. Both plots show that dice at the center of the wafer perform better than dice on the edges.

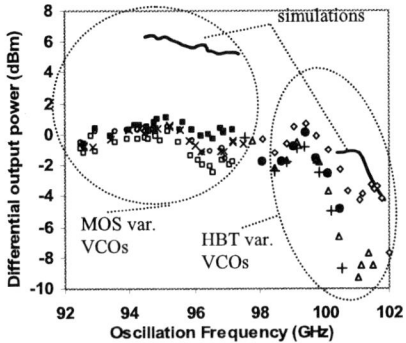

Fig. 11. Output power versus oscillation frequency for BiCMOS9 VCOs on different wafers.

■ VCO not present	■ VCO not present
■ Die not tested	■ Die not tested
▦ 104.5-105.0 GHz	▦ -98 – -100dBc/Hz
▦ 104.0-104.5 GHz	▦ -95 – -98 dBc/Hz
▦ 103.5-104.0 GHz	▦ -92 – -95 dBc/Hz
▦ 103.0-103.5 GHz	▦ > -92 dBc/Hz

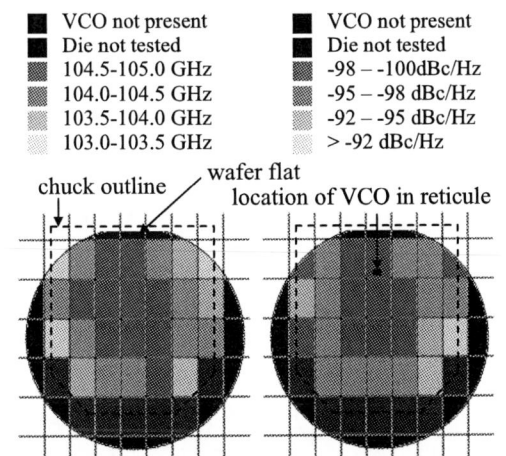

Fig. 12. BipX wafer maps of oscillation frequency (left) and phase noise at 1 MHz offset (right).

TABLE 2: SUMMARY OF BICMOS9 MOS VAR. VCOS.

Wafer	1	2	3	4
Center freq. (GHz)	94.7	94.9	94.9	95.0
Tuning range (GHz)	4.6	4.6	4.6	4.6
Output power (dBm)	0.2	0.7	0.6	0.8
DC power (mW)	133.8	133.2	137.3	132.6

TABLE 3: SUMMARY OF BICMOS9 HBT VAR. VCOS.

Wafer	1	2	3	4
Center freq. (GHz)	99.6	100.5	100.1	100.5
Tuning range (GHz)	3.4	3.6	3.6	3.7
Output power (dBm)	-1.1	-1	-1.4	-0.9
DC power (mW)	133.0	133.0	136.2	132.8

V. CONCLUSIONS

W-band low-voltage VCOs have been presented with record phase noise for SiGe VCOs above 80 GHz.

Experimental results indicate that transistors in W-band VCOs should be biased at peak f_T/f_{MAX} to minimize phase noise because NF_{MIN} current density approaches peak f_T/f_{MAX} current density in the W-band. Additionally, MOS varactors are shown to be superior to HBT varactors for achieving low phase noise. Furthermore, while VCO performance improves when SiGe technology is scaled, the oscillation frequency – determined by passive components – is insensitive to scaling. Wafer mapping and temperature data show that SiGe HBTs with over 200 GHz f_T/f_{MAX} are required to obtain production-quality W-band VCOs.

REFERENCES

[1] M. Laurens et al, "A 150 GHz f_T/f_{MAX} 0.13μm SiGe:C BiCMOS Technology," *Proc. IEEE BCTM*, pp.199-202, Sept. 2003.

[2] P. Chevalier, *et al.* "300 GHz f_{max} self-aligned SiGeC HBT optimized towards CMOS compatibility" *Proc. BCTM 2005*, pp. 120-123.

[3] G. Jagannathan et al., "Self-aligned SiGe NPN transistors with 285 GHz f_{MAX} and 207GHz f_T in a manufacturable technology," *IEEE Electron Device Ltrs.*, vol. 23, no. 5, 2002.

[4] J. Bock et al., "SiGe Bipolar Technology for Automotive Radar Applications," *Proc. IEEE BCTM*, pp. 84-87, Sept. 2004.

[5] M. Racanelli "SiGe BiCMOS Technology for RF Circuit Applications," *IEEE Trans. on Electron Devices*, vol. 52, no. 7, July 2005. pp. 1259-1270.

[6] H. Rucker et. al., "SiGe:C BiCMOS Technology with 3.6 ps Gate Delay," *IEDM Techn. Digest*, pp. 5.3.1-5.3.4, Dec. 2003.

[7] L. Dauphinee et al., "A Balanced 1.5 GHz voltage controlled oscillator with an integrated LC resonator," *IEEE ISSCC Dig.*, pp. 390-391, Feb. 1997.

[8] H. Li, and H.-M. Rein, "Fully Integrated SiGe VCOs With Powerful Output Buffer for 77-GHz Automotive Radar Systems and Applications Around 100 GHz," *IEEE JSSC*, vol. 39, pp. 1650-1658, Oct. 2004,

[9] R. Wanner, et al, "A SiGe Monolithically Integrated 75 GHz Push-Push VCO," *Si Monolithic ICs in RF Sys. Digest*, 2005, pp. 375-378.

[10] W. Perndl, et al, "Voltage-Controlled Oscillators up to 98 GHz in SiGe Bipolar Technology," *IEEE JSSC*, vol. 39, no. 10, Oct. 2004, pp. 1773-1777.

[11] B.A. Floyd, "V-band and W-band SiGe Bipolar Low-noise Amplifier and Voltage-Controlled Oscillators," *RFIC Symposium Digest*, June 2004, pp. 295-298

[12] C. Cao and K.O, "A 90-GHz Voltage-Controlled Oscillator with a 2.2-GHz Tuning Range in a 130-nm CMOS Technology," *Sym. on VLSI Cir. Digest*, Jun 2005, pp. 242-243.

[13] C. Lee, et al, "SiGe BiCMOS 65-GHz BPSK Transmitter and 30 to 122 GHz LC-Varactor VCOs with up to 21% Tuning Range," *IEEE CSICS*, 2004. Page(s):179 – 182.

[14] T.O. Dickson, and S.P. Voinigescu, SiGe BiCMOS Topologies for Low-Voltage Millimeter-Wave Voltage Controlled Oscillators and Frequency Dividers," *SiGe Monolithic ICs in RF Sys. Digest*, 2005, Pp. 273-276.

[15] K.H.K. Yau, and S.P. Voinigescu, " Modeling and Extraction of SiGe HBT Noise Parameters from Measured Y-Parameters and Accounting for Noise Correlation," *Si Monolithic ICs in RF Sys. Digest*, pp. 226-229, Jan. 2006.

2006 Bipolar/BoCMOS Circuits and Technology Meeting Proceedings

Turn-On Voltage Control in BSCR and LDMOS-SCR by the Local Blocking Junction Connection

V.A. Vashchenko and P.J. Hopper

National Semiconductor Corporation 2900 Semiconductor Drive, M/S E-155, Santa Clara, CA 95052

Abstract — **The problem of local ESD protection of power arrays is addressed at the device level. A wide voltage range of the pulsed dV/dt turn-on is achieved using a local blocking junction connection. The approach is experimentally validated on both examples of Bipolar SCR and NLDMOS-SCR devices and implemented in a 0.5μm 24V BiCMOS process.**

Index Terms — **Silicon bipolar/BiCMOS process technology, ESD, power devices, snapback.**

I. INTRODUCTION

Emerging analog power management applications require new ESD (electrostatic discharge) protection solutions. In the case of fast switching outputs, whereby the speed of the output signal is comparable to the ESD pulse rise time, a problem exists to implement a local clamp approach. Due to the constraints of multiple voltage domains; high voltage output and excessive IC area consumption, a conventional local clamp ESD protection approach, based upon snapback devices, is not always easy to implement. Among the major challenges is the protection of the output pins for switching voltage regulator circuit applications.

In the case of a high voltage output array circuit, a difficulty is derived from the similarity of the arrays snapback characteristics to both the array output and ESD devices. Since both devices may be based upon similar blocking junctions, the resultant ESD current flow direction depends on the dynamic turn-on characteristics of the ESD snapback cell, the dynamic coupling of the output array control electrode to the internal circuit and the equivalent capacitance of the array's output node.

Under these circumstances the precise targeting of the snapback characteristics for local snapback devices becomes a challenging design task both in the output voltage and in the rise time domains.

Typical an ESD protection solution for high-voltage BiCMOS or BCD processes is based upon the use a silicon controlled rectifier (SCR) based ESD device [1]. To avoid additional process steps the ESD protection devices are based upon those supported in the process as NPN BJT (Fig.1a) or NLDMOS (Fig.2a) devices.

The ESD device architecture is achieved by the embedding of an additional parasitic PNP structure be means of an additional P-emitter region. The simplified cross-section for a Bipolar SCR (BSCR) [2, 3] is presented in Fig.1b, where the "B", "E", "P", "C" are the base, the N-emitter, the P-SCR-emitter and the collector electrodes, respectively. Similarly, the cross-section for an LDMOS-SCR device [4-6] is presented in Fig.2b, where the "B", "S", "G", "P", "D" are the bulk, the source, the gate, the P-SCR-emitter and the drain electrodes, respectively.

Both BSCR and NLDMOS-SCR devices are based upon the blocking junction and diffusion regions from the supported devices NPN BJT and NLDMOS devices respectively.

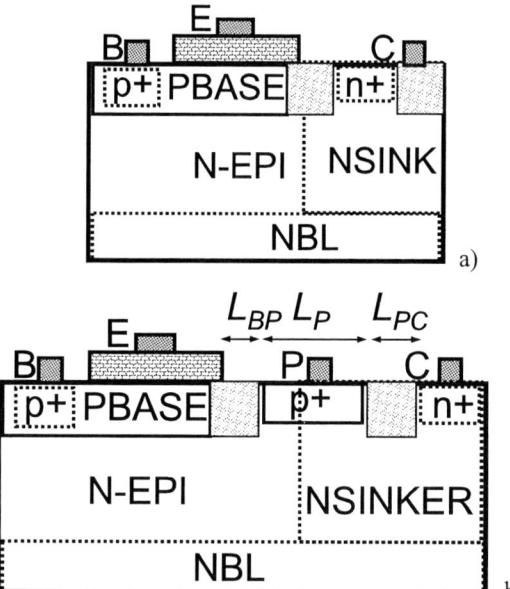

Fig. 1. Simplified cross-sections for NPN BJT (a) and the Bipolar SCR ESD protection device (b)

Since obtaining a proper reference voltage becomes a challenge, an alternative way to control the pulsed turn-on characteristics is to implement the dV/dt triggering using a displacement current.

In some cases the desired change in the triggering characteristics can be achieved by a simple scaling of the device parameters that effect on the blocking the P-SCR emitter and N-emitter isolation level.

1-4244-0458-4/06/$20.00 ©2006 IEEE   146

However, in general, the desired turn-on voltage range might not be achieved.

The purpose of this study is to suggest and experimentally validate a new principle for ESD protection cell design. A local blocking junction connection is suggested to amplify the dV/dt effect in ESD protection cells and thus reduce the turn-on voltage with a considerably wide voltage range.

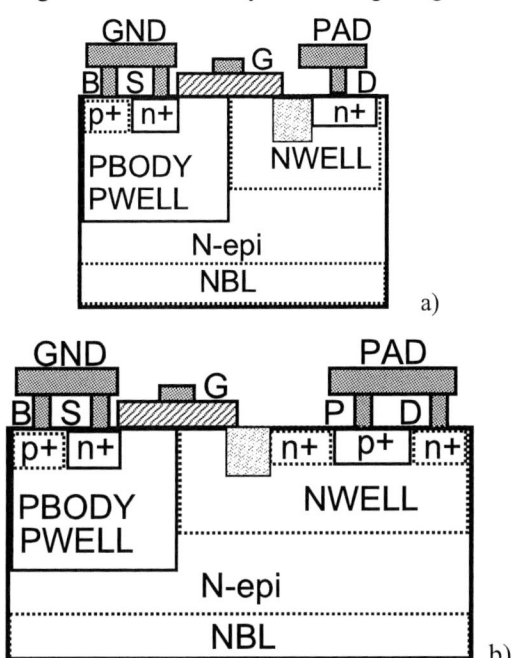

Fig. 2 Simplified cross-section for NLDMOS (a) and NLDMOS-SCR ESD protection device (b)

II. THE PRINCIPLE OF THE TURN-ON VOLTAGE CONTROL USING LOCAL BLOCKING JUNCTION CONNECTION

The process used in this study is a complementary 24V 0.5 μm BCD process. The devices have been tested using transmission line pulse (TLP), human body model (HBM) and machine model (MM) ESD tests.

The simplified equivalent circuit for the BSCR (Fig.1b) and LDMOS-SCR (Fig.2b) are presented in Fig.3.

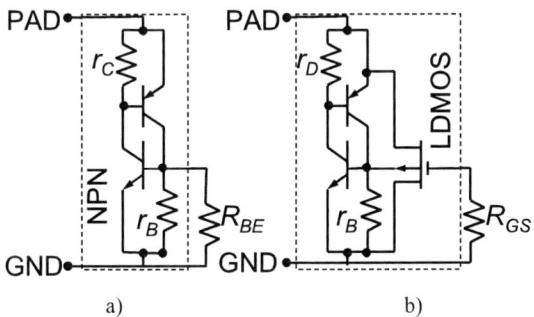

Fig. 3 Equivalent circuit diagrams for the BSCR (a) and LDMOS-SCR (b) devices.

To achieve high-voltage tolerance at submicron dimension the blocking junction electrode in the ESD protection devices are shorted to the corresponded anode P-SCR emitter and cathode Emitter ("E") or Source ("S") electrodes. The internal resistance of the drift blocking junction regions r_D, r_C, r_B, (Fig.3) controls the BSCR turn-on [7].

Usually the change in the internal resistance can be achieved by the change in the dimension of the anode region (L_{BP}, L_P, L_{PD}.) and similar parameters in the cathode region. However in the case of the given process this technique has not provided the proper impact to the turn-on characteristics. The experimental TLP characteristics for different structure parameters for the BSCR device are presented in Fig.4a. The pulsed turn-on voltage of the device is a rather weak function of the structure parameters (Fig.4b).

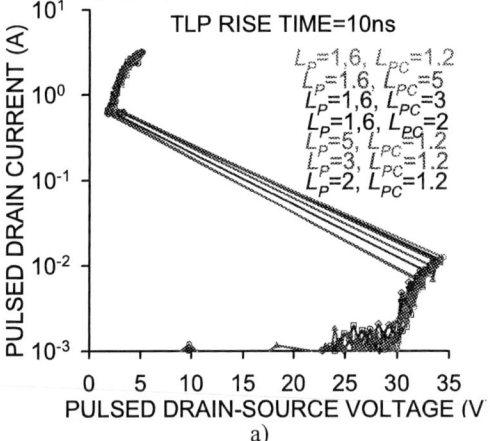

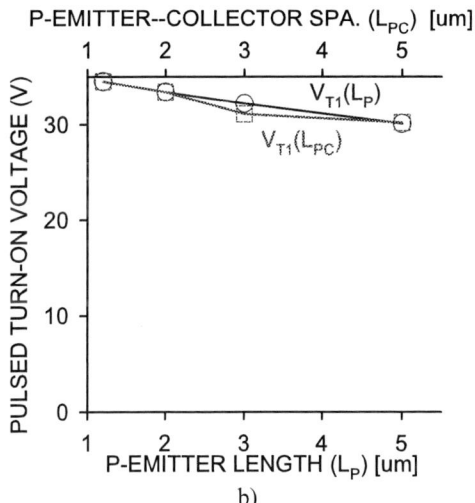

Fig.4 TLP characteristics for BSCR with different L_P and L_{PC} parameters (a) and the plotted dependence for the turn-on voltage (b)

To overcome the effect of excessive blocking junction isolation, a local blocking junction connection has been implemented. In order to gain additional internal resistance r_C the collector region

147

has been reduced in width from the full size W of the device finger (Fig.5a) down to a local region of a variable width W_C. Different versions of the local collector region have been studied for placement in the center or in the side of the finger (Fig.5 b, c).

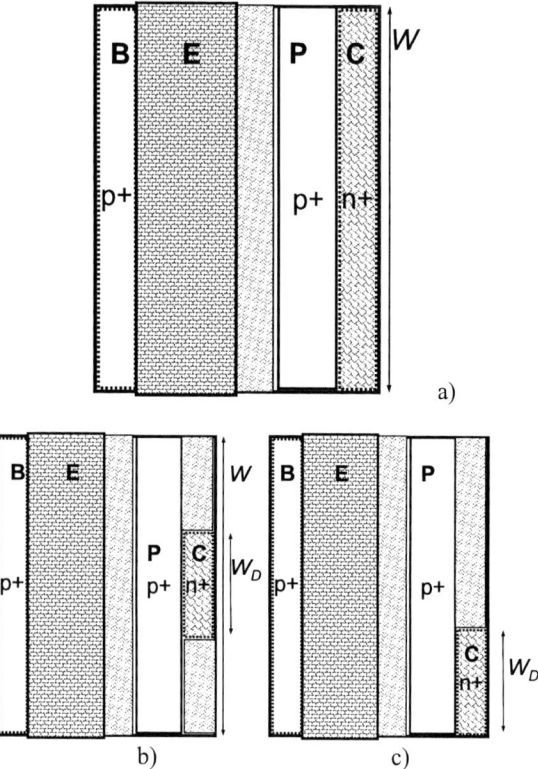

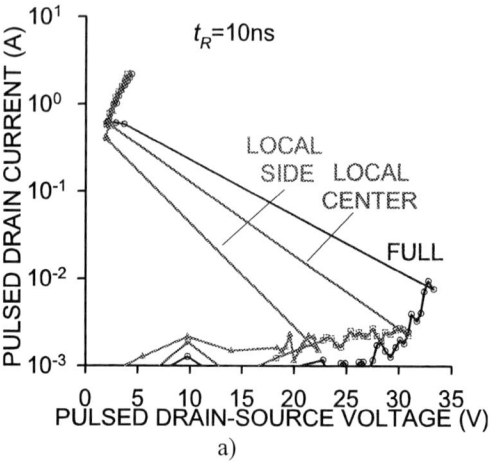

a)

Fig. 5   Simplified topology for the reference BSCR device (a) and the versions of the BSCR with local center (b) and the one side (c) collector contacts

II. EXPERIMENTAL RESULTS FOR THE LOCAL BLOCKING JUNCTION CONNECTION

Experimental TLP characteristics of the BSCR devices with local blocking junction connection are presented in Fig.6a. As it was assumed, the pulsed turn-on voltage becomes a strong function of the contact dimension and position.

Similarly the device with a local blocking junction connection produces the wide voltage range of the turn –on characteristics from 7V to 24V depending on the TLP pulse rise time. (Fig.6 b, c)

The implementation of the similar approach for local blocking junction diffusion in the NLDMOS-SCR devices with a local drain diffusion region has demonstrated similar regularities. Pulsed snapback turn-on voltage control has been obtained similar to the BSCR devices in the range from 50V to 20V. Experimental validation of the approach as implemented with high voltage NLDMOS SCR devices is presented in Fig.7.

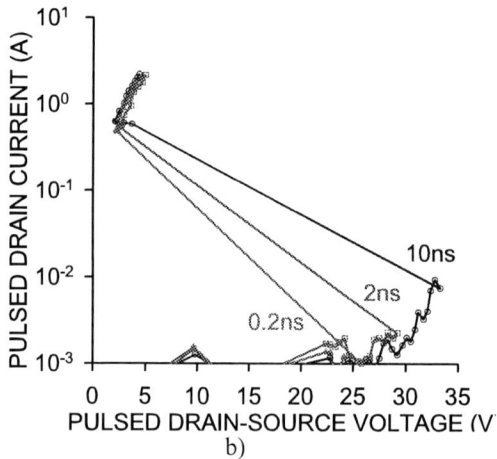

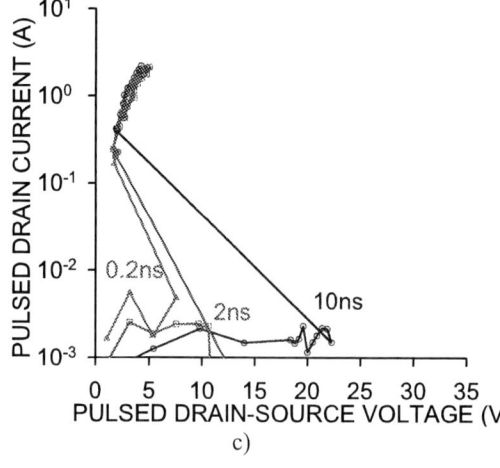

Fig. 6   TLP characteristics for the BSCR devices: the effect of the local collector on the turn-on voltage (a) and comparison for full (b) and the side local contact BSCR (c) devices for different rise time.

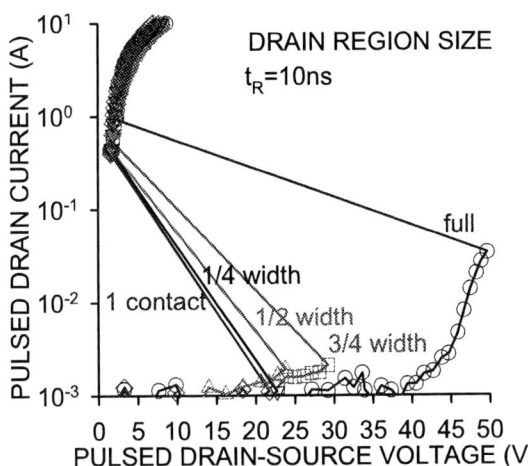

Fig. 7 Effect of local drain connection in NLDMOS-SCR ESD protection devices: TLP characteristics for different local drain.

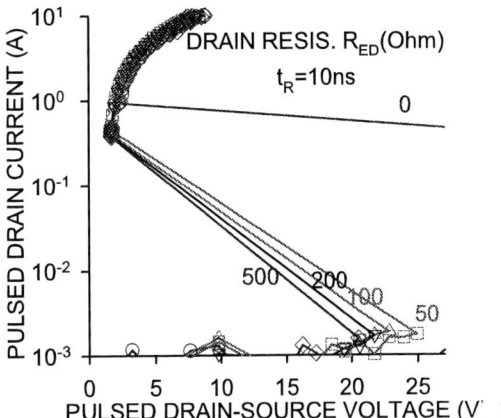

Fig. 8 The effect of the additional external resistor in the N-collector – P-emitter circuit; the equivalent circuit with the P-emitter-drain resistor R_{PD} and pulsed TLP characteristics for different R_{GS} value.

IV. BLOCKING JUNCTION CONNECTION WITH SUB-CIRCUIT ELEMENTS

For a more precise tuning the change in the turn-on characteristics can be further achieved by both the separation of the blocking junction and the adding of a subcircuit element. A simple additional external resistor provides for the dV/dt effect amplification. The change in the equivalent circuit (Fig.3b) for the NLDMOS-SCR device, with the adding of the drain-P-emitter resistor R_{PD} , is shown in Fig, 8a. With increasing R_{PD} value a significant reduction in the turn-on voltage is observed (Fig.8b)

The correlation between the HBM and MM test and maximum TLP current has been characterized and verified.

V. CONCLUSION

Device level control of the pulsed turn-on characteristics of ESD devices has been experimentally demonstrated using a local blocking junction diffusion region or an additional sub-circuit resistor for blocking junction connection.

An advantage of the newly proposed approach is the realization of triggering characteristics in a desired voltage range with only a minor deviation from the supported device architecture.

This suggested approach is a valuable methodology for ESD protection design of analog circuits.

REFERENCES

[1] L. Avery, "Using SCR's as transient protection structures in integrated circuits", In Proceed. *EOS/ESD Symposium* 1983, pp.27-29, 1983.
[2] J.Z. Chen, A. Amerasekera and T. Vrotos "Bipolar SCR ESD protection for high speed submicron Bipolar/BiCMOS Circuits frequency integrated circuits", *IEDM.*, 1995, pp.337-340, 1995.
[3] V. A. Vashchenko and P. Hopper, "Bipolar SCR ESD Devices," *Microelectronics Reliability Jour..*, pp. 457-471, 2005.
[4] S. Pendharkar, R. Teggatz, J. Devore, J Carpenter, T. Efland, C-Y Tsai, *"SCR-LDMOS □ A novel LDMOS device with ESD robustness,"* in Proceed. *ISPSD* 2000, pp.341-344., 2000
[5] A. Concannon, V. A. Vashchenko, M. ter Beek, P. Hopper, "ESD Protection of Double-Diffusion Devices in Submicron CMOS Processes", in Proceed. *ESSDERC*, 2004. pp.261-264, 2004
[6] V. A. Vashchenko, M. ter Beek, "ESD Protection window targeting using LDMOS-SCR devices with PWELL-NWELL super-junction", in Proceed. *IRPS* 2005, pp.612-613, 2005.
[7] S.M. Sze, *Physics of semiconductor devices*, Wiley, 1981,

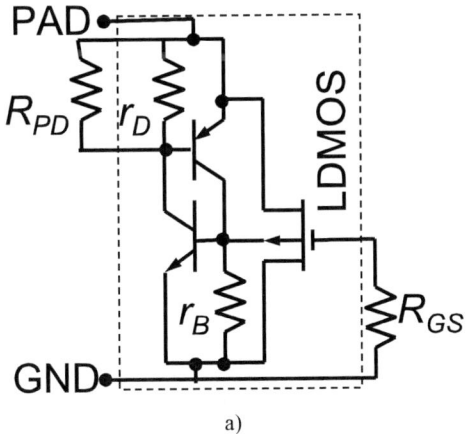

a)

2006 Bipolar/BoCMOS Circuits and Technology Meeting Proceedings

Deep Trench NPN Transistor for Low-R_{ON} ESD Protection of High-Voltage I/Os in Advanced Smart Power Technology

A. Gendron[1,2], C. Salamero[2], N. Nolhier[2], M. Bafleur[2], P. Renaud[1], P. Besse[1]

[1] *Freescale Semiconductor, 134 Av du Général Eisenhower 31023 TOULOUSE Cedex, FRANCE*
[2] *LAAS-CNRS, 7 Av. du Colonel Roche 31077 TOULOUSE Cedex 4, FRANCE*

Abstract: An innovative self-biased NPN transistor dedicated to the ESD protection of high voltage I/Os is presented. To fulfil a high clamping voltage / low on-state resistance specification, we have taken benefit of specific technology features, as deep insulation trenches, low-doped epitaxy and high-doped buried layer. First, the guidelines allowing the increase of the clamping voltage and the lowering of the on-state resistance are defined, based on an accurate description of the physical mechanisms involved during an ESD stress. Then, the proposed NPN transistor is described, and the results of measurements and TCAD simulations are presented. Excellent capabilities as 40 Volt clamping voltage, zero on-state resistance and I_{t2} higher than 5 Ampere have been achieved.

I. Introduction

The main feature of smart power technologies is the integration of command and power components on the same chip. They generally address operating voltages up to 100 V and are dedicated to applications such as automotive or power over ethernet. Protecting the high voltage I/Os against ESD is very challenging. In addition to high operation voltages, a high ESD robustness ranging from 4 kV up to 8 kV HBM becomes a standard customer request. Finally, to be cost-effective, a minimum silicon area of the protection structure is mandatory. All these requirements induce high power dissipation in a small region, and lead to high local self-heating.

Besides, the high current clamping voltage of the protection must be higher than the application operating voltage and maximum rating to avoid any unwanted triggering upon a fast transient or during system level EMC/ESD testing. Particularly, the major risk is latch-up triggering during the IC operation. For many applications, this results in clamping voltages close to the circuit destruction voltage (usually oxide breakdown), and therefore a narrow voltage window for the design of the ESD protection. Given these application constraints, a standard "snapback ESD protection" (Figure 1) should exhibit both a high holding voltage, V_H and a low on-state resistance R_{ON}. Thus, SCR cannot be used because of their very low holding voltages. Another ESD protection, currently used for its good efficiency in terms of on-resistance (R_{ON}) versus area and robustness, is the vertical NPN transistor. However, its holding voltage is defined by the technology parameters such as the epi-layer thickness and the N-type buried layer profile that cannot be easily adjusted. For the technology considered in this work, the holding voltage of the vertical NPN is too low (<20 V) to match most of the ESD specifications. An existing solution consists in stacking several transistors [1] so that the total

holding voltage, equal to the sum of each individual one, matches the specification. The major drawbacks are an important circuit footprint and a degraded on-resistance.

In this work, we propose a new NPN-based ESD protection structure that exhibits both a high holding voltage and a very low on-resistance, compatible with the specifications of 40 V / 80 V I/Os. The originality of the proposed protection consists in taking advantage of technology features such as deep isolation trenches and low-doped epitaxy.

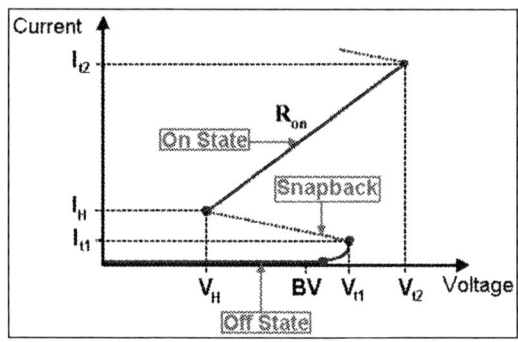

Figure 1: Standard current-voltage TLP characteristic of a "snapback ESD protection". BV is the static breakdown voltage, V_{t1} the triggering voltage, V_H the holding voltage, V_{t2} the thermal breakdown voltage and R_{ON} the on-state resistance.

II. Design of high holding voltage, low on-state resistance ESD protections based on self-biased bipolar

A. High holding voltage V_H

A self-biased NPN transistor (Figure 2) behaves as a diode when undergoing a negative ESD stress on its collector. Under a positive ESD stress, the base-collector voltage increases up to the avalanche breakdown and the induced current triggers on the

1-4244-0458-4/06/$20.00 ©2006 IEEE

bipolar transistor. The bipolar turn-on is associated with a decrease of the base-collector voltage or snapback. Using some simplifications (negligible recombination, no access resistance), a simple analytical study allows describing the on-state of a self-biased bipolar transistor [2]:

$$\Box M = 1 + \frac{I_B}{I_E} \qquad (1)$$

Where $\Box$ is the common base gain (in self-biased configuration), M the avalanche multiplication factor, I_B the current at base contact, I_E the current at emitter contact. It is noteworthy to outline that M is an increasing function of the voltage V.

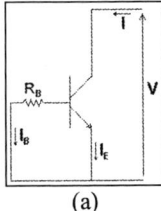 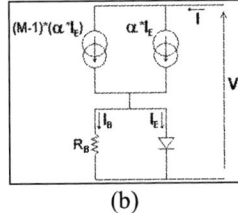

(a)   (b)

Figure 2: Electrical schematic (a) and equivalent circuit (b) of a self-biased NPN.

The ratio $\dfrac{I_B}{I_E}$ is a decreasing function of the current and rapidly tends to zero, which corresponds to an open-base configuration. Thus, a first approximation of the holding voltage V_H is the open-base collector emitter breakdown voltage BV_{CEo}, given by the implicit equation:

$$\Box M = 1 \qquad (2)$$

As a result, the avalanche multiplication factor M and the bipolar gain $\Box$ are the two relevant parameters for controlling the holding voltage. To increase the holding voltage value, M should increase and $\Box$ decrease. In practice, $\Box$ can be more easily controlled. In particular, the base transfer coefficient can be lowered by increasing the internal base length.

B. Low on-state resistance, R_{ON}

Equation (1) is not sufficient to determine the self-biased NPN bipolar transistor R_{ON}, which is closely related to the effects of high current densities (approximately 10^5-10^6 A.cm^{-2}) and high temperatures, induced by an ESD event. To accurately describe the R_{ON} three contributions have to be taken into account: the access resistances, the impact of the electric field modification at the base-collector junction and the impact of the electric field building up in the low-doped collector region.

B.1 Access Resistances

The access resistances correspond to the ohmic resistances in the emitter and the collector that are inherent to the structure. Generally, their values are determined by the ballast resistance necessary to obtain a uniform current conduction and thus a high robustness.

B.2 Field modification at the base-collector junction

a. High current effects

In a bipolar transistor, the carriers injected by the emitter modulate the base and the collector doping levels. This results in a modification of the base-collector space charge region, which leads to a shift of the effective base-collector junction into the collector region and a widening of the effective base. Kirk reported for the first time this effect [3]. It is possible to take advantage of this behaviour to reduce R_{ON}. If the base doping level is low and the collector doping profile abrupt, only the base undergoes doping modulation, thus reducing the space region extension. This induces a voltage decrease, beneficial for R_{ON}.

b. Temperature effects

During an ESD stress, the device self-heating modifies the physical parameters of the silicon. Concerning R_{ON}, the most significant thermal effects are the decrease of both the impact ionization coefficients and the carrier saturation velocities. On the one hand, a higher electric field, i.e. a higher voltage, is needed to compensate the impact ionisation decrease; this tends to degrade R_{ON}. On the other hand, the decrease of saturation velocities enhances the high injection effects. As a result, no general R_{ON} trend can be determined. If the protection is designed to take advantage of these effects, R_{ON} improvement could be expected. However, in practice, most of the ESD protections exhibit an increase of R_{ON} with the temperature.

B.3 Electric field building-up in low-doped collector regions

At high current densities, an electric field could build up in the low-doped collector region (Figure 3) By instance, this is the case in a vertical NPN bipolar structure with a graded collector composed of a N well and a N$^+$ buried layer. As long as the carrier velocity grows linearly with the electric field (ohmic conduction), the electric field remains quite low (region 1 in (Figure 3)) However, when the carrier velocity saturates (approximately at 2 10^4 V.cm^{-1}), the electric field rapidly rises to maintain the increase of the current density (region 2 in (Figure 3))[4]. Under such conditions, the voltage also rapidly increases, until the high electric field generates impact ionisation (approximately at $3 \cdot 10^5$ V.cm^{-1}), which induces a voltage snapback (region 3 in (Figure 3)) Generally, the resulting over-voltage prevents to develop ESD protections with a narrow design window. To avoid this problem, the solution consists in increasing the collector doping and its surface, so that the current generated by the ESD stress is not sufficient to lead to the over-voltage.

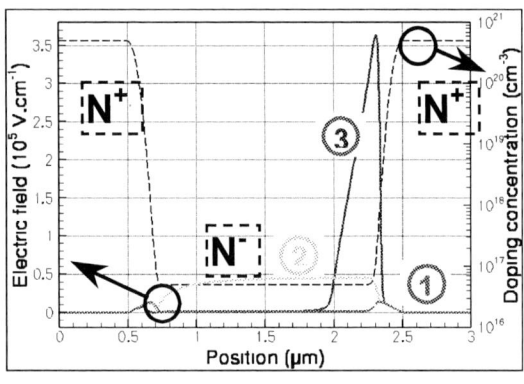

(a)

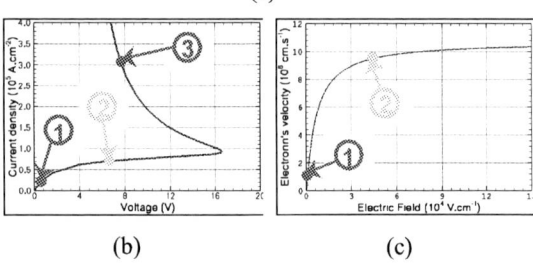

(b)   (c)

Figure 3: Description of the behavior of a simple N$^+$-N$^-$-N$^+$ doping profile under high current density: electric field distributions at different current densities (a), current-voltage characteristic (b) and electron velocity versus electric field (c). The third state is not represented on figure (c): the current is no more limited by the saturation velocity as a large amount of carrier is generated by impact ionization.

III. High holding voltage and low R_{ON} self- biased NPN transistor using a trench isolated smart power technology

The considered technology is a 0.25 μm smart power technology featuring a low-doped P$^-$-epitaxy, a high-doped P$^+$-substrate, a high-doped N$^+$-buried layer and deep trench insulation. The objective consists in developing a high holding voltage and low R_{ON} ESD protection by taking advantage of the technology features to implement some of the design guidelines defined above.

A. Device description

The proposed device (Figure 4) is a NPN bipolar transistor in which a deep trench separates the collector and the emitter-base regions. The P$^-$-epitaxy forms the base region. By forcing the current to flow into the silicon bulk, the trench widens the

base with minimum supplementary area. Therefore, the bipolar gain is greatly reduced, which in turn increases the holding voltage. Concerning the R_{ON}, the base-collector junction, formed by the P$^-$-epitaxy and the N$^+$ buried layer, fulfils the low base doping / abrupt collector profile condition, which tends to reduce R_{ON}. However, there is a low-doped region in the collector, between the buried layer and the deep well and electric field building up could be expected and then a R_{ON} degradation.

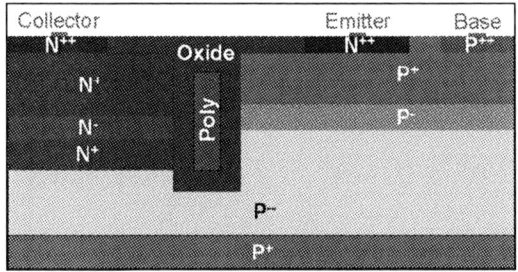

Figure 4: Schematic cross-section of the NPN under study.

B. Measurement data

The structure under study is composed of four 100 μm-long fingers, the total surface is approximately $100*100$ μm^2 and an 850 Ω polysilicon resistance ($50*13$ μm^2) connects the base to the emitter. The behaviour under ESD conditions was characterized by Transmission Line Pulse (TLP) measurements.

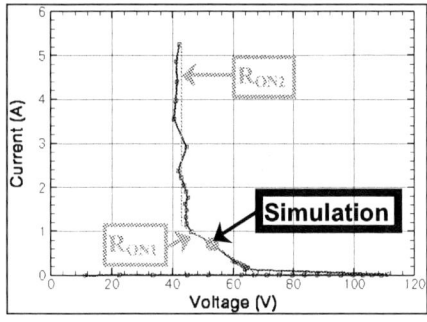

Figure 5: Measured TLP characteristic and the simulated point used for the device behavior analysis.

The TLP IV characteristic (Figure 5) (Table 1) exhibits a high clamping voltage (>40 V) and a current-voltage characteristic almost vertical above one Ampere. These two features are excellent for developing a high-voltage / low R_{ON} ESD protection. Concerning the robustness, the TLP set-up limit was reached before destruction, and at least an 8 kV HBM robustness is expected (corresponding to 5A current). The main drawback is the high triggering voltage (110V). However, to cope with this problem, in an input protection scheme, this component can be implemented as the primary element of a two-stage protection circuit [5]. If used as a power clamp, active triggering of the base of the transistor [5] [6] should allow a direct turn-on into snapback then avoiding the drawback of the high voltage triggering.

Table 1: Measured electrical parameters. V_{t1} is the triggering voltage, V_{Sb} the voltage of the first point after snapback, V_{min} the minimum voltage after snapback, R_{ON1} the on-state after below 1 Ampere, R_{ON2} the the on-state after above 1 Ampere and I_{t2} the destruction current

V_{t1} (V)	V_{Sb} (V)	V_{min} (V)	R_{ON1} (Ω)	R_{ON2} (Ω)	I_{t2} (A)
109	63	40	22	0	>5

C. Simulation analysis of the device behaviour under ESD conditions

A simulation analysis was carried out using ISE-TCAD simulator [7]. First, the component was described by a process simulation (DIOS), and then the behaviour under TLP pulses was simulated (DESSIS). Calibration difficulties were encounter to describe the N^+-buried layer/P-substrate junction: this doping profile is not well characterized since not critical for the active devices. As the transistor bias current is provided by the avalanche at this N^+-buried layer/P-substrate junction, TCAD simulations could be only used for qualitative analysis. We provide in (Figure 6) the current density and field distributions at the end of the pulse, for a simulated TLP point sufficiently close to measurement. At the base-collector junction, the electric field modification along the current path shows a strong base push-out. This effect tends to reduce R_{ON} but is not sufficient to induce the observed extremely low R_{ON} value. In a previous study [8] and corroborated by several authors [9], NPN devices having a strong snapback exhibit an inhomogeneous current flow along the length of the emitter. This was experimentally demonstrated by Pogany [10] using laser interferometry. The ESD robustness and low on-resistance results from the movement of the current filament due to the impact of temperature on the avalanche generation.

IV. Conclusion

The physical mechanisms, which rule the holding voltage and the on-resistance of an ESD protection, were described. Based on this analysis, design guidelines were defined to achieve high holding / low R_{ON} specifications. These guidelines were applied to develop an innovative self-biased NPN transistor in an advanced smart power technology by using a deep trench to widen the base. TLP characterizations have shown a high clamping voltage (40 V), an excellent high current R_{ON} and a very good robustness. The drawback is the over-voltage at triggering. However, simple methods, as two stages circuits or active triggering, are available to overcome this over-voltage issue.

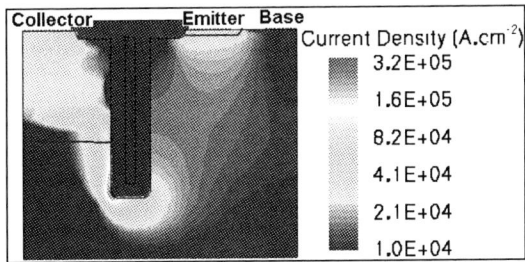

(a)

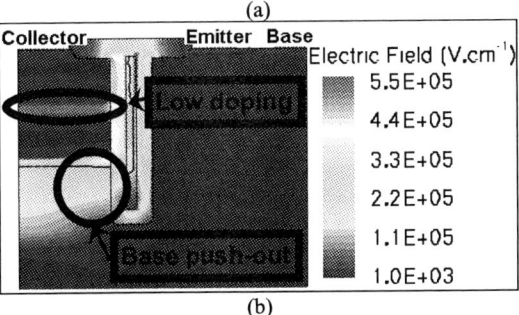

(b)

Figure 6: Current density (a) and electric field (b) distributions at TLP point (V=53 V ; I=700 mA).

References

[1] M-D. Ker, K-H. Lin, "The Impact of Low-Holding-Voltage and the Design of Latch-up Free Power-Rail ESD Clamp Circuit for LCD Drivers ICs ", IEEE Journal of Solid State Circuit, vol. 40, n° 8, August 2005.

[2] S. Joshi, R. Ida, P. Givelin, E. Rosenbaum, "An analysis of bipolar breakdown and its application to the design of ESD protection circuits", 39th IRPS, Orlando, Florida 2001.

[3] C. T. Kirk,"A theory of transistor cutoff frequency (fT) falloff at high current densities", IRE Trans. on electron devices, pp 164-173, 1961.

[4] G. Notermans, "On the Use of N-Well Resistors for Uniform Triggering of ESD Protection Elements", in Proceedings EOS/ESD Symp, 1997, pp.221-229.

[5] A. Amerasekera, C. Duvvury, ESD in silicon integrated circuits, John Wiley & sons, 1995.

[6] M.D. Ker, K.C. Hsu, Substrate-Triggered SCR Device for On-Chip ESD Protection in Fully Silicided Sub-0.25-m CMOS Process, IEEE Trans. on Electron Devices, vol. 50, n°2, Feb. 2003, pp. 397-405.

[7] ISE/Synopsys TCAD Manuals, Release 10.0, 2004

[8] D. Trémouilles, G. Bertrand, M. Bafleur, N. Nolhier, L. Lescouzères, Design guidelines to achieve a very high ESD robustness in a self-biased NPN, EOS/ESD Symposium, Charlotte (USA), October 6-10th, 2002, pp.281-288.

[9] R. M. Steinhoff, J.B. Huang , P. L. Hower, J. S. Brodsky, Current Filament Movement and Silicon Melting in an ESD-Robust DENMOS Transistor, EOS/ESD Symposium, October ??, 2003, pp.??.

[10] D. Pogany et al, Single-Shot Nanosecond Thermal Imaging of Semiconductor Devices Using Absorption Measurements, IEEE Transactions On Device And Materials Reliability, vol. 3, N°. 3, September 2003, pp.85-88.

GaAs HBT ESD Diode Layout and its Relationship to Human Body Model Rating

Douglas A. Teeter[1], Kathy Muhonen[2], David Widay[1], Mike Fresina[2]

RF Micro Devices   [1]Billerica, MA, 01821, USA   [2]Greensboro, NC, 27409, USA

Abstract — **Different GaAs HBT ESD diode configurations are analyzed and compared using Human Body Model (HBM) and Transmission Line Pulse (TLP) measurements. A common diode stack test circuit consisting of 3 series forward and reverse diodes is used to analyze ESD ruggedness dependence on layout and area. Both base emitter and base collector junction ESD ruggedness is compared and analyzed. To our knowledge, this is the first time data has been presented for GaAs HBT diodes that calculates the ratio of HBM voltage to maximum TLP current.**

Index Terms — **Electrostatic Discharges, ESD, heterojunction bipolar transistors, HBT, TLP**

I. INTRODUCTION

Electrostatic discharge (ESD) is becoming increasingly important in handset manufacturing as the pressure to reduce manufacturing costs demands higher assembly yields and the use of low cost assembly houses. Increased ESD ruggedness allows for reduced ESD precautions in the factory, higher yields, and an overall reduced manufacturing cost.

While extensive research has been and continues to be performed for Silicon integrated circuits, relatively little work has been published regarding ESD protection of GaAs circuits [1]-[7]. ESD diodes are a common building block for many protection circuits. This paper presents a detailed study of GaAs HBT ESD diode layout tradeoffs using a simple diode stack circuit as a demonstration vehicle.

Transmission Line Pulse (TLP) measurements are commonly used in evaluation of Si ESD protection circuits [8]-[10]. We demonstrate that, as with Silicon, a direct relationship exists between maximum TLP current and Human Body Model (HBM) failure voltage. While there has been some prior reporting of TLP measurements on GaAs devices [4, 5], we believe this is the first time data has been presented for GaAs HBT diodes that calculates the ratio of HBM voltage to maximum TLP current.

II. DEVICE DESCRIPTION

All results presented in this paper make use of RF Micro Devices advanced 3G GaAs HBT technology. The devices are grown using MBE and have AlGaAs emitters with a Be doped base layer [11]. Typical process Beta is 330 at 20 kA/cm^2 with a base-collector breakdown voltage (V_{bco}) of 30 V. F_t

of the process is 29 GHz at 25 kA/cm^2 for Vbc=1.7 V and associated f_{max} is 66 GHz.

III. ESD MEASUREMENTS

Human Body Model (HBM) and Transmission Line Pulse (TLP) Testing are two ESD simulation techniques commonly used to analyze ESD in devices. Figure 1 provides a simplified comparison of the two approaches. During HBM testing, a 100 pF capacitor is charged to a specified voltage and then discharged through a 1500 ohm resistor. After ESD zapping the device, the leakage current is measured by applying 0.5 V to the circuit. A higher ESD voltage is then applied to the device and the process is repeated until failure. TLP measurements utilize a transmission line instead of a capacitor. The transmission line is used to transfer a square voltage pulse to the DUT while the voltage and current are measured at the DUT. The DUT voltage and current are then recorded, usually about 2/3 through the pulse. Typically, a 100 nS pulse with 10 nS rise and fall times are used. After each transmission line discharge, the leakage current of the circuit is measured with a 0.5 V source. Then the pulse voltage is increased and the process is repeated until failure. References [8]-[10] provide a more detailed description of TLP measurements.

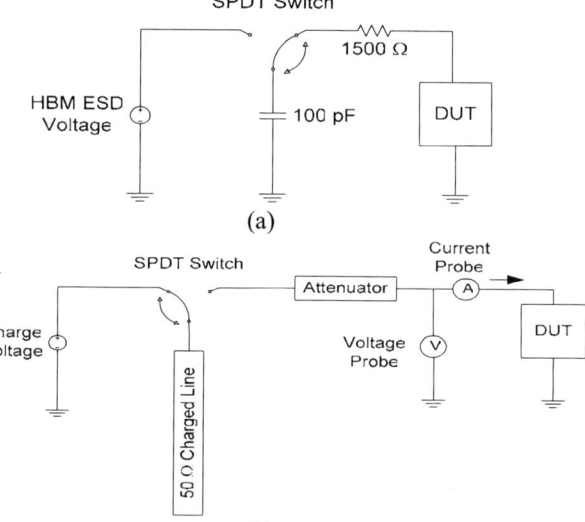

Fig. 1: Simplified block diagram comparing a) standard Human Body Model and b) Transmission Line Pulse measurement

IV. ESD DIODE STUDY

For simplicity and consistency, a simple 3 series forward and reverse diode test circuit was used to compare each ESD device performance. Figure 2 shows the basic circuit used. This circuit is typical of what one might use on a 2.8 V DC control or bias pin of a GaAs RFIC.

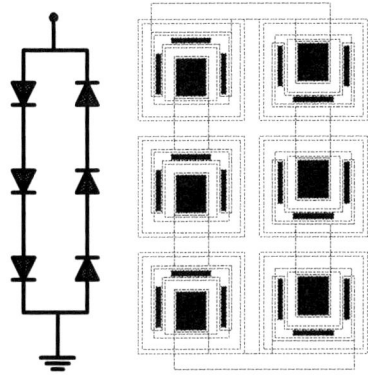

Fig. 2: Test circuit used to evaluate each diode layout considered in this paper. All devices studies were ESD tested in a 3 series forward, 3 series reverse circuit.

A. HBT Transistor Base-Emitter versus Base-Collector Junction Diodes

Before analyzing the diode stack circuit shown in figure 2, we compared the ESD ruggedness of single base-collector and base-emitter junctions of our HBT device. TLP measurements confirmed that the reverse bias failure characteristics of the base-collector junction (60 mA @ 34 V for 160 um² HBT) exceeded that of the base-emitter junction (15 mA @ 9 V for 160 um² HBT). However, we were surprised to discover that the onset of "soft failure" during *forward* bias, defined by a gradual but increasing change in low level leakage, was present for the base-collector junction but not for the base-emitter junction. The improved forward bias BE junction ruggedness compared to the forward BC junction may be attributed to lower series resistance from higher emitter doping and better heat sinking from the emitter contact metal [4]. When a human body model pulse is applied to the ESD clamp circuit in figure 2, it is the current handling capability of the forward biased arm in the clamp circuit that will limit the overall ESD voltage rating. The forward biased arm of the clamp circuit keeps the voltage on the reverse biased arm below the breakdown voltage of the junction. Thus, it is most relevant to compare ruggedness of the base-collector and base-emitter junction under forward bias conditions when designing symmetric clamp circuits.

Using quad 2x20 um² emitter (160 um² total emitter area) HBT transistors as test vehicles, we TLP tested the base-collector and base-emitter junction on discrete devices to failure. Figure 3 shows typical data under forward bias conditions for a single base-emitter and base-collector junction. Multiple devices were tested, some with AlGaAs emitters and others with InGaP emitter, all showing similar characteristics shown in figure 3. During each TLP pulse, the current through the device is measured about 60 ns into the pulse. After each pulse, the leakage through the test circuit was measured by applying 0.5 V to the circuit. Figure 3 plots the leakage current as a function of the peak current through the circuit during the TLP pulse. For the base-collector diode, the leakage current gradually starts increasing once the current through the junction exceeds about 450 mA, indicating the onset of soft failure. Leakage current in the junction continues to increase with further current passing through the diode. The maximum current that the base-collector junction can withstand is about 1.3 A, as noted by the rapid increase in leakage current. The associated junction voltage at soft failure is 3 V and at hard failure it is 11 V. For the base-emitter junction, the leakage current stays low, then abruptly increases once the current in through the DUT reaches 1.2 Amps, indicating hard failure. The associated DUT voltage at hard failure is 4.5 V. Notice the absence of soft failure for the base-emitter junction. For this reason, the remainder of the work discussed in this paper focuses on base-emitter diode stacks since we concluded the ESD clamp circuit shown in figure 2 would be more robust with base-emitter junction diodes than base-collector junction diodes.

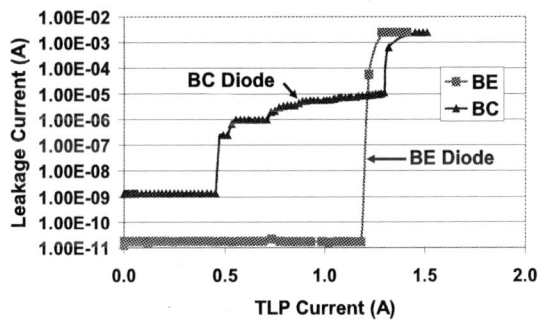

Fig. 3: Leakage current comparison for single HBT base-emitter and base-collector junctions.

B. Diode layout study

We also studied ESD performance of different base-emitter diode layouts. Figure 4 summarizes the diode layouts studied. Two basic trends were investigated. First, four diodes, (all with 160 um² of emitter area), with differing layouts were compared. The objective was to determine whether one large 160 um² finger or smaller parallel fingers with 160 um² total area performed better. Second, rectangular diodes of differing area but the same basic layout were compared.

These diodes were each used in a 3 series forward and reverse diode stack configuration (see figure 2) and tested to failure via TLP. Figure 5 compares the maximum TLP current handling of each of the diode stacks. As will be illustrated later in this paper, higher maximum TLP current corresponds to higher HBM ESD rating. In addition to the device layouts illustrated in figure 4, we also include data on base-emitter diode stacks made from 80 um^2 (4x2x10 um^2) and 160 um^2 (4x2x20 um^2) emitter area HBT transistors. Several interesting points can be made based on these results. First, diode connected transistors of the same emitter area have the highest maximum TLP current handling capability compared to standard diodes of the same emitter area. Thus, connecting HBT transistors in a base-emitter diode configuration will result in a more rugged ESD protection circuit. Second, it is better to use a multi-finger junction rather than a single, large area junction. This finding is actually consistent with our first finding in that as you move from layout XEB160 to XEB160C, (refer to figure 4), the diode layout becomes more like a standard HBT transistor. Our results are consistent with those reported in [4].

Figure 6 compares the layout area for the test circuit utilizing device XEB160 to that of the 160 um^2 HBT base-emitter diode. While the circuit utilizing the 160 um^2 HBT base-emitter diode provides 76% higher current handling, it also occupies 76% more die area.

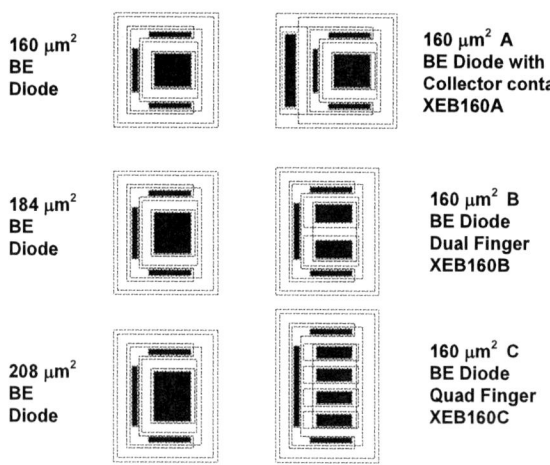

Fig. 4: Comparison of different ESD diode layouts evaluated in this work

C. Diode Area Study

Diode stacks (figure 2) utilizing 80 um^2, 160 um^2, and 208 um^2 base-emitter junction diodes were measured using transmission line pulse testing. As shown in figure 7, the maximum current handling of the diode is proportional to the area of the diode.

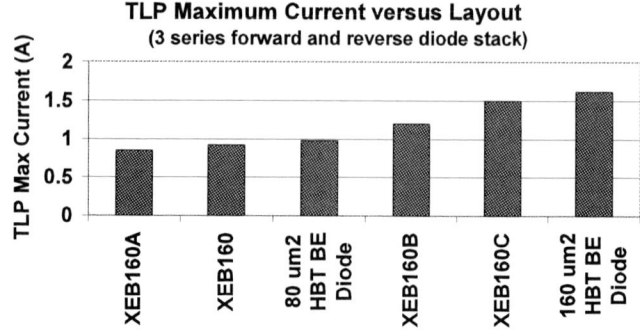

Fig. 5: Comparison of TLP Maximum Current for different ESD diode layouts. The 3 series forward, 3 series reverse test circuit shown in figure 2 was used for all the results in this figure.

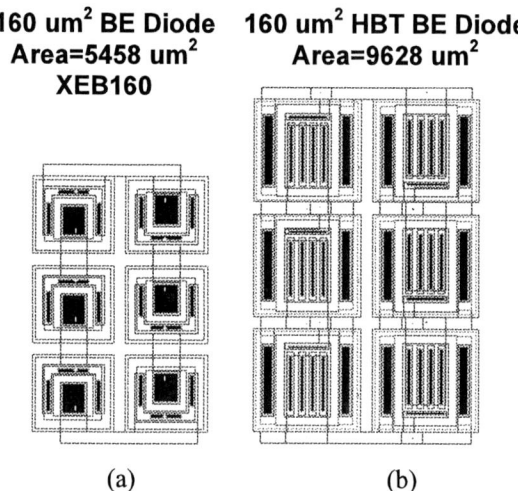

(a)   (b)

Fig. 6: Comparison of layout area for two of the diode stacks analyzed in figure 5. The 160 um^2 HBT base-emitter diode (b) handles 76% more current compared to the XEB160 diode (a), but requires 76% more layout area.

V. TLP VERSUS HBM RATING

Next, 11 different diode structures were measured using both TLP and HBM tests. The ratio of the HBM failure voltage to maximum TLP current was calculated, as shown in figure 8. Ideally, this ratio should be the HBM resistance of 1500 ohms. The computed average value is 2156. Si devices typically have a ratio between 1000 and 3000 that is constant within a particular technology node [8]. Our data clearly shows a consistent ratio between maximum TLP current and HBM ESD rating. This illustrates TLP data is a valuable tool in designing and evaluating ESD protection for GaAs circuits.

VI. CONCLUSION

A common test circuit was used to evaluate ESD performance of different device layouts. HBT base-emitter junctions were found to have better forward bias current handling capability compared to base-collector junctions for a given transistor. Diode layouts utilizing multiple fingers have better transient current handling and hence ESD robustness than diodes of the same area constructed a single rectangular junction. In other words, diode layouts that look like transistor layouts have better ESD current handling capability. We noticed a consistent ratio between the maximum TLP current and the corresponding HBM rating of various diode stack structures. The ratio was constant and independent of the specifics of the diode stack circuit. Thus, one can accurately estimate Human Body Model performance of a circuit based on its Transmission Line Pulse current handling capability.

ACKNOWLEDGEMENTS

The authors wish to acknowledge the assistance and support Jena Antonell and Ann Rhan as well as Larry Gatewood for helpful discussions. We also wish to thank Barth Electronics, Oryx Instruments, and SQP Products for allowing us to use their TLP test equipment to collect data.

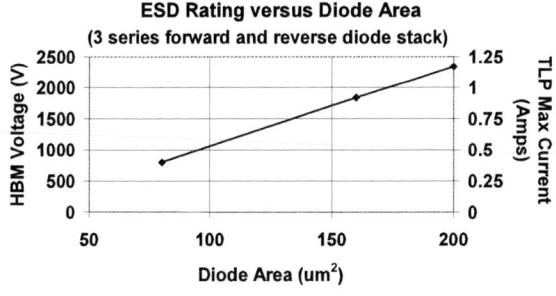

Fig. 7: Maximum TLP Current and HBM Voltage rating as a function of diode area for devices XEB80, XEB160, and XEB208.

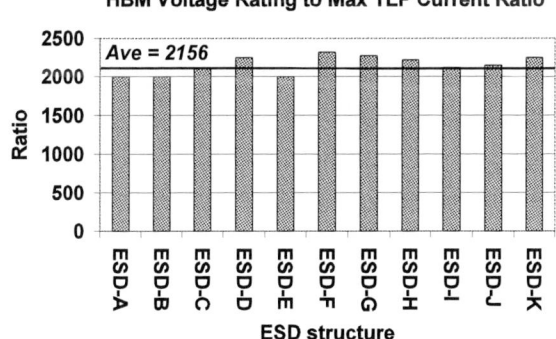

Fig. 8 Ratio of HBM ESD Voltage Rating to the maximum TLP current capability for 11 different diode stack structures. Detailed descriptions for each structure are in Table I.

TABLE I – Circuits Used to Calculate Ratio in Fig. 8

Circuit	Diode Stack Description
ESD-A	3SF 8x18um^2 BC, 3SR 8x18um^2 BC
ESD-B	3SF 8x18um^2 BC, 2SR 8x18um^2 BC
ESD-C	3SF 160um^2 HBT BE Diode, 3SR 160um^2 HBT BE Diode
ESD-D	3SF 160um^2 BE XEB160, 3SR 160um^2 BE XEB160
ESD-E	3SF 22x22um^2 BC, 3SR 22x22um^2 BC
ESD-F	3SF 160um^2 BE (XEB160), 2PR 22x22um^2 BC
ESD-G	3SF 160um^2 BE (XEB160), 1R 22x22um^2 BC
ESD-H	3SF 160um^2 BE XEB160, 2PR 160um^2 BE XEB160
ESD-I	3SF 160um^2 BE XEB160, 3SR 160um^2 BE XEB160
ESD-J	3SF 160um^2 BE XEB160, 2SR 160um^2 BE XEB160
ESD-K	3SF 160 um^2 BE XEB160, 1R 160 um^2 BE XEB160

Notes: 3SF = 3 Series Forward diodes
3SR = 3 Series Reverse diodes
2PR = 2 Parallel Reverse diodes
1R = 1 Reverse diode
BC = Base-Collector junction
BE = Base-Emitter junction

REFERENCES

[1] T. Henderson, "Effects of Electrostatic Discharge on GaAs-Based HBTs," *1997 GaAs IC Symposium Digest*, pp. 147-150, October 1997.

[2] K. Bock, "ESD Issues in Compound Semiconductor High Frequency Devices and Circuits," *1997 EOS/ESD Symposium*, pp. 1-12, October 1997.

[3] F. M. Yamada, et. al., "ESD Sensitivity Study of Various Diode Protection Circuits Implemented in a Production 1 um GaAs HBT Technology," *Proc. GaAs Reliability Workshop*, pp. 139-146, 1999.

[4] Y. Ma, and G. Li, "ESD Protection Design Considerations for InGaP/GaAs HBT RF Power Amplifiers," *IEEE Trans. On Microwave Theory and Tech.*, vol. 53, no. 1, pp. 221-228, January 2005.

[5] Y. Ma and G. Li, "On the Road to ESD Safe HBT MMICs," *2005 IEEE CSIC Digest*, pp. 272-275.

[6] Y. Ma and G. Li, "InGaP/GaAs HBT RF Power Amplifier with Compact ESD Protection Circuit," *2004 IEEE MTT-S Digest*, pp. 1173-1176.

[7] M. Sun and Y, Lu, "A New ESD Protection Structure for High-Speed GaAs RFICs," *IEEE Trans. On Electron Device Letters*, vol. 26, no. 3, pp. 133-135, March 2005.

[8] J. Barth, et. al., "TLP Calibration, Correlation, Standards, and New Techniques," *IEEE Trans. On Electronics Manufacturing*, vol. 24, no. 2, pp. 99-108, April 2001.

[9] T. J. Maloney and N. Khurana, "Transmission Line Pulsing Techniques for Circuit Modeling of ESD Phenomena," *1985 ESD/EOS Symposium*, pp. 49-54, September 1985.

[10] L. G. Henry, et. al., "Transmission Line Pulse Testing of the ESD Protection Structures of ICs – A Failure Analysts Perspective," *Proceeding from the 26th International Symposium for Testing and Failure Analysis*, pp. 203-215, November 2000.

[11] M. Fresina, et. al., "Reliable AlGaAs/GaAs HBTs grown by MBE with increased Beryllium Doping and Aluminum Concentration", *2001 GaAs Reliability Workshop*, Baltimore, MD

2006 Bipolar/BoCMOS Circuits and Technology Meeting Proceedings

Human Body Model ESD Protection of RF Bipolar Circuits

Paul Davis[1] and Brian Horton[2]

1. Consultant, Reading, PA   2. Agere systems, Inc., Allentown, PA

Abstract — **RF and other high-frequency bipolar circuits operating at 1-5 GHz have been ESD protected using the Human Body Model test to 2kV. The described design and modeling techniques minimize the operating parasitics, chip area, and number of process turn-arounds.**

Index Terms — **Silicon bipolar/BiCMOS process technology, bipolar modeling and simulation, power devices, analog or digital circuits, RF circuits, device physics.**

I. INTRODUCTION

Electrostatic Discharge (ESD) protection is normally required for all circuits, but is more difficult for cellular RF circuits because of the small paracitics required for 1 to 5 GHz operation.

Although there are other types of ESD tests, the Human Body Model (HBM) is well defined [1] and for small area bipolar circuits, the solutions seem to suffice for other tests. The Charged Device Model (CDM) testing [2] involves currents that are >10 times as high and pulses that are <1/100 the width of the HBM. Significant parasitics for analysis in CDM are therefore very different from HBM.

The HBM test is a circuit representation of a charged person touching a single pin of an IC with any or all other pins grounded. The capacitance to ground of the idealized person has been standardized to 100pF. The skin resistance has been standardized to a 1500 Ohm resistor between the capacitor and the IC device under test (DUT). The touching of the pin, with appropriate pins grounded, is represented by the closing of a switch. This simple (single loop) circuit is shown in Fig. 1.

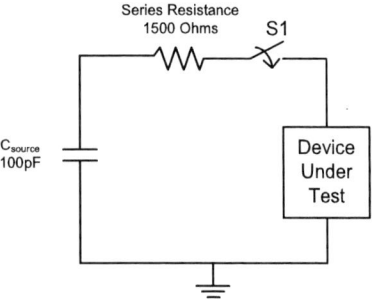

Fig. 1 ESD Test Circuit, Human Body Model

The (open) 100pF capacitor is charged to a test voltage, such as 1500V. One test consists of a single discharge of the capacitor, with the resulting one

ampere going through the IC (or its protection circuit) between the input pin and the grounded pin(s). The quality of the protection against ESD is defined by the highest test voltage that does not damage the circuit, with all pins tested and with multiple ground configurations. The breakdown voltage (BV) of most RF IC's is very much less than the test voltage. As a result the shape of the current pulse through the circuit can be easily calculated. The current pulse has a fast rise (<1ns) and an exponential fall with a time constant of 150 ns. Suitable current sources to aid in modeling elements of the protection circuits can be built in the laboratory using high powered (even if rather antique) rectangular-wave pulse generators.

II. PROTECTION TECHNIQUES FOR MEDIUM VOLTAGE RF CIRCUITS

Under test, an example current of one ampere tries to enter the RF IC on the tested pin. With no protection, the voltage will rise until some element of the circuit breaks down. This is usually a MOS capacitor, which is then permanently damaged. Also, small sized transistors may break down and can be permanently damaged even if the pulse duration is only 150ns. The general mode of the protection circuit is to steer the one ampere to ground without bringing the internal circuit voltages above damaging voltage. The basic elements we use for RF protection are steering diodes and voltage clamp circuits (Zener equivalents). There are special tricks for RF input and output circuits [3,4] explained in another section, but most of the pins are protected as shown in Fig. 2.

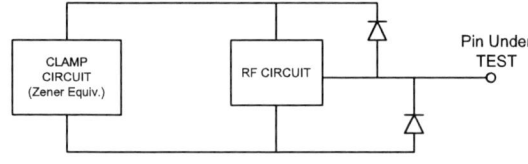

Fig. 2 Basic ESD Protection Scheme for RF

Several positive supply and ground connections are assumed to power the circuit. A small-sized, reverse-biased PN diode is between the pin and ground. Another (reverse-biased) diode is between the pin and a positive supply. Between each positive supply and ground is a "clamp circuit", capable of carrying the full discharge current without allowing the supply bus to go above a damaging voltage. One of the steering diodes (depending on the polarity of the input test voltage) carries the test current (one ampere) either directly to ground, or to a positive supply lead (which is open during test). If the current reaches the positive supply lead, the clamp-circuit conducts like a Zener in breakdown with a voltage across it. This BV must be low enough to avoid damage to the RF circuit it is protecting. Notice that inside the RF circuit (connected to the positive supply pin) there are usually resistors in series with small transistors, which help to avoid damaging currents even if these small devices go into breakdown. The ESD current only lasts for (equivalent) 150ns, therefore the protection devices can be designed to meet the following criteria: 1. Carry the current of the specified HBM ESD test current pulse without damage to itself or to the RF circuitry. 2. The steering diodes (connected to the input/output test pins) must have a small parasitic capacitance, while maintaining adequately low forward voltage at the test current. 3. The clamp-device (across the supplies) must not maintain a high-current latched condition across the supply, even if triggered there accidentally by a glitch or turn-on ramp. 4. Minimize the silicon area of the clamp. RF IC's tend to have several positive supplies for isolation. There is a required clamp for each, on the relatively small RF chip.

III. DESIGN OF THE STEERING DIODES AND CLAMPS

Since two steering diodes are connected between each pin and AC ground, the parasitic capacitance must be kept as small as possible, consistent with the protection requirements, i.e. carrying one ampere (for example) with an acceptable (1-2 Volt) voltage drop. Even though Schottky diodes have a lower voltage drop at low current, p-n junctions have an advantage at high currents. The built-in voltage decreases significantly as the temperature rises. With high level injection, the series resistance drops at high currents. The diode must have a BV greater than the clamp voltage (in operation) plus a low total forward drop at high currents. The area chosen is a compromise between parasitic capacitance and forward drop at the specified high current.

The design does not have to be totally trial and error. A look at the thermodynamics of the device under stress might help. High power is present only for a short time (150ns), so the temperature can certainly approach the melting point of aluminum. A worst case design assumption, reasonably supported by experimental data, is an adiabatic process. This means that for silicon protection elements, the temperature of silicon can be estimated as if the active volume were hit with an impulse of power. The volume can be estimated as the area of the active junctions and the thickness of material where peak

power occurs. Matrix or row structures have been used, and make a lot of sense for the clamp devices, and/or if heat spreading is included with very small geometries.

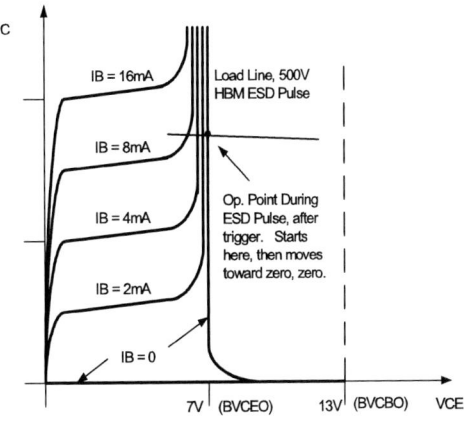

Fig. 3
Idealized Large Signal Transistor Output Characteristics

The clamp device of choice is a large NPN transistor connected between the two supplies. When operating as a protection device it is in sustain break-down. DC characteristics are illustrated in Fig. 3. However, since the sustain (BV) voltage in RF circuits is often very close to the supply voltage, a resistor of 1 to 2k is added across the base-emitter junction. A capacitor of a few tens of pico-Farads supplements the normal collector-base capacitance. With fast-rise pulses, the capacitor current (collector to base) drives ON the transistor, putting it directly in

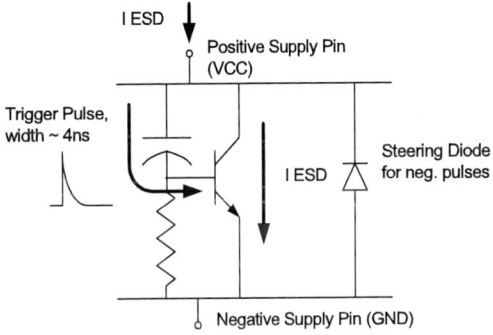

Fig. 4
Active Clamp Circuit Using Transistor Sustain Voltage

the sustain mode. The full circuit of the clamp device, in Fig. 4, includes a normally reverse biased steering diode, in case the test voltage across the supplies should be reversed.

IV. DESIGN OF TEST CIRCUITS

ESD protection is sometimes an afterthought after thousands of man-hours are spent in optimizing the RF circuit. The resulting speculative layouts (of ESD protection) result in a series of trial-and-error processing runs that are inefficient at best and ineffective at worst. Fortunately ESD protection can become fairly standard for a large class of bipolar high-frequency circuits. Designing an adequate ESD protection circuit can be similar to designing a high performance RF circuit. The technique requires one "modeling" and/or performance pass for the elements (such as clamp transistors and steering diodes) prior to the design of the protection circuits. If the elements are modeled (tested) correctly, the protection circuit can then be designed, including adjusting the size of the elements) and laid out with a reasonable chance of first-time success.

To enhance modeling, several varieties and sizes of steering diodes and clamp transistors (large NPN's optimized for low series resistance) are laid out with pads that can be wire-bonded to pins in a package. The devices are being tested at full (pulsed) current, and probes simply have too much resistance (and probably inductance). The devices will be tested for their I-V characteristics and capabilities at full current (up to two amps). The modeler is really doing a curve trace under low-frequency pulsed conditions, such that the device does not overheat (with false results) and/or burn out. Although it might be tedious and time consuming to obtain, the resulting I-V characteristics look very similar to low-current ones. Pulse measurements overlay idealized DC characteristics in Fig. 5. The steering diodes look like p-n diodes with a low forward drop and a series resistance of one to two ohms. The clamp transistors look like Zener diodes with a resistance in breakdown of less than one ohm, which starts to increase rapidly close to burn-out.

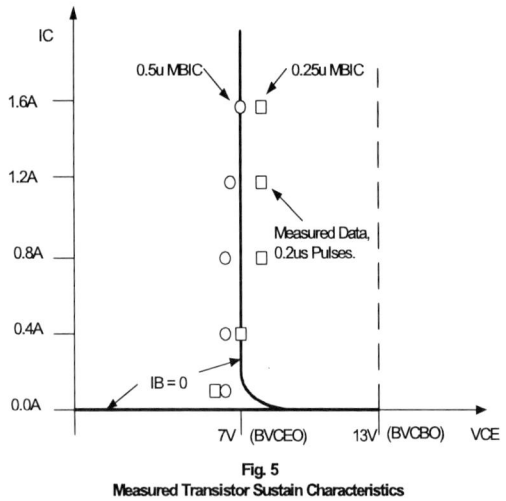

Fig. 5
Measured Transistor Sustain Characteristics

The circuit technique we have found most convenient is a printed circuit board with a package clamp or socket. We used a rectangular 100V pulse generator with 2Amp capability. The current is measured through a CT-2 (Tektronix) current probe, thereby simulating a constant current with relatively low source resistance (50 Ohms). The experimental results are shown in Fig. 5 are for two designed clamp transistors in BiCMOS technologies with the line widths shown.

V. INPUT AND OUTPUT PROTECTION FOR RF CIRCUITS

The input LNA of an RF bipolar circuit is usually a common emitter NPN transistor, with the base tied to the input pin and the emitter to ground. The above suggested ESD circuit would allow an arbitrary open pin to go one diode drop above the clamp BV (~Vcc). The standard LNA is one forward biased base-emitter to ground, shunting the protection circuit. Although the input transistor is usually large to reduce base resistance, it may not be designed to carry 1.5 Amps ESD pulse, even for 150 ns. One technique is to put a custom clamp across the input pin of two (or more) series forward biased "steering" diodes. Whether an additional series resistor in needed (between the diodes and the input base) depends on the size of the input transistor. In addition, a reverse-biased steering diode to ground is required to avoid breaking down the base-emitter junction of the LNA transistor. (ref. 3)

The output pin of an RF circuit is often the collector of a large transistor, which is connected to the supply through an external inductor. The output RF signal swings above the supply every cycle and would be distorted by a single steering diode between the output pin and Vcc. One solution to this problem is to add more steering diodes in series to Vcc to avoid RF distortion. [4] There is the possibility, during an ESD pulse, that the output transistor will go into sustain. Fortunately, most output transistors can survive undamaged, becoming clamp devices themselves. If the output transistor is too small, the RF current is small and a series resistor can be added without degrading the RF signal.

VI. OVERALL HIGH FREQUENCY PROTECTION SCHEME

Bipolar transistors are often used in high frequency circuits, such as RF and fiber optic conditioners, where parasitics are important. A generalized shunt and clamp protection scheme can be applied with minimal effort, after qualified devices are designed and tested. It is as follows: Each positive supply is connected to ground through one or more clamp devices in parallel. Every pin except supply pins (and possibly RF input pins) is connected to Vcc and GND through steering diodes. There is often a layout convenience if any Vcc or GND rail can be used for this connection. If all Vcc's are the same voltage, but require isolation, parallel steering diodes can connect

160

the Vcc's. The schematic is checked for every conceivable ESD condition, such that the ESD current is shunted around the RF circuit. Special cases, such as an LNA and power output stages are handled separately, using the suggestions above. Obviously operating circuit simulations must include the capacitive and nonlinear parasitics of the steering diodes. The mask layout is monitored to insure that high resistance does not accumulate in any of the paths. This must be resolved either with wider paths or multiple clamp devices.

VII. RESULTS AND CONCLUSION

Using the above techniques, the ESD capability of bipolar RF circuits has increased from about 200 Volts to 2kV. The technology was changed from 1u to 0.25 u [0.4 u emitters] BiCMOS. The supply voltage was 5V max.

Two steering diodes with 0.26pF capacitance (pin to ac ground) has a current carrying capability of 1.5 A at <2V each. See Fig 6. A clamp transistor has 1.5 A capability at 8V. The clamp circuit, including a large NPN with a matrix of emitters, collector-base capacitor, and steering diode were laid out in an area of about 100 X 100 u^2. Experimentally, pins that use the steering diodes shown in Fig. 6 and the clamps of Fig. 5 have passed 2KV on production RF circuits.

A design and testing technique for bipolar ESD protection of RF (and other high frequency) circuits has been described. A technique for applying the protection devices to a large class of high frequency circuits has been described and tested. Three

possible problems areas (input and output circuits and long supply paths) have also been mentioned.

The presently accepted protection level of 2kV has been achieved a few years ago for RF circuits. Obviously higher current clamps with their increased area can be designed. The real problem is the steering diodes, where doubling the protection means doubling the operating signal parasitics on the pins. However, since the need for external components in the signal path is reduced, the sensitivity to parasitics is also reduced. Higher level ESD protection is certainly achievable.

ACKNOWLEDGEMENT

The authors wish to acknowledge the device design of Vance Archer.

REFERENCES

[1] JEDEC STANDARD, Electrostatic Discharge (ESD) Sensitivity Testing Human Body Model (HBM), JESD22-A114D, March 2006.

[2] JEDEC STANDARD, Field-Induced Charged-Device Model "Test Method for Electrostatic-Withstand Thresholds of Microelectronic Components, JESD22-C101C, December 2004.

[3] P. C. Davis, "Input Stage ESD Protection for an Integrated Circuit" US Patent #6400204, Jun. 4, 2002,

[4] P. C. Davis, "Output Stage ESD Protection for an Integrated Circuit" US Patent #6529059, Mar. 4, 2003.

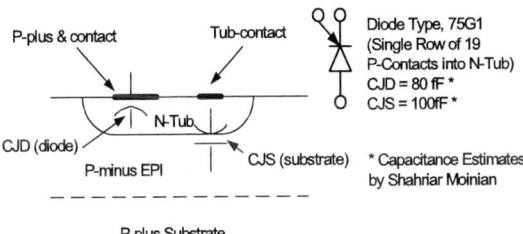

Fig. 6
ESD Diode on 0.5 u MBIC 3V

2006 Bipolar/BoCMOS Circuits and Technology Meeting Proceedings

Characterization and Modeling of Intermodulation Linearity in a 200 GHz SiGe HBT Technology

Guofu Niu, Ying Li, Zhiming Feng, Jun Pan, and David C. Sheridan[1]

ECE Department, 200 Broun Hall, Auburn University, Auburn, AL 36849, USA
Tel: 334 844-1856 / Fax: 334 844-1888 / E-mail: niuguof@auburn.edu
[1]IBM Microelectronics, Essex Junction, VT 05452, USA

Abstract — **This paper presents experimental characterization and modeling of intermodulation linearity in a 200 GHz SiGe HBT technology. The impact of biasing current, voltage, and breakdown voltage is examined. IP3 is simulated using the VBIC, HICUM and Mextram model to evaluate the linearity capability of these models. The impact of avalanche and self-heating on IIP3 are examined. A weak avalanche is shown to have significant impact on IIP3.**

I. INTRODUCTION

Intermodulation linearity is an important figure-of-merit for RF devices, as it relates to the selectivity of a RF receiver and the spectral purity of a RF transmitter. Various theories, simulations, and experimental investigation of linearity have been reported for Si, SiGe and III-V bipolar transistors [1]-[6]. Accurate simulation and modeling of linearity is challenging, as it requires accurate description of higher order derivatives of all nonlinear current and charge relations in the device. This calls for systematic experimental characterization to better understand the intermodulation linearity behavior of bipolar transistors, which practically does not exist for SiGe HBTs. From a circuit design standpoint, it is desirable to understand how linearity changes as a function of biasing current and voltage, as well as device breakdown voltage, as biasing point and breakdown voltage are design variables. Another practically important issue is how accurate the compact models are in simulating linearity. Model validation is typically done using DC I-V data and small signal s-parameters, but not linearity data because of the complexity involved and time consuming nature of linearity measurement.

The purpose of this work is to experimentally characterize intermodulation linearity, the most relevant linearity for RF applications, in a 200 GHz SiGe HBT technology [7] and evaluate the linearity simulation capability of the VBIC, HICUM and Mextram models using harmonic balance. The results provide insight into the device physics underlying linearity behavior, guidelines to optimal biasing, device selection (e.g. high breakdown versus low breakdown versions), as well as quantified simulation to data comparison of linearity useful for designers and modelers. The input 3rd order intercept point, IIP3, is used as a figure-of-merit for intermodulation linearity. IIP3 is measured on $I_C - V_{CE}$

plane for devices of various size and breakdown voltage. The S-parameters are also measured, from which f_T is extracted.

II. EXPERIMENTAL SETUP

Fig. 1 shows the experimental setup used. Both the source and load are terminated with 50 Ω. The broadband nature of the 50 Ω system is desired as it does not filter out the second order harmonics which may remix with the fundamental output to produce 3rd order intermodulation. Devices are probed on-wafer using GSG probes. Two Agilent signal sources are used to provide the two tone input. The attenuators are used to reduce the intermodulation between the two sources. The automatic level control in the sources is turned off to minimize intermodulation generated within the sources. A spectrum analyzer is used to measure the output spectrum. Power meters are used for calibration. The Agilent intermodulation utility is used to choose the optimum analyzer setting for minimum analyzer IM3 and maximum signal to noise ratio. An in-house HP-VEE program is written to automate the measurement. For each bias point and frequency, the input power is swept and the third order intercept is obtained by extrapolation, as shown in Fig. 2. The analyzer setting is optimized dynamically for each power level. The measurement system intermodulation is verified to be well below the device intermodulation. The upper and lower IM3 are the same in our measurements. Measurements are made at 2, 5 and 15 GHz, using a two tone spacing of 1 MHz. The results are similar. Below we present the 5 GHz data. Due to space limitation, we will present only devices with an emitter area of $0.12 \times 18 \ \mu m^2$.

III. EXPERIMENTAL RESULTS

A. Standard Breakdown Devices

Fig. 3 shows the output curves for a standard breakdown device with an emitter area of $0.12 \times 18 \ \mu m^2$. S-parameters and IIP3 are measured as a function of I_C from 0 to 60 mA, for V_{CE}=1.0, 1.3, 1.6 and 1.9 V. At higher $I_C V_{CE}$, self heating is visible. For lower I_C, avalanche multiplication is weak, but noticeable for the highest V_{CE} used (1.9V). Avalanche measured using RF pulse I-V, however, shows that avalanche is much

1-4244-0458-4/06/$20.00 ©2006 IEEE

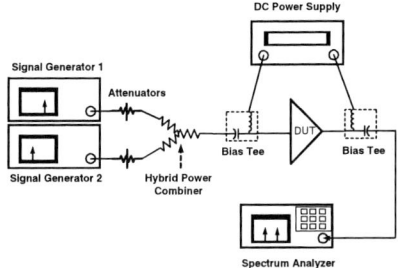

Fig. 1. A schematic of the IIP3 measurement setup.

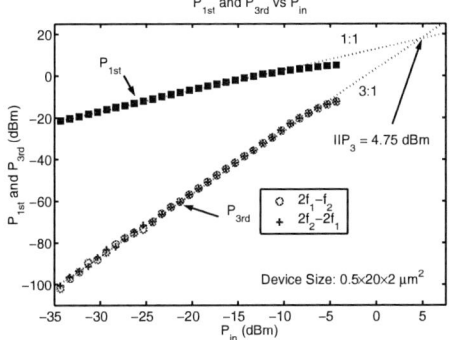

Fig. 2. Illustration of the IIP3 extrapolation.

weaker at higher I_C, even at V_{CE}=1.9 V, because of reduced internal V_{CB} due to voltage drop across the collector resistance, and reduced field due to charge modulation by mobile carriers in the CB junction [8].

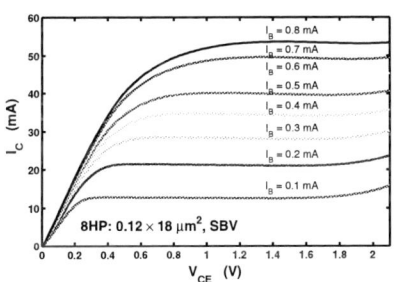

Fig. 3. Measured output curves of a standard breakdown HBT.

Fig. 4 (a) shows the measured IIP3 versus I_C for various V_{CE}. At a given V_{CE}, IIP3 first increases with I_C, reaches a peak, and then drops slightly down to a relatively constant level. At higher I_C, another rapid drop of IIP3 occurs, followed by another increase. Both the peak IIP3 value and the peak IIP3 I_C increase with increasing V_{CE}. Before the peak, IIP3 is nearly independent of V_{CE}. A comparison of Fig. 4 with the f_T curves in Fig. 4 (b) shows that the peak IIP3 I_C is well below the peak f_T I_C. The second IIP3 decrease occurs at a higher I_C for higher V_{CE}. V_{CE} affects the CB junction

capacitance, avalanche multiplication, and high injection f_T rolloff.

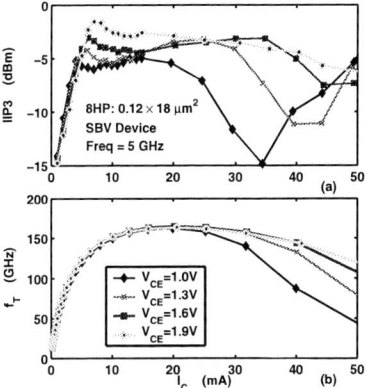

Fig. 4. (a) IIP3–I_C (b) f_T–I_C. Standard breakdown HBT.

Below 20 mA, the impact of V_{CE} on f_T is negligible, however, its impact on IP3 is significant. Additional measurements show that for V_{CE} below 1.6 V, avalanche is weak, the primary impact of V_{CE} on IIP3 is likely through the CB capacitance, including both its absolute value and its derivatives with respect to both V_{CB} and I_C. The peak of IIP3 is a result of complicated cancellation between various nonlinearities, and highly depends on the details of not only the absolute value, but also the higher order derivatives of the involved physical quantities (such as C_{CB} and I_C), making its accurate modeling challenging, as detailed below. We also note that Early effect is negligible for these HBTs, and can be safely excluded.

B. Medium and High Breakdown HBTs

Typical SiGe processes offer devices with multiple collector profiles to provide devices with multiple breakdown voltages. Fig. 5(a) and (b) show the IIP3 and f_T of the medium breakdown HBT respectively. For a meaningful comparison, a higher V_{CE} range and a smaller I_C range is used with increasing breakdown voltage. Like f_T, the IIP3 peak shifts to a higher I_C as V_{CE} increases, The peak IIP3 value increases as well. The IIP3 values shifts to a higher I_C as well. Interestingly, f_T is still fairly close to, or right at its peak when IIP3 reaches the valley point. After the valley point, IIP3 starts to rapidly increase again while f_T continues to drop.

Fig. 6(a) and (b) show the IIP3 and f_T of the high breakdown HBT. Again, IIP3 shows a clear peak and a clear valley for all V_{CE}. Both the peak and valley I_C values increase with increasing V_{CE}. The peak IIP3 value increases with V_{CE}, the valley IIP3 value, however, is nearly independent of V_{CE}.

C. Comparison of Breakdown Versions

Fig. 7(a) and (b) compares the IIP3 and f_T of the three breakdown versions. V_{CE}=1.9 V is chosen to avoid quasi-saturation in the high breakdown HBT. With increasing breakdown voltage (decreasing collector doping), peak IIP3 pushes out to a higher I_C. For the standard breakdown HBT, a nice high IIP3 plateau is observed from 7 to 30 mA, over which f_T is high as well. For the medium and high breakdown HBTs, however, such IIP3 plateau does not exist, and the biasing I_C needs to be kept the peak IIP3 point, which is below the peak f_T point, to achieve both high f_T and high IIP3. For the high breakdown HBT, at the peak f_T point, IIP3 is at its minimum. Therefore, the peak f_T point would be a bad bias point from a linearity standpoint.

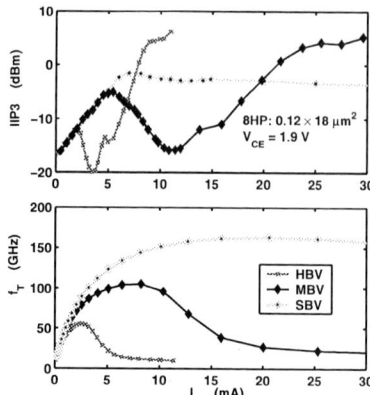

Fig. 7.   (a) IIP3 and (b) f_T comparison for the standard, medium and high breakdown HBTs.

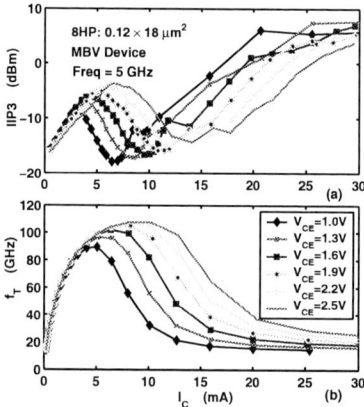

Fig. 5. (a) IIP3 and (b) f_T for medium breakdown HBTs.

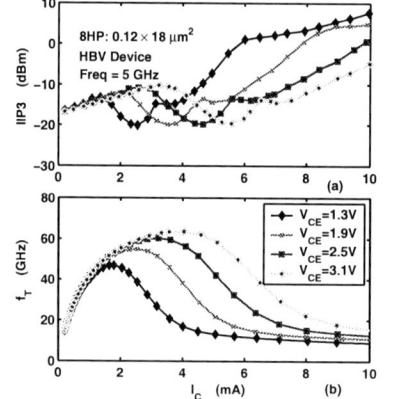

Fig. 6. (a) IIP3 and (b) f_T for high breakdown HBTs.

IV. DATA AND MODEL CORRELATION

We now compare simulated and measured IIP3. For the standard breakdown HBT, we use VBIC, Mextram and HICUM models from the design kit developed for the same HBT used in this work. All of the three models produce reasonable fitting of I-V and small signal s-parameters at lower current levels. The main difference is high current behavior. A detailed comparison of the high current $I - V$ and S-parameters as a function of bias shows that for all models, accuracy degrades at higher I_C (even below peak f_T). No model is available

for the medium breakdown HBT, and only the VBIC model is available for the high breakdown HBT. ADS as opposed to Cadence SpectreRF is used, because of its accuracy and efficiency for IP3 simulation. The Cadence design kit is accessed through the ADS dynamic link in the Cadence tools.

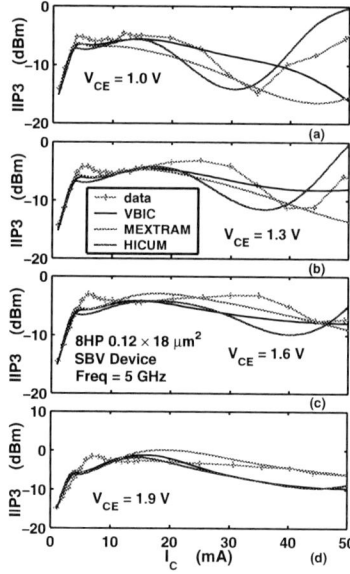

Fig. 8.   Comparison of simulated and measured IIP3 at V_{CE}=1.0, 1.3, 1.6 and 1.9V. Standard Breakdown HBT.

A. Standard Breakdown HBT

Fig. 8 (a)-(d) compare simulated and measured IIP3 as a function of I_C for the standard breakdown HBT. We first examine the V_{CE}=1.0V case. All of the 3 models do a good job at $I_C < 5$ mA. For $I_C > 5$ mA and $I_C < 20$ mA, all of the models are off by approximately 2 dB. At $V_{CE} = 1.0$ and 1.3V, only HICUM is able to model the high current IP3 rolloff and rise behavior (near 30 mA), because of its improved high current re-

164

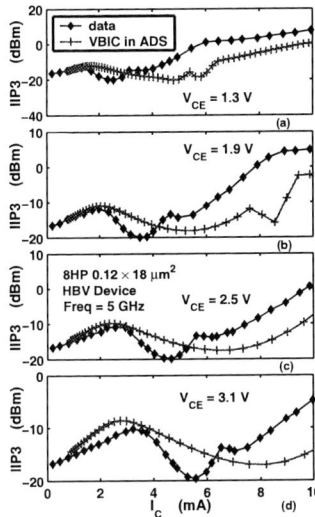

Fig. 9. Comparison of simulated and measured IIP3 at V_{CE}=1.3, 1.9, 2.5 and 3.1V. High Breakdown HBT.

gion modeling. However, in terms of the IIP3 value discrepancy, the VBIC model is the most accurate below 25 mA. Model comparison at V_{CE}=1.3V is similar to that at V_{CE}=1.0V. At V_{CE}=1.6 and 1.9V, HICUM and Mextram do not show clear advantages over VBIC.

B. High Breakdown HBT

Fig. 9 (a)-(d) compares simulated and measured IIP3 as a function of I_C for V_{CE}=1.3, 1.9, 2.5 and 3.1V, using VBIC, the only model available. The simulation agrees with measurement at lower I_C and lower V_{CE}. The simulation does capture the overall variation of IIP3 with I_C and V_{CE}. Quantitative agreement, however, is not satisfactory at higher I_C or higher V_{CE}.

C. Impact of Avalanche and Self-heating

Volterra series based analysis in [3] showed that cancellation of CB capacitance nonlinearity and avalanche nonlinearity can result in improvement of IIP3, or they can enhance each other to further degrade IIP3. In simulation, we can turn on and off avalanche, as well as self-heating, to examine their impact on IIP3. Fig. 10 compares IIP3 simulated with various options for the standard breakdown HBT, at V_{CE}=1.9V. The M-1 value in simulation (and measurement) is 0.002 at 5 mA and V_{CE} = 1.9V, which is by all means weak. Below 25 mA, the IIP3 simulated with avalanche is considerably higher, both with and without self-heating. Above 25 mA, the IIP3 simulated without avalanche is higher. In present comapct models, M-1 is only a function of the internal V_{CB}. Therefore, only the I_C dependence of M-1 due to the $I_C R_C$ voltage drop is included, while the I_C dependence of avalanche due to charge modulation in the CB junction is not accounted for. This results in an overestimation of M-1 at higher I_C in simulation. Including the decrease of M-1 with increas-

ing I_C would bring the high I_C region IIP3 closer to data in the high I_C region. Given that a weak avalanche can have a large impact on IIP3, accurate measurement and modeling of the I_C dependence of avalanche are clearly necessary. Self-heating shows only a weak effect on IIP3.

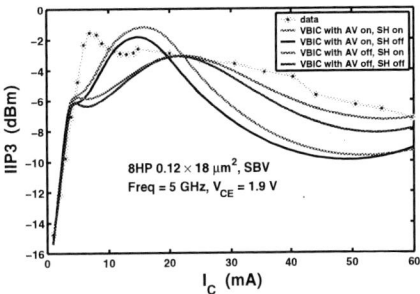

Fig. 10. Comparison of IIP3 simulated using different combinations of avalanche and self-heating settings and measured IIP3 at V_{CE}=1.9V. Standard Breakdown HBT.

V. CONCLUSIONS

We have presented systematic characterization results of intermodulation linearity in a 200 GHz SiGe HBT technology, and compared the I_C and V_{CE} dependences of IIP3 and f_T. At relatively lower I_C and V_{CE}, VBIC, HICUM, and Mextram all yield reasonably accurate IIP3. At relatively higher I_C (still well below peak f_T for standard breakdown HBT) and V_{CE}, significant deviations between simulation and measurement occur. In some cases, HICUM better captures the trend of IIP3 variation with I_C in the high current region. The effect of avalanche on IIP3 is shown to be significant, while the effect of self-heating on IIP3 is shown to be weak.

ACKNOWLEDGMENTS

This work was supported by NSF under ECS-0119623 and ECS-0112923, and IBM under an IBM Faculty Partnership Award. The wafers were fabricated at IBM Microelectronics, Essex Junction, VT. We would like to thank A. Joseph, D. Harame, and the IBM SiGe team for their contributions.

REFERENCES

[1] J. Reynolds, *IEEE Trans. On Electronic Devices*, no. 11, pp. 595-599, 1965.

[2] S. A. Maas et al., *IEEE Trans. Microwave Theory and Tech.*, vol. 40, no. 3, pp. 442-448, 1992.

[3] G. Niu et al., *IEEE Trans. Microwave Theory and Tech.*, vol. 49, no. 9, pp. 1558-1565, 2001.

[4] M.P. van der Heijden et al., *IEEE Journal of Solid State Circuits*, vol. 37, no. 9, pp. 1176-1183, 2002.

[5] M. Iwamoto et al., *IEEE Trans. Microwave Theory and Tech.*, vol. 48, no. 12, pp. 2377-2388, 2000.

[6] M. Iwamoto et al., *IEEE Trans. Microwave Theory and Tech.*, vol. 50, no. 12, pp. 2954-2962, 2002.

[7] S. J. Jeng et al., *IEEE Electron Device Letters*, vol. 22, no. 11, pp. 542-544, 2001

[8] J. Pan and G. Niu, *Proceedings of ECS Symposium on SiGe Material, Devices and Applications*, pp. 429–436, 2004.

2006 Bipolar/BoCMOS Circuits and Technology Meeting Proceedings

Impact of Collector-Base Space Charge Region on RF Noise in Bipolar Transistors

Kejun Xia and Guofu Niu

Electrical and Computer Engineering Department

200 Broun Hall, Auburn University, Auburn, AL 36849, USA

Tel: 334 844-1861 / Fax: 334 844-1888 / E-mail: xiakeju@auburn.edu

Abstract — **This paper examines the impact of collector-base space charge region (CB SCR) on RF noise in scaled bipolar transistors by solving the Langevin equation of electron noise transport. The van Vliet model, which was derived for intrinsic base only, was evaluated for scaled bipolar transistors in which CB SCR transit time is most significant. An improved noise model accounting for CB SCR effects is derived.**

I. INTRODUCTION

Accurate modeling of intrinsic RF noise is important for RF noise modeling of bipolar transistors. The most fundamental noise model developed so far is the the van Vliet model [1], which solved the microscopic noise transport equation for base minority carrier (electrons for npn considered here). Van Vliet's derivation of base and collector current noise power spectrum densities (PSDs) assumed adiabatic boundary condition i.e., $\Delta n = 0$ or zero electron density flucutation at both ends of the base, and did not consider electron transport in the CB SCR. For scaled bipolar transistors, e.g. SiGe HBTs of 200 GHz peak f_T, CB SCR electron transport becomes more significant than base electron transport, calling for an investigation of its impact on transistor noise.

An extremely useful result of van Vliet's derivation is that the base and collector current noise and their correlation can be related to the Y-parameters due to intrinsic base electron transport:

$$
\begin{aligned}
S_{ib}^{van} &= 4kT\Re(Y_{11}') - 2qI_B, \\
S_{ic}^{van} &= 4kT\Re(Y_{22}') + 2qI_C, \\
S_{icib*}^{van} &= 2kT(Y_{21}' + Y_{12}'^* - gm_0).
\end{aligned}
\tag{1}
$$

gm_0 is the transconductance at low frequency. We emphasize that Y' in van Vliet's derivation refers to the Y-parameters of the intrinsic base Y_B only, see Fig. 1. In the literature, the van Vliet model is often applied using Y-parameters of the whole intrinsic transistor, e.g. in [2] and [3], as opposed to Y_B, for which the model was derived. Physically speaking, both the Y-parameters and the noise parameters are modified by electron transport through the CB SCR, it is not clear at all what the relation between Y-parameters and transistor noise should be when the CB SCR is accounted for.

This paper investigates the impact of CB SCR on transistor noise and derives an improved noise model including such impact. The CB SCR affects electron transport (and hence noise transport) in two ways. First, a velocity saturation boundary condition, i.e., $\Delta n = -J/qv_{sat}$ should be applied at the end of neutral base, as opposed to the adiabatic boundary condition i.e., $\Delta n = 0$ used by van Vliet. Its

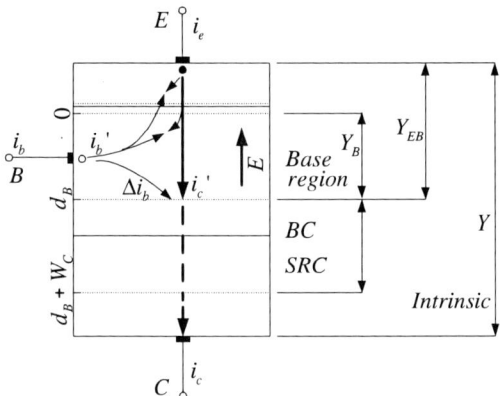

Fig. 1. Illustration of the region structure of a NPN intrinsic transistor with corresponding Y-parameters. The carrier flow signs will be used in Section III.

effect on dc currents and base transit time is well known, but its effect on noise remains to be examined, which we address in Section II. Second, electron transport through the CB SCR modifies both the Y-parameters and the noise parameters. We will address the second effect in Section III, and further derive a new set of relationship between noise currents and Y-parameters in presence of CB SCR. Such relationship is extremely useful for device physics based noise analysis and noise modeling, because Y-parameters can be readily traced to transistor design (e.g. Ge gradient), and readily measured as well.

II. VELOCITY SATURATION BOUNDARY

We assume here uniform base built-in field (e.g. from Ge grading or doping gradient). The neutral base width is denoted as d_B, as shown in Fig. 1. The built-in field E is measured by a parameter η as $E = -\eta\frac{kT}{qd_B}$. The minus sign indicates an acceleration field. We solve the DC, AC continuity equations and Langevin equation to obtain Y-parameters and the PSDs of the intrinsic base and collector noise currents. Velocity saturation boundary condition is forced at the end of the base for all three cases. We briefly describe only the noise analysis.

A. Noise Analysis

The Langvin equation based on Drift-Diffusion model for the base electron noise is

$$
D_n\frac{\partial^2}{\partial x^2}\Delta n + \mu_n E\frac{\partial}{\partial x}\Delta n - \frac{\Delta n}{\tau_n} - j\omega\Delta n + \xi(\omega) = 0, \tag{2}
$$

where μ_n is the electron high field mobility described by

$$\mu_n = \frac{\mu_{n0}}{\sqrt[\beta]{1 + \left(\frac{\mu_{n0}E}{v_{sat}}\right)^\beta}}.$$

and the noise source

$$\xi(\omega) = \zeta(\omega) + \frac{1}{q}\frac{\partial}{\partial x}\gamma(\omega), \quad \gamma(\omega) = 4q^2 D_n N(x), \quad (3)$$

where $\zeta(\omega)$ is GR noise source, $\gamma(\omega)$ is diffusion noise source and $N(x)$ is the DC electron concentration. We only consider diffusion noise here as the GR noise is negligible at RF. Einstein relation is used. The boundary condition for (2) is

$$\Delta n|_{x=0} = 0, \quad \Delta n|_{x=d_B} = -\frac{J_n}{qv_{sat}}. \quad (4)$$

To solve (2)-(3) we use the Green function method [4]. Due to space limitation, we do not show the lengthy derivation details here and give below only the solution of S_{icib^*}.

$$S_{icib^*} = A_E\{-C(A^* + C^*)\bar{K}\frac{d_B}{-\alpha_1 - \alpha_1^* + \bar{\alpha}_1}\left(e^{-\alpha_1 - \alpha_1^* + \bar{\alpha}_1} - 1\right)$$

$$-D(B^* + D^*)\bar{K}\frac{d_B}{-\alpha_2 - \alpha_2^* + \bar{\alpha}_1}\left(e^{-\alpha_2 - \alpha_2^* + \bar{\alpha}_1} - 1\right)$$

$$-D(A^* + C^*)\bar{K}\frac{d_B}{-\alpha_1^* - \alpha_2 + \bar{\alpha}_1}\left(e^{-\alpha_1^* - \alpha_2 + \bar{\alpha}_1} - 1\right)$$

$$-C(B^* + D^*)\bar{K}\frac{d_B}{-\alpha_1 - \alpha_2^* + \bar{\alpha}_1}\left(e^{-\alpha_1 - \alpha_2^* + \bar{\alpha}_1} - 1\right)$$

$$+ first\ 4\ lines,\ substitute\ \bar{\alpha}_1 \rightarrow \bar{\alpha}_2\ and\ \bar{K} \rightarrow \bar{L}$$

$$+ first\ 4\ lines,\ substitute\ \bar{\alpha}_1 \rightarrow \eta\ and\ \bar{K} \rightarrow \overline{KL}\} \quad (5)$$

where

$$A = -\frac{\alpha_1}{d_B\left(1 - \kappa e^{-2\theta}\right)}, \quad B = \frac{\kappa e^{-2\theta}\alpha_2}{d_B\left(1 - \kappa e^{-2\theta}\right)},$$

$$C = A\frac{(\alpha_1 - \kappa\alpha_2)e^{\alpha_2}}{\alpha_2 - \alpha_1}, \quad D = -C\frac{\alpha_2}{\alpha_1},$$

$$\bar{K} = \frac{4q^2 D_n n_{00}e^{\frac{V_{BE}}{V_T}}}{1 - \bar{\kappa}^{-1}e^{2\theta}}, \quad \bar{L} = \frac{4q^2 D_n n_{00}e^{\frac{V_{BE}}{V_T}}}{1 - \bar{\kappa}e^{-2\theta}}, \quad \overline{KL} = 4q^2 D_n n_{00},$$

$$\kappa = \frac{D_n\alpha_1 - v_{sat}d_B}{D_n\alpha_2 - v_{sat}d_B}, \quad \alpha_1 = \frac{\eta}{2} + \theta, \quad \alpha_2 = \frac{\eta}{2} - \theta,$$

$$\theta = \sqrt{\left(\frac{\eta}{2}\right)^2 + \frac{d_B^2}{D_n\tau_n} + \frac{j\omega d_B^2}{D_n}},$$

$\bar{\alpha}_1 = \alpha_1|_{\omega=0}, \quad \bar{\alpha}_2 = \alpha_2|_{\omega=0}, \quad \bar{\kappa} = \kappa|_{\omega=0}.$

The last two lines in (5) represent eight terms obtained by a parameter substitution procedure performed on the first four lines in (5). n_{00} is the DC electron concentration at $x = 0$ when $V_{BE} = 0V$. The velocity saturation boundary condition is involved through parameter κ. $\kappa = 1$ if such a condition is neglected.

B. Evaluation of van Vliet Model - Base Only

Using our Langevin equation solution, we find that without including velocity saturation in the boundary condition,

the van Vliet model (1) gives the base electron noise exactly, as expected. With the velocity saturation boundary condition, however, the van Vliet model deviates from the analytical results as the base width is narrowed down, for example, to 20nm. The van Vliet model overestimates S_{ib} and $|\mathfrak{R}(S_{icib^*})|$ while underestimates $|\mathfrak{S}(S_{icib^*})|$. However, with a strong base built-in field, typically the case in graded SiGe HBTs, the deviation is significantly reduced, as shown in Fig. 2, where $\eta = 5.4$. A careful inspection of whole solution process does not yield an intuitive explanation for this observation. However, calculations show that this is generally true for all practical values of built-in field found in modern SiGe HBTs with graded base. Therefore, we can continue to use the van Vliet model for describing the relationship between noise and Y-parameters of the intrinsic base for graded SiGe HBTs. This is our starting point for Section III.

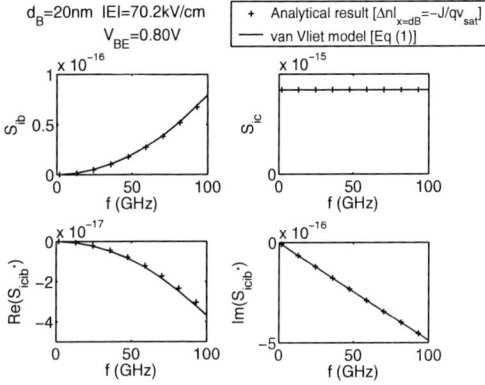

Fig. 2. Evaluation of van Vliet model for base region noise under $d_B = 20nm$, $\eta = 5.4$ ($|E|$=70.2 kV/cm), V_{BE}=0.8V.

III. CB SCR TRANSPORT EFFECT

A. Derivation

As shown in Fig. 1, the AC electron current injected into SCR is denoted as i_c' and the AC collector current is denoted as i_c. The electrons inside the CB SCR induce base hole accumulation at the SCR side of the base region and electron depletion at the SCR side of the collector region. The first part results in an extra base hole current Δi_b, which is $i_c' - i_c$. Note that i_c and i_b take positive signs when they flow into the electrodes. Physics analysis shows that i_c and i_c' can be related [5] by

$$\lambda(\omega) \equiv \frac{i_c}{i_c'} = \frac{1 - e^{-2j\omega\tau_c}}{2j\omega\tau_c}. \quad (6)$$

where $\tau_c = W_c/(2v_{sat})$ to first order. The total base and collector currents can be derived as

$$i_b = i_b' + (1 - \lambda)i_c', \qquad i_c = \lambda i_c'. \quad (7)$$

With (7) and by neglecting Y_{12}' and Y_{22}', we obtain the Y-parameters of the whole intrinsic transistor including CB SCR as

$$y_{11} = y_{11}' + (1 - \lambda)y_{21}', \qquad y_{21} = \lambda y_{21}'. \quad (8)$$

Here Y' is Y_B.

The noise PSDs including CB SCR transport are derived from (7) as

$$S_{ib} \equiv < i_b i_b^* > = S'_{ib} + 2\Re[(1-\lambda)S_{icib^*}] + |1-\lambda|^2 S'_{ic},$$

$$S_{ic} \equiv < i_c i_c^* > = |\lambda|^2 S'_{ic}.$$

$$S_{icib^*} \equiv < i_c i_b^* > = \lambda S'_{icib^*} + \lambda(1-\lambda^*)S'_{ic}. \qquad (9)$$

Here S' can be obtained from (1) with $Y'=Y_B$ for reasons discussed in Section II. Fig. 3 shows the effect of τ_c on noise, $\tau_c=0$ for the dash lines and $\tau_c=0.75\tau_{tr}$ for the solid lines. Here $\tau_{tr} = \tau_c + \tau_b$, with τ_b being the base transit time. Clearly the base current noise is significantly *enlarged* due to CB SCR electron transport, particularly with increasing frequency.

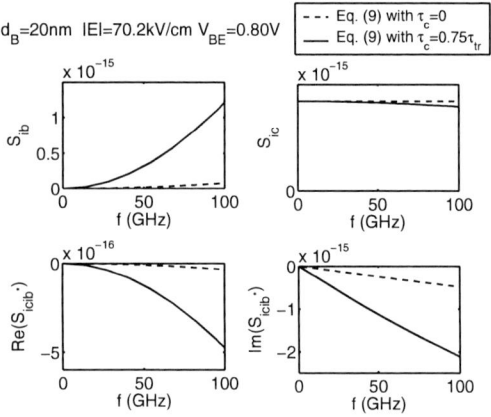

Fig. 3. Comparison the intrinsic noise with $\tau_c=0$ and $\tau_c=0.75\tau_{tr}$. For $\tau_c=0.75\tau_{tr}$, $f_T=174$ GHz.

For various reasons discussed above, it is highly desirable to express the noise PSDs in (9) in terms of the Y-parameters for the whole intrinsic transistor Y. A set of such expressions are derived below

$$S_{ib} = 4kT\Re(Y'_{11}) - 2qI_B + 4kT\Re[(1-\lambda)Y'_{21}]$$
$$- 4kT\Re(1-\lambda)gm_0 + |1-\lambda|^2 S'_{ic}$$
$$= 4kT\Re(Y_{11}) - 2qI_B - 4kT\Re(1-\lambda)gm_0 + |1-\lambda|^2 S'_{ic}$$
$$\approx \{4kT\Re(Y_{11}) - 2qI_B\} - (1-|\lambda|^2)2kTgm_0.$$

$$S_{ic} = \{2qI_C\}|\lambda|^2.$$

$$S_{icib^*} = \lambda 2kTY'_{21} - \lambda 2kTgm_0 + \lambda(1-\lambda^*)S'_{ic}$$
$$= 2kTY_{21} - \lambda 2kTgm_0 + \lambda(1-\lambda^*)S'_{ic}$$
$$\approx \{2kT(Y_{21} - gm_0)\} + (1-|\lambda|^2)2kTgm_0. \qquad (10)$$

We illustrate the S_{ib} derivation as an example. The first step is obtained directly from (9) and (1). The second step is obtained using (8). The third step is obtained noticing that $S'_{ic} = 2qI_C \approx 2kTgm_0$. Note that the terms enclosed by {} in (10) are the noise expressed by (1) with $Y' = Y$, a brutal force application of van Vliet model (using Y despite that it needs Y' – often used without justification). The additional terms in our new model, (10), represent the error introduced by using the van Vliet model with the overall transistor Y-parameters.

Fig. 4 compares the improved model, the brutal use of van Vliet model and the exact result, that is, Langevin equation solution used with (9). The improved model works very well, and gives results nearly identical to the exact result. S_{ib} and $|\Re(S_{icib^*})|$ are overestimated by the brutal use of van Vliet model, while $|\Im(S_{icib^*})|$ is correctly modeled. The inconsistent modeling of S_{ib} and $|\Im(S_{icib^*})|$ results in an overestimation of NF_{min} for the brutal use of van Vliet model. The magnitude of derivation depends on the ratio τ_c/τ_{tr}, which increases with scaling. Even though the differences look small on the plots shown, the resulting differences in noise parameters of the intrinsic transistor (NF_{min}, R_n, and Y_{opt}) are significant, making them important to model. For transistors in which base resistance is large, the final impact on overall transistor noise parameters is smaller, simply because of the less importance of intrinsic transistor noise.

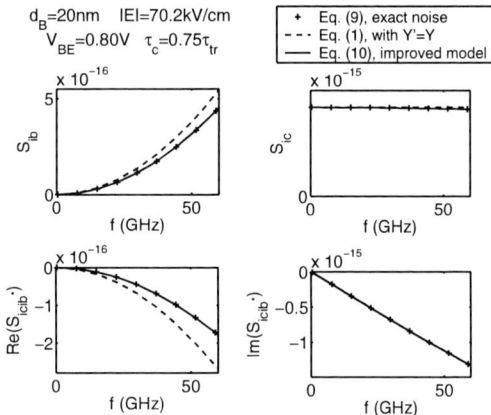

Fig. 4. Comparison between the brutal used van Vliet model and the improved model under $\tau_c=0.75\tau_{tr}$. $f_T=174$ GHz.

B. Verification

To verify our derivations, we examine the new model using hydrodynamic DESSIS noise simulation. The device has 184 GHz peak f_T with effective $d_B=20$nm. At $V_{BE} = 0.79V$, $\tau_c=0.75(\tau_b+\tau_c)$, $f_T=155$ GHz. Fig. 5 compares the improved model with the extracted τ_c, the brutal use of van Vliet model and the extracted intrinsic base electron noise. The new model improves S_{ib} and $\Re(S_{icib^*})$ modeling. The DESSIS simulated $S_{ic} < 2qI_C$ is a direct result of hydrodynamic simulation.

Fig. 6 shows the results of the improved model and van Vliet model using overall Y-parameters, and measured data for a SiGe HBT having 200 GHz peak f_T at $V_{BE} = 0.92V$. The solid line is the result of the improved model with $\tau_c = \tau_{tr}$. The dash line is the result of van Vliet model using overall Y-parameters. The new model improves NF_{min} modeling. The experimental data from [6] is from 2 to 26 GHz due to measurement frequency range limitation. The calculation is made over a much wider frequency range, as the device is capable of much higher frequency operation. Clearly the difference between the improved model and the van Vliet model with overall Y becomes larger at higher frequencies. The τ_c, with a phyical meaning of collector transit time, can be used as a fitting parameter to improve noise

168

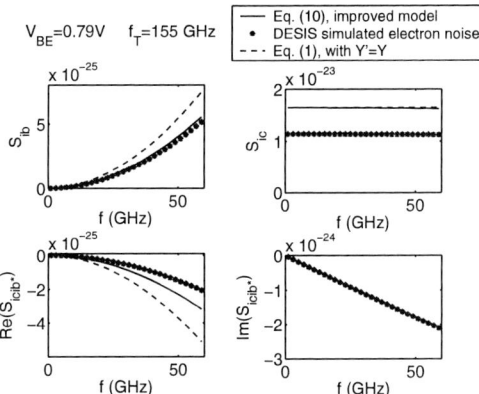

Fig. 5. Comparison between van Vliet model, new model and the extracted intrinsic noise from DESIS simulation results. $\tau_c=0.75(\tau_b+\tau_c)$ is used in the new model. Effective d_B=20nm, η=5.4, $|E|$=70.2 kV/cm.

data fitting. We note that the difference between the two approaches will significantly increase when the base resistance is reduced.

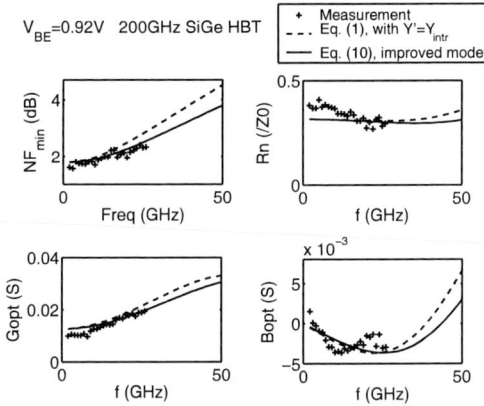

Fig. 6. Comparison of noise parameters between brutal-used van Vliet model and improved model with $\tau_c=\tau_{tr}$ using experimental data.

Now we consider an extreme case, i.e. $\tau_c \gg \tau_b$. This eventually becomes the physical scenario described by the transport noise model [7] [8]. Under such condition, both Y-parameters and λ are expressed with τ_c. With approximations up to the second order of ω, we obtain that

$$S_{ib} \approx 2\omega^2\tau_c^2 kTgm_0, \quad S_{ic} \approx S_{ic}^{van}(1 - 1/3\omega^2\tau_c^2),$$
$$S_{icib^*} \approx -2/3\omega^2\tau_c^2 kTgm_0 - j\omega\tau_c 2kTgm_0. \quad (11)$$

Comparing with the Taylor expression of transport model equations in [2], we found that

$$S_{ib} \approx S_{ib}^{tran}, \quad S_{ic} \approx S_{ic}^{tran} + 2/3\Re(S_{icib^*}^{tran}),$$
$$S_{icib^*} \approx S_{icib^*}^{tran} - 2/3\Re(S_{icib^*}^{tran}). \quad (12)$$

This shows that under $\tau_c \gg \tau_b$ condition, the transport noise model does not well model the intrinsic noise. However, it is a good approximation as S_{ib} and $\Im(S_{icib^*})$ have been correctly modeled. The improved model thus provides

a means of "bridging" the van Vliet model and the transport noise model.

IV. CONCLUSIONS

We have examined the impact of CB SCR on intrinsic transistor noise in scaled bipolar transistors. The van Vliet model, which relates noise to Y-parameters due to base electron transport, has been examined in presence of CB SCR effects. A new set of expressions relating noise and intrinsic Y-parameters accounting for CB SCR are derived. Improved noise modeling is shown using hydrodynamic noise simulation and measured data on a 200 GHz HBT.

ACKNOWLEDGMENTS

This work was supported by SRC under #2003-NJ-1133 and IBM under an IBM Faculty Partnership Award. We thank Yan Cui, D. Sheridan, S. Sweeney, J. Cressler and the IBM SiGe team for their contributions.

REFERENCES

[1] K. M. van Vliet, *Solid state electronics*, 15(10), 1033, 1972.
[2] K. Xia, et al., *IEEE TED*, 53(3), 515, 2006.
[3] J. Paasschens, et al., *Proc. of the IEEE BCTM*, pp. 221-224, 2003.
[4] F. Bonani, and G. Ghione, *Noise in Semiconductor Devices: Modeling and Simulation –Springer Series in Advanced Microelectronics*. Berlin, Germany: Springer Verlag, 2001.
[5] R.L. Pritchard, *Electrical Characteristics of Transistors*, McGraw-Hill, New York, 1967.
[6] K. Xia, et al., *Proc. of the IEEE BCTM*, pp. 180-183, 2005.
[7] M. Rudolph, et al.,*IEEE EDL*, 20(1), 24, 1999.
[8] G. Niu, et al., *IEEE TED*, 48(3), 2568, 2001.

2006 Bipolar/BoCMOS Circuits and Technology Meeting Proceedings

SiGe Profile Optimization for Improved Cryogenic Operation at High Injection

Yan Cui, Guofu Niu, Yun Shi, Chendong Zhu[1], Laleh Najafizadeh[1]
John D. Cressler[1], and Alvin Joseph[2]

Alabama Microelectronics Science and Technology Center
Electrical and Computer Engineering Department
200 Broun Hall, Auburn University, Auburn, AL 36849, USA
Tel: 334 844-1856 / Fax: 334 844-1888 / E-mail: cuiyan1@auburn.edu
[1] School of ECE, Georgia Institute of Technology, Atlanta, GA 30332-0250, USA
[2] IBM Microelectronics, Essex Junction, VT 05452, USA

Abstract — **This paper explores SiGe profile optimization for improved cryogenic operation at high injection. Through analyzing distributive transit time profiles, the bottle neck limiting high injection f_T is identified and then eliminated in an optimized profile design. The fabricated profile indeed shows considerably improved f_T and β at high injection.**

Index Terms — **SiGe HBT device physics, SiGe profile optimization, cryogenic electronics, extreme environment electronics.**

I. INTRODUCTION

The ability to maintain high f_T and β at high current density (J_C) is highly desired for many analog, digital, and RF circuits. With cooling, the f_T of a SiGe HBT rolls off at a higher J_C, because of increased saturation velocity, as shown in Fig. 1 for a 50 GHz SiGe HBT, which we will refer to as the "control". Heterojunction barrier effect [1], on the other hand, becomes worse with cooling. Observe that at 162 K, f_T rolls off rather quickly by a significant 35 GHz as J_C increases from 3 to 4 mA/μm^2 at 162 K. This is undesired for high J_C operation of digital circuits, RF oscillators, and power amplifiers.

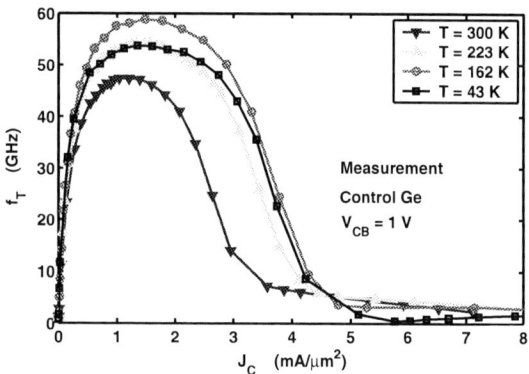

Fig. 1. Measured f_T vs J_C for the control Ge profile at 300 K, 223 K, 162 K, and 43 K. $V_{CB} = 1$ V.

This paper aims to improve high injection f_T (and β)

performance through SiGe profile optimization, particularly at cryogenic temperatures of interest to space exploration. We first identify the local position at which undesired high injection transit time buildup occurs using device simulation. The results are then used to optimize the SiGe profile to eliminate or minimize such high injection transit time buildup. DESSIS is used for device simulation. New physics models for Si and SiGe are implemented through a C++ physical modeling interface (PMI) [2], and model coefficients are calibrated to fit measured $I - V$ data.

II. CRYOGENIC PROFILE DESIGN

In simulation, the device structure is constructed based on device layout. The doping and Ge profiles were determined using SIMS. Without loss of generality, we have sketched the shape of the control Ge profile in Fig. 2, together with another Ge profile that extends deep into the collector, which will be referred to as profile I. The main disadvantage of profile I is the much higher total Ge content, and thus less stable SiGe film. Such a profile is traditionally used to improve high injection f_T performance, as detailed in [3]. However, as shown in Fig. 3, for the HBT considered here, according to simulation, the improvement from using deep Ge extension occurs at a very high J_C where f_T has already rolled off considerably. This calls for new profile design approaches to improving high injection f_T.

A. Identifying High Injection f_T Limiting Factor

To find out what limits high injection f_T, we perform distributive transit time analysis as a function of J_C. Details of distributive transit time analysis can be found in [1]. The spatial distribution of the total transit time is simulated, in terms of the so called differential transit time τ_{diff}. In an ideal 1-D bipolar transistor, at any position x, $\tau_{diff}(x) \cdot \Delta x$ represents the local contribution to the total transit time due to minority carrier charge storage from depth x to $(x+\Delta x)$. τ_{diff} has a unit of ps/μm, and its integration from emitter to collector gives the total transit time τ_{ec} [1]. The cutoff frequency

1-4244-0458-4/06/$20.00 ©2006 IEEE   170

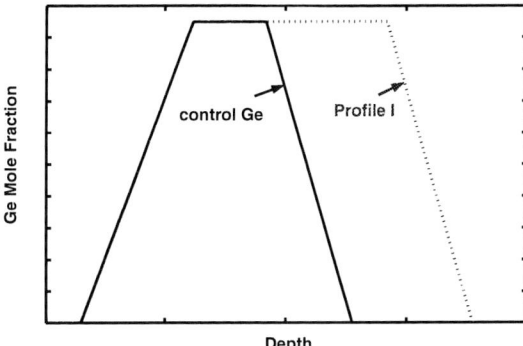

Fig. 2. Sketch of the control Ge profile and Ge profile I.

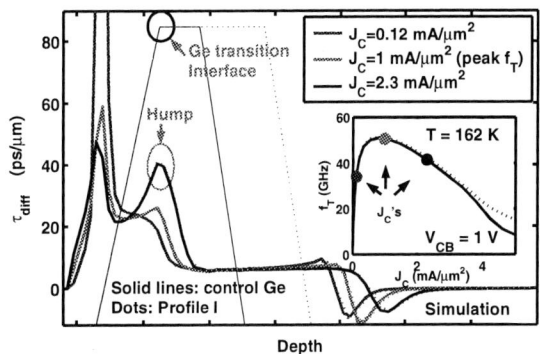

Fig. 4. τ_{diff} for the control Ge profile and profile I at 162 K. $J_C = 0.12$, 1, and 2.3 mA/μm^2. $V_{CB} = 1$ V.

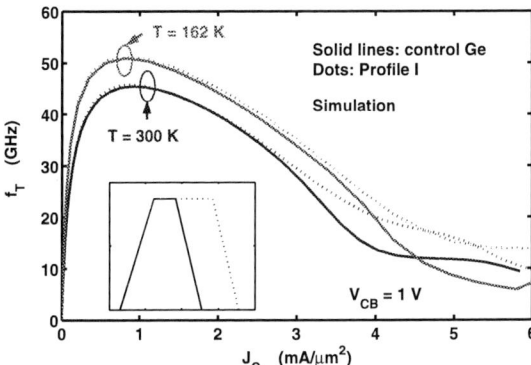

Fig. 3. Simulated f_T vs J_C for the control Ge profile and profile I at 300 K and 162 K. $V_{CB} = 1$ V.

f_T is related to τ_{ec} by $f_T = 1/2\pi\tau_{ec}$.

Fig. 4 shows the simulated differential transit time τ_{diff} profile for the control Ge profile and Ge profile I, at 162 K, for representative J_C values. As J_C increases from 0.12 to 1 mA/μm^2 (near peak f_T), a small hump of τ_{diff} begins to appear at the point of transition from linear Ge grading to constant Ge. As J_C further increases to 2.3 mA/μm^2, the differential transit time hump increases rapidly, thus increasing τ_{ec} and degrading f_T. At such J_C, Ge profile I shows little improvement over the control Ge profile. From the above analysis, the bottle neck that limits high injection f_T performance occurs at or near the point of Ge transition from linear grading to constant, as indicated in Fig. 4. To eliminate such differential transit time "hump", we need to create an accelerating field through additional forward Ge grading, as detailed below.

B. New Cryogenic Profile Design

A simple way to eliminate or minimize the differential transit time hump is to keep on grading Ge after the "hump" position. This, however, inevitably increases the total Ge content if the front side Ge gradient is to be kept. Alternatively, we can decrease the Ge gradient on the front side, and then use the extra Ge content to provide the Ge gradient needed after the first Ge transition

point of the control Ge profile. For simplicity and easier SiGe epi growth development, we choose to use a triangle profile that has not only the same start and end point as the control Ge profile, but also the same total Ge content. The SiGe film stability can then be kept the same. The problem is then narrowed down to determining where the peak Ge point should be. Dozens of such profiles are simulated, from which the optimized profile is selected. The result is shown in Fig. 5 as profile II. Interestingly the optimum peak occurs at the second transition point of the original Ge control profile. Because of the higher peak Ge mole fraction, Ge profile II has a steeper Ge retrograding into the collector than the control profile. This, however, does not lead to worsened high injection f_T roll-off.

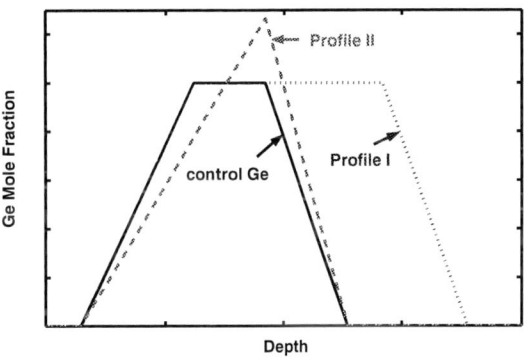

Fig. 5. Sketch of the control Ge profile, SiGe profile I, and SiGe profile II.

Fig. 6 compares the high injection differential transit time τ_{diff} profiles for the three Ge profile designs at 162 K. The J_C used is 2.3 mA/μm^2, just above peak f_T. Observe that the differential transit time "hump" disappears for the new cryogenic profile II, as expected. The result is improved high injection f_T, as shown in Fig. 7. The high injection f_T improvement becomes more significant with further cooling.

Note that profile II is not the best in terms of RF noise performance, as detailed in [4], because the lo-

171

cal noise density is the highest near the EB junction. However, for SiGe HBTs operating at cryogenic temperatures, noise is less an issue, because of the natural reduction of thermal noise with cooling. Even for room temperature application, the associated noise degradation can be negligible or justified, particularly given the low noise capability of SiGe HBTs.

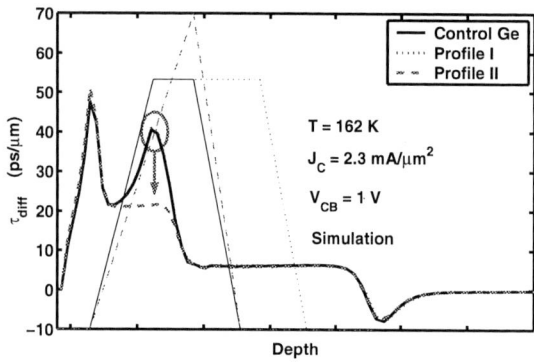

Fig. 6. τ_{diff} for the control Ge profile, profile I and profile II at 162 K. $J_C = 2.3$ mA/μm^2. $V_{CB} = 1$ V.

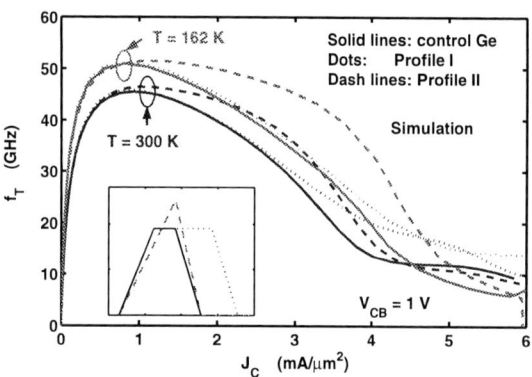

Fig. 7. Simulated f_T vs J_C for the control Ge profile, profile I, and profile II at 300 K and 162 K. $V_{CB} = 1$ V.

III. EXPERIMENTAL RESULTS

The three profiles were fabricated in the same wafer lot under identical processing conditions. SIMS shows that the fabricated profiles are approximately the same as intended. The devices are measured on wafer using a customized cryogenic microwave probing system. Standard open/short de-embedding is used to remove pad and interconnect parasitics. Fig. 8 shows measured f_T vs J_C for the optimized profile II at 300 K, 223 K, 162 K, and 43 K. Compared to the Ge control profile (Fig. 1), the optimized profile II has much better high injection f_T performance. At 162 K, as J_C increases from 3 to 4 mA/μm^2, f_T drops by only 10 GHz as opposed to 35 GHz in the control profile. The high f_T operating current density range is clearly widened in the optimized profile. For instance, at 43 K, the J_C at

which f_T drops to 10 GHz is 4.2 mA/μm^2 in the control Ge profile, while in profile II, this occurs at $J_C = 7.5$ mA/μm^2, a much higher J_C.

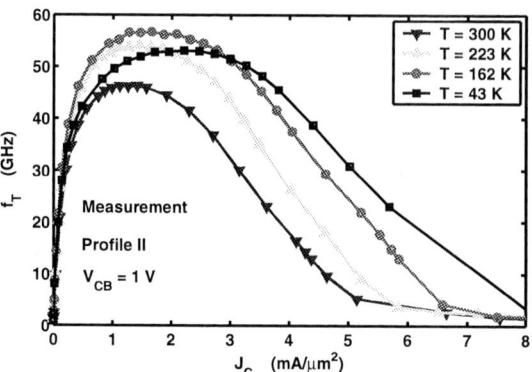

Fig. 8. Measured f_T vs J_C for the optimized profile II at 300 K, 223 K, 162 K, and 43 K. $V_{CB} = 1$ V.

Fig. 9 compares the measured f_T and β vs J_C at 162 K for the three profiles. At $J_C = 4.5$ mA/μm^2, the f_T of the control Ge profile and profile I have dropped to well below 10 GHz, while the f_T of the optimized profile II is still above 30 GHz. At lower J_C, β is lower in the optimized profile II, as expected. However, at $J_C > 5$ mA/μm^2, β is higher in the optimized profile II, because of reduced high injection charge storage. The slightly lower low injection β is not important as the β value remains high enough. For instance, at 1 mA/μm^2, β is 70 in the optimized profile II, and 80 in the control profile. For typical circuit applications that benefit from high f_T at high J_C, the f_T improvement is expected to be most significant in enabling circuit operation at higher J_C.

The high injection f_T and β improvements from using the optimized Ge profile are more significant at 43 K, as shown in Fig. 10. At $J_C = 4.5$ mA/μm^2, where f_T of the control Ge profile and profile I have already decreased to below 10 GHz, the f_T of the optimized profile II is still above 40 GHz. Similarly, β is much higher above 5 mA/μm^2 in profile II. At $J_C = 6$ mA/μm^2, both β and f_T of the control Ge profile and profile I have fully degraded. On the other hand, f_T is greater than 20 GHz, and β is greater than 35 in the optimized profile II at 6 mA/μm^2.

We note in passing that quantitative calibration of high injection f_T and β characteristics are very challenging, even at room temperature, and are thus not shown. Improved modeling of minority carrier mobility, saturation velocity, 2-D doping profile, as well as process specific non-uniform trap distribution are needed to achieve better agreement between simulation and measurement data. The relative comparisons between profiles, however, are found to be consistent between simulation and measurement. This demonstrates the utility of such simulation, even though the calibration is only done at low injection.

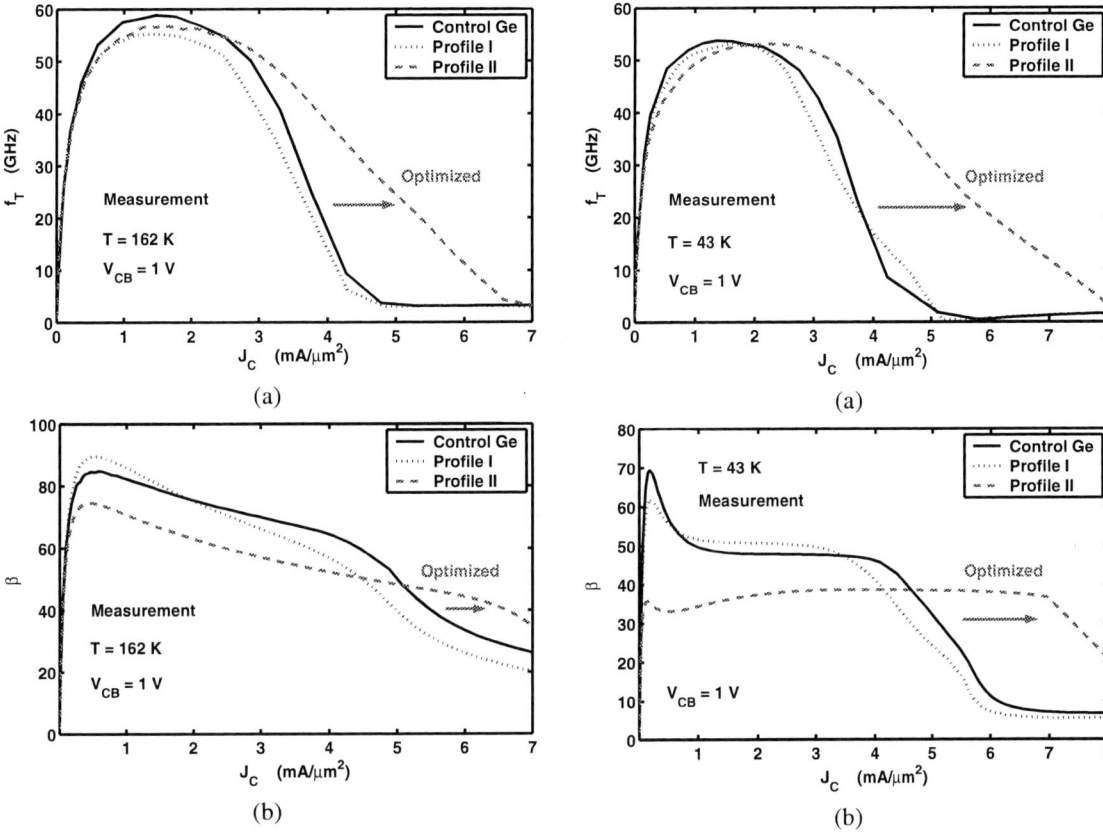

Fig. 9. (a) Measured f_T, and (b) Measured β vs J_C for the control Ge profile, profile I, and profile II at 162 K. $V_{CB} = 1$ V.

Fig. 10. (a) Measured f_T, and (b) Measured β vs J_C for the control Ge profile, profile I, and profile II at 43 K. $V_{CB} = 1$ V.

IV. CONCLUSIONS

We have presented a new approach to SiGe profile design optimization for improved high injection performance, particularly in cryogenic environment. Differential transit time profiles are first analyzed as a function of current density to reveal local transit time buildup position limiting high injection f_T. The results are then used to optimize SiGe profile and eliminate such high injection transit time buildup. Experimental results indeed show significant improvements in high injection f_T, as well as high injection β. These improvements are more significant at lower temperatures.

ACKNOWLEDGMENT

This work was supported by NASA, under grant NNL05AA37C, and the Georgia Electronic Design Center at Georgia Tech. We are grateful for the support of C. Moore, M. Watson, M. Beatty, L. Nadeau of NASA, E. Kolawa of JPL, and the IBM SiGe development group, as well the many contributions of the SiGe ETDP team, including: M. Mojarradi, B. Blalock, W. Johnson, R. Garbos, R. Berger, F. Dai, L. Peltz, A. Mantooth, P. McCluskey, M. Alles, R. Reed, A. Joseph, and C. Eckert.

REFERENCES

[1] J.D. Cressler and G.F. Niu, *Silicon-Germanium heterojunction bipolar transistors*, Artech House, pp. 261-301, Jan. 2003.
[2] DESSIS, 2-D Device Simulator, version 10.0, Synopsys.
[3] A.J. Joseph, J.D. Cressler, D.M. Richey, and G. Niu, "Optimization of SiGe HBTs for operation at high current densities," *IEEE Transactions on Electron Devices*, vol. 46, pp. 1347–1354, 1999.
[4] Y. Cui and G. Niu, "SiGe profile optimization for low noise using microscopic noise simulation," *IEEE Bipolar/BiCMOS Circuits and Technology Meeting*, pp. 268-271, Oct. 2005.

Influence of the Ge profile on V_{BE} and current gain mismatch in Advanced SiGe BICMOS NPN HBT with 200 GHz f_T

M. Dahlström, K. Walter, S. Von Bruns, R.M. Malladi, Kim M. Newton, A.J. Joseph

IBM Microelectronics Division, IBM, Bldg. 972D, Essex Junction, VT 05452, 802-769-4228, email: mattias@us.ibm.com

Abstract — **Transistor mismatch is a key parameter for the design and operation of advanced analog circuits. We present for the first time data from several generations of BiCMOS technology nodes for V_{BE} and current gain (β) mismatch. We show that the 0.12 μm BiCMOS has a 3-σ V_{BE} mismatch of 0.63 mV-μm and β mismatch of 0.24 %-μm. CBE NPNs have essentially the same but slightly lower mismatch than CBEBC NPNs. Very small and very long devices have increased mismatch, especially at high currents. We also present a physical model and experimental data showing the influence of the emitter-base Ge slope on the device mismatch.**

Index Terms — **Bipolar/BiCMOS analog circuits, device mismatch, bipolar modeling and simulation, Ge profile, SiGe, HBT, NPN, mismatch, V_{BE}, base emitter voltage**

I. INTRODUCTION

Device matching is an important consideration in analog design [1]. Circuit elements such as the differential pair at the input of an amplifier require low V_{BE} mismatch to minimize DC offset voltage. A bank of transistor current sources can provide well-matched collector currents if the V_{BE} mismatch between the devices is small. Generally speaking, limiting the variation in DC gain (β) between devices is important for controlling the amplification in different parts of a circuit.

The mismatch in V_{BE} or β is typically expressed as (1):

$$\sigma^2 = \frac{b_1^2}{W_E L_E} + \frac{b_2^2}{W_E^2} + \frac{b_3^2}{L_E^2} + b_4^2 s^2 \qquad (1)$$

where W_e and L_e are the emitter dimensions, s is the separation between devices, and b_N are technology dependent parameters [2-3]. Often only b_1 is reported, and for earlier generations of NPNs it is on the order of 0.3-0.5 mV-μm. Eq. (1) captures the fundamental variation in V_{BE} or β due to microscopic differences in device dimension, doping and base SiGe profile. We expect this variation to average out for devices with larger emitter areas, so small devices will have worse mismatch than larger devices. Any variations that affect I_b and I_c in the same way, such as emitter opening size or emitter resistance at high currents, will not show up in the β mismatch [3].

As pointed out in [2], Eq. (1) does not model the increase in V_{BE} mismatch observed at high current densities, where the effects of device and parasitic resistances become more prominent. Hardware measurements provide data on the external V_{BE} and not the internal V_{BE} of the emitter-base junction. By taking emitter and base resistance into account, the current dependence of V_{BE} can be expressed as:

$$V_{BE} = V_{BE,i} + R_{ex} I_E + R_b I_B . \qquad (2)$$

where R_{ex} and R_b are the emitter and base resistances, including wiring resistance. Natural variations in these resistances – especially the emitter resistance – will add a current dependence to the V_{BE} mismatch [2].

A multiple emitter device could have increased mismatch due to current hogging in individual fingers. Variations in emitter and wiring resistance as well as self-heating are likely reasons for this. The model also does not account for current crowding along the length of a device or in a very wide device.

Monitoring the device mismatch of a process line over time is a useful tool for tracking non-ideal device behavior due to lithography, epitaxy, wiring resistance, etc.[4-5].

II. THEORY

The base-emitter potential V_{BE} is linked to the collector current I_c (3):

$$I_C = A_E J_C = \frac{q n_i^2 A_E e^{\frac{q V_{BE}}{n_s k_B T}}}{G_B} \qquad (3)$$

where G_B is the base Gummel number (4) and $n_{t,eB}$ contains the current enhancement originating from the heterojunction [6].

$$G_B = \int_0^{W_B} \frac{n_i^2}{n_{i,eB}^2} \frac{p_p}{D_{nB}} dx \qquad (4)$$

I_c can be rewritten as:

$$I_C \approx I_{Co} \frac{\Delta E_g(W_b)}{kT} \frac{1}{1 - \exp\left(\frac{-\Delta E_g(W_b)}{kT}\right)} \exp\left(\frac{\Delta E_g(0)}{k_B T}\right) \quad (5)$$

where I_{co} is the collector-base leakage current with the emitter open, $\Delta E_g(W_b)$ is the effective bandgap difference over the base, and $\Delta E_g(0)$ is the emitter-base bandgap difference at the start of the base. As seen in Eqs (3) and (5), variation in the Ge slope in the base or variation in the position of the emitter-base junction will determine I_C and V_{BE}, and these variations can lead to increased mismatch. The last term in (5) shows I_c and V_{BE} are exponentially related to the actual position of the emitter-base junction, which determines the effective bandgap. We expect that for a small shift in emitter-base junction position, a steeper Ge slope will lead to a larger I_c or V_{BE} change, of special concern for advanced SiGe NPNs that tend to have very aggressive base profiles.

To understand how β mismatch compares to V_{BE} mismatch, the effects on I_B need to be understood. In general, I_B mismatch follows I_C mismatch in the linear region, just as I_B follows I_C. Emitter dimension variations and base doping variation provide an area dependent term. I_B has the same relation to parasitic emitter and base resistances as I_C (1) through V_{BE} [2], providing a current dependent mismatch. However, emitter interface properties have a strong effect on I_B, and I_B mismatch has been used to identify non-ideal emitter silicidation [4]. β mismatch can therefore be expected to show little current dependence but larger sensitivity to emitter properties than V_{BE} mismatch.

III. EXPERIMENTAL SETUP

For measurement of V_{BE} mismatch, the collector terminal of each of four npn HBTs is shorted to its corresponding base terminal with a wide metal line, and the resultant four base-emitter diodes are connected in a bridge configuration. The transistors are separated by 30 μm. A current $2 \cdot I_E = 2 \cdot J_E A_e$ is passed through the outer terminals and the voltage at each of the inner terminals as well as the difference voltages V_{12} and V_{21} between them are monitored using sensitive Digital Voltmeters. The measured potentials V_{12} and V_{21} represent bridge imbalance, and the bridge tracking ΔV_{BR} and resulting 3-σ HBT V_{BE} mismatch is expressed as:

$$\Delta V_{BR} = V_{12} - \left(\frac{V_{12} + V_{21}}{2}\right) \quad (6)$$

$$3\sigma_{\Delta VBE} = 3\sqrt{2}\sigma_{\Delta VBR}$$

The voltage across the metal interconnects is kept negligibly small compared to ΔV_{BE} by keeping them

wide. Increasing the width of the metal interconnects also limits the voltage drop due to Joule heating at high current densities.

A transistor pair with the emitters shorted together is used for beta mismatch measurements. A current I_E is passed through each of the transistors in two separate tests and the currents I_C and I_B are monitored. The beta difference between the two tests corresponds to the beta mismatch [7].

The measurements take place on a temperature-controlled shielded wafer prober, and are done at different currents for several different emitter dimensions. A large number of sites per wafer are measured on several wafers from each of the available lots. The 3-σ mismatch is calculated for each wafer, and then averaged over all wafers. Data where the measured V_{BE} of a transistor in the bridge is well outside the range of expected operation are screened out.

Node (nm)	180	180	130	130
Version	7WL	7HP	8WL	8HP
F_T (GHz)	60	120	100	200
F_{max} (GHz)	85	100	150	280
J_{peak} (mA/μm²)	1.65	5.5	9	12.5
β		140	200	165	400
BV_{CEO} (V)	3.3	2.0	2.5	1.8

Table 1. Properties of different BiCMOS technologies

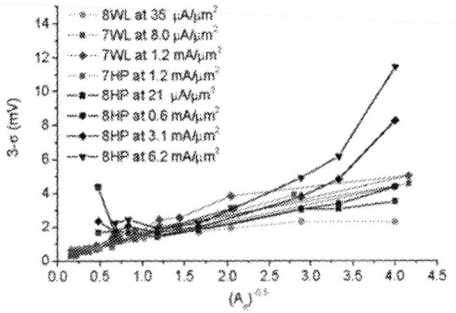

Fig. 1. Measured V_{BE} mismatch for several IBM BiCMOS technologies at different current densities.

IV. RESULTS

Fig.1 shows the measured V_{BE} mismatch in the BiCMOS7HP, BiCMOS7WL, BiCMOS8HP and BiCMOS8WL technologies [8]. BiCMOS8HP devices are characterized at the highest current levels, up to 6.2 mA/μm², and the impact of device and parasitic resistances can be seen in the high values of mismatch for devices with the smallest emitter dimensions. At lower current levels the mismatch is comparable to

BiCMOS7HP or BiCMOS7WL. The BiCMOS8HP results also show a distinct increase in mismatch for very long experimental devices (L_e=36 μm), probably due to non-uniform current injection caused by a base potential drop along the length of the device.

The 3-σ V_{BE} mismatch parameter b_1 for 8HP was extracted as 0.63 mV-μm at a low current density of 21 μA/μm^2, rising to 1.12 mV-μm at 6.2 mA/μm^2.

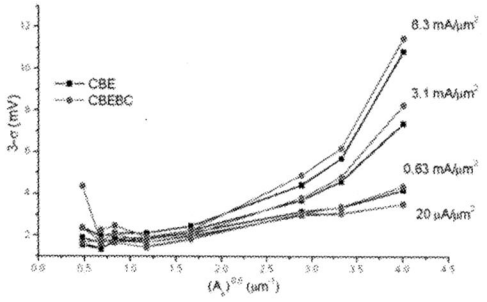

Fig. 2. Measured V_{BE} mismatch for BiCMOS8HP comparing CBE and CBEBC layout.

Fig 2 shows a comparison between CBE and CBEBC layout. The CBE layout features a wrap-around silicided base, contacted on one side only. The CBEBC layout is contacted to the first metal level on both sides of the emitter. The mismatch is essentially identical for either layout, despite the differences in base and collector resistance.

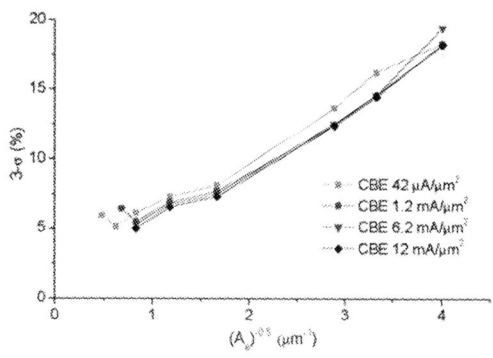

Fig. 3. Measured β mismatch for BICMOS8HP technology at different current densities.

Fig. 3 shows the measured β mismatch for BiCMOS8HP. The beta mismatch parameter b_1 for 8HP was extracted to 0.24 %-μm and is essentially unchanged with current.

Fig. 4 shows the current dependence of the V_{BE} mismatch for different emitter dimensions. The current dependence is significantly stronger for the smallest emitter dimensions, where the emitter resistance R_{ex} is the highest. The average emitter resistance for each emitter size is displayed.

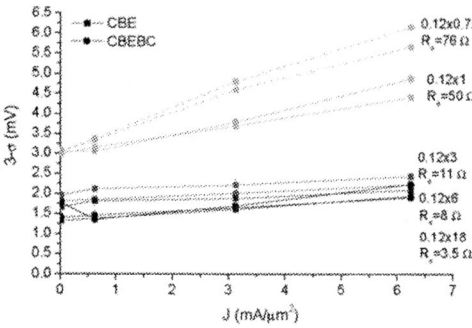

Fig. 4. Measured V_{BE} mismatch for BiCMOS8HP as function of current density.

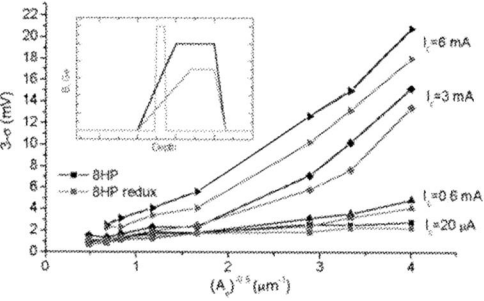

Fig. 5. Measured V_{BE} mismatch for BiCMOS8HP NPNs with original Ge slope (8HP) and reduced Ge slope (8HP redux). Inset: B and Ge profile for 8HP and 8HP redux.

In order to understand the impact of the Ge profile on mismatch, we processed six wafers with a baseline Ge slope (8HP) and six wafers with a reduced Ge slope (8HP redux) in the emitter-base junction (fig 5 inset). They were otherwise identical and processed together. Device DC performance is reasonably comparable, but the 8HP redux wafers show a drop in β from 433 to 166 and a drop in f_T of 27%. This is an expected effect of the increased base transit time. Current at peak f_T is essentially constant at 12 mA/μm^2 (8HP) and 11 mA/μm^2 (8HP redux). Fig. 5 shows the mismatch data for these wafers, and the mismatch for a shallower Ge ramp is smaller, in agreement with Eq. (5). Future work will investigate the impact of sub-collector design on the mismatch.

V. CONCLUSION

The V_{BE} and β mismatch for IBM's 200 GHz BiCMOS technology are found to be similar to that of previous technologies of the same family, except at very high current densities where the V_{BE} mismatch increases dramatically for the smallest devices. This is primarily attributed to the emitter resistance. Comparing the results for CBE and CBEBC layouts show essentially the same mismatch. Very long devices ($L_e > 12$ μm) seem to carry a V_{BE} mismatch penalty.

We also show that the V_{BE} mismatch of wafers with reduced Ge ramp is lower, in accordance with Eq. (5).

It is important that these effects are accurately accounted for in the device model, and that analog circuit designers are familiar with them.

ACKNOWLEDGEMENT

The authors wish to acknowledge the assistance of Maurice LaCroix and Jay Rascoe for measurements.

REFERENCES

[1] P.R. Kinget, "Device mismatch and tradeoffs in the design of analog circuits," *IEEE Journal of Solid-State Circuits*, vol.40, no.6, pp. 1212-1224, June 2005.

[2] S. Bordez et al, "Study of Bipolar Transistor Matching at High current level with Various Test Configurations Leading to a new Model Approach," *Proceedings of the 2005 Bipolar/BiCMOS Circuits and Technology Meeting*, section 4.2, Oct. 2005.

[3] P.G. Drennan, "Device mismatch in BiCMOS technologies," *Proceedings of the 2002 Bipolar/BiCMOS Circuits and Technology Meeting*, Oct. 2002.

[4] H.P. Tuinhout, "Improving BiCMOS technologies using BJT parametric mismatch characterization," *Proceedings of the 2003 Bipolar/BiCMOS Circuits and Technology Meeting*, pp. 163-170, Sept. 2003.

[5] H.P. Tuinhout and W.C.M Peters, "Measurement of lithographical proximity effects on matching of bipolar transistors," *Proceedings of the 1998 Microelectronic Test Structures*, pp.7-12, March 1998.

[6] Y. Taur and T. Ning, *Fundamentals of Modern VLSI Devices*, Cambridge University Press, 1998.

[7] H.P. Tuinhout, "Design of Matching Test Structure," *Proc. 1994 Int. Conference on Microelectronic Test Structures*, vol. 7, pp. 21-27, March 1994.

[8] N. Feilchenfeld *et al*, "High Performance, Low Complexity 0.18 um SiGe BiCMOS Technology for Wireless Circuit Applications," *BCTM Proceedings*, p. 197, 2002.

[9] B. A. Orner *et al*, "A 0.13 um BiCMOS Technology featuring a 200/280 GHz (fT / fmax) SiGe HBT," *BCTM Proceedings*, p. 203, 2003.

[10] L. Lanzerotti *et al*, "A Low Complexity 0.13 μm SiGe BiCMOS Technology for Wireless and Mixed Signal Applications," *BCTM Proceedings* 2004.

2006 Bipolar/BoCMOS Circuits and Technology Meeting Proceedings

Substrate Transfer:
an Enabling Technology for System-in-Package Solutions

R. Dekker, M. Dümling [1], J.-H. Fock [2], J.R. Haartsen [3], H.G.R. Maas, T.M. Michielsen,
H. Pohlmann [2], W. Schnitt [2] and A.M.H. Tombeur.

Philips Research, Prof. Holstlaan 4 (WAG-02), 5656 AA Eindhoven, The Netherlands, ronald.dekker@philips.com;
[1] Technical University Hamburg-Harburg, Germany; [2] Philips Semiconductors, Hamburg, Germany;
[3] Technical University Eindhoven, Eindhoven, The Netherlands.

Abstract — **A simple and robust technology for the transfer of circuits, processed on normal silicon wafers, to alternative substrates is presented. Substrate Transfer Technology (STT) is a post-processing technology based on adhesive bonding using a UV curing adhesive. We demonstrate that STT can be used for the elimination of RF substrate losses, for double sided device processing and for the fabrication of flexible circuits and thermal sensors and actuators. The successful industrialization of the process demonstrates that STT is indeed a viable option for high-yield, low-cost mass production.**

Index Terms — **Semiconductor device fabrication, Semiconductor device bonding, bonding, Silicon on insulator technology, Transducers, Thermoelectric devices**

I. INTRODUCTION

Although silicon has proven to be an excellent material for the fabrication of active devices and micro-systems, it is often not realized that the structures of interest are usually confined in the top few microns of the silicon wafer. The remaining 99.5% of the silicon wafer just serves to keep the structures together during processing and in some cases to conduct heat and electricity to the package. Fortunately, the material properties of silicon are adequately suitable for these purposes. Compared to for instance compound semiconductor substrates it is quite strong and has a good thermal and reasonable electrical conductivity. Depending on the application, however, some material properties of silicon are less desirable:

Resistive losses at RF frequencies. It is well known that the resistive nature of the silicon substrate causes dissipative losses at RF frequencies, resulting in a degradation of the quality factor of passive components [1] and an increased power consumption [2]. Additionally, the capacitive coupling of different circuit components through the resistive substrate results in cross-talk.

Thermal Conductivity. The good thermal conductivity of silicon has enabled the integration of high speed processors and efficient (RF) power devices. There are, however, numerous applications, especially in the area of sensors and Lab-on-Chip concepts, where the high thermal conductivity is a significant drawback. Many sensors like bolometers, thermal flow sensors, IR detectors and micro-calorimeters rely on the sensing of small temperature gradients across the device [3]. The high thermal leakage through the silicon substrate in these devices results in a strongly reduced sensitivity. Similarly, in many Lab-on-Chip concepts accurate temperature control is required to promote specific reactions e.g. for PCR (Polymer Chain Reaction), a technique for in-vitro amplification of DNA strands, which requires a carefully controlled sequencing of temperature steps. As a result, many of these devices are fabricated on free standing membranes to create the required thermal isolation. The fabrication and subsequent processing of these free standing membranes, however, is far from trivial and has hampered high-volume, low-cost industrialization.

Rigidity. Silicon substrates are relatively thick, hard, brittle, not bendable and certainly not stretchable. Yet, there is a growing interest in ultra thin and flexible integrated circuits [4]. The main application is RF-tags e.g. for security and authentication (value paper, passports, tickets, vouchers, etc.), for tracking (medicines, medical samples, etc.), and for data storage (digital version of a printed document). A second important application is in-vivo implantable electronic devices e.g. cochlear implants, artificial retinas etc.

Opaqueness. In an optical sensor array usually only a part of the chip area is optically sensitive, with the remaining chip area containing readout amplifiers, pixel address logic, interface, etc. The fill-factor is defined as the optical sensitive silicon area divided by the total chip area. The aim is of course to maximize the fill-factor and accordingly optimize the price/sensitivity trade-off. This holds for optical image sensors, but also for medical (X-ray) detectors. A way to realize a fill-factor of almost 100% is to use a backside illuminated buried junction underneath the front-side circuit layer. To reduce dark-current and pixel smear, the silicon substrate needs to be thinned down to the proximity of the buried junction and replaced by a transparent substrate [5].

Double-sided Device Processing. The simple fact that the silicon substrate is present implies that it blocks double-sided device processing. In device fabrication, we have been conditioned to think of silicon devices as being only accessible from the front-side of the wafer. Only in rare cases, e.g. in the fabrication of free-standing membranes, the possibility to access devices or structures from the backside of the wafer is

1-4244-0458-4/06/$20.00 ©2006 IEEE

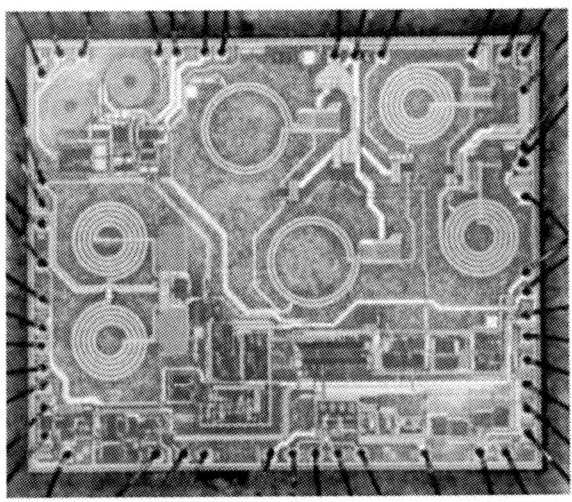

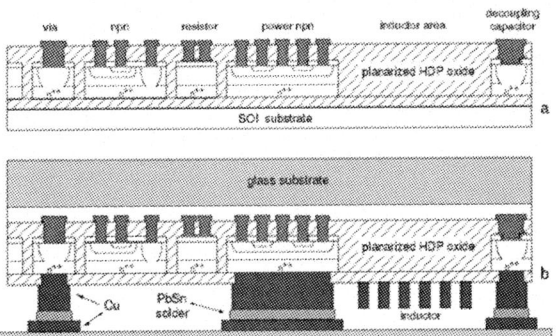

Figure 2: Schematic cross-section of the surface mounted RF IC technology: (a) fully processed SOI wafer with typical devices, (b) final device soldered to a PCB after transfer-to-glass, etching of silicon, backside processing and dicing.

Figure 1: Fully integrated GPS receiver frontend. The circuit was processed in an RF-SOI technology and after processing transferred to glass. The complete process has been industrialized and the part complies with normal reliability standards.

used. In the substrate transfer process discussed in this paper, devices are first processed from the front side. Next, the wafers are glued, top-down, to a supporting substrate, followed by a removal of the original substrate. This procedure gives full access to the backside of the devices. Given this possibility many new device concepts in the area of MEMS, sensors, actuators and System-in-Package (SIP) have been envisaged.

The aim of Substrate Transfer Technology (STT) is to tailor the properties of the substrate to the specific requirements of the application. Although it is a relatively simple and straightforward procedure, STT has so far - with a few exceptions - remained a research curiosity. A reason for this is perhaps that most process architects are unaware of the powerful and fascinating possibilities that STT offers. Moreover, if one is not trained in thinking in terms of STT, these possibilities can easily be overlooked. Since we have developed STT more than a decade ago, we have experienced repeatedly that it can be a very useful process tool for the fabrication of integrated devices for a broad spectrum of applications. It is the aim of this paper to give an overview of these applications, and to convey an idea of the versatility of STT.

II. SUBSTRATE TRANSFER TECHNOLOGY

STT was developed to combine the ease and advantages of standard silicon processing with a complete freedom of substrate choice [6]. The technology is based on the gluing of a fully processed silicon wafer, top down, to an alternative substrate. Usually, a commonly available alkaline-free, high barium content glass like AF45 from Schott is used because it combines excellent RF properties with a low price. If needed, many other types of substrates may be used as is demonstrated by the successful transfer of circuits to AlN, Al_2O_3, ferrite and glass-fiber epoxy printed circuit board [6]. The transparency of

glass additionally allows for UV curing of the adhesive.

In literature many different adhesives have been proposed for permanent wafer bonding such as: bisbenzocyclobutene (BCB) from Dow Chemical, SU-8 epoxy based photo resin from Microchem Corp., positive and negative photo-resists, polymethylmethacrylate (PMMA) and polyimides [7]. The adhesive we have selected is a solvent free monomer to which a UV-sensitive initiator has been added. On exposure to UV-light the adhesive immediately cross-links to form a very hard acrylic. Since curing is performed at room temperature, it does not result in bowing or warping of the substrate sandwich due to differences in thermal expansion coefficients. The main advantages of the adhesive selected by us are: easy to work with, a water like viscosity, accommodates many microns of topography on the wafer, allows for room temperature UV-curing, stability to a temperature of 300 °C, transparency and resistance against 100 °C KOH etch for several hours. The major drawbacks are the relatively low glass transition temperature (< 100 °C) and the relative large shrinkage during curing (≈ 10 %). To bring STT to an industrial level, fully automatic batch-to-batch equipment was developed for 150 mm and 200 mm wafers. In our equipment only a crude wafer alignment is used based on wafer flat recognition. More accurate wafer alignment is possible with more advanced commercially available adhesive bonding equipment [8, 9].

Once the wafer has been attached to the new carrier, the wafer sandwich is flipped and processing continues on the backside of the original silicon wafer, now the new front side. Depending on the application or product, different process routes can be followed here. STT was originally developed for the transfer of an RF-SOI process to glass to allow for high quality passives. In this case the complete silicon substrate can be removed. This is done by a combination of standard DISCO grinding followed by silicon etching in KOH stopping on the buried oxide layer. Figure 1 depicts a fully integrated GPS frontend circuit fabricated in this technology which has been transferred to Philips Semiconductors in Hamburg. The gluing of wafers to glass and KOH etching are done in a separate "gray-

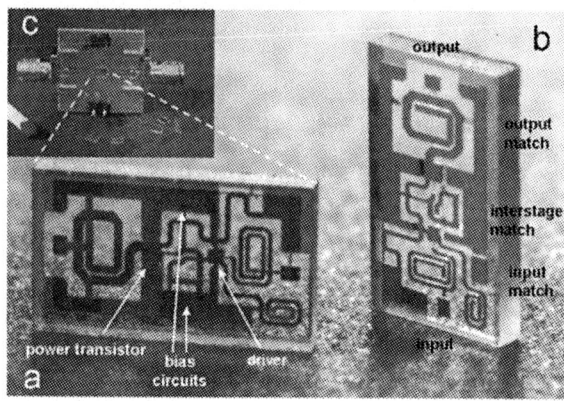

Figure 3: Design study for a fully integrated 2 stage 1.8 GHz, 3 V, 2 W power amplifier line-up (3.1x1.8 mm²). (a) view through the glass carrier showing (left-to-right) the output-, interstage- and input-matching circuits, (b) view on the backside showing the backside processed copper layer used for the matching circuits and SMD soldering, (c) the device soldered to a PCB in a 50 Ω environment.

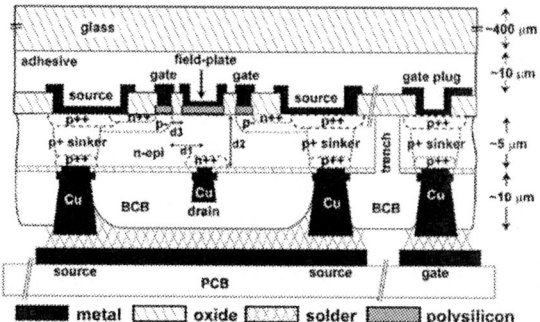

Figure 4: Schematic cross-section of a VDMOSFET on glass. The device was fabricated on SOI and subsequently transferred to glass. The backside processing includes: fine-pitch lithography, laser annealed contacts and a dual copper interconnect. (after Nenadović [20])

room" area. After transfer to glass, the wafers are cleaned and clean-room processing is continued for final PECVD Si_3N_4 passivation and opening of bondpads. Extensive reliability and lifetime tests were carried out on the transferred products to ensure that these parts fully comply with normal standards.

Complete substrate removal is an option for all SOI processes since the buried oxide layer provides a convenient etch stop. In fact any type of process that uses a closed SiO_2 or Si_3N_4 layer underneath the device layers may be transferred this way e.g.: flow-sensors, bolometers, micro-calorimeters, passive RF devices, etc. In the absence of a suitable etch stop the silicon substrate can be removed partially using masked KOH etching or deep RIE etching. This enables the intriguing possibility to sculpture the silicon substrate by etching mesas, cavities and (fluidic) micro-channels which form a functional part of the device.

III. DOUBLE-SIDED DEVICE PROCESSING

After the transfer of the device layer to an alternative substrate, there are many possibilities for double-sided device processing. However, the decomposition of the adhesive used (320 °C) places an upper limit on the maximum allowed process temperature. Apart from this, most of the remaining process restrictions are related to wafer handling and can easily be solved.

The most elementary backside processing step is passivation and opening of the bondpads as in the case of the RF-SOI process mentioned in the previous section. Note, that this will address the underside of the first metallization layer! In the IC-Fab Ultratech steppers with "in-die" alignment markers are used for photolithography. The transparency of the glass wafers initially caused some problems since some process equipment uses optical flat-finding to align the wafer. This was solved by sputter coating the backside of the glass wafer with a thin layer of titanium. The PECVD Si_3N_4 passivation layer is deposited

at 280 °C.

A somewhat more elaborate double-sided process scheme is depicted in Fig. 2. Starting point here is a thick-film SOI process with conventional vertical devices. After opening of contacts to the heavily doped buried layer, a 10 μm thick layer of copper is plated. The copper layer is used to reflow-solder the devices to the PCB, but also for the realization of low-loss inductors and striplines (Fig. 3). The possibility to connect the collector of an RF-power transistor directly to the output match circuit not only reduces losses, but additionally strongly reduces parasitic interconnect capacitances normally present in a large 3-terminal device which is contacted only from one side [10]. What has been realized here is infact a wafer-scale packaged device, illustrating the present trend towards a fading boundary between "front-end processing" and assembly. Very high efficiency RF-power transistors have been demonstrated using this kind of device concepts [11]

Without the possibility for high temperature anneals after transfer, the highly doped regions required for a low contact resistance for the backside metal have to be fabricated during front-end processing. This can be difficult, especially for p-type contacts. Nanver et al. have demonstrated that 5 keV implants with a dose of 2×10^{15} cm^{-2} of both As$^+$ and B$^+$, activated by excimer laser annealing results in a low contact resistance of 1.5×10^{-7} Ωcm^2 and 5×10^{-7} Ωcm^2, respectively [12]. It was found that with respect to laser annealing, a 1 μm thick silicon layer transferred to glass will behave as bulk silicon.

A combination of the backside copper interconnect and laser annealed contact implantations was used by Buisman et al. for the fabrication of high-performance substrate transferred varactor diodes [13]. By sandwiching the 1.6 μm thick varactor diode between two metal layers a MIM like performance is obtained. Quality factors at 2 GHz ranging from 150 (zero bias, 6 pF)) to 600 (6 V reverse, 2 pF) were realized.

For more advanced double-sided device structures fine pitch lithography with small alignment errors is required. In case the alignment markers are visible after substrate transfer, these markers can be used for Back-To-Front Alignment (BTFA). Since the markers are flipped during transfer, mirror symmetric

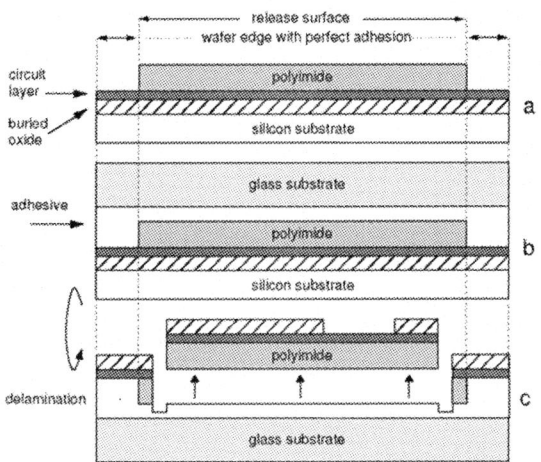

Figure 5: Processing sequence for the transfer of single-crystalline SOI circuits to polyimide foil. The area with perfect adhesion at the edge of the wafer protects the foil against the aggressive chemicals used to etch the silicon substrate and allows for a controlled delamination.

Figure 6: Cutting of the foil with a razor blade, enables easy delamination of the RF-ID circuits.

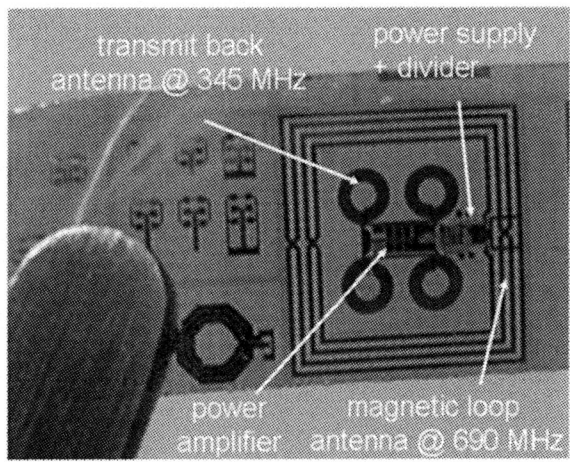

Figure 7: The 10 μm thick RF-ID tag circuit after delamination (chip area: 3x3 mm^2).

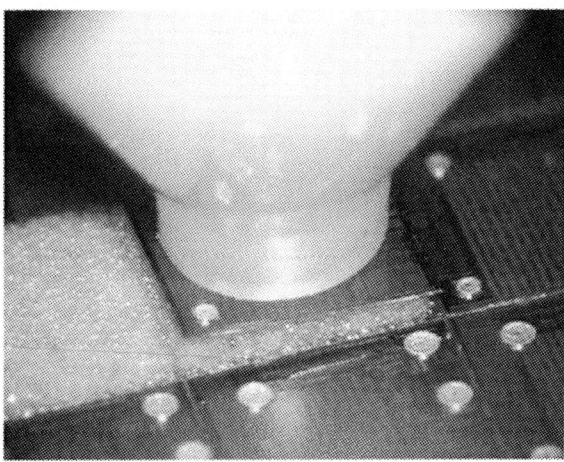

Figure 8: Picking and placing of 10 μm thick RF-ID circuits from a dicing foil (with permission Mühlbauer, High Tech International).

alignment markers for an ASML PAS5000 stepper have been developed by Van Zeijl et al. [14, 15]. These mirror symmetric alignment markers used in combination with standard overlay test procedures indicated a minimum feature size of 1 μm with an overlay of 0.3 μm. It was found that layer distortions and adhesive thickness variations were to a large extent responsible for the overlay errors. Recently, Marinier et al. have demonstrated that by using more than two alignment markers and die-by-die leveling on a PAS5500 stepper, a feature size and overlay accuracy of 300 nm and 70 nm, respectively are possible [16]. In case the alignment maskers are not visible after transfer, the recently introduced 3DAlign^{TM} option for ASML PAS5000/PAS5500 steppers may be used [17]. The 3DAlign^{TM} option is easily installed on an existing system, and without interference with normal stepper operation, allows for any arbitrary mixing of front- and back-wafer expo-

sures with a specified alignment accuracy better than 250 nm.

A combination of fine pitch backside lithography and laser annealed contact implantations was used by Nanver et al. for the fabrication of a backside contacted collector bipolar transistor process [18]. The backside contact improved electrical isolation and reduced the collector-base capacitance, collector resistance and collector-substrate capacitance. This work also clearly demonstrated the limitations of using a thermally isolating glass substrate. Unless the process is designed to operate at very low power levels, the circuits will suffer from excessive self-heating resulting in thermal breakdown. A similar device structure was proposed by Rodwell et al. for the fabrication of > 100 GHz MMICs in InP [19].

The RF Power Silicon-on-Glass VDMOSFET presented by Nenadović combines all of the features discussed above, complemented with a dual copper layer backside interconnect with a BCB intermetal dielectric (Fig. 4) [20, 21]. The de-

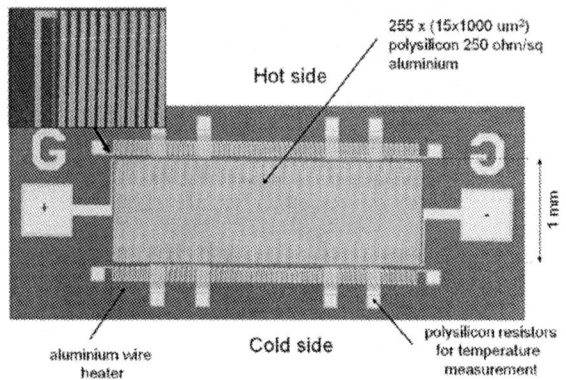

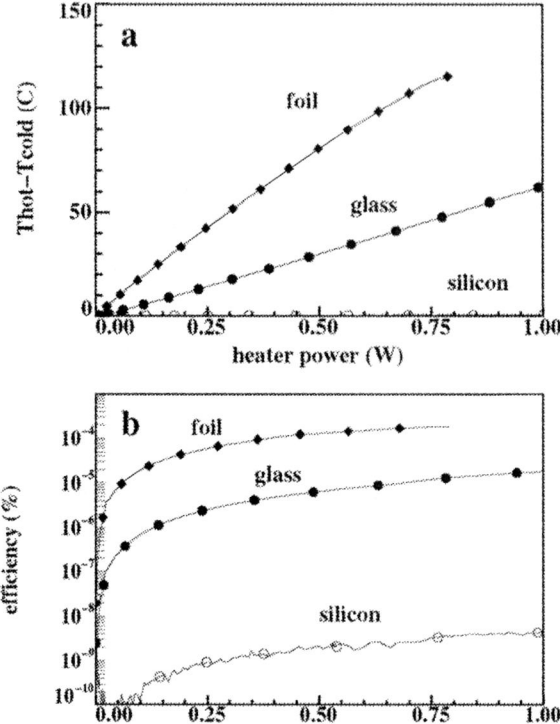

Figure 9: Thermopile test structure with aluminium heaters and polysilicon temperature sensors.

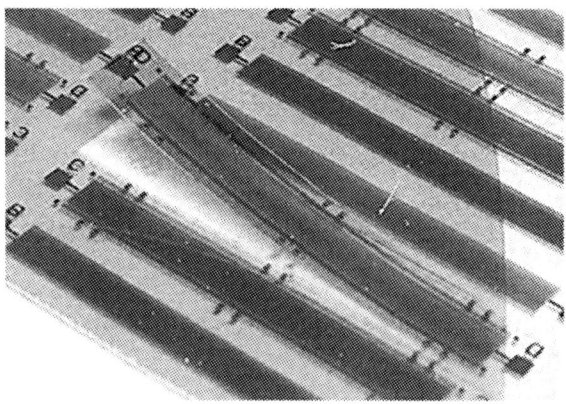

Figure 10: Delamination of the thermopile structure.

Figure 11: Temperature difference between "hot" and "cold" junctions (a) and conversion efficiency (b) of the thermopile test structure depicted in Fig. 9 as a function of heater power.

vice, intended for use in base stations, is designed for optimal electrical- and thermal-grounding resulting in a high gain and excellent thermal stability.

The Resonant-Cavity-Enhanced Photodiode proposed by Sinnis et al. shows the potential of STT for other than RF applications [22]. The photodiode is a 1 μm thick epitaxially grown pin structure on SOI, with a semitransparent front side mirror and an aluminium backside mirror. Since the light passes more than once through the same thin silicon layer, an improvement in response at the same speed is achieved. Moreover, the cavity thickness can be designed such that at the desired wavelength the absorption can be enhanced. This was demonstrated by the large Fabry-Pérot oscillations present in the photoresponse as a function of wavelength.

IV. FLEXIBLE DEVICES

In our recently proposed CIRCONFLEX technology, circuits fabricated on SOI wafers are transferred to a 10 μm thick polyimide carrier [23]. CIRCONFLEX is an extension of the STT technology discussed in the previous section (Fig. 5). In this case the glass is only used as a temporary carrier to allow for handling of the flexible circuits. Starting point are again

fully processed wafers fabricated in the industrialized thin-film RF-SOI process mentioned before. The transfer process starts with the deposition of a 10 μm thick polyimide layer by spin-coating. By careful primering an optimal adhesion between the polyimide and the wafer surface is achieved. Since the polyimide is attacked by the KOH etch used to remove the substrate, it is removed at the edge of the wafer by etching in TMAH (Fig. 5a). The surface of the polyimide is provided with a releasing surface, resulting in a moderate adhesion to the adhesive. Next, the standard STT process flow is resumed, so the wafers are glued to the temporary glass substrate and the silicon substrate is removed. At this stage the circuit layer attached to the polyimide layer remains fully intact, without buckling or cracking thanks to the moderate adhesion to the glass and the suspension of the circuit layer to the edge of the wafer. After cutting the foil with a razor blade (or laser), the circuits can now be easily delaminated, and the temporary glass substrate can be recycled (Fig. 6).

Measurements on individual bipolar transistors both after and during bending, of the foil to radii of less than 1 mm showed only minor and non-systematic changes in the characteristics. Since this process is ideally suitable for ultra-thin flexible RF-ID tags in paper, a fully integrated RF-ID tag demonstrator circuit was designed and fabricated (Fig. 7) [23]. The 3×3 mm^2 circuit operates at 700 MHz and has a reading dis-

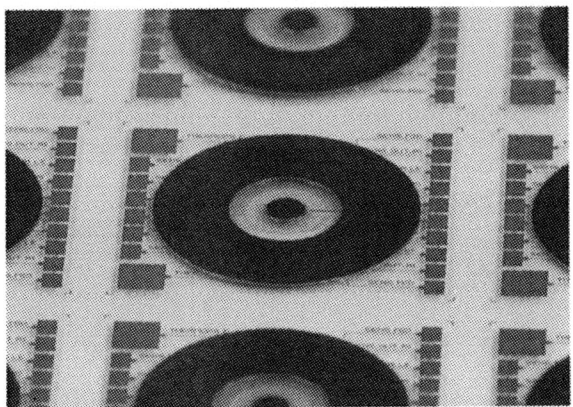

Figure 12: Transferred substrate micro-calorimeter. The high thermal resistivity of the glass substrate results in a sensitive device without having to resort to membrane processing. This photograph is taken with the supporting glass plate facing down, so that the 50 μm patterned silicon substrate is visible. This test device measures 9x10 mm^2.

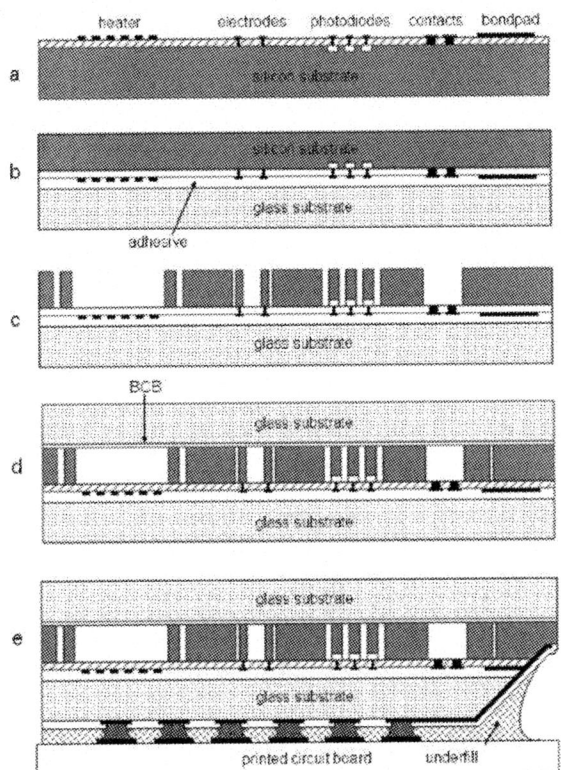

Figure 13: Schematic process sequence for the transferred substrate Lab-on-Chip platform.

tance of 1.5 cm above the antenna of the 0.5 W transmitter, corresponding to a received power level of 1 mW. During bending the tag circuit surprisingly remains fully functional, even down to a bending radius of 0.7 mm!

Extensive tests have been carried out to study the cracking behavior of the ceramic circuit layers as a result of bending [24]. In compressive mode no cracks were observed even for a radius of 0.5 mm. However, in tensile mode cracks are already formed at a radius of 4 mm. Typically these cracks start at the edge of the foil, probably as a result of the propagation and growth of microcracks formed during cutting of the foil. Fortunately, the experiments also revealed that if the ceramic circuit layers are placed under compressive stress, crack formation in tensile mode can be completely suppressed. The compressive stress can be built in into the ceramic layers during deposition, or can be supplied externally, e.g. by making use of the shrinkage of the polyimide layer during curing.

In the German Government funded SITRA project a slightly different approach was followed for the fabrication of ultra-thin ICs. Again a polyimide layer is used to provide strength to the circuits after thinning. In this approach the wafer is attached to a temporary glass support wafer using a 90 °C thermal releasing foil. After the wafer is thinned to 50 μm by grinding, the remainder of the substrate is etched in HF/HNO$_3$/H$_2$O using a spin etcher since the thermal releasing foil would not withstand the KOH etch. Next, the wafer is attached to a frame carrier containing a 120 °C thermal releasing dicing foil, and released from the temporary substrate at 90 °C. Figure 8 shows the automatic picking of the 10 μm thick RF-ID circuits (including solderbumps) from the dicing foil.

V. THERMAL ISOLATION

Many sensor and Lab-on-Chip concepts rely on the detection or generation of temperature gradients across the chip. On silicon this is difficult due to the high thermal conductivity of the substrate. In many cases STT may offer a solution to this problem without the need to resort to fragile membrane processing.

A thermopile is a frequently used temperature difference sensor in flow-meters, microcalorimeters, biosensors, IR detectors etc [3, 25]. It basically consists of a large number of thermocouple pairs in series to obtain a large response. By nature thermopiles are offset free: no temperature difference results in zero output voltage. Silicon microfabrication technology is ideally suitable for the fabrication of these thermopiles, since microlithography enables the fabrication of large numbers of tiny individual thermocouples, while at the same time (poly-) silicon exhibits a high Seebeck coefficient [26]. To investigate the advantage of STT for the fabrication of these thermopiles, a test structure was fabricated consisting of 255 thermocouple pairs in series (Fig. 9). Each pair consisted of an aluminium and polysilicon leg (n-type, 250Ω/sq each with a width of 15 μm , a length of 1 mm and a spacing of 5 μm). To facilitate on-wafer testing the thermopile was provided with aluminium heaters and polysilicon temperature sensors. The thermopile structures were transferred to glass, as well as to 10 μm polyimide foil using the procedure presented in the previous section (Fig. 10).

Figure 11a shows the measured temperature difference across the thermopile as a function of heater power. The temperature difference for the reference device on silicon was

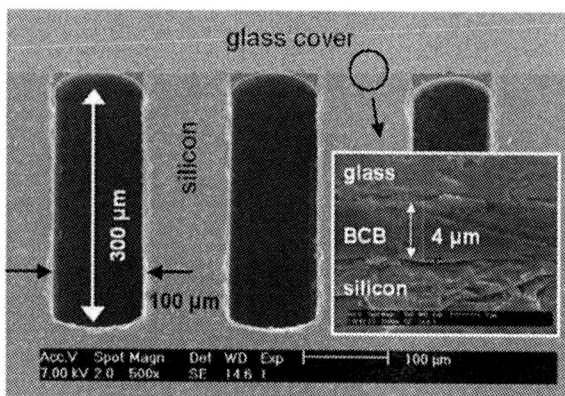

Figure 14: The sealing of micro-fluidic channels by using a BCB adhesive.

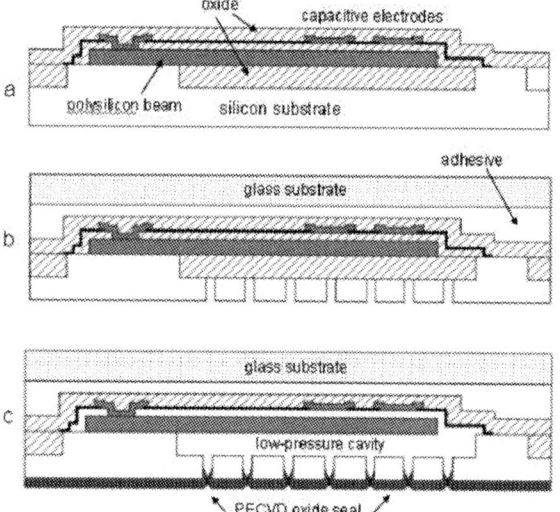

Figure 15: Schematic process sequence for the waferscale vacuum package based on substrate transfer.

so small that it could not be determined accurately with the polysilicon temperature sensors. From the slope of the thermopile output voltage as a function of temperature difference across the device on glass, the sensitivity was determined to be $200\ \mu V/K$ per couple. Figure 11b finally shows that by transferring the device to glass, and subsequently to foil an increase in conversion efficiency by 4 and 5 orders of magnitude respectively, can be obtained. A combination of the RF-ID tag discussed in the previous section and a flexible thermopile is at the moment under investigation for wireless in-vivo blood-flow sensing.

Figure 12 depicts an STT based microcalorimeter which is currently being investigated. The thermopile legs in this device have been designed as spokes in a wheel. In this case not all the silicon is removed. By using deep RIE etching a disc of silicon is left on top of the "hot" side of the thermopile and a doughnut shaped piece of silicon is left on the "cold" side.

Figure 16: A non-conformal deposition of PECVD oxide is used to vacuum seal MEMS resonator devices.

These pieces of silicon serve as thermal shunts to ensure that all junctions have the same temperature. It is the intention to use the microcalorimeter for the detection and measurement of concentrations of agents like glucose and cholesterol. This can be done by attaching enzymes to the central disk which react with these agents [27, 28]. The reaction of the agent with the enzyme will generate minute amounts of heat which can be detected by the sensitive thermopile.

VI. SUBSTRATE SCULPTURING

The micro-calorimeter device presented in the previous section illustrates how STT enables the possibility to "sculpture" the silicon substrate so that it becomes a functional part of the device. This feature can be used for the realization micro-fluidic devices. In Fig. 13 a schematic process flow for a Lab-on-Chip platform is sketched. Starting point is again a normally processed silicon wafer, containing devices such as heaters, temperature sensors, photodetectors, etc (Fig. 13a). After gluing of the wafer, the silicon is thinned to $100\ \mu m$ (Fig. 13b). Using deep RIE etching any arbitrary fluidic channel pattern may now be etched with channels as narrow as $5\ \mu m$ (Fig. 13c). To seal the channels a second glass substrate is spin-coated with a $10\ \mu m$ thick layer of non-photosensitive BCB, and pressed against the channel structures (Fig. 13d). During curing the soft baked BCB becomes liquid at 170 °C and finally cures at 200 °C thereby sealing the channels. The advantage of this procedure is that the channels remain open (Fig. 14). An elegant way to electrically contact the device is to combine it with the Shellcase waferscale packaging concept (Fig. 13e) [29, 30].

Figure 15 finally shows how STT is being used in the development of a waferscale vacuum package for MEMS devices [31]. The MEMS device is sandwiched between a glass wafer and the $50\ \mu m$ thick silicon substrate. The MEMS structure is released by HF (vapor) etching through narrow trenches in the silicon. The vacuum cavity is sealed with a non-conformal PECVD oxide (Fig. 16).

VII. DISCUSSION AND CONCLUSIONS

The examples in this paper illustrate how Substrate Transfer Technology (STT) can be used to overcome the limitations of silicon as a substrate while retaining the attractive properties of silicon.

It can be used to eliminate RF substrate losses, enable double sided device processing, fabricate flexible circuits and thermal sensors. It is based on a simple and straightforward adhesive bonding procedure using a UV-curing adhesive. The industrialization of the transfer process, and the qualification of transferred devices shows that STT is suitable for high-yield, high-volume production.

Nevertheless, STT has so far not been accepted by the process community and is still regarded as an "exotic" process step. This may be due to several reasons. First of all STT-based solutions are often not in line with existing roadmaps, and once roadmaps have been defined and development world-wide has gained momentum in a certain direction it is almost impossible to introduce alternatives. Secondly, the implications of STT - such as the possibility for double sided device processing - are so unconventional, that they require a different way of thinking with respect to process integration. It takes a specific mindset to use the possibilities of STT to their full advantage. Moreover, the semiconductor industry tends to be conservative. The investments involved and risks are enormous. As a result existing solutions and process steps are squeezed to the very limit before resorting to a revolutionary alternative.

Finally, it should be recognized that STT is balancing on the borderline between "normal" cleanroom processing and assembly. In fact neither one is adequately suitable for a process like STT. Standard cleanrooms are hesitant to process materials like glass or adhesives, while assembly centers are not equipped to process devices. It is highly plausible that in the future more and more System-in-Package solutions will enter this gray area between processing and assembly. The introduction of specialized System-in-Package foundries might fill the increasing gap between both disciplines.

References

[1] J.N. Burghartz, et al., "On the Design of RF Spiral Inductors on Silicon," *IEEE Transactions on Electron Devices*, vol. 50, no. 3, pp. 718-729, March 2003

[2] R. Dekker, "Silicon Process Technology Innovations for Low- Power RF Applications," in *Proc. European Solid-State Device Research Conf. (ESSDERC)*, 1998, pp.71-80

[3] A.W. van Herwaarden, et al., "Integrated Thermopile Sensors," *Sensors and Actuators*, A21-A23, pp. 621-630, 1989

[4] M. Usami, et al., "An SOI-Based 7.5 μm Thick 0.15x0.15 mm² RFID Chip," in Proc. IEEE International Solid-State Circuits Conf. (ISSCC), 2006, pp. 308-309

[5] L. Strüder, et al. "Backside Illuminated Pixel Sensors," *Optical Detectors for Astronomy II*, P. Amico and J.W. Beletic (eds.), Kluwer Academic Publishers, 2000, pp. 245-269

[6] R. Dekker, et al., "Substrate Transfer for RF Technologies," *IEEE Transactions on Electron Devices*, pp. 747-759, March 2003

[7] F. Niklaus, et al., "Low-Temperature Full Wafer Adhesive Bonding," *Journal of Micromechanics and Microengineering*, vol. 11, no. 2, pp. 100-107, 2001

[8] see: www.evgroup.com

[9] see: www.suss.com

[10] F. van Rijs, et al., "Influence of output impedance on power added efficiency of Si-bipolar power transistors," in *IEEE Int. Microwave Symposium Dig. MTT-S*, vol. 3, 2000, pp. 1945-1948

[11] R. Dekker, et al., "77% Power added efficiency surface-mounted bipolar transistors for low-voltage wireless applications," in *Proc. IEEE Bipolar/BiCMOS Circuits and Technology meeting (BCTM)*, 2000, pp. 191-194

[12] L.K. Nanver, et al., "Ultra-low- temperature low-ohmic contacts for SOA applications," in *Proc. IEEE Bipolar/BiCMOS Circuits and Technology Meeting (BCTM)*, 1999, pp. 137-140

[13] K. Buisman, et al., "High-Performance Varactor Diodes Integrated in a Silicon-on-Glass Technology," in *Proc. European Solid-State Device Research Conf. (ESSDERC)*, 2005, pp. 117-120

[14] H.W. van Zeijl, et al., "Front- to backwafer overlay accuracy in substrate transfer technologies," in *Proc. Int. Conf. on Semiconductor Technology*, Proc. vol. 2001-17, Editor M. Yang, The Electrochemical Society, pp. 356-367

[15] H.W. van Zeijl, "Bipolar Transistors with Self-Aligned Emitter-Base Metallization and Back-Wafer-Aligned Collector Contacts," PhD Thesis, Technical University Delft, ISBN 90-8559-045-0, 2005 (download: www.library.tudelft.nl > Resource Guide > TU Delft Publications > Search > Zeijl)

[16] L. Marinier, et al., "Front- to back-side overlay optimization after wafer bonding for 3D integration," in *Proc. MNE Conference*, Elsevier, 2005

[17] see: www.asml.com

[18] L.K. Nanver, et al., "A Back-wafer Contacted Silicon-On-Glass Integrated Bipolar Process, Part I - The Conflict Electrical versus Thermal Isolation," *IEEE Transactions on Electron Devices*, vol. 51, no. 1, pp. 42-50, Jan. 2004

[19] M.J.W. Rodwell, et al., "Submicron scaling of HBTs," *IEEE Transactions on Electron Devices*, vol. 48, no. 11, pp. 2606-2624, Nov. 2001

[20] N. Nenadović, et al., "RF Power Silicon-On-Glass VDMOSFETs," *IEEE Electron Device Letters*, vol. 25, no. 6, pp. 424-426, June 2004

[21] N. Nenadović, "Electrothermal Behavior of High-Frequency Silicon-On-Glass Transistors," PhD Thesis, Technical University Delft, ISBN 90-6464-114-5, 2004 (download: www.library.tudelft.nl > Resource Guide > TU Delft Publications > Search > Nenadovic)

[22] V.S. Sinnis, et al., "Resonant-cavity-enhanced photodiode using silicon-on-anything technology," in *Proc. European Solid State Device Research Conf. (ESSDERC)*, 1999, pp.524-527

[23] R. Dekker, et al., "A 10 μm Thick RF-ID Tag for Chip-in-Paper Applications," in *Proc. IEEE Bipolar/BiCMOS Circuits and Technology meeting (BCTM)*, 2005, pp.18-21

[24] R. Dekker, et al., "CIRCONFLEX: an ultra-thin and flexible technology for RF-ID tags," in *Proc. 15th European Microelectronics and Packaging Conference & Exhibition (IMAPS)*, June 12-15, Brugge Belgium, 2005, pp. 268-271

[25] A.W. van Herwaarden, et al., "Thermal Sensors based on the Seebeck effect," *Sensors and Actuators*, 10, pp. 21-346, 1986

[26] J. Schieferdecker, et al., "Infrared thermopile sensors with high sensitivity and very low temperature coefficient," *Sensors and Actuators*, A46-A47, pp. 422-427, 1995

[27] P. Bataillard, E. Steffgen, S. Haemmerli, A. Manz and H.M. Widmer, "An integrated silicon thermopile as biosensor for the thermal monitoring of glucose, urea and penicillin," *Biosensors & Bioelectron.*, 8 (2), pp. 89-90, 1993

[28] A.W. van Herwaarden, et al., "Liquid and gas micro-calorimeters for (bio)chemical measurements", *Sensors and Actuators*, A43, pp. 24-30, 1994

[29] A. Badidi, "Ultrathin Wafer Level Chip Size Package," *IEEE Transactions on Advanced Packaging*, vol. 23, no. 2, pp. 212-214, May 2000

[30] see: www.tessera.com

[31] M. Duemling, et al., "Properties of PECVD-Oxide Sealing layers in Vacuum Packages Utilizing Substrate Transfer," in *Proc. 4th European Microelectronics and Packaging Symposium*, Terme Čatež, Slovenia, 22-24 May 2006, pp.191-196

Above IC RF MEMS and BAW filters:
fact or fiction?

P. Ancey

STMicroelectronics, 850 rue Jean Monnet, 38926 Crolles, France, pascal.ancey@st.com

Abstract — In the Above IC integration, components are directly built monolithically upon or within the chip itself. The paper describes the works relative to Bulk Acoustic Wave filter and RF MEMS (micro-switch and electromechanical resonator). The paper demonstrates the feasibility of a fully-integrated RF Front-End using Above-IC BAW integration technique for W-CDMA applications. A MEMS micro-switch, processed on top of a standard BiCMOS wafer including the driver, is also reported here. Finally, the paper presents a new low voltage micromechanical resonator based on suspended-gate MOSFET operating at 16 MHz which have a great potential for CMOS co-integrated reference oscillator applications.

Index Terms — Bulk Acoustic Wave filter, RF micro-switch, RF-MEMS, electromechanical resonator, Resonant Suspended Gate MOSFET.

I. INTRODUCTION

In the race towards still higher integration of RF technologies, two main strategies are today competing. The first is System in Package (SiP) and aims at providing in a modular way all necessary components within the package. The current module Passive Device Integration (PDI) offers a promising platform for RF front-end integration. The second strategy is System on Chip (SoC) and constitutes the focus of this present paper. According to SoC integration, components are directly built monolithically upon or within the chip itself, and can be considered as options of a technology platform, as DRAM or imagers.

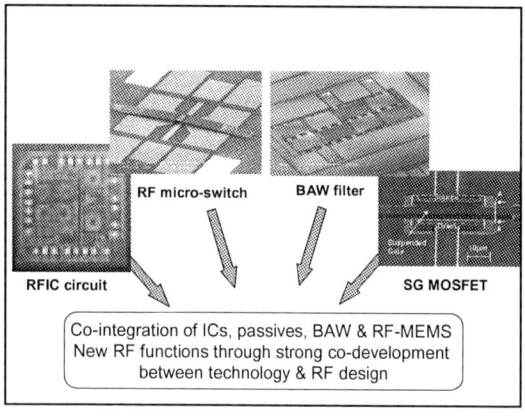

Fig. 1. Above IC integration of BAW & RF MEMS

The mobile phone industry which requires cheaper, smaller and integrated RF components has generated many developments in the field of radio frequency micro electro mechanical system (RF-MEMS). RFMEMS technology is relatively new, but has generated a tremendous amount of excitement because both the performances enhancement and the manufacturing cost reduction are evident characteristics of the technology. RF-MEMS devices include RF switches, tunable capacitors, high Q inductors, high Q electromechanical resonators and Bulk Acoustic Wave (BAW) filter. [1]

This paper will focus on three components (BAW filter, micro-switch, electromechanical resonator) and will demonstrate the interest of Above IC integration. SoC approach opens the door for innovative architectures & RF front-end.

II. RF FRONT-END USING AN ABOVE-IC BAW FILTER

Bulk Acoustic Wave (BAW) filters and duplexers are determined to replace conventional RF SAW filters in the mobile phone as they have reached excellent performances and can be manufactured in a very cost competitive way using standard IC manufacturing. BAW piezoelectric resonator are very promising devices for meeting the needs of modern wireless communication standards. They can be used advantageously in RF systems operating between 1 and 10 GHz because they feature Q factor much higher than on-chip LC tanks, a large power capability a low volume, a low cost and a reasonable coupling coefficient. These properties are all essential when portability is at stake, since they contribute to reduced power consumption and a greater compactness. Such resonators can be used to create passive high performances filters [2]-[4]. In comparison with surface acoustic wave (SAW) devices, BAW exhibit a lower frequency drift and a better power handling. Furthermore, BAW devices have a tremendous advantage over other technologies such as ceramic or surface acoustic wave filters, which is the possibility of being fabricated above integrated circuits since their materials and thermal budget are compatible with a post-processing approach. Such a co-integration can reduce even further the size and the cost of high performance RF front-ends, and opens the way toward a SoC radio.

1-4244-0458-4/06/$20.00 ©2006 IEEE

The European project MARTINA [5] demonstrated the feasibility of an above-IC BAW technology with a filter designed for the Rx chain of a WCDMA mobile phone. Filter was fabricated at the wafer level above 0.25μm BICMOS SiGe:C wafers, from ST BICMOS7RF technology. Film Bulk Acoustic Resonator (FBAR) architecture has been chosen, since it is the most appropriate choice for above-IC integration. Surface micro-machining can create an excellent isolation with a thin air gap under the resonator (Fig.2).

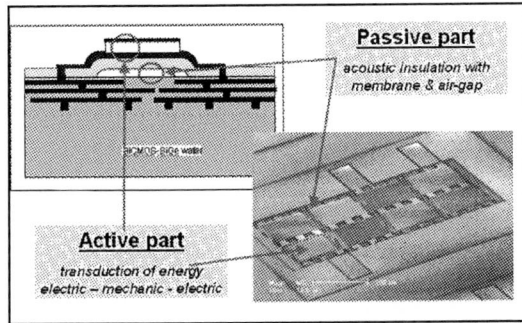

Fig. 2. Above IC integration of FBAR resonator & filter

The BAW technology is based on the piezoelectric thin film. The aluminum nitride (AlN) is grown on platinum (Pt) electrode, since the use of Pt promote the growth of AlN film with excellent piezoelectric properties. This material enables the fabrication of individual FBAR with a coupling coefficient k_{eff}^2 and a quality factor Q higher than 6.5% and 900, respectively. FBAR resonator and filters are post-processed over BiCMOS SiGe:C wafers. First, silicon oxide is deposited over the passivated wafers and planarized by CMP in order to get a smooth and flat surface. A polymer sacrificial layer is deposited and patterned for defining each resonator's position, and a dielectric encapsulation layer is applied to protect the polymer. Next, the active part of the devices is built up, with the subsequent deposition and patterning of a Pt bottom electrodes, the piezoelectric AlN layer and an Al top electrode. Via holes are then etched through the different dielectric layers until the last metal level of the integrated circuit (M5), and a thick Al film is deposited and patterned to create interconnections between the BAW filter and the IC. Finally, the sacrificial layer is etched for releasing the membranes.

The circuit proposed is a simplified implementation of a zero-IF front-end for the evaluation of the impact of an above-IC post-LNA filter on the system performances (Fig. 3). It is implemented in a SiGe:C BiCMOS process that exhibits very high RF performances on both actives and passives components [6]. From the system point of view, the single-ended LNA is followed by an on-chip single to

differential balun with moderate insertion loss to drive in the differential mode the post-LNA band pass filter at 2.14GHz.

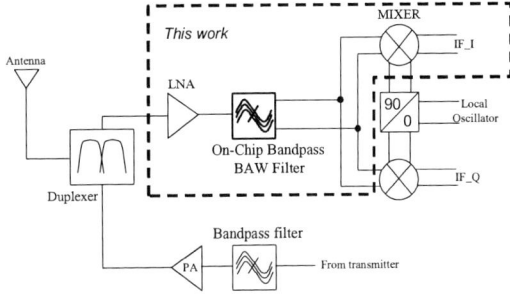

Fig. 3. Simplified block diagram of WCDMA-ZIF transceiver

The down conversion is built on a classical Gilbert fully differential mixer for IP2 improvement. The effective chip area for the complete front-end is 2.44mm², including the BAW filter (Fig.4).

In order to fulfill the filter specifications for a WCDMA receiver (60 MHz pass band, very low insertion loss and high rejection), a two stage lattice topology was selected. The filter shows good insertion loss of -3dB and excellent rejection is observed for all the out-of-band frequencies, with a value of -55dB at Tx central frequency (1.95GHz). This figure is about 10dB better than the most up-to-date SAW filter. Moreover, the constant 50dB wide-stop-band attenuation is very attractive for RF front-end mobile communication circuits.

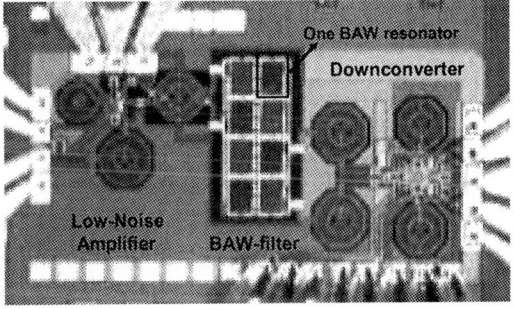

Fig. 4. BiCMOS and above IC BAW filter RF front-end micrograph

The demonstration of FBAR filter integrated between a LNA and a mixer has been demonstrated and presents the major advantage of the ultimate integration [5]. By reducing the number of external components, the mobile terminal bill-of-materials and board area are drastically reduced, resulting in a shorter product development cycle and cost. This FBAR technology enables also the design of new circuit such as filtering-LNA comprising two broadband amplifiers and a FBAR filter [7].

These results demonstrate that the FBAR technology is compatible with advanced BiCMOS technology, and that RF high-Q passive and active devices can be integrated on a single chip. The design of a fully integrated RF front-end for wideband CDMA applications, or of a filtering LNA, that both include a double lattice BAW filter integrated above-IC, are only two examples of the enormous potential of this technology.

We are also developing at STMicroelectronics and at CEA-LETI, another BAW technology, without surface micromachining, which offers excellent mechanical robustness, simplifies the packaging and is fully realized in 8" wafers with tools & processes compatible with semiconductor fab. This architecture is called Solidly Mounted Resonator (SMR) in which the acoustical insulation is realized by a Bragg mirror made of a set of quarter wavelength sections of several layers with alternating values of high and low acoustic impedance. The conventional Bragg materials (W, SiO2) [4] have showed limitation in the integration (patterning). We propose to replace them by materials such silicon nitride and organo-silicate glass (SiOC:H) showing simpler integration (no patterning) and already qualified for the interconnection of the advanced CMOS processes.

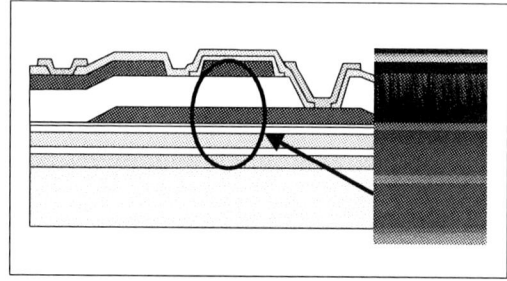

Fig. 5. SMR BAW technology

All the process steps of the SMR devices are realized on 8" wafers with qualified industrial equipment and a first W-CDMA duplexer has been realized & flip-chipped on PDI according a SiP integration. Investigation of the SoC integration of this SMR technology is currently under investigation: the validation of vias through the Bragg mirror will bring new perspectives for integration.

III. Above RF MEMS Micro-switch

Micro Electro Mechanical System (MEMS) are used in a wide spectrum of applications. Based on their physical properties, they may not only interface electronic circuit with their environment – where they are used as sensors and actuators – but also bring a major breakthrough in specific Radio Frequency (RF) transceivers blocks.

The paper focuses on the RF MEMS switch. Our objective was to develop a band switching solution for the 2 GHz signal. Since the switch was to be inserted in the RF signal path, a low insertion loss was required (0.5dB) for preserving a low global noise factor on one side and for limiting the consumption of amplifiers on the other. In the OFF state, an isolation of 40 dB was specified. Switching time and reliability was fixed at 300μs and 10^8 cycles [8].

The RF micro switch presented is this paper is an ohmic type (metal-to-metal contact) and the structural deformable part is made of a silicon nitride clamped beam. This beam includes titanium nitride heating resistors, electrostatic holding electrodes, and two aluminum blocks disposed at the fixed end of the beam. The RF lines and the contacts are made of a thick gold layer, and the air gap is defined by a polymer sacrificial layer. The actuation is based on the combination of thermal actuation and electrostatic latching.

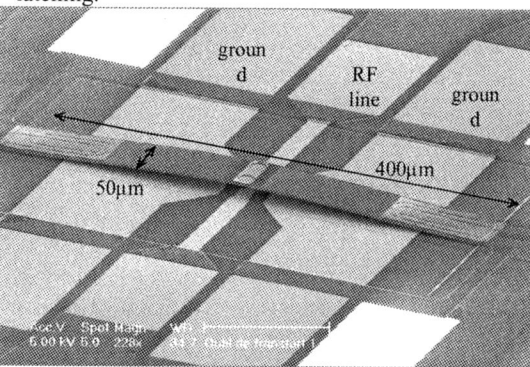

Fig. 6. RF micro switch micrograph

We demonstrated the possibility to realize the monolithic integration: the switch was processed on top of the 0.25μm BICMOS SiGe:C wafers, from ST BICMOS7 technology, including the driver and we demonstrated that the MEMS process has no influence on the IC performances.

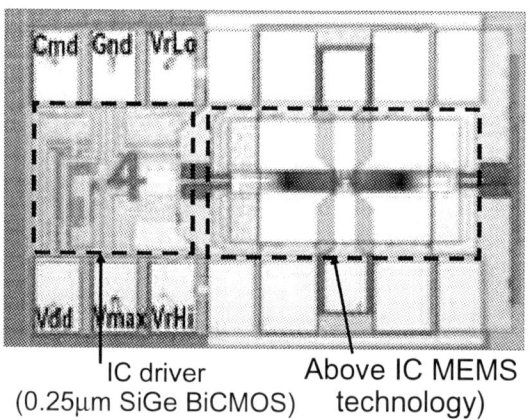

Fig. 7. Above IC integration of RF micro switch

188

The switch exhibits at 2GHz a 0.18dB insertion loss and a 57dB isolation level. The measured switching time was commonly 250µs when switching from the OFF state to the ON state, and 30µs when switching from the ON state to the OFF state. The device was switched over 10^9 cycles (cold switching) without failure. The targeted specifications were completely achieved and such device could be used within a matching network in order to address the reconfiguration of devices such as Power Amplifier.

IV. MEMS RESONATOR BASED ON RSG-MOSFET

MEMS resonators have been demonstrated to have a great potential for CMOS co-integrated reference oscillator applications, ranging from few MHz to 1.5 GHz with quality factors up to Q = 180 000. Among presented major drawbacks of current MEMS capacitive resonator are the very high DC voltage needed to actuate the structure, between 10V & 100V, and the low level of the output signal [9].

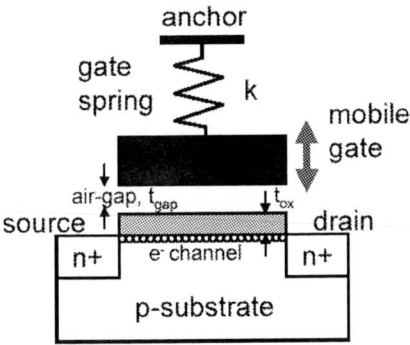

Fig. 8. Schematic of the RSG MOSFET including the mobile conductive gate over air/oxide/semiconductor

The paper describes a 16MHz micromechanical resonator based on the Resonant Suspended-Gate MOSFET (RSG-MOSFET) which brings low actuation voltage and larger output signal due to the transistor gain. The principle of the suspended-gate MOSFET is depicted in Fig. 8: a metal gate is suspended over the oxide/semiconductor channel of a MOS transistor by supporting arms (with equivalent k mechanical spring constant). By applying a voltage on the gate, a combined electro-mechanical action results: while the inversion channel is forming, the gate deflects vertically the gate-to-channel capacitance. This results in a super-exponential dependence of the inversion charge on the gate voltage in the sub threshold regime. Non-equilibrium between electric and elastic forces occurs at a value that is slightly large than 1/3 of the gap (compared to a pure metal-metal MEMS capacitor) because of the

in-series connection of the oxide and depletion capacitance.

The structure of the suspended-gate MOSFET is shown in Fig. 09. The resonator were designed and fabricated using the CMOS compatible process flow with an original metal-over channel MOSFET design. Active areas are etched through 100nm oxide and 200nm nitride isolation layers. A 40nm gate oxide is grown, and source & drain are phosphorous implanted. Polysilicon planarized by CMP serve as sacrificial layer (100 to 200nm) and AlSi layer (2µm) processed over the channel defines the suspended gate. The vibrating clamped-clamped beam acts as the suspended-gate of the MOSFET transistor (Fig. 9)

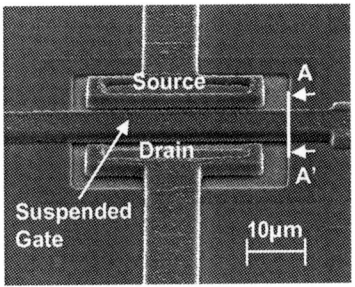

Fig. 9. SEM picture of a clamped-clamped beam SG-MOSFET with W=34µm and L=6µm

At the mechanical frequency which was designed at 16MHz, a maximum of impedance is measured in vacuum and an equivalent quality factor of 650 was extracted. The very low actuation DC voltage (below 1V) is compatible with low-voltage CMOS technology [10].

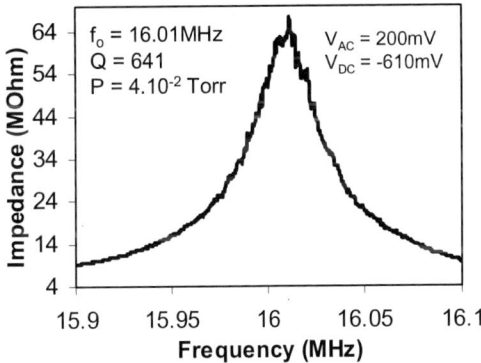

Fig. 10. Measured impedance spectrum of a 16.01MHz RSG-MOSFET under vacuum

We are also investigating at STMicroelectronics an innovative process in order to realize integrated thin mono-silicon membrane or beam. The Silicon On Nothing (SON) process, well known for advanced CMOS application, consists of an epitaxy of a mono-SiGe layer followed by an epitaxy of a mono-Si layer.

189

By an isotropic plasma etching with high selectivity, it is possible to realize mono-Si beam or disks [11]. The very good mechanical properties of mono-Si should allow reaching very high quality factor. A clamped-clamped flexural beam resonator can achieve a resonance frequency from 10 to 100MHz, depending on the structure dimensions, material and stress. Disk structures could be also investigated in order to reach GHz range in the future.

V. CONCLUSION

We presented in this paper innovative BAW filter & RF MEMS devices which show good performances when realized in above IC integration. SoC approaches offers a higher integration solution answering to the tremendous pressure to reduce the number of external passive components in transceivers and other communication circuits. Moreover, new emerging devices as RSG-MOSFET show promising results and a strong potential for a future convergence of silicon CMOS and MEMS technologies, bringing new functionalities and innovative architectures.

SOC integration of BAW and RF-MEMS is becoming a reality. Production stage can be expected in 3 or 4 years especially for BAW which are more mature. Meanwhile, SiP is rather to provide the immediate integration required for cellular terminals. At middle term, and once monolithic integration has been well developed, SoC will allow for a cheaper integration when present components will be proposed as options of standard technology.

ACKNOWLEDGEMENT

The author would like to kindly acknowledge the contribution of their colleagues at CEA-LETI & EPFL STMicroelectronics and European Commission for the funding of MARTINA IST-2001-37362 project.

REFERENCES

[1] J.J. Yao, "RF MEMS from a device perspective", *Journal of Micromechanical & Micro engineering*, vol. 10, pp. R9-R38, October 2000.

[2] K.M Lakin, "Thin film resonator and filter", *1999 IEEE Ultrasonics Symposium Dig.*, vol. 2, pp. 895-906, October 1999.

[3] R.C. Ruby, P. Bradley, J. Larson, Y. Oshmyansky, and D. Figueredo "Ultra-miniature high-Q filter and duplexers using FBAR technology", *2001 IEEE International Solid-State Circuit Conference Dig.*, pp. 120-121, February 2001.

[4] R. Aigner et al, "Bulk acoustic wave filter: performances optimization and volume manufacturing", *2003 IEEE MTT-S Int. Microwave Symp. Dig.*, pp. 2001-2004, June 2003.

[5] J-F Carpentier et al, "SiGe:C BiCMOS WCDMA Zero-IF RF Front-End Using An Above-IC BAW Filter", *2005 IEEE International Solid-state Circuits Conference Dig.*, pp. 394-395, February 2005

[6] P. Garcia et al, "Fully-integrated WCDMA direct conversion SiGe:C BiCMOS receiver", *2004 IEEE Bipolar/BiCMOS Circuits and Technology Meeting*, Montreal, Canada, 2004

[7] M-A Dubois, C. Billard, C. Muller, G. Parat, and P. Vincent, "Integration of high-Q BAW resonators and filters above IC", International Solid-state Circuits Conference, pp. 392-394, February 2005

[8] D. Saias et al, "An above IC MEMS RF switch", *IEEE Journal of Solid-State Circuits*, Vol.39, N°12, pp. 2318-2324, December 2003

[9] M.U. Demirci, C.T.-C Nguyen, "Single-resonator fourth-order micromechanical disk filter", *MEMS 2005 Dig.*, pp. 207-210, January 2005

[10] N. Abelé et al., "Suspended-Gate MOSFET: bringing new MEMS functionality into solid-state MOS transistor", *IEEE International Electron Devices Meeting (IEDM), Late News*, pp. 1075-1077, December 2005

[11] S. Monfray, Design & characterization of CMOS devices based on SON technology, PhD thesis, Nov 2003

2006 Bipolar/BoCMOS Circuits and Technology Meeting Proceedings

Lithography for the 32-nm Node and Beyond

J. N. Burghartz[a], M. Irmscher[a], F. Letzkus[a], J. Kretz[b], and D. Resnick[c]

[a] Institute for Microelectronics Stuttgart (IMS CHIPS); Allmandring 30a; 70569 Stuttgart; Germany;
Phone: +49 711 21855 200; FAX: +49 711 21855 222; E-Mail: burghartz@ims-chips.de;

[b] Qimonda Dresden GmbH & Co. OHG, Koenigsbruecker-Str. 180, 01099 Dresden; Germany;

[c] Molecular Imprints Inc., 1807C West Braker Lane, 78758 Austin, TX, USA.

Abstract — **This review article presents, discusses and compares three emerging photo lithography techniques for use in future BiCMOS fabrication processes: EUV lithography, e-beam direct write, and nano imprint. Specific challenges are discussed and the state-of-the-art is illustrated with respect to bipolar device scaling.**

Index Terms — **Silicon bipolar/BiCMOS process technology, bipolar scaling, photo lithography, EUV, E-beam, nano imprint.**

I. INTRODUCTION

Since Gordon Moore's famous 1965 prediction about the trends of miniaturization in silicon technology [1] this vision could be fulfilled up to this day with strong confidence in its continuation until at least 2020 [2]. In the history of microelectronics the two candidate devices, the MOS and the bipolar transistor, have been in a severe competition about the leading role, with CMOS now being the dominant device technology. The enabling factor for this dominance is the scaling principle of CMOS [3], which is not strictly transferable to a bipolar transistor due to the non-scalable turn-on voltage of the emitter-base diode [4]. Recently, however, a similar issue has arisen for CMOS since the threshold voltage of MOS cannot be reduced as enforced by the scaling rules. This may put the focus back on bipolar technology, and in particular SiGe HBT's, for which a given frequency limit can be reached at a comparably larger minimum feature size and at a higher supply voltage. Nevertheless, in radio-frequency (RF) applications, which is a field still dominated by bipolar [5], CMOS emerges as a serious contender as well [6].

CMOS performance has always been strictly depending on the capabilities of lithography in terms of minimum feature size and overlay. For the bipolar technologies of the 70's the situation was similar because back then transistor structures were non-self-aligned. With the invention of the self-aligned double-polysilicon bipolar transistor structure [7] the performance of bipolar transistors was greatly improved as far as device parasitics go, thus exposing the limitations in intrinsic doping profile design [8]. A change back to a lithography-dominated performance of bipolar technology came with the demonstration of the SiGe HBT [9], [10] in the late 80's and even more so with the SiGeC HBT in the mid-90's [11]. With these advancements in the intrinsic doping profile design cut-off and maximum oscillation frequencies of bipolar transistors could rapidly be elevated to about 200 GHz [12], [13], [14]. Recent improvements of the transistor structure, particularly the use of a raised extrinsic base [10], [15], [16], and further advancements in photolithography have ramped up the bipolar performance figures to 350 GHz. Since in the development of current BiCMOS technologies the CMOS part, and thus the lithographic capability, is typically adopted from the preceding CMOS generation, it can be expected that SiGeC HBT's may even reach performance levels of around 500 GHz, if the most advanced lithographic processes can be employed [17].

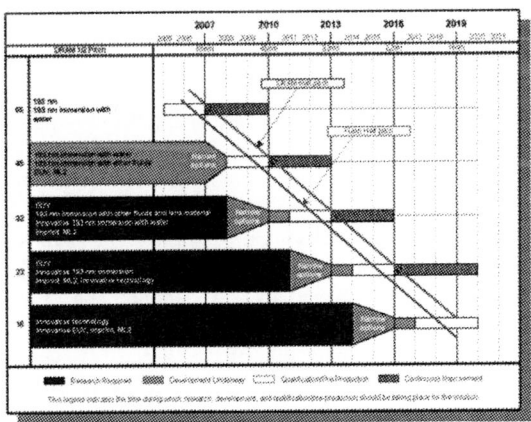

Fig. 1   Illustration of research, development and pre-production for the lithography options considered on the 2005 ITRS roadmap.

In this paper we will therefore evaluate the status of three photolithographic techniques that give promise to considerably further advancing the art to the 32 nm node and beyond. The techniques we address here are deep-EUV, e-beam direct write, and nano imprint, each representing a distinct set of advantages and challenges in the prototyping and manufacturing phases of the technology life cycle. They are all in the category 'next generation lithography' that has the potential to break the 100-nm barrier in IC manufacturing.

Fig.1 illustrates, at which DRAM half-pitch node these techniques will likely be introduced in the research, development and pre-production phases according to the International Technology Roadmap for Semiconductors (ITRS) [2]. It should be noted though that e-beam direct write will have to be brought to next generation multi-beam systems (ML2) in order to be considered. Nevertheless, efforts in e-beam direct writing remain relevant as long as coordinated progress in the development of ML2 systems is underway.

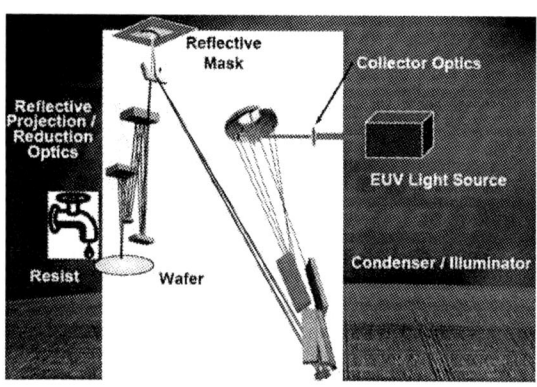

Fig. 2   Concept of EUV lithography

II. EUV LITHOGRAPHY

EUV lithography, in contrast to the classical transmission based optical i-line or DUV lithography, is running in reflection. The basic configuration of an EUV system is shown in Fig. 2. The 13.4 nm exposure wavelength from an EUV source [18] is collected by a condensor mirror. The light beam of the condensor illuminates a reflective mask and the mask pattern is imaged by a reduction lens optics comprised of several mirrors onto the resist coated wafer. Since the foundation of the EUV LLC [19] in the US in 1997 serious research activities have been started worldwide in the field of tool, source, optics and mask blank related issues to identify possible showstoppers of this challenging technology for the semiconductor industry.

One key component of EUV lithography is the reflective mask (Fig. 3). The EUV radiation is reflected by a stack of 40 Mo/Si bilayers, which serve as a bragg mirror at this given wavelength. The mirror is covered by a thin Si or Ru capping layer for environmental protection [20] followed by a buffer layer for repair purposes and an absorber layer that absorbs the EUV radiation during exposure. Since the reflectivity of the EUV radiation at an incident angle of 6° is limited to ~ 68% the thermal input into the mask is not negligible. LTEM (Low Thermal Expansion Material) substrates with thermal expansion coefficients well below 10 ppb/K have to be used to avoid any pattern distortion during exposure.

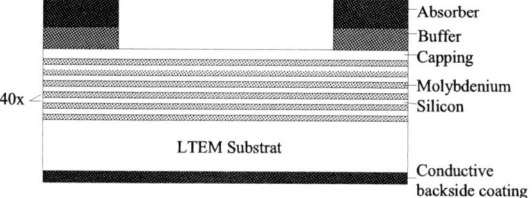

Fig. 3   Cross section of a patterned EUV mask

In 2001 a MEDEA+ project "EXTUMASK" was started in Europe with the goal to establish a production infrastructure for experimental EUV masks down to the 45 nm half-pitch node and beyond. The project was first headed by the Infineon mask house in Munich and later on by the AMTC (Advanced Mask Technology Centre) in Dresden. IMS Chips in Stuttgart was taking care of the development of a complete patterning process with tight CD uniformity and linearity specs for final EUV test masks. This included the e-beam lithography of complex mask pattern in the 100 nm feature size range with chemically amplified resists [21], dry etching of new Ta based absorbers [22] and etching of different buffer materials like SiO_2 or Cr [23]. The developed single processes have been integrated into a full mask flow process and finally applied for the

Fig. 4   ASML α-tool setup mask

192

fabrication of exposure test masks for printing experiments at the small-field Micro Exposure Tool (MET) in Berkeley [24] and the first full field setup mask for the ASML α-tool [25, 26]. Fig. 4 shows a picture of this setup mask. The mask pattern defined an area of 141x141 mm^2 with an opening density of ~ 6 %. 100 nm lines/spaces. Isolated lines in x- and y-direction were realized with the entire mask patterning process completed. SEM top down pictures of the final mask features are shown in the Fig. 5. First printing results of the α-tool with this mask have been published [27]. Dense 35 nm lines have been realized showing the potential of the tool and the mask technology.

Fig. 6 Leica SB350 E-beam direct write tool

Fig. 5 100 nm dense and isolated lines after TaN absorber and SiO$_2$ buffer etch

III. E-BEAM DIRECT WRITE SYSTEMS

E-beam direct writing of critical layers on silicon wafers is currently discussed as an option for early device and technology development as well as for fast prototyping. Due to the tremendously increasing mask costs for advanced lithography nodes, prototyping and small volume applications using optical steppers become more and more expensive and time-consuming. In addition, lead time of the mask set reduces the technological margin. As an alternative e-beam direct write allows the overall cycle time of circuit designs to be drastically shortened. For logic devices this approach has been successfully demonstrated recently by integration of variable shaped beam (VSB) e-beam direct writing into productive optical lithography [28]. The ITRS predicts the 32 nm node to be realized in 2013. For early device and technology development, using this maskless approach, a 32nm e-beam technology should be available in 2008-2010 (Fig. 6).

Due to throughput and stability requirements, only resists with a high sensitivity are applicable. Chemically amplified resists (CARs) require doses up to two orders of magnitude lower compared to the currently applied non-chemically amplified alternatives (e.g. PMMA). In order to cover all applications for different layers positive and negative tone resists are necessary. Therefore, the availability of high-performance and high resolution resists is one of the most crucial issues and thereby thus the key for enabling this technology for future device generations. Unfortunately, there are only a few specifically designed e-beam resists commercially

available because of the small market opportunity. The current challenges are to identify e-beam sensitive pCARs and nCARs, as well as appropriate process conditions for the direct write at the 50 nm node and beyond by, e.g., using the Leica variable shaped beam writer SB350. Clearly, e-beam direct write will have to be expanded to the ML2 concept in order to make it a viable contender of EUV and nano imprint lithographies.

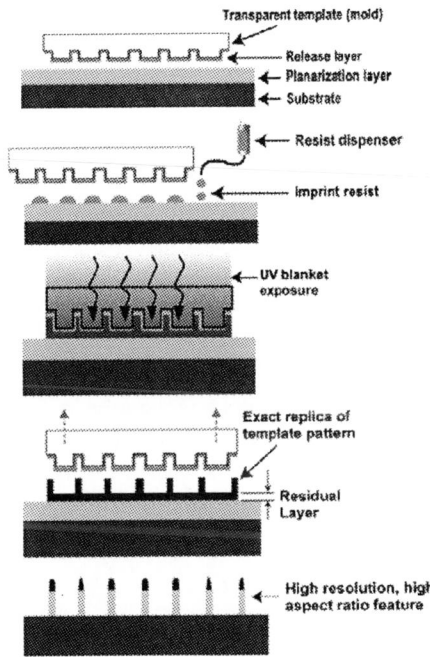

Fig. 7 The S-FIL process steps

IV. NANO IMPRINT TECHNOLOGY

Escalating costs of next generation lithographies with complex sources and optics drives the search for cost-effective sub-100 nm alternative printing techniques. Nano imprint lithography or specifically "Step and flash imprint lithography (S-FIL)", as called by Molecular Imprints (Fig. 7), is one promising alternative among the emerging lithography technologies. During the last years imprint tools have been steadily developed and improved, such as the MII's "Imprio" series, and no showstopper could be identified so far.

Fig. 8 Imprinted 65nm grating covering the entire template area of 25 x 25 mm^2

The imprint lithography has now been included on the ITRS lithography roadmap at the 32 and 22 nm nodes. Step-and-flash imprint lithography is a unique method for printing sub-100 nm geometries [29, 30]. Relative to other imprinting processes, S-FIL has the advantage that the template is transparent, thereby facilitating conventional overlay techniques. S-FIL might provide sub-100 nm feature resolution fairly inexpensively compared to optical lithography. However, since imprint technology does not reduce dimensions the quartz template fabrication for the monomer printing appears presently to be the most critical part. The imprinting process exactly replicates the template feature onto the wafer. Parameters like resolution, uniformity, linearity, placement accuracy and defectivity are to a first order determined by the quartz template.

Accordingly, progress and acceptance as well as possible applications of the nano imprint technology depend on the performance and availability of related templates. At present, the realization of defect free quartz templates with a complex pattern of feature sizes down to 32 nm and below appears challenging. By employing Gaussian beam pattern generators, the targeted resolution can easily be realized. The throughput, however, is not yet acceptable. In addition, present tools suffer from placement inaccuracy. In contrast, variable shaped e-beam (VSB) writers, primarily used as photo mask pattern generators, provide the required placement accuracy and throughput, but most of them cannot meet the ultimate resolution target. The Leica electron optical system implemented in VSB mask and direct writers of the SB3xxx series, however, enables an outstanding resolution of 50nm dense lines and below. At IMS Chips a nano imprint template technology has been developed by using a Leica SB350 writer as well as a state-of-the-art mask tool

set [31]. The template quality has been demonstrated by various imprinting tests at Molecular Imprints, Austin, TX, USA (Fig 8).

V. CONCLUSIONS

Obviously, a first thought may be that those sophisticated post-optical lithography techniques are applied to CMOS long before they are adopted for bipolar technologies; this, at least, is the situation we experience today. A second thought, however, may point to the need for bipolar to also exploit the most advanced lithography in order to cope with both CMOS and III-V RF technologies in the future. This is, because, on the one hand, there is a clear chance that CMOS will grasp much of the consumer communication markets, in which SiGe bipolar technologies still are viable today. Only, if bipolar can follow to the nodes far below 90 nm, it can show its superiority in providing ultra-high speed at adequately high supply voltages and thus sufficiently high device impedances. On the other hand, the comparably lower breakdown voltage of SiGe in comparison to III-V semiconductor technologies may also require the exploitation of the most advanced lithography at the critical mask levels in order overcome that disadvantage by a considerably higher operating frequency. With this possible scenario in mind it is advisable for bipolar process developers to gain early experience with those post-optical lithography techniques.

ACKNOWLEDGEMENT

The authors wish to acknowledge the support of the coworkers of the Lithography Department at IMS CHIPS who are not part of the author list. The EUV work has been supported by MEDEA$^+$, the German Federal Ministry for Education and Research, the Ministry of Economic Affairs of Baden-Wuerttemberg. The AMTC is a joint venture of AMD, Inc., Infineon Technologies AG and Toppan Photomasks, Inc. The authors alone are responsible for the content.

REFERENCES

[1] G.E. Moore, "Cramming more components onto integrated circuits", *Electronics*, Vol. 38, No. 8, 1965.
[2] http://www.itrs.net/Common/2005ITRS.
[3] R.H. Dennard, F.H. Gaennsslen, H.N. Yu, V.L. Rideout, E. Bassous, A.R. LeBlanc, "Design of implanted MOSFETs with very small physical dimensions", *IEEE J. Sol. St. Circ.*, Vol. 9, pp. 256-268, 1974.

[4] P.M. Solomon and D.D. Tang, "Bipolar circuit scaling", *Dig. Int. Sol. St. Circ. Conf. (ISSCC)*, pp. 86-87, 1979.

[5] L.E. Larson, "Integrated circuit technology options for RF-ICs – present status and future directions", *IEEE J. Sol. St. Circ.*, Vo. 33, No. 3, pp. 387-399, 1998.

[6] J.N. Burghartz, M. Hargrove, C. Webster, R. Groves, M. Keene, K. Jenkins, D. Edelstein, R. Logan, E. Nowak , "RF potential of a 0.18-μm CMOS logic technology", *Techn. Dig. IEEE Internat. El. Dev. Mtg (IEDM)*, 5-8 December 1999, pp. 853-856, 1999.

[7] T.H. Ning et al., "Self-aligned bipolar transistor for high performance and low power delay VLSI", *IEEE Trans. El. Dev.*, Vol. 28, pp. 1010-1013, 1981.

[8] G. Zimmer, B. Höfflinger, B. Schneider, „A fully-implanted NMOS, CMOS, bipolar technology for VLSI of analog-digital systems", *IEEE J. Sol. St. Circ.*, Vol. 14, No. 4, pp. 312-318, 1979.

[9] G.L. Patton, J.H. Comfort, B.S. Meyerson, E.F. Crabbe, G.J. Scilla, E. de Fresart, J.M.C. Stork, J.Y.-C. Sun, D.L. Harame, J.N. Burghartz, "75 GHz f-sub-T SiGe-Base Heterojunction Bipolar Transistors", *IEEE Electron Device Letters*, vol. 11, no. 4, pp. 171-173, 1990.

[10] J.N. Burghartz, J.H. Comfort, G.L. Patton, J.D. Cressler, B.S. Meyerson, B.S. Meyerson, J.Y.-C. Sun, G. Scilla, J. Warnock, "Sub-30 ps ECL Circuits Using High-f-sub-T Si and SiGe Epitaxial Base Transistors, *Technical Digest of the IEEE Internat. El. Dev. Mtg (IEDM)*, pp. 297-300, 1990.

[11] H.J. Osten, D. Knoll, B. Heinemann, G. Lippert, G. Schley, „Increasing process margin in SiGe heterobipolar technology by adding carbon", *IEEE Trans. El. Dev*, Vol. 46, pp. 1910-1916, 1999.

[12] A.J. Joseph, D.L. Harame, B. Jagannathan, D. Coolbaugh, D. Ahlgren, J. Magerlein, L. Lanzerotti, N. Feilchenfeld, S. St.Onge, J. Dunn, E. Nowak, "Status and direction of communication technologies – SiGe BiCMOS and RF CMOS", *IEEE Proceedings*, Vol. 93, No. 9, pp. 1539-1558, 2005.

[13] B. Heinemann; R. Barth; D. Bolze; J. Drews; P. Formanek; O. Fursenko; M. Glante; K. Glowatzki; A. Gregor; U. Haak; W. Hoppner; D. Knoll; R. Kurps; S. Marschmeyer; S. Orlowski; H. Rucker; P. Schley; D. Schmidt; R. Scholz; W. Winkler; Y. Yamamoto; "A complementary BiCMOS technology with high speed npn and pnp SiGe:C HBTs", *Techn. Dig. IEEE Internat. El. Dev. Mtg.*, pp. 521-524, 2003.

[14] M. Racanelli and P. Kempf, "SiGe BiCMOS technology for RF circuit applications", *IEEE Trans. El. Dev.*, Vol. 52, No. 7, pp. 1259-1270, 2005.

[15] J.N. Burghartz, J.D. Cressler, K.A. Jenkins, J.Y.-C. Sun, J.M.C. Stork, J.H. Comfort, T.A. Brunner, C.L. Stanis, "Device Design Issues for a High-Performance Bipolar Technology with Si or SiGe Epitaxial Base", *Proc. European Solid-State Dev. Res. Conf. (ESSDERC)*, pp. 11-14, 1991.

[16] J.-S. Rieh, B. Jagannathan, H. Chen, K.T. Schoenenberg, D. Agnell, A. Chinthakindi, J. Florkey, F. Golan, D. Greeberg, S.-J. Jeng, M. Khater, F. Pagette, C. Schnabel, P. Smith, A. Stricker, K. Vaed, R. Volant, D. Ahlgren, D. Freeman, K. Stein and S. Subbanna, "SiGe-HBT's with cut-off frequency of 350 GHz", *Dig. IEEE Internat. El. Dev. Mtg.*, pp. 771-774, 2002.

[17] J.W. Slotboom, "The Shrinking Bipolar Transistor", *IEEE Bipolar/BiCMOS Circ. Techn. Mtg.*, Keynote Speech, 2003.

[18] T. Tomie, T. Aota, Y. Ueno, G. Nimi, H. Yashiro, J. Lin, I. Matshusima, K. Komiyama, D. Lee, K. Nishigori, H. Yokota, "Use of tin as a plasma source material for high conversion efficiency", *Proc. SPIE*, Vol. 5037, pp. 147-155, 2003.

[19] Industry consortium with the member companies: Intel, Motorola, AMD, Micron Technologies, Infineon Technologies , IBM.

[20] P. Y. Yan, G. Zhang, S. Chegwidden, E. Spiller, P. Mirkarimi, "EUVL masks with Ru ML capping," Photomask Technology and Management, *Proc. SPIE*, Vol. 5256, p. 1281, 2003.

[21] J. Butschke, D. Beyer, C. Constantine, P. Dress, P. Hudek, M. Irmscher, C. Koepernick, C. Krauss, J. Plumhoff, P. Voehringer, "90 nm mask making processes using the positive tone chemically amplified resist FEP171," *Proc. SPIE Int. Soc. Opt. Eng.*, Vol. 5256, p. 344, 2003

[22] F. Letzkus, J. Butschke, M. Irmscher, C. Koepernik, F. M. Kamm, J. Mathuni, J. P. Rau, G. Ruhl „Dry Etch Processes for the Fabrication of EUV Masks" *Microelectronic Engineering 73-74, pp. 282-288, 2004*

[23] F. Letzkus, C. Köpernik, J. Butschke, C. Holfeld , J. Mathuni, L. Aschke, F. Sobel „SiO2 Buffer Etch Process with a TaN Absorber for EUV Mask Fabrication", *Proc. of SPIE, Vol. 5567, pp. 1407-1416*, 2004.

[24] S. Schwarzl, F.-M. Kamm, S. Hirscher, K. Lowack, W.-D. Domke, M. Bender, S. Wurm, A. R. Pawloski, B. La Fontaine, C. Holfeld, U. Dersch, F. Letzkus, J. Butschke, "Comparison of EUV Mask Architectures by Process Window Analysis", *Proc of SPIE, Vol. 5751, pp 1107-1115, 2005*

[25] F. Letzkus, J. Butschke, M. Irmscher, H. Sailer, U. Dersch, C. Holfeld „EUVL Mask Manufacturing Technologies and Results", *Proc. of SPIE, Vol. 5992, 59922A, 2005*

[26] U. Dersch, R. Buettner, C. Chovino, T. Heins, H. Herguth, J. H. Peters, T. Rode, F. Letzkus, J. Butschke, M. Irmscher „Manufacturing of the first EUV full field scanner mask" to be published at *Proc. BACUS 2006*, Monterrey, USA, 2006.

[27] H. Meilin, H. Meijer, P. Kürz, N. Harned, „First Performance Results of the ASML Alpha Demo Tool", *Proc. SPIE Conf. on Microlithography*, February 19 – 24, San Jose, CA, USA, 2006.

[28] J. Todeschini, L. Pain, S. Manakli, B. Icard, V. Dejonghe, B. Minghetti, M. Jurdit, D. Henry, V. Wang "Electron beam direct write process development for sub 45nm CMOS manufacturing", *Proc. SPIE Conf. on Microlithography*, Santa Clara, CA, USA, 2005.

[29] M. Colburn, S. Johnson, M. Stewart, S. Damle, T. Bailey, et al., *Proc. SPIE Emerging Lithographic Technologies III*, p. 379, 1999.

[30] M. Colburn, T. Bailey, B.J.Choi, J.G.Ekerdt, S.V. Sreenivasan, C.J. Wilson, "Development and Advantages of Step-and-Flash Lithography", *Solid State Technology*, 67, July 2001.

[31] M. Irmscher, J. Butschke, G. Hess, C. Koepernik, F. Letzkus, M. Renno, H. Sailer, H. Schulz, A. Schwersenz, and E. Thompson, "NIL Template Manufacturing Using a Variable Shaped E-Beam Writer and a new pCAR, *Proc. SPIE Conf. on Microlithography*, Vol. 6151, p.615115, April 19 – 24, San Jose, CA, USA, 2006.

2006 Bipolar/BoCMOS Circuits and Technology Meeting Proceedings

Applications of SiGe Material for CMOS and Related Processing

S.Monfray[1], T.Skotnicki[1], P.Coronel[1], S.Harrison[2]

D.Chanemougame[1], F.Payet[1], D.Dutartre[1], A.Talbot[1], S.Borel[3]

[1])STMicroelectronics, [2])Philips Semiconductor, 850 rue J.Monnet, 38920 Crolles, France.
[3]) CEA-LETI,, rue des Martyrs, 38054 Grenoble, France
ph: +33 4 76 92 53 44, fax : +33 4 76 92 50 71
e-mail: stephane.monfray@st.com

ABSTRACT — Despite the fact that the main commercial heterostructure based on SiGe is the HBT, the CMOS R&D has also focused on the use of SiGe material for boosting the performances. SiGe alloys are indeed extensively used for their electrical properties (band gap engineering by adjusting the Ge content, carriers' mobility improvement...). New applications for CMOS are emerging that are based on the crystalline properties of the SiGe material: strained devices, high-mobility SiGe channel transistors or architectures where it is used as a sacrificial layer.

The selective SiGe/Si etching can be used in several advanced microelectronics architectures that require a thin single-crystal Si film isolated from the substrate by a cavity. The capability to perform sustained mono-Si areas over an empty tunnel opens a wide range of applications for this technique, in particular for the realization of localized single-gate fully depleted transistors, double-gate or multi-channel devices, MEMS in nanoscale dimensions ...

Index Terms — SiGe, Ge, Strain, heterostructures, selective SiGe etching, Silicon On Nothing, Fully depleted devices, Double Gate devices

Introduction

In order to maintain the circuit speed enhancement at the rate of 1.2 times per year or even higher, the field of CMOS devices is investigating the use of silicon heterostructures for boosting the performances. In particular, SiGe alloys are extensively used for their impact on the electrical properties of the device: for example, the band gap can be modulated by the adjustment of the Ge content in the conduction channel, or the drive current can be enhanced thanks to the carriers' mobility improvement. In consequence, with the progress in epitaxial growth, new applications are emerging that are based on the crystalline properties of the material: strained devices,

high-mobility SiGe channel transistors or architectures where it is used as a sacrificial layer.

The objective of this paper is to give an overview of the different applications of SiGe in the nanometer CMOS world. We will investigate in the first part how the SiGe is used to induce strain (biaxial and unixial) into Si-conduction devices, and we will also present the opportunity of using the SiGe itself as a conduction-channel material

From pure Si (x=0) to pure Ge (x=1), any $Si_{1-x}Ge_x$ alloy can be deposited on Si with the same crystal structure and the same lattice constant as the substrate (if it does not exceed the critical thickness for plastic relaxation). In consequence, the SiGe can be buried between mono-Si layers without modifying their conduction properties. In the second part, we will illustrate how this SiGe layer can be used as a sacrificial layer. In this approach, the selectivity to Si is crucial because both materials are exposed to the complete etching process whereas only one has to be removed.

We will describe here the mainstream of the selective SiGe etching and its applications. In particular, we will present the SON (Silicon-On-Nothing) process in the third part and its possible technological variants for multiple devices applications, such as single-gate fully-depleted devices and its extension to double and multiple gate transistors.

I/ SiGe for Heterostructure FETs

CMOS logic circuit speed is mainly determined by the current drive of the MOSFETs and the load capacitance. One booster used to enhance the drivability of the devices is the modification and the improvement of carrier transport properties of silicon by the use of strain.

1-4244-0458-4/06/$20.00 ©2006 IEEE   196

I.1/ Biaxial strained Si for CMOS devices

In a strained-Si MOSFET, the surface channel is formed at the gate oxide/Si interface. When a thin layer of Si is pseudomorphically grown on a relaxed SiGe thick layer (figure I-1), the lattice mismatch induces a biaxial tensile strain in this silicon layer. The theory predicts a preferential occupation of the two-fold valleys in the tensile layer, where the in-plane conduction mass is lower, and a suppression of the intervalley phonon scattering. In consequence, the electron mobility is increased, and this effect has been widely reported in the literature (see the figure I-2 as an example).

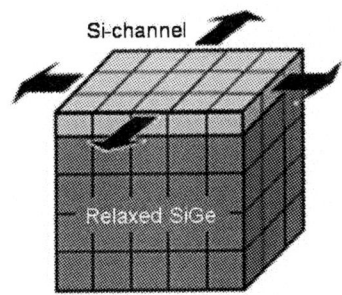

Fig.I-1:Tensile-strained Si on relaxed Si1-xGex

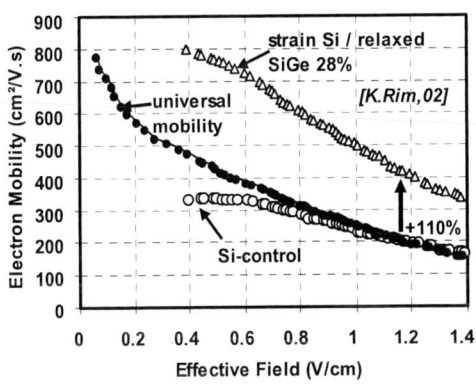

Fig.I-2: Effective electron inversion mobility versus Effective Field for different strained-Si MOSFETs [K.Rim,02]

Different groups have reported on the demonstration of short channel MOSFETs obtained on strained Si grown on a relaxed SiGe layer. The figure I-3 gives an example of a 42nm device fabricated on SiGe Relaxed Buffer (SRB) [Boeuf, 04]. SRBs consist in a step graded buffer, followed by a $Si_{0.8}Ge_{0.2}$ epitaxy for this demonstration. In this work, after a specific STI formation, a 150Å Si epitaxy was made to define the strained-Si channel. The figure I-4 demonstrates the NMOS device saturation current improvement as a function of gate length Lg. At same overdrive (open circles), the measured gain is ~15% for L=55nm. At same threshold voltage (black circles), the gain is found to be close to 100% for long channel devices but it diminishes with Lg. This degradation with L_{gate}

is fundamentally explained by the device physics (critical lateral field). In addition, because of the difference of conductivity between the SiGe layer and the Si layer, a self heating effect appears in static measurement that is responsible for a reduction of the saturation current by 5-10%. Even if this technology can make a direct impact on the Si-properties, some critical challenges have also to be taken into consideration for the integration: dislocation defects in epitaxy layers, thermal stability of the layers, dopants diffusion in SiGe…

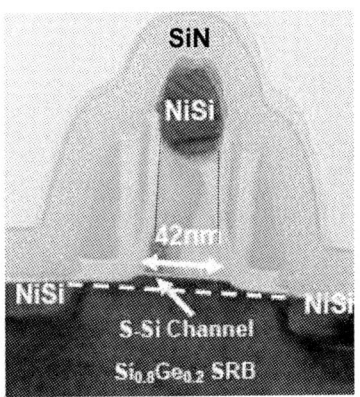

Fig.I-3: TEM cross section of a strained-Si MOSFET [F.Boeuf,04]

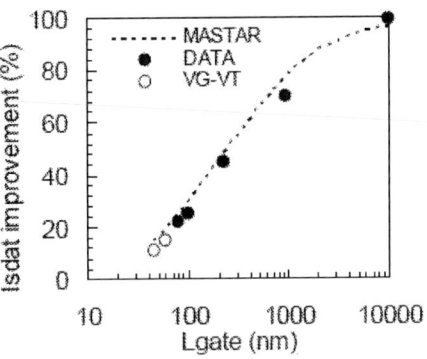

Fig.I-4: NMOS device saturation current improvement as a function of gate length Lgate [F.Boeuf,04]

I.2/ Uniaxial strained Si for PMOS devices

For PMOS devices, the hole mobility enhancement experimentally observed tends to reduce near zero at high vertical electric field for biaxial stress. However, the improvement is maintained for uniaxial compressive stress (figure I-5). This can be explained by the band curvature calculation: the relative magnitude of the out-of-plane light and heavy hole is more favorable at high vertical field for uniaxial stress (strain and surface confinement band splitting are becoming additive) compared to the biaxial case (strain effect is canceled by the confinement band splitting).

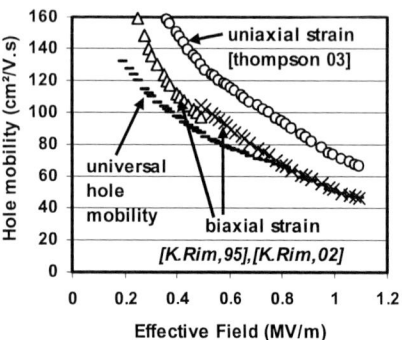

Fig.I-5: Hole mobility versus vertical electric field for uniaxial and biaxial stress [Thompson,03].

Different ways have been studied to introduce uniaxial strain in the device (stress capping liners deposition, STI impact, SiGe in SD areas...), and it seems that the heteroepitaxy approach offers the greatest potential to induce high level of stress. As shown in figure I-6, heteroepitaxy of $Si_{1-x}Ge_x$ creates uniaxial compressive strain in the Si channel in the direction of the current flow. This effect is induced by the mismatch in the lattice constant between the SD-region material and the Si-substrate.

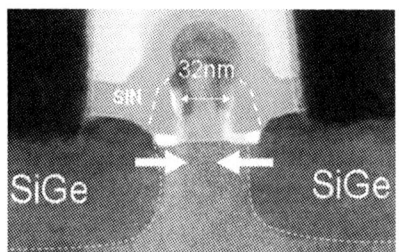

Fig.I-6: TEM cross section of 45nm devices with uniaxial compressive stress induced by the SiGe SD

The I_{on} vs I_{off} plot (figure I-7) gives an example of 20% improvement in the drive current induced by the SiGe SD on standard wafers compared to standard PMOS devices integrated on rotated substrates (with the transport along the <100> direction).

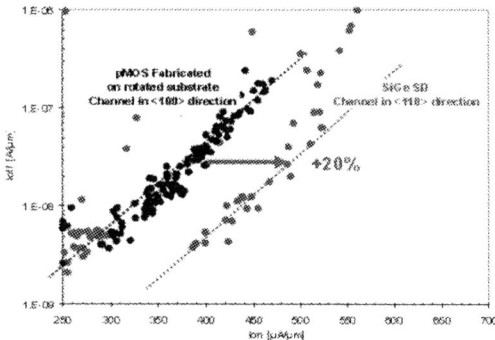

Fig.I-7: Ion vs Ioff for PMOS bulk devices integrated on rotated <100> channel (black) and conventional <110> channel with embedded SiGe S/D (red). 20% of improvement in the drive current is measured.

I.3/ SiGe-Channel MOSFETs

In parallel to the investigation of strained-Si channel devices, the intrinsic higher mobility of SiGe material itself always constitutes a promising solution for the mobility issue [Fischetti, 96]. Today, compressively strained SiGe (or strained Ge) channel appears to be one of the best candidate for hole mobility enhancement (figure I-8), and one might consider dual channel architectures for CMOS applications (tensile Si for NMOS and strained SiGe for PMOS, [M.L.Lee, 2003]). Using a SiGe channel with an appropriate high-K gate dielectric makes it possible to get rid of a historically used Si-cap and thus take full advantage of a surface channel operation [Weber, 05].

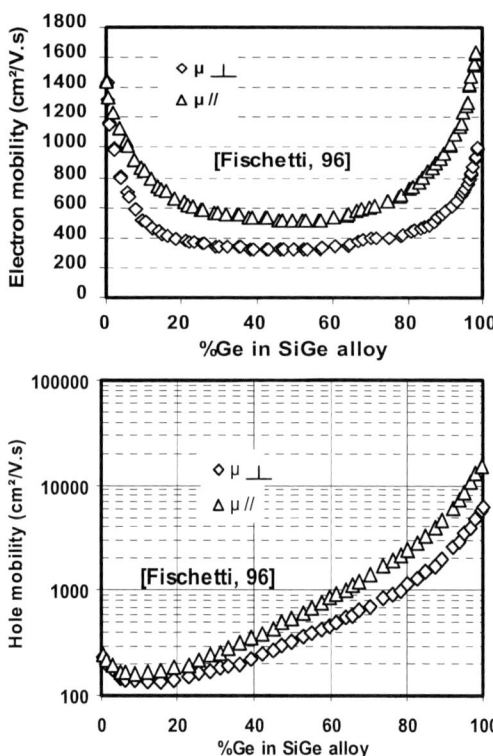

Fig.I-8: Calculated phonon-limited 300 K electron and hole low field mobility in $Si_{1-x}Ge_x$ alloys grown on (001) Si substrates, [Fischetti, 96].

In the examples given in figures I-9&I-10, the benefits of an optimized HfO_2/SiGe interfacial layer are exposed. The comparison of Hole mobility vs the Si cap thickness clearly shows that the interface quality (determined by the interface state density and the interface roughness) is greatly improved with a sacrificial silicon cap. Therefore, the integration of strained SiGe channel pMOSFETs with HfO_2 gate dielectric remains possible, but the entire potential of improvement due to the use of this high-mobility material strictly depends on its integration with a low defect gate dielectric.

198

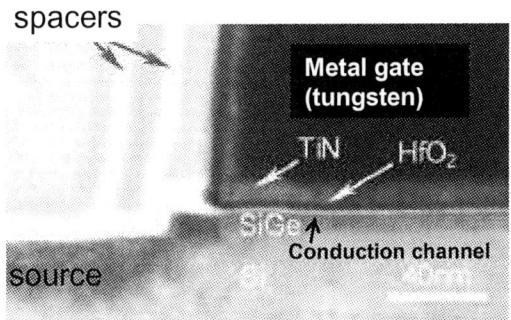

Fig.I-9:Exemple of SiGe/HfO2/TiN gate stacked-transistor [O.Weber,06]

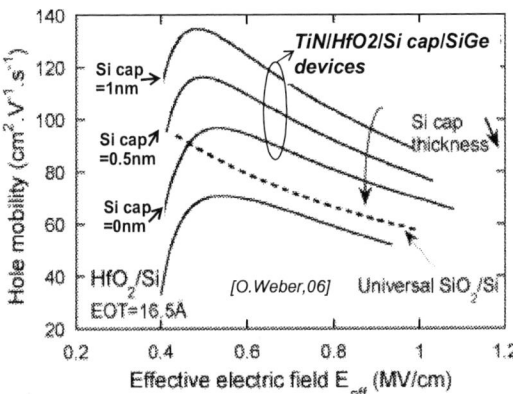

Fig.I-10:Examples of measured hole mobility for SiGe pMOSFETs with different Si cap thicknesses [O.Weber,06]

II/ SiGe as a sacrificial material: selective SiGe/Si etching

With the progress in epitaxial growth, a new application for SiGe is emerging, based on the crystalline properties of this material. Indeed, the SiGe can be buried between mono-Si layers without modifying their conduction properties. This method is used in several advanced microelectronics architectures that require a thin single-crystal Si-film isolated from the substrate by a cavity: the SiGe layer is used as a sacrificial layer for selective lateral etching. In this approach, the selectivity is crucial because both materials are exposed to the etching process whereas only one has to be removed.

Getting good selectivity between $Si_{1-x}Ge_x$ alloys and pure Si is challenging because of the similarity between the two materials. Indeed, they have the same diamond crystal structure, their lattice constants are not very different (<1% for $Si_{1-x}Ge_x$ with x <25) and their components (Si and Ge) are very close in term of reactivity to halogen-based chemistries (due to the same external electronic configuration).

In order to discriminate the Si and the $Si_{1-x}Ge_x$ in highly selective etching processes, the gases must contain elements which have a spontaneous affinity for the material to etch, and they must form volatile products with it [Borel, 04]. Silicon and Germanium are commonly etched with halogen-based chemistries because of the easy formation of volatile SiX_4 and GeX_4 (X = F, Cl, Br…) in Reactive Ion Etching. In the case of Chemical Dry Etching, there is no ion bombardment to assist the reaction, and all halogens are not of equal efficiency. Fluorine has a spontaneous action on Si and Ge, this is the reason why fluorinated compounds are the basic gases for Si or SiGe isotropic etching. The Figure II-1(a) shows an example of 350nm tunnel in a $Si_{70}Ge_{30}$ layer etched selectively to Si. Figure II-1(b) shows a "Tunnel depth"/Si ratio around 50, which means that when a 100nm-deep tunnel is etched in the SiGe layer, only 2nm of Si are consumed beneath and underneath the entrance of the tunnel.

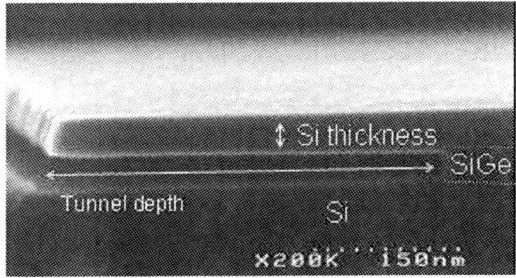

(a) S.E.M. cross-section micrograph of the tunnel

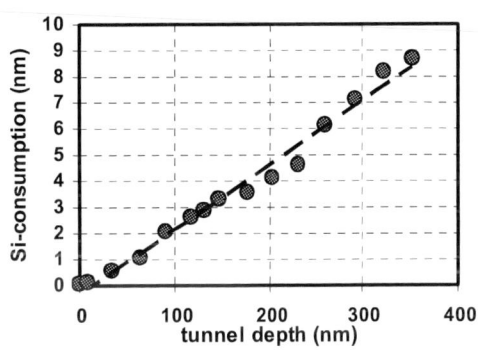

(b) Example of Si consumption all along the tunnel

Fig.II-1:Characterization of a dry-etching process for the tunnel etching

III/ Application of SiGe selective removal to thin-film devices: SON (Silicon-On-Nothing) technology

III.1/ Single gate devices

Nanometric films of mono-crystalline Silicon on insulator are extensively studied for their potential for end-of-roadmap CMOS transistors. In particular, it is

199

recognized that the best electrostatic control is obtained for the thinnest Si-film and buried oxide [C.Fenouillet-Beranger, 2004]. Different fabrication technologies (SOI, SIMOX, etc.) have been developed for continuous down scaling of the Si film and the buried dielectric, but the commercial availability of extremely thin films as 5-10nm still faces problems of feasibility and dispersion.

In this context, the use of a sacrificial buried-SiGe layer has been developed to overcome such limits, allowing for the fabrication of a nanometric-scale silicon film on a buried insulator [Monfray, 01,02,04]. The silicon-film and the buried dielectric thicknesses are both defined by the epitaxial process. The main advantage is that this process (named SON for "Silicon-On-Nothing") opens access to extremely thin films (the silicon channel and the buried insulator) at the same time offering the thickness control as fine as the resolution of the epitaxy process. In addition, a single bulk-type wafer is needed to fabricate the SON layers, in contrast to two wafers required in wafer-bonding-type techniques. Another advantage of the introduction of extremely thin fully depleted Si-films is their compatibility with single-metal gate (mid-gap) due to their intrinsically low threshold voltage [Monfray, 02].

The SON process can be easily used for fabricating thin film Fully-Depleted SOI-like devices. For this purpose, two selective epitaxial growth (SEG) steps are performed in the conventional CMOS flow just after the STI realization. The first SEG step provides a $Si_{70}Ge_{30}$ layer, and the second a Si-cap layer, selectively on active areas, and the process follows a conventional CMOS flow until the gate stack patterning and the spacers formation (figure III-1). Trenches are etched in active areas to give access to the buried SiGe layer that is then selectively etched thanks to the previously described removal process (figure III-2). The gate stack becomes sustained over the entire active area, giving the name (Silicon-On-Nothing) to the technology.

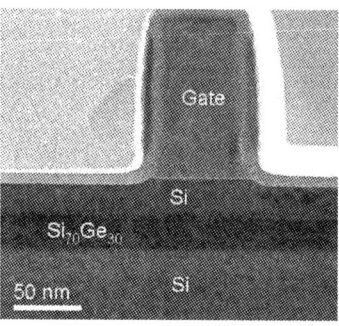

Fig.III-1: In the SON process flow, the gate stack is patterned on Si/SiGe selective epitaxial layers.

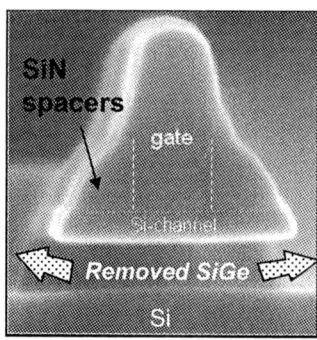

Fig.III-2: Trenches are opened in the Si SD areas to open the access to the buried SiGe layer that is then selectively removed.

The stability of the gate stack at this step is owed to its bridging across the active area supported by the STI. The tunnel is then cleaned and filled with a dual-dielectric layer structure that can be etched with isotropic plasma, everywhere except inside the tunnel. Finally, a second selective Si-epitaxy is grown in order to reunify the SD areas with the conduction channel (figure III-3a). Optionally, the SiGe can be removed directly from the borders of the active area, leading to a non rupture of the Si-film of the entire active area (this approach is named "Localized-SOI" on figure III-3(b)). This selective SiGe removal can also be performed below the SD areas only (see the figure III-3(c)).

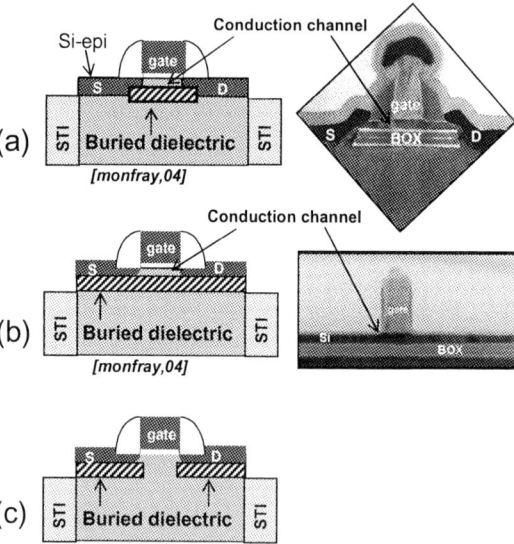

Fig.III-3(a): The tunnel is filled with dielectric and SD areas are cleaned before the selective epi-growth. Source and drain area becomes contacted with the conduction channel. [Monfray,04]
Fig.III-3(b): Concept of SON for localized SOI: the buried SiGe layer is etched from STI edges so that the buried dielectric becomes located below the entire active area.
Fig.III-3(c): in this approach, the SiGe is wet-etched only below the SD areas.[S.H.Kimg, 05]

III.2/ Gate All Around (GAA) and Multi-channel devices

Double (or multi-) gate structures are recognized for providing the best electrostatic integrity control for ultimate CMOS devices. These kinds of devices are also renowned for their improved transport efficiency. The selective SiGe-removal process may also used to ease the fabrication of such double (or multi-) gate devices [Harrison, 03], [Cerutti, 05], [Y.S.Lee, 05]. The main problems remaining to be solved are related to: (1) - the thickness control of the conduction channel, (2) - its surface roughness-, (3) – the top and bottom gate alignment.

In the first GAA example given in this section, the SON-based processes start with a SEG of SiGe on STI isolated wafers. The SiGe/Si stack is next patterned by RIE into a strip expanding above the active area and overlapping on STI (future S/D). Then, the buried SiGe is selectively removed, as in the process described before (see figure III-4). This future transistor channel is then covered with the gate dielectric and the gate material.

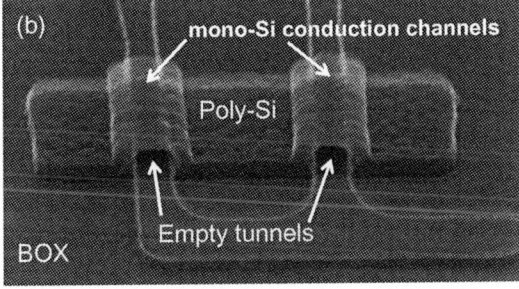

Fig.III-4(a): Sustained silicon conduction channel after the SiGe removal. [Harrison,03].
(b): Sustained silicon conduction channel in circuits areas.

The stack is then patterned into a gate in a masked RIE step. Highly performant devices with Lgate down to 40nm and Si-channel films of 16nm have also been achieved (figure III-5) with electrical results among the best ever reported. The process was recently extended to allow for top gate and the bottom gate to be the same size (figure III-6). Finally, with the use of multiple SiGe/Si epitaxial steps, multi-bridges can also be obtained after the SiGe removal and in consequence multi-channel devices can be implemented (figure III-7).

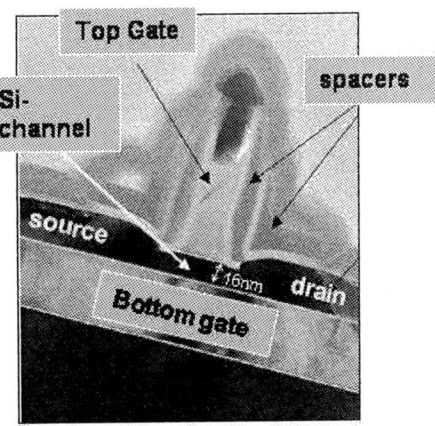

Fig.III-5: Along-channel TEM Cross section of a GAA device with a large overlapped bottom gate. Lgate is 40nm. The measured Si-film thickness is 16nm [Harrison,03].

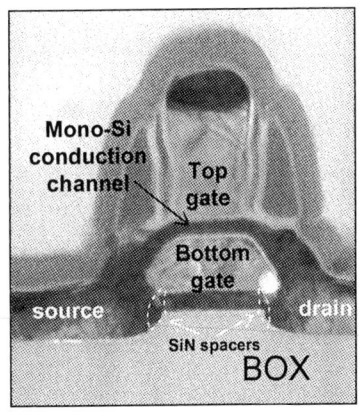

Fig.III-6: Cross section of a GAA device with a small overlapped bottom gate. [Cerutti,05].

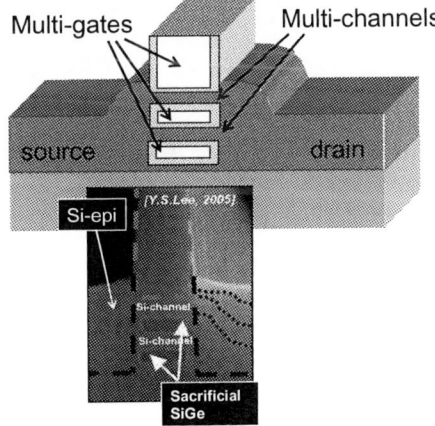

Fig.III-7: Cross section in a multi-channel device after the SD epi growth [Y.S.Lee, 2005].

IV/ Application of SiGe selective removal to NEMS (Nano-Electro-Mechanical Structures)

The fabrication of mono-silicon bridges over active areas can be adapted into electromechanical micro switch, where the flexure Si- beam can be obtained thank to a selective SiGe etching process. This opens new opportunities for such applications, for example the integration of nano-electro-mechanical structure at the front-end level (figure IV-1).

Considering the state of the art, the maximal vibration frequencies that can be obtained with flexure beam are in the range of 10 to 100 MHz. With the introduction of monocristalline Si beams with very small gap between the detection electrode and the flexure beam (figure IV-2), frequencies in the range of several GHz can be reached in theory. The possibilities that are proposed with this technology in this domain are in consequence of great interest.

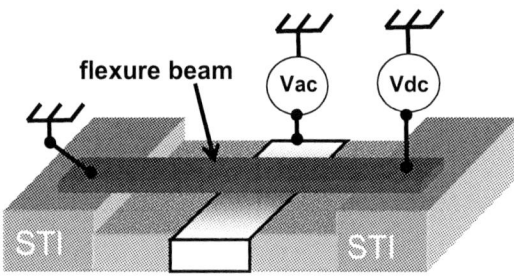

Fig.IV-1: -Electromechanical Micro-switch: the flexure beam is realized with thin monocristalline Si.

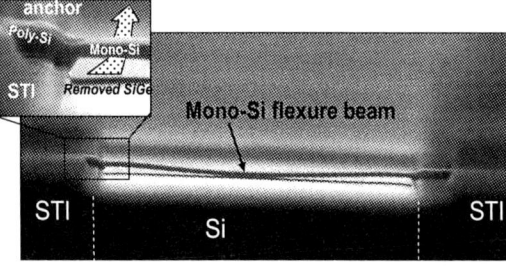

Fig.IV-2: -Example of morphological realization of a thin monocristalline flexure beam after the SiGe removal.

Summary

SiGe alloys are extensively used in the CMOS nanometer world for their wide range of applications such as high mobility conduction channels, uniaxial and biaxial stressors. We also demonstrated that this material can also be used as a sacrificial layer for a lateral etching selectively to silicon. This approach can be used in several advanced microelectronics architectures that require a thin single-crystal Si film isolated from the substrate by a cavity. The capability to perform sustained mono-Si areas over an empty tunnel opens a wide range of applications for this technique, in particular for the realization of localized single-gate fully depleted transistors, double-gate or multi-channel devices, MEMS in nanoscale dimensions … Today, the SiGe is considered as a fundamental material for highly performant logic applications, but also for new and innovative approaches, demonstrating in this way a large potential for future applications.

References

-K.Rim et al, Proc. Symp.VLSI Tech., pp. 98-101, 2002
-F.Boeuf et al, proceedings SSDM 2004
-Thomson et al, IEDM tech. Dig. p.978, 2003
-M.L.Lee et al, IEDM Tech.Dig, pp429-432, 2003
-Fischetti et al, J. Appl. Phys. 80 (4), 15 August 1996
-O.Weber et al, IEEE TED, VOL. 53, No. 3, March 2006
-C.Fenouillet-Beranger et al, SSE vol.48, pp961-967, 2004
-S.Borel et al, Japanese JAP, vol 43 (6B), pp 3964-3966, 2004.
-S.Borel et al,Microelectronic Engineering, vol 73-74, pp 301, 2004.
-S.Monfray et al, IEDM Tech.Dig, pp.645-648, 2001
-S.Monfray et al, IEDM Tech.Dig, pp.635-638, 2004
-S.Monfray et al, IEDM Tech.Dig., pp.263-266, 2002.
-S.H.Kim et al, proceedings of IEEE SOI conference, pp-174-175, 2005
-S.Harrison et al, IEDM tech. Dig. pp449-452, 2003
-Cerutti et al, proceeding of SNW, 2005
-S.Y.Lee et al, Proc. Symp.VLSI Tech, p.154, 2005

2006 Bipolar/BoCMOS Circuits and Technology Meeting Proceedings

Design and Modeling of mm-Wave Monolithic Transformers

Tak Shun D. Cheung and John R. Long

Electronics Research Laboratory/DIMES, Delft University of Technology, the Netherlands

Abstract — Monolithic transformers operating near self-resonance for mm-wave circuit applications are presented. The proposed transformer circuit models and analytical equations accurately capture impedance transformations predicted by electromagnetic simulation to within ±5% and power loss within ±0.5dB. These models enable circuit design with load impedances ranging from 50Ω to just a few Ohms.

Index Terms — silicon monolithic transformers, mm-wave passive modeling, self-shielded passive components.

I. INTRODUCTION

On-chip transmission lines, inductors and transformers are indispensable for system-on-a-chip implementation of matching networks, feedback loops, and resonant tanks in multi-stage amplifiers, mixers and oscillators operating at millimeter-wave (mm-wave) frequencies. These passives must have controlled impedance, good isolation and low losses to complement the high frequency performance of modern silicon transistors. Their application could potentially enable migration towards unlicensed (ISM) bands at 24 and 60GHz in order to reduce spectral congestion.

SiGe transistors with transit frequencies (f_T) over 70GHz facilitate the implementation of mm-wave circuits, however, breakdown is typically $V_{CEO} < 2V$. Common-base gain stages offer superior breakdown and bandwidth compared to common-emitter amplifiers, but they are rarely used because of the impedance transformation needed to interface a typical RF source (e.g., 50Ω) to a transistor input impedance of 2-5Ω (see Fig. 1). Recently, self-shielded interstage matching transformers have been demonstrated that minimize losses in the silicon substrate and due to the skin-effect at 24GHz [1]. These transformers are designed to maximize power transfer between windings by exploiting the thick top-metal layers available in modern BiCMOS technologies. However, electromagnetic simulation of the self-shielded transformer is time-consuming because a relatively large number of metal layers are required (i.e., compared to a spiral inductor). Therefore, accurate circuit simulation models that allow fast computation of the transformer's electrical characteristics are required for design.

In this paper, compact models suitable for mm-wave applications of n:1 silicon monolithic transformers are presented. Despite their simplicity, these models offer fast and accurate calculation of impedance transformation (±5% error) and power loss (±0.5dB error). They are applicable to loads ranging from 50Ω down to 1Ω, making the model suitable for use with

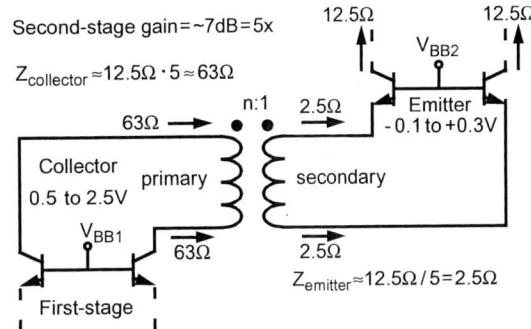

Fig. 1: Interstage matching with a step-down transformer.

low input impedance (e.g., common-base) amplifiers.

II. THE SELF-SHIELDED TRANSFORMER

A differential common-base amplifier (Fig. 1) with 7dB gain per stage at 24GHz is used as an example to illustrate transformer design and modeling. Fig. 2 shows a self-shielded 2:1 transformer that transforms the 2.5Ω second-stage emitters to 63Ω loads for the first gain stage. Self-shielding [2] is applied to the transformers in order to minimize losses in the silicon substrate. A low-voltage shield winding (secondary in Fig. 2) is configured so that it physically surrounds the winding carrying a relatively high voltage (e.g., primary winding in Fig. 2). The shield winding is connected to nodes in the RF circuit that have relatively low voltage and impedance levels, such as the emitters of the second gain stage in Fig. 1 (i.e., -0.1 ~ 0.3V), thereby minimizing the electric field coupled to the

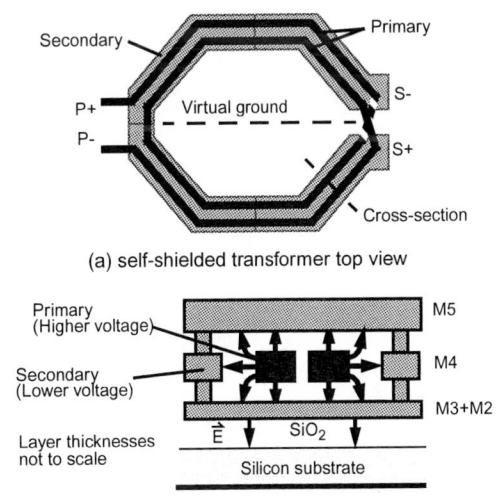

(a) self-shielded transformer top view

(b) cross-section of the self-shielded structure

Fig. 2: Self-shielded monolithic transformer.

substrate. In this way, the potential on the low-voltage winding is defined by the circuit, instead of relying on an explicit ground connection (as used in patterned ground shielding). The electric field coupling to the substrate is reduced without using a separate shield layer or ground reference. Self-shielded passive components are also well-isolated from each other, because there is no common ground connection.

III. COUPLED-LINE TRANSFORMER MODEL

In a 1:1 transformer, the primary and secondary form a pair of coupled transmission lines. For a 2:1 transformer (Fig. 3a), the equivalent half-circuit model of Fig. 3b uses two pairs of coupled-lines which are isolated from each other. The longer primary winding is represented by 2 coupled-line sections in series, while the secondary is realized as coupled-line sections in parallel. A phase inversion accounts for the connection to the other half of the transformer winding (absent in a half-circuit model) and is not needed in a full-circuit model. The model in Fig. 3b is reciprocal. Therefore, a 1:2 step-up transformer model can be constructed by swapping the primary and secondary terminals in the 2:1. Models for n:1 step-up or step-down ratios can also be constructed in a similar way.

Secondary terminal S- is connected to load Z_s. Mutual magnetic coupling between windings causes currents I_{odd} to flow in opposite directions on the windings, which have a characteristic impedance Z_o. The self-resonant frequency (beyond which coupling between the lines is mainly capacitive) is f_{SRo}.

In general, the two windings have unequal length and width. In such cases, flux leakage is modeled by the two grounded transmission line (i.e., short-circuited) stubs with self-resonant frequencies f_{SRp}, f_{SRs} and characteristic impedances Z_{op}, Z_{os}. Larger Z_{op}/Z_o and Z_{os}/Z_o ratios indicate greater magnetic coupling. Ohmic losses in the coupled-lines are modeled by DC loss coefficient α_{oo}, and α_{fo} models losses proportional to frequency. Similarly, α_{op}, α_{os} are the DC losses, and α_{fp}, α_{fs} are the frequency-dependent losses of the primary and secondary stubs.

The propagation coefficients of the coupled, primary and secondary stub transmission lines are γ_o, γ_p, γ_s, respectively, which are defined by Eq. 1. The impedances of the primary and secondary stubs, Z_{tp} and Z_{ts}, are given by Eq. 2. The impedance seen at the primary (Z_p) due to a secondary impedance (Z_s) is given by Eq. 3 and 4. The power loss (i.e., the ratio of transformer input power and output power) for a 2:1 transformer is given by Eq. 5 in the unit of dB.

$$\gamma_{o,p,s} = \alpha_{oo,p,s} + \alpha_{fo,p,s} \cdot f + \frac{j\pi}{2}\left(\frac{f}{f_{SRo,p,s}}\right) \quad (1)$$

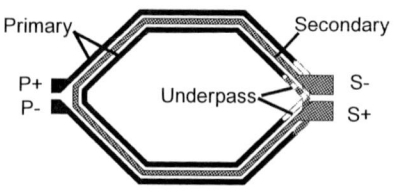

(a) 2:1 planar transformer layout

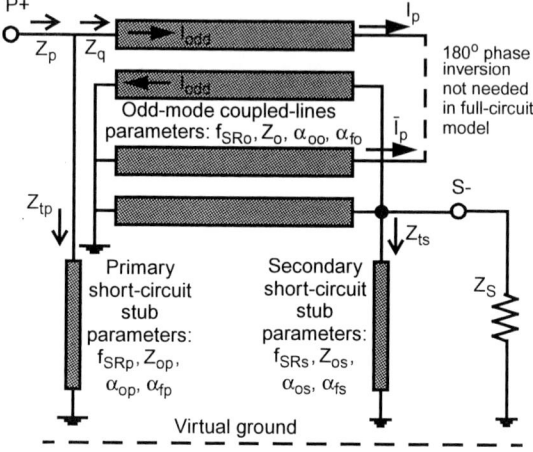

(b) 2:1 transformer half-circuit model

Fig. 3: Monolithic transformer with 2:1 turns-ratio.

$$Z_{tp,s} = Z_{op,s}\left(\frac{e^{\gamma_{p,s}}-1}{e^{\gamma_{p,s}}+1}\right) \quad (2)$$

$$Z_p = Z_q \,/\!/\, Z_{tp} \quad (3)$$

$$Z_q = \left(\frac{2bZ_o\left(a\left(e^{2\gamma_o}-1\right)Z_o + \left(e^{2\gamma_o}+1\right)(Z_s \,/\!/\, Z_{ts})\right)}{a\left(e^{4\gamma_o/a}+1\right)Z_o + \left(e^{4\gamma_o/a}-1\right)(Z_s \,/\!/\, Z_{ts})}\right) \quad (4)$$

where a=2, b=1 for a 1:1 transformer, and

a=1, b = $e^{2\gamma_o}+1$ for a 2:1 transformer.

$$Loss = 10\log\left(\left|\frac{\left(\left(e^{4\gamma_o}+1\right)Z_o + \left(e^{4\gamma_o}-1\right)(Z_s \,/\!/\, Z_{ts})\right)Z_s Z_q}{2e^{\gamma_o}\left(e^{2\gamma_o}+1\right)Z_o(Z_s \,/\!/\, Z_{ts})Z_p}\right|^2 \frac{Re(Z_p)}{Re(Z_s)}\right) \quad (5)$$

A lossless 1:1 transformer with Z_o of 25Ω and resonant frequency of 10GHz (i.e., $f_{SRo,p,s}$=10GHz) is used to illustrate typical transformation characteristics. For near perfect magnetic coupling, Z_{op} and Z_{os} are chosen as 1000·Z_o. The impedances seen at the primary from Eq. 3 for secondary loads (Z_s) of 25Ω and 100Ω are plotted in Fig. 4. In Region 1 (f<<0.001f_{SRo}), there is negligible coupling and the primary behaves like an inductor (i.e., ∠ Z_p≈90°). In Region 2 (0.01f_{SRo}<f<0.1f_{SRo}), transformer action is seen (i.e., Z_p∝Z_s). In Region 3 (near self-resonance), Z_p∝1/Z_s, which is behavior similar to a λ/4 transmission line. A 25Ω load is therefore transformed to 25Ω in Region 2 and 100Ω in Region 3 (despite the 1:1 turns ratio). Note that the rising impedance seen at the transformer primary at

204

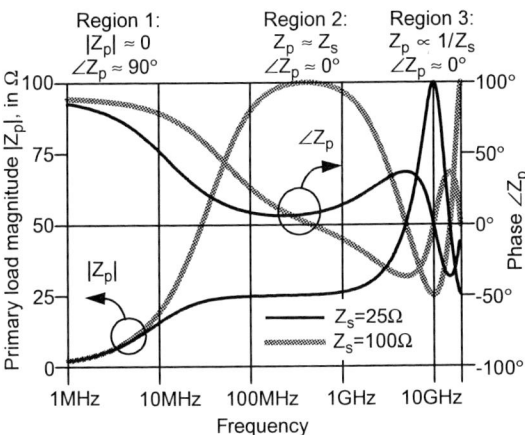

Fig. 4: Ideal 1:1 impedance transformation.

self-resonance can be used to equalize for transistor bandwidth limitations.

IV. EFFECTS OF IMPERFECT MAGNETIC COUPLING

Fig. 5 plots the primary impedance Z_p for a 1:1 transformer with imperfect magnetic coupling (i.e., $Z_{op}=Z_{os}=10 \cdot Z_o$). The secondary load Z_s is 25Ω, Z_o is 25Ω and the self-resonant frequencies are 10GHz as assumed in the previous example. The self-inductance dominates the primary impedance ($\angle Z_p > 60°$) in Region 2, causing Z_p to deviate from the ideal 1:1 transformation (ideally $Z_p=Z_s$). When the load is tuned by shunt capacitors to remove the inductive effect, the resulting load impedance $Z_{p\text{-tuned}}$, increases (see Eq. 6). However, the transformation in Region 3 is unaffected by lower Z_{op} and Z_{os}. In this example, $Z_{p\text{-tuned}}$ in Region 2 is increased by 44% (i.e., 36Ω vs. 25Ω) compared to Z_p for perfect magnetic coupling as seen in Fig. 4 for $Z_s=25\Omega$.

$$Z_{p-tuned} = \begin{cases} Z_s \left(\dfrac{Z_{op} + (Z_o // Z_{op})}{Z_{op} - (Z_o // Z_{op})} \right)^2 & \text{Region 2} \\[2em] \dfrac{(2Z_o)^2}{Z_s} & \text{Region 3} \end{cases} \quad (6)$$

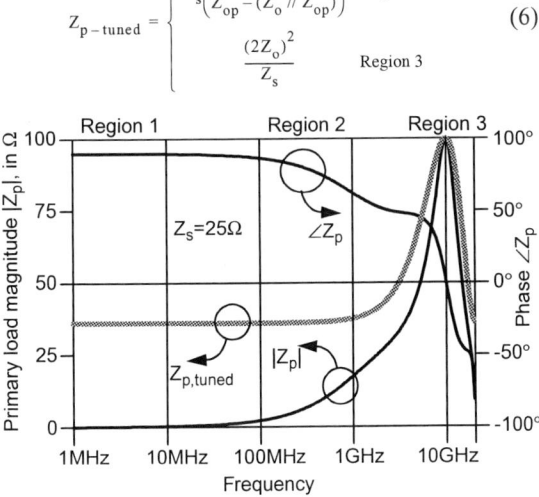

Fig. 5: 1:1 impedance transformation with leakage.

A transformer designed to operate in Region 3 maximizes the operating frequency (for a given winding length), and a longer winding (compared to that required for Region 2) can be used to increase the number of turns and thereby achieve higher magnetic coupling. By contrast, operating in Region 2 requires a large tuning capacitors to reduce the effects of leakage, resulting in some power loss and an unintended additional step-up of the impedance transformation.

V. MODEL COMPARISON

For comparison, a lumped-element equivalent based on the coupled-line model was developed. Fig. 6 shows one subsection of a lumped-element model. Six of these subsections are distributed in a cascade to form a half-circuit model of a 2:1 transformer with similar accuracy to the coupled-line model proposed in this paper. Thus, the advantages of the coupled-line model are simplicity, and a close link between physical parameters (α, Z_o, etc.), analytical expressions (Eqs. 1-5), and the actual model.

In this section, the coupled-line and lumped-element models are compared to electromagnetic (EM) simulations of a practical self-shielded transformer (Fig. 2a) with an outer-dimension of 270µm x 200µm. A 2.5-D EM simulator (Agilent Momentum) is used to compute the transformer s-parameters. The model parameters are extracted by fitting the impedance transformation and power loss to the models. The transformation for secondary loads (Z_s) of 1Ω, 2.5Ω, 10Ω, and 50Ω computed from Eq. (1) to (5) are shown in Fig. 7, whereas the lumped-element model and EM simulations are compared in Fig. 8. A 0.13pF shunt capacitor representing a typical output capacitance for a BJT transistor is added to the primary winding, but it has no effect on the extraction of the parameters. The self-shielded transformer matches a 2.5Ω to about 60Ω at 25GHz with 1.5dB loss. Both the coupled-line and lumped-element models exhibit excellent fits to the EM data over the entire 5-40GHz bandwidth, and also over a large range of secondary impedance, Z_s. It should be noted that the transformation of load impedance Z_s down to near 1Ω is very sensitive to Ohmic losses in the transformer windings

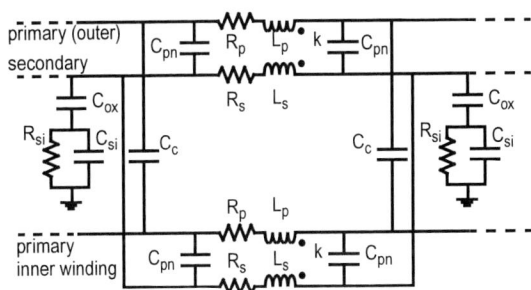

Fig. 6: Single lumped-element section for 2:1 transformer.

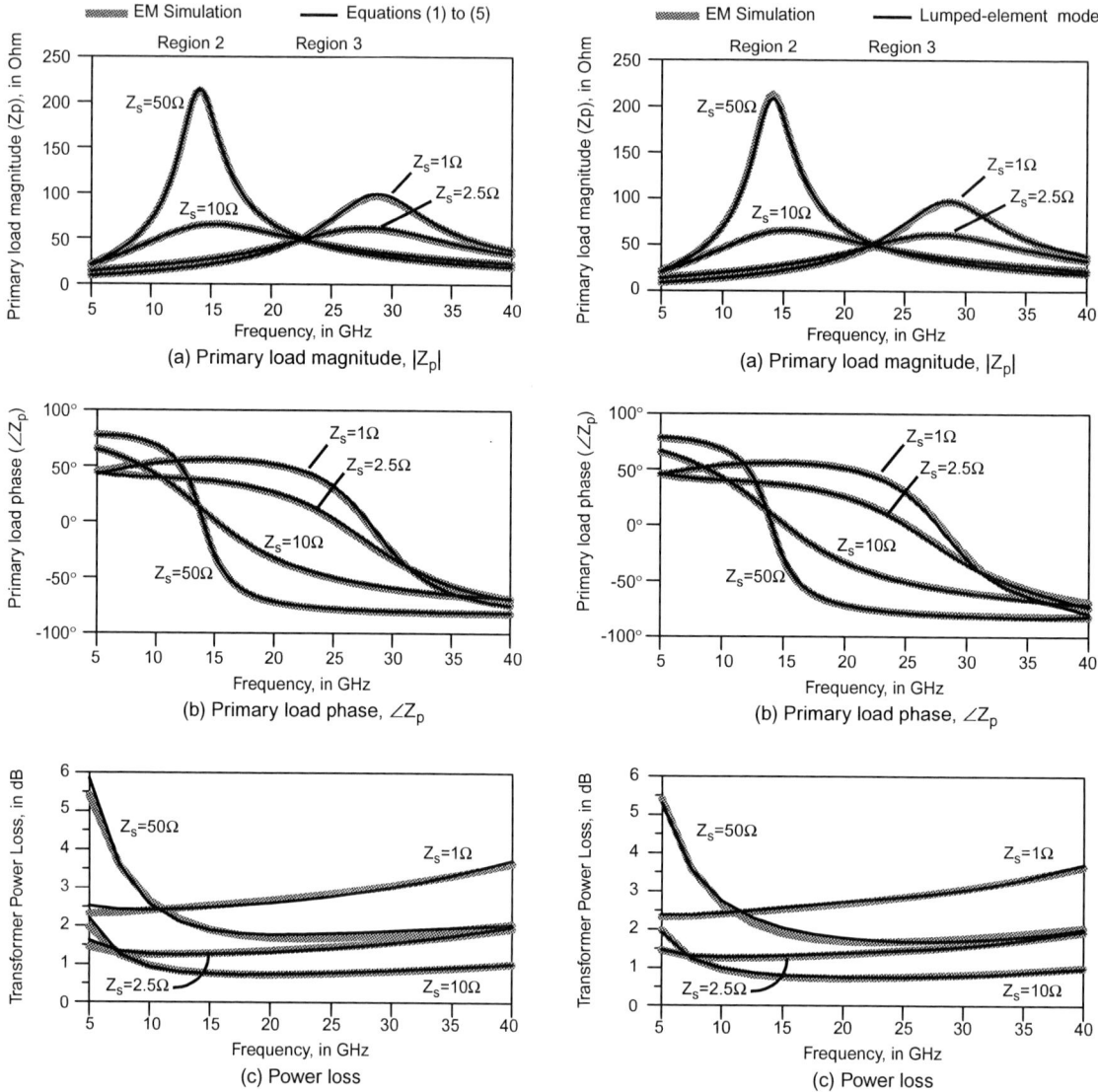

Fig. 7: Coupled-line model vs. simulation.

Fig. 8: Lumped-element model vs. simulation.

and errors in the model parameters. However, both models are still accurate for small load impedances. Therefore, the coupled-line model and equations can be used in applications such as interstage matching between common-base stages, where the load impedance is just a few Ohms.

A 2:1 transformer lumped-element model consists of 200+ elements, and a design change affects many parameters simultaneously. The high accuracy of the coupled-line model is achieved with only 8 elements (i.e., 4 coupled-line sections and 4 stubs) using fewer and more intuitive parameters. From the equations derived for the model, a designer can modify one or two parameters to preview results of a design change, and thereby quickly optimize a new circuit to meet design goals before resorting to time-consuming EM simulations. The coupled-line model is also expected to maintain accuracy at frequencies beyond available simulation data because it is a distributed model.

VI. CONCLUSION

Self-shielded transformers enable interstage matching in low impedance circuits such as multi-stage common-base amplifiers. Equations derived for a relatively simple distributed model with parameters extracted from EM simulation allow fast and accurate calculation of impedance transformation and power loss for a large range of loads.

VII. REFERENCES

[1] T. S. D. Cheung, J. R. Long, D. L. Harame, "A 21-26GHz SiGe Bipolar Power Amplifier MMIC," *IEEE Journal of Solid-State Circuits,* Vol. 40, No. 12, pp. 2583-2597, December, 2005.

[2] T. S. D. Cheung, J. R. Long, Y. V. Tretiakov, D. L. Harame, "A 21-27GHz Self-Shielded 4-way Power-Combining PA Balun," *IEEE Custom Integrated Circuits Conference,* pp. 617-620, October, 2004.

2006 Bipolar/BoCMOS Circuits and Technology Meeting Proceedings

Lumped Modeling of Differentially Driven Symmetric Inductors for RF IC Design

E. Ragonese[1], A. Scuderi[1], T. Biondi[2], A. Italia[1], and G. Palmisano[1]

[1]Università di Catania, Facoltà di Ingegneria, DIEES, Viale A. Doria 6, 95125 Catania, Italia,
[2]STMicroelectronics, Stradale Primosole 50, 95121 Catania, Italia

Abstract — This paper presents a lumped scalable model of integrated inductors for RF IC design. The model is validated up to 20 GHz by comparison with on-wafer measurements of both single-ended and differentially driven spirals covering 0.2- to 8.3-nH inductance values. Average errors on inductance, self-resonance frequency and quality factor are 2%, 5%, and 7%, respectively. High accuracy, full geometrical scalability and compactness promote the proposed model as a reliable time-saving design tool. As an example of model exploitation, an image-reject down-converter for 5-GHz wireless local area networks is presented. In this circuit differentially driven symmetric inductors are adopted to implement monolithic filters, which provide an image-rejection ratio as high as 56 dB.

Index Terms — Bipolar technology, image-reject ratio, integrated inductors, modeling, RF circuit design.

I. INTRODUCTION

Nowadays, RF circuit performance are strictly related to the quality of the available passive devices, i.e. spiral inductors and transformers, as demonstrated by the huge number of works present in literature, [1]-[3]. Since RF ICs are widely based on differential topology to exploit the common node benefits, symmetrical coils can be profitably used to drive and/or load differential pairs. Instead of two separate spirals, a symmetric coil consumes lower area providing the full electrical symmetry. Indeed, inductor excited by a differential source presents inherently benefits in terms of quality factor (Q) and self-resonance frequency (f_{SR}) with respect to the case of a single-ended excitation [4].

The use of integrated inductors in RF design requires simple, accurate and geometrically scalable models [5]-[7]. Despite the spread of symmetric inductors in present-day circuits, most models are developed for single-ended spirals without facing the issue of differential topologies. On the other hand, the few models concerning symmetric inductors are not validated for single-ended structures.

In this work a lumped scalable model, previously extracted and validated by comparison with measurements of a large number of single-ended coils [8], is here extended to symmetric devices. The validity of this model is demonstrated up to 20 GHz by comparison with experimental data of a wide set of geometrically scaled symmetric inductors. Thanks to its compactness and geometrical scalability, the model

is well suitable for device optimization in the design of RF circuits. As an example, the paper reports the design and measured performance of an image-reject (IR) down-converter for 5-GHz wireless local area networks (WLANs), which largely exploits both single-ended and differential inductors and transformers. In particular, in this circuit differentially driven symmetric inductors are adopted to implement monolithic filters, allowing high on-chip image-rejection ratio (IRR) performance to be achieved.

II. DIFFERENTIALLY DRIVEN INDUCTOR MODELING

In this work a 46-GHz-f_T 50-GHz-f_{max} silicon bipolar process, provided by STMicroelectronics, was used. This is a low-cost technology requiring only 17 mask steps featuring oxide trench isolation, poly resistors, metal-insulator-metal (MIM) capacitors with 0.7 fF/μm^2. The process allows three metal layers to fabricate inductors and transformers. The highest metal level is used for the coils, while the more resistive layers for connection paths and ground planes.

A wide set of symmetric devices with turn number (N) from 1 to 5, width (w) from 6 to 14 μm, and inner diameter (D_{in}) from 50 to 150 μm, was fabricated and experimentally characterized. Five-step de-embedding technique was used to avoid test fixture parasites. The buried layer was profitably exploited to build a patterned ground shield. To this aim, an oxide trench pattern was used to cut the buried layer under the spiral ortogonally to magnetically induced current loops. This technique provides a low-impedance path to ground for the radial displacement currents, thus minimizing the energy loss in the substrate.

Measurements of these symmetric inductors were used to extend the validation of the model in Fig. 1, whose accuracy was previously demonstrated only for single-ended coils [8]. Closed-form expressions of model components are geometrically scalable and reported in [8]. The accuracy of the proposed model is demonstrated in Figs. 2 and 3. Fig. 2 compares estimated and measured inductance and Q for 2.5- and 4.7-nH symmetric coils. The model closely predicts the frequency behavior of both inductance and Q up to the f_{SR}. The maximum errors calculated at

1-4244-0458-4/06/$20.00 ©2006 IEEE

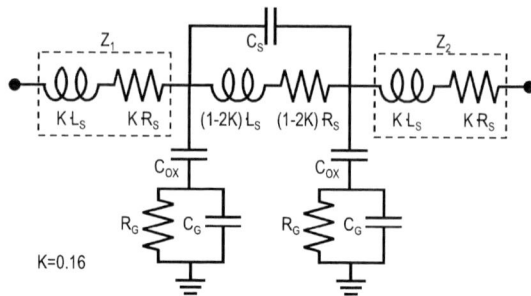

Fig. 1. Proposed inductor model.

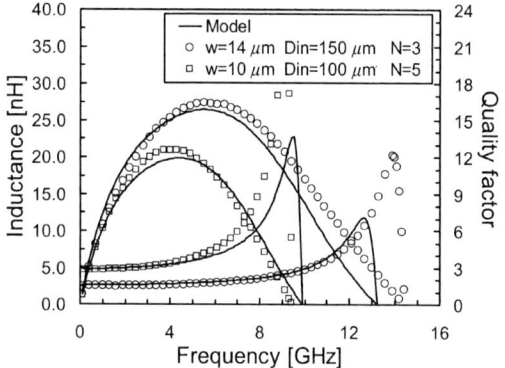

Fig. 2. Comparison between measured (symbols) and simulated (solid lines) inductances and quality factors.

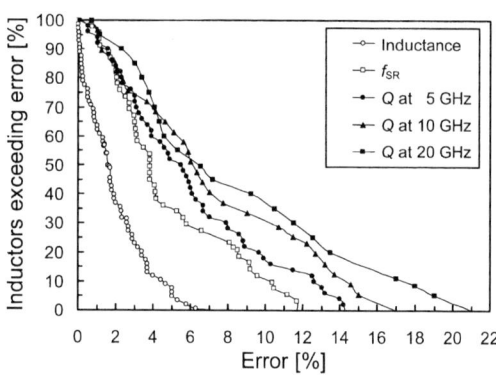

Fig. 3. Error distributions of low-frequency inductance, f_{SR}, and Q at three working frequencies.

Q-peak frequency are lower than 3 % and 5% for inductance and Q, respectively. Fig. 3 reports the error distribution curves of low-frequency inductance, f_{SR}, and Q at three working frequencies for both single and differential coils (covering 0.2- to 8.3-nH inductance values). The inductance is predicted with median and maximum errors of 2% and 7%, respectively, for all integrated inductors. Moreover, the Q error distribution curves highlight the soundness of both the topology and closed-form expressions describing series and substrate losses. Indeed, the Q of more than 50% of measured devices is predicted with error lower than 7% up to 20 GHz. Finally, the f_{SR} is estimated

with maximum and average errors of 12% and 5%, respectively. Displayed results demonstrate the accuracy and full scalability of the model over the whole range of geometrical parameters for both single and differential devices and promote this model as reliable design tool.

III. DESIGN EXAMPLE: 5-GHz IR DOWN-CONVERTER

The proposed model was exploited in the design of an IR down-converter for 5-GHz WLANs. The down-converter is part of a double-conversion receiver adopting a sliding-IF approach with the local oscillators running at $4f_{RF}/5$ and $f_{RF}/5$, respectively.

The simplified schematic of the proposed circuit is shown in Fig. 4. It consists of a two-stage variable-gain LNA and a double-balanced mixer. Both stages LNA1 and LNA2 adopt a cascode topology with resonant load, which allows high reverse isolation, power gain and linearity to be achieved. The LNA1, which uses a single-input differential-output configuration, was designed to achieve simultaneous noise / input impedance matching, by using optimum transistor sizing and bonding wire base and emitter inductances (L_B, L_{E1}). To improve the linearity performance when a large RF signal drives the down-converter, a single-bit gain control was implemented by means of the voltage V_{CTRL}. The LNA1 is loaded by an integrated stacked transformer (T_1), which implements both inter-matching network (with capacitors C_{C1}, and C_{S1}) and single-ended-to-differential conversion to the LNA2 input. Indeed a fully differential topology is adopted for the second stage and the subsequent mixer to assure higher rejection of common-mode spurious signals and reduce the LO feed-through. The LNA2 was designed to trade-off power gain and linearity performance, using single-ended degeneration inductors (L_{E2}) of 2 nH and a differential load inductor (L_C) of 2.7 nH. The down-conversion mixer adopts a double-balanced topology and consists of a voltage-to-current converter and a Gilbert switching quad. To achieve better transconductor linearity single-ended degeneration inductors (L_{E3}) of 2 nH are exploited. An integrated stacked balun (T_2) is adopted to perform the single-ended-to-differential conversion at the LO port. Finally, at the IF output an off-chip discrete network (not in figure) is used to provide both output matching and differential-to-single-ended conversion for testing purpose.

As highlighted in the dashed boxes of Fig. 4, both LNA2 and mixer include an image rejection filter (IRF) to provide on-chip image rejection. The adopted IRF is a third-order passive filter, consisting of a differentially driven inductor (L_F) and MIM capacitors, which produce a notch in the power gain frequency response corresponding to the image band [9]. Since MIM capacitors have negligible losses, the

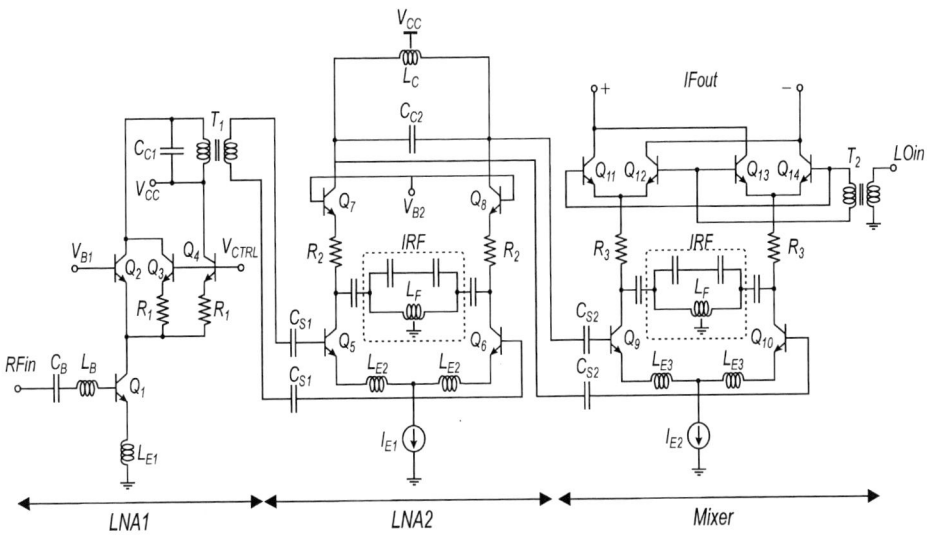

Fig. 4.   Simplified schematic of 5-GHz IR down-converter.

notch depth and hence the image rejection depends on the Q of L_F inductor. To further improve the IRR, the current partitioning between the active circuits and the IRFs was optimized by means of additional resistances R_2 and R_3, whose values were properly chosen to maximize the notch depth with a tolerable degradation of the noise figure and power gain.

All the passive devices of the IR down-converter were designed with the model sketched in Fig. 1, which was used to set optimum spiral geometrical parameters, thus restricting electromagnetic (EM) simulations only to post-layout analysis. Since the model was validated for both single-ended and differential topologies, it was also exploited to choose the most advantageous configuration, as trade-off between inductor performance and area consumption.

IV. EXPERIMENTAL RESULTS AND DISCUSSION

To validate the use of the proposed model in the design flow of the IR down-converter, on-wafer test structures of both symmetric inductors L_C ($N = 3$, $w = 12\,\mu$m, $D_{in} = 140\,\mu$m), L_F ($N = 3$, $w = 16\,\mu$m, $D_{in} = 82\,\mu$m), and single-ended inductors $L_{E2,3}$ ($N = 3$, $w = 7\,\mu$m, $D_{in} = 98\,\mu$m) were fabricated and measured. Fig. 5 reports the comparison between modelled and measured inductance and Q for L_C and L_F as a function of frequency. Maximum errors calculated at 5.5 GHz are lower than 3 % and 10% for inductance and Q, respectively. Similar errors were also found for single-ended inductors $L_{E2,3}$, (in complete agreement with results in [8]) and are not reported here for the sake of brevity.

The die micrograph of the IR down-converter is shown in Fig. 6. The two-stage LNA is placed on the left and the mixer on the right hand side, while the IRFs are arranged on the top. The chip measures 1.4×2.6 mm^2 and was assembled in a $5 \times 5 \times 0.9$-mm leadless plastic package.

At high-gain setting the down-converter provides a 21-dB power gain and a noise figure of 4.3 dB, as shown in Fig. 7. The circuit provides an input 1-dB compression point (P_{1dB}) of -21 dBm. When the LNA1 gain is reduced of 11 dB (low-gain setting), the P_{1dB} increases up to -11 dBm with a 8.5-dB noise figure. The current consumption is 22 mA from a 3-V supply voltage.

The adopted image-filtering approach mainly suffers from absolute fabrication tolerances for capacitors (δ_C). Fig. 8 shows the measured IRR compared with typical and worst-case simulated curves ($\delta_C = \pm 10\%$). In the frequency range 5.3–5.6 GHz an IRR higher than 56 dB was measured, while the down-converter maintains an IRR higher than 40 dB under worst-case conditions. Unlike other image filtering techniques reported in literature [10]-[13], such a performance has been achieved without using any notch frequency tracking or Q-enhancement of the IRFs. As demonstrated in Fig. 8, measured results are in agreement with simulations, thus confirming the reliability of the proposed model as a simple tool for the design of integrated inductors in RF circuits.

V. CONCLUSION

The capabilities of a lumped scalable model for both single-ended and differential monolithic inductors have been demonstrated. The model accuracy has been validated by comparing simulated and measured inductance and Q of more than 70 devices up to 20 GHz. The design and experimental measurements of a 5-GHz IR down-converter has been also reported to demonstrate the reliability of the proposed model as a time-saving accurate design tool for inductive devices in RF circuits.

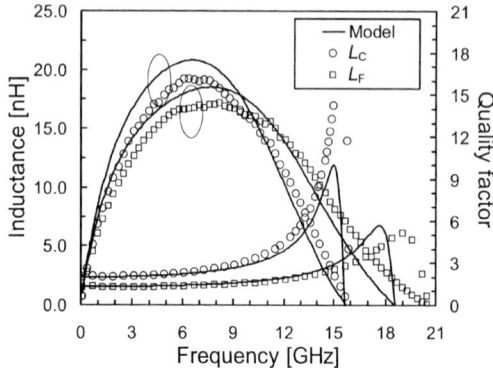

Fig. 5. Measured (symbols) and simulated (solid lines) inductances and quality factors of L_C and L_F inductors.

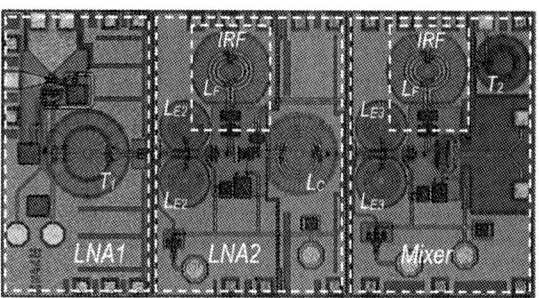

Fig. 6. Die micrograph of the IR down-converter.

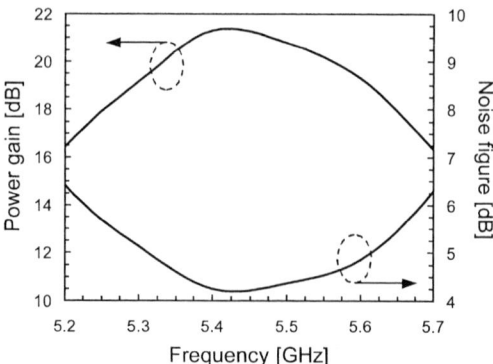

Fig. 7. Measured power gain and noise figure.

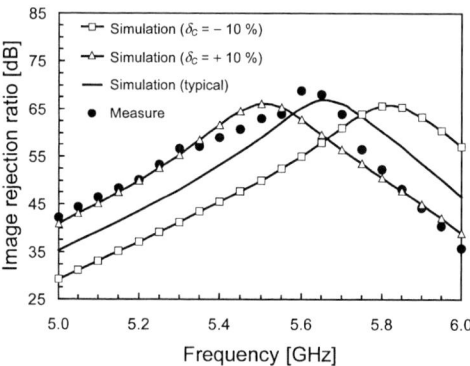

Fig. 8. Effect of the fabrication tolerances on the IRR.

REFERENCES

[1] Y. K. Koutsoyannopoulos, and Y. Papananos, "Systematic analysis and modeling of integrated inductors and transformer in RF IC design," *IEEE Trans. Circuits Syst. II*, vol. 47, no. 8, pp. 699-712, Aug. 2000.

[2] J. N. Burghartz, and B. Rejaei, "On the design of RF spiral inductors on silicon," *IEEE Trans. Electron Devices* vol. 50, no. 3, pp. 718-729, Mar., 2003.

[3] F. Rotella *et al.*, "A broad-band lumped element analytic model incorporating skin effect and substrate loss for inductors and inductor like components for silicon technology performance assessment and RFIC design," *IEEE Trans. Electron Devices*, vol. 52, no. 7, pp. 1429-1441, Jul., 2005.

[4] M. Danesh, and J. R. Long, "Differentially driven symmetric microstrip inductors", *IEEE Trans. Microwave Theory Tech.*, vol. 50, no. 1, pp. 332-341, Jan. 2002.

[5] C. P. Yue, and S. S. Wong, "Physical modeling of spiral inductors on silicon," *IEEE Trans. Electron Devices*, vol. 47, no. 3, pp. 560-568, Mar. 2000.

[6] J. Sieiro, J. M. López-Villegas, J. Cabanillas, J. A. Osorio, and J. Samitier, "A physical frequency-dependent compact model for RF integrated inductors, *IEEE Trans. Microwave Theory and Tech.*, vol. 50, no. 1, pp. 384-392, Jan. 2002.

[7] A. C. Watson, D. Melendy, P. Francis, K. Hwang, and A. Weisshaar, "Comprehensive compact-modeling methodology for spiral inductors in silicon-based RFICs," *IEEE Trans. Microwave Theory and Tech.*, vol. 52, no. 3, pp. 849-857, Mar. 2004.

[8] A. Scuderi, T. Biondi, E. Ragonese and G. Palmisano, "A lumped scalable model for silicon integrated spiral inductors," *IEEE Trans. Circuits Syst. I*, vol. 51, no. 6, pp. 1203-1209, Jun. 2004.

[9] A. Italia, E. Ragonese, L. La Paglia, and G. Palmisano, "A 5-GHz silicon bipolar radio transceiver front-end," in *Proc. IEEE Bipolar/BiCMOS Circuits and Technology Meeting*, Sep. 2004, pp. 120-123.

[10] M. A. Copeland, S. P. Voinigescu, D. Marchesan, P. Popescu, and M. C. Malerpard, "5-GHz SiGe HBT monolithic radio transceiver with tunable filtering," *IEEE Trans. Microwave Theory and Tech.*, vol. 48, no. 2, pp. 170-180, Feb. 2000.

[11] T. H. Lee, H. Samavati, and H. R. Rategh, "A 5-GHz CMOS wireless LANs," *IEEE Trans. Microwave Theory and Tech.*, vol. 50, no. 1, pp. 268-280, Jan. 2002.

[12] J. W. M Rogers, and C Plett, "A 5-GHz radio front-end with automatically Q-tuned notch filter and VCO," *IEEE J. Solid-State Circuits*, vol. 38, no. 9, pp. 1547-1554, Sep. 2003.

[13] T.-K. Nguyen *et al.*, "Image-rejection CMOS low-noise amplifier design optimization techniques," *IEEE Trans. Microwave Theory Tech.*, vol. 53, no. 2, pp. 538-547, Feb. 2005.

210

2006 Bipolar/BoCMOS Circuits and Technology Meeting Proceedings

Demonstration of three-dimensional 35nF/mm² MIM Capacitor integrated in BiCMOS Circuits

JC. Giraudin [1], F. Badets [1], JP. Blanc [1], E. Chataigner [1], A. Bajolet [1,2],
T. Jagueneau [1], C. Rossato [1] and P. Delpech [1]

[1] STMicroelectronics, 850 rue Jean Monnet, F-38926 Crolles Cedex, France
[2] IMEP, 3 parvis Louis Neel, B.P. 257, F-38016 Grenoble Cedex, France

Abstract — This paper summarizes the electrical characterization of MIM capacitor realized in three-dimensional. High density of 35nF/mm² is obtained with low leakage current. Its integration in BiCMOS technology is demonstrated and three circuits are characterized.

Index Terms — BiCMOS integrated circuits, MIM devices, Voltage controlled oscillators, Frequency synthesizers, Filters, Capacitors.

I. INTRODUCTION

Today's trend of technologies for mobile phones and personal computers is to minimize their size, volume and weight, while improving their performances. In this context, integrated circuits must be scaled down. The down-scaling of transistors is largely studied today and follows the Moore's law. On the other hand, passive components did not follow the same progress: most of them are not integrated on the circuit but are discrete components. Consequently, they occupy a larger area and mass fraction of devices. Applications like mobile phones need about twenty passive components for one integrated circuit, and most of them are capacitors [1]. It seems then essential to integrate them in the circuit. This is yet realized with well known dielectric materials, like SiO_2, Si_3N_4, Al_2O_3 or Ta_2O_5, which lead respectively to capacitance density of 1, 2, 3.5 and 5nF/mm², while respecting reliability specifications [2]-[3].

To integrate decoupling capacitors while occupying reasonable areas, the needed capacitance density is about 35nF/mm², which is not reachable with planar geometry and the classical insulators cited before. Indeed, to reach this goal, the relative permittivity of the dielectric should be around 175. Materials like TiO_2, STO, PZT or BST [2] present such a relative permittivity, but their integration in CMOS technology is not totally under control today. In order to use conventional insulator materials, the realization of three-dimensional (3D) capacitors is then mandatory to obtain 35nF/mm² [4]: it allows increasing the capacitor developed area without increasing the area occupied on the circuit.

In this paper, we describe the realization of 3D Metal Insulator Metal (MIM) capacitor called "HiDTC" (High Density Trench Capacitor) and its

electrical performances. Then, we demonstrate for the first time its successful integration in three different BiCMOS circuits and discuss the gain obtained.

II. HiDTC MANUFACTURING AND ELECTRICAL CHARACTERIZATION

A. HiDTC Manufacturing

HiDTC is realized in the Inter-Metallization Dielectric (IMD) layers of standard BiCMOS interconnect levels. Trenches of 4μm height are etched in the three IMD layers localized between Metal1 and Metal4. Then the MIM stack is deposited by Chemical Vapor Deposition (CVD) for TiN metal and Atomic Layer Deposition (ALD) for Al_2O_3 dielectric. A tungsten layer deposited by CVD fills the trenches. Finally, electrodes are patterned using classical photo-sensitive resist and dry etching. Standard vias connect the top electrode through the tungsten layer, whereas bottom electrode access is realized directly on the Metal1 layer also used as etching-stop layer for the trench patterning. A schematic cross-section, as well as a scanning electron microscopy (SEM) view is presented in Fig. 1.

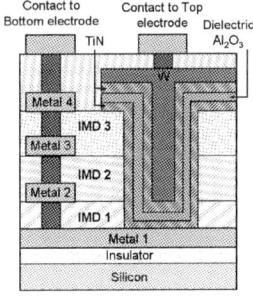

Fig. 1. Schematic (left) and SEM (right) cross view of HiDTC.

Typical width of trenches is 0.4μm for a trench height of 4μm. This high aspect ratio enables to increase by a factor of 10 the capacitance density. Indeed, we reach the value of 35nF/mm² using 200A of Al_2O_3 dielectric, whereas the capacitance density is only 3.5nF/mm² in planar configuration.

1-4244-0458-4/06/$20.00 ©2006 IEEE

B. HiDTC Electrical Characterization

Statistical electrical measurements have been performed on 1300 HiDTC realized on 86 wafers processed in 6 runs, and compared to planar capacitors. The results are reported in table I and presented in Fig. 2.

TABLE I.

Capacitance density for planar MIM and HiDTC

Capacitance density per projected area	Mean value (nF/mm²)	Standard Deviation (%)	Minimum value (nF/mm²)	Maximum value (nF/mm²)
Planar MIM	3.43	1.6	3.30	3.56
HiDTC	33.41	3.8	30.05	36.68

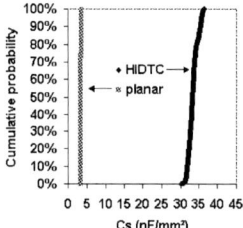

Fig. 2. Cumulative probability of capacitance density for planar MIM and HiDTC.

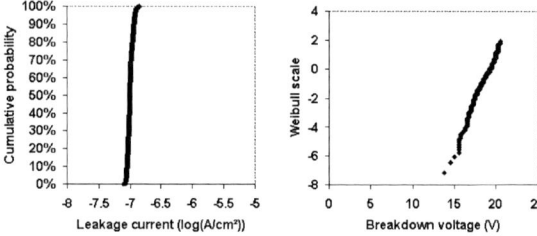

Fig. 3. Cumulative probability of leakage current at 5.5V and 25°C for HiDTC (left graph) and Weibull scale of breakdown voltage for HiDTC (right graph).

As expected, the mean capacitance value is about ten times higher for HiDTC than for planar capacitor. The standard deviation is also increased for the HiDTC. This is due to the dispersion on the IMD layers thickness, while the dispersion of the planar devices depends only on the dispersion of the Al_2O_3 thickness. However, this result stays satisfactory for decoupling applications without improving IMD dispersion.

On the same devices, statistical measurements of leakage current at 5.5 volts (see left graph on Fig. 3) and breakdown voltages (see right graph on Fig. 3) have also been performed.

At 5.5V, leakage current is very low: about 100nA/cm² with extremely low dispersion, which corresponds to less than 100pA/nF. Breakdown voltages are higher than 14V, corresponding to a minimum breakdown electric field of 7MV/cm, in

agreement with literature describing results for planar capacitor [3]. Thus, no degradation is seen with the 3D architecture. Moreover, concerning leakage current, there is no dependence with the temperature, as represented in Fig. 4, where leakage current has been evaluated for HiDTC as a function of voltage at 25°C and 175°C.

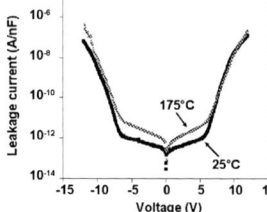

Fig. 4. Leakage currents for HiDTC as a function of voltage at 25°C and 175°C.

This excellent result allows using HiDTC in any kind of functions, including power amplifier module, where the temperature can reach 125°C or even 150°C, without increasing the power consumption which is a key factor in nowadays mobile electronic devices such as phones and personal computers.

III. HiDTC INTEGRATED IN BiCMOS CIRCUITS

The high density of 35nF/mm² obtained with HiDTC allows integrating high value capacitors such as series capacitors or decoupling capacitors in BiCMOS circuits. This integration has been realized in three circuits designed in a 0.25μm SiGe:C BiCMOS technology that addresses wireless application needs [5]. Based on a 0.25μm CMOS process, this technology includes active and passive devices such as isolated NMOS, 15V BV_{DS} NLDEMOS, 2.5V and 5V SiGe:C HBT, isolated vertical PNP BJT, 1kohms Poly resistor, MOS & junction varactors, 5nF/mm² MIM capacitor and high Q inductor. The three circuits and their electrical characterizations are described in the following parts.

A. Series capacitor

Direct current (DC) biasing is a key factor in analog base band signal processing for mobile phones. Offset voltages may be generated locally because of component mismatches and then amplified. As a result, some stages may be biased in an unwanted region, causing non linearity or distortion.

Classical solutions provide DC level adjustments at different points of the analog chain so that every stage operates in its nominal region.

Another approach consists in using a series capacitor between two blocks (see Fig. 5). This capacitor allows the DC voltage of the second block to be fixed thanks to Vbias independently of the output voltage of the first block. If low frequency signals must be fed, the capacitance must be quite large, typically few nanofarads. Up to now, such capacitance could not be integrated on silicon without penalizing chip area.

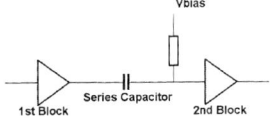

Fig. 5.   Simplified design using series capacitor.

A test chip has been realized in BiCMOS technology integrating HiDTC. The design provides an alternate current (AC) coupling between the Variable Gain Amplifier (VGA) and the Output Buffer (see Fig. 6). The cut off frequency of the high pass filter is set to 4.4kHz thanks to the 2.18nF capacitor. This cut off frequency conforms to GSM standard requirements.

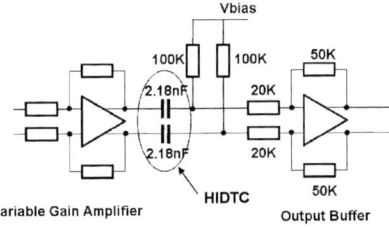

Fig. 6.   Simplified schematic of the test chip.

The effective area needed to implement the two capacitors including all the connections on the silicon is 0.15mm², as shown on the layout in Fig. 7. This corresponds to an effective density of 30nF/mm², which is more than 6 times higher than the standard planar MIM capacitors.

The frequency response obtained is presented in Fig. 7 (right graph). Moreover, the cut off frequency has been measured on several devices: average value is 4.9kHz with a very low dispersion. The shift of 11% in the cut off frequency is explained by the sum of the shift of the poly resistor (4%) and HiDTC (7%) on the dices tested.

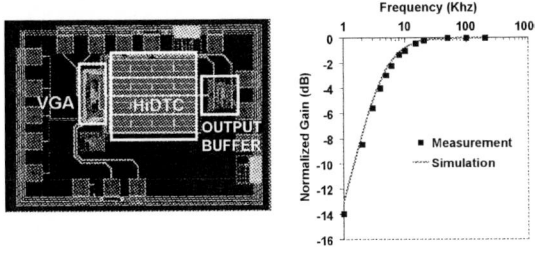

Fig. 7.   Test chip layout (on the left) and Frequency response -measured and simulated- (on the right).

This good result puts forward the ability of using HiDTC as series capacitor without penalizing chip area while keeping electrical performances.

B. Decoupling capacitors and filters for VCO filters

In modern telecommunication transceiver the sensitive blocks such as Voltage Controlled Oscillators (VCO) are integrated on the same chip as noisy blocks such as digital functions or output buffers. Avoiding any coupling between them is crucial in order to maintain the overall performance of the chip. Decoupling the supplies and the biasing voltages with capacitors is one of the solutions used to achieve this goal. The best results are obtained when the capacitance is large and positioned as close as possible to the sensitive node, hence the need for on-chip high density capacitors.

The 35nF/mm² HiDTC has been used in a 4-GHz LC VCO originally designed with the standard 5nF/mm² planar MIM capacitors. The whole block features a BiCMOS Colpitts oscillator, an amplitude automatic control (AAC) loop, a variable attenuator and an output buffer. All the planar MIM capacitors used for decoupling and filtering purposes have been replaced by HiDTC. A total of 670pF were concerned by this change: 300pF for the filtering inside the AAC, 200pF for the core oscillator biasing voltages, 40pF for the input tuning voltage and 130pF for the supply filtering.

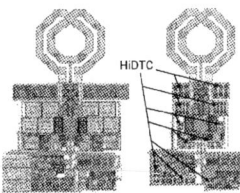

Fig. 8.   VCO layout without (left) or with (right) HiDTC.

The area of the whole VCO is reduced by 25%, from 0.4 to 0.3mm². Taking into account only the active part of the circuit, i.e. excluding the inductor whose size depends on the targeted phase noise, the area reduction is as high as 42%, from 0.24 to 0.14mm². Moreover, as it can be seen in Fig. 8, the new VCO has a rectangular form factor which greatly facilitates its integration with other blocks. The phase noise performance of the VCO was kept unchanged (see Fig. 9) with values of -96dBc/Hz at 100kHz, -120dBc/Hz at 1MHz and -152dBc/Hz at 40MHz offsets for a core oscillator current of 6mA.

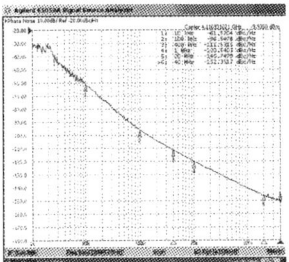

Fig. 9.   VCO phase noise.

Same results were obtained on several tenths of chips, demonstrating the high quality of the HiDTC and its ability to replace planar MIM capacitor with a significant area reduction without any loss of performance in a complex design such as a VCO.

C. Fractional PLL with HiDTC based loop filter

The third tested circuit integrating HiDTC is a Phase Locked Loop (PLL). Frequency synthesizers are key building blocks of RF transceivers. They have to provide a stable programmable frequency with a step that equals the application channel bandwidth. Usually, frequency synthesizers are made around a Phase Locked Loop depicted in Fig. 10.

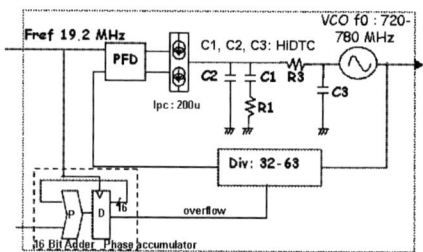

Fig. 10.   Fractional synthesizer schematic.

It consists in a Phase Frequency Detector (PFD), a Loop Filter, a VCO, and a programmable feed back frequency divider. The PFD compares the VCO frequency through the frequency divider to a pure reference frequency that comes usually from a crystal oscillator. The PFD delivers a signal proportional to the phase difference and corrects the VCO frequency through the loop filter. As in any feed back controlled system, the loop filter is calculated in order to ensure stability.

Usually, the PLL loop filter remains out of chip, as the capacitors values are about some nano farads and are almost incompatible with integration. This implies the use of two additional pads and some discrete components and so, implies a global cost increase.

Fig. 11.   Fractional PLL layout.

HiDTC have been used in order to fully integrate a 760MHz fractional synthesizer involved in a multimode GSM/DCS/WCDMA frequency synthesizer. The values of the three capacitors are respectively 3.5nF, 175pF and 125pF. The layout with the HiDTC based loop filter is depicted in Fig. 11.

Measurement results are shown in Fig. 12. The overall loop bandwidth is almost 100kHz as it had been predicted by system simulation, ensuring thus a good agreement with expected capacitors values. Moreover the loop filter takes advantage of the very low leakage current of HiDTC leading to a low reference spurious tone (about -90dBc). This result ensures the very good performance of the fully integrated PLL loop filters with saving assembly cost of the system.

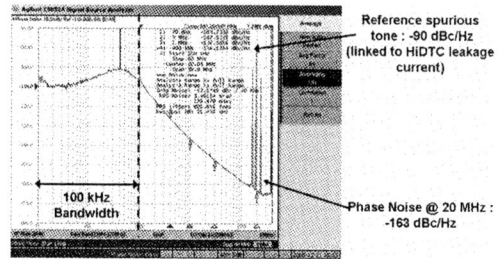

Fig. 12.   Output PLL phase noise.

V. CONCLUSION

This paper gives the electrical results of HiDTC, the first three-dimensional 35nF/mm² MIM capacitor integrated in BiCMOS technology. This high density has been reached with a statistical standard deviation of 3.8%. Breakdown voltages are above 14V and leakage current stays below 100pA/nF at 5.5V whatever the polarization and the temperature. Three circuits with integrated HiDTC used as series capacitor or as decoupling capacitor in fractional PLL and VCO have been evaluated. All the three circuits give excellent results thanks to HiDTC performances with good correlation with simulation, and with significant area and assembly cost savings. This demonstrates the ability of this new component to be integrated in complex system like system-on-chip used for mobile applications, where the key factor is the compromise between cost, size, weight and performances.

REFERENCES

[1] R.K. Ulrich *et al*, "Integrated passive component technology",*IEEE Press, Copyright 2003.*
[2] H. Singh Nalwa, Handbook of thin film materials, Vol.3.
[3] K.H. Allers et al, "Dielectric reliability and material properties of Al_2O_3 in metal insulator metal capacitors (MIMCAP) for RF bipolar technologies in comparison to SiO_2, SiN and Ta_2O_5", *in BCTM Tech. Dig*, 2003, pp. 35-38.
[4] A. Bajolet et al, "Three-dimensional 35 nF/mm² MIM capacitors integrated in BiCMOS technology",*in ESSDERC Tech. Dig.*, 2005, pp. 121-124.
[5] H. Baudry et al, "BiCMOS7RF: a highly-manufacturable 0.25-µm BiCMOS RF-applications-dedicated technology using non selective SiGe:C epitaxy*", in BCTM Tech. Dig.*, 2003, pp. 207-210.

2006 Bipolar/BoCMOS Circuits and Technology Meeting Proceedings

Integrated BiCMOS 10 GHz S-Parameter Module

Jangsup Yoon, Robert M. Fox, and William R. Eisenstadt

Department of Electrical and Computer Engineering/ University of Florida, Gainesville, FL, 32611-6130

Abstract — **The paper presents integrated BiCMOS 10 GHz s-parameter measurement circuits for performing RF/Microwave signal gain and phase detection. Potential applications include embedded IC test and production IC test.**

Index Terms — **Silicon bipolar/BiCMOS process technology, System-on-chip, Automatic Test Equipment.**

I. INTRODUCTION

Modern integrated systems include dense ICs containing diverse circuits such as microprocessors, memory, digital processing blocks, analog functional blocks, and RF functional blocks. As systems-on-a-chip (SoCs) become more advanced and complex, it is mandatory to reduce manufacturing costs, especially testing costs.

Mixed signal/RF IC production test is becoming increasingly costly for the IC industry and may require multi-million dollar production ATE testers. SoCs with wireless, baseband analog and dense digital functionality compound the production IC test complexity and greatly increase test cost. State-of-the-art 90 nm CMOS technologies are outperforming the test electronics of many ATE systems in microwave/millimeter-wave frequencies and Gb/s I/O speeds. Unfortunately, the commercial marketplace tightly constrains IC test costs and may not permit manufacturers to pass on test cost increases despite the increased performance of integrated systems.

This work employs 10 GHz integrated peak and phase detector circuits, directional couplers, and passive RF IC elements [1]-[5] to implement on-chip circuits for vector power measurements (s-parameters). The circuits were implemented using IBM 7WL (180 nm SiGe/CMOS) technology.

II. ON-CHIP S PARAMETER MEASUREMENT

The s-parameter measurement modules contain integrated subcircuits capable of measuring gain and phase at frequencies near 10 GHz; these include directional couplers, switches, dividers, peak detectors and phase detectors. S-parameter measurement circuits connected to a DUT are shown in Fig. 1. In this figure, the directional couplers sample signals on the transmission lines and switches

direct the sampled data to microwave power detectors and to phase detectors. These s-parameter measurement modules can be integrated on a chip for built-in self-test. Alternatively, they could be put on an ATE load board or in a microwave probe to greatly extend test capability at very low cost. The modules enable multi-port test and parallel test on the load board.

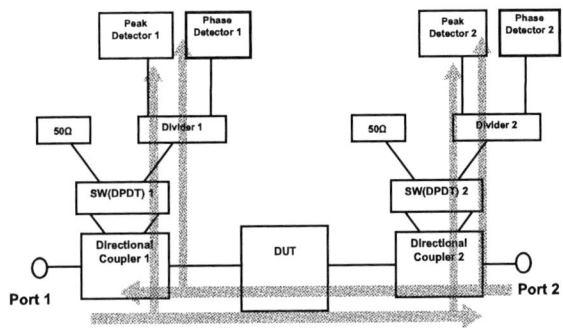

Fig. 1.   S-parameter Test Block Diagram.

III. SUB-CIRCUIT DESIGN FOR BUILT-IN S PARAMETER MEASUREMENT

As shown in Fig. 1, the s-parameter measurement system includes directional couplers, switches, dividers, peak detectors and phase detectors. A lumped-element directional coupler [5] is used, as shown in Fig. 2.

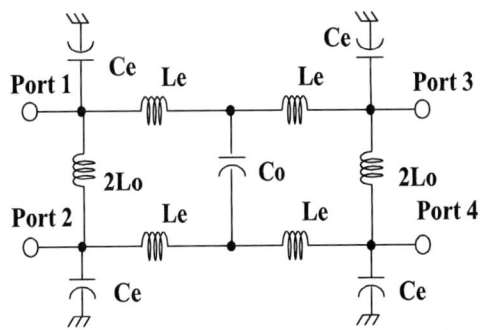

Fig. 2.   Schematic of the lumped passive directional coupler.

A double-pole double-throw (DPDT) RF switch is employed as shown in Fig. 3. The switch includes four n-channel MOSFETs and six poly resistors. The

1-4244-0458-4/06/$20.00 ©2006 IEEE   215

MOSFET gate length and width are the dominant factors controlling RF switch insertion loss [6]. Minimum length (180 nm) is used. 30 fingers are used, with total gate width of 4230 nm for each FET.

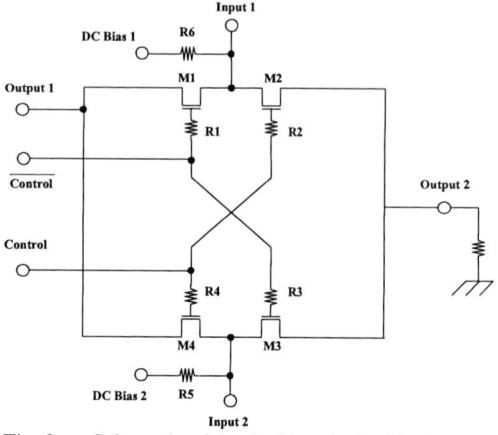

Fig. 3.   Schematic of the double pole double throw (DPDT) switch.

Two types of Si-Ge BJT-based phase detector circuits were designed [7][8]. One is an S-R flip-flop and the other is a Gilbert cell. The phase detector compares the phase of the input signal to that of a reference signal. The S-R flip-flop uses single-ended reference and signal inputs. The Gilbert cell phase detector uses differential signals. Simulations and measurements showed that the Gilbert phase detector output is highly amplitude dependent. A limiting amplifier using a four-stage cascade of Cherry-Hooper amplifiers [11] was designed to minimize variations of the inputs to the phase detectors. The schematic of one stage of the amplifier is shown in Fig. 4. The biasing is chosen to provide strong impedance mismatch and emitter-follower feedback is used to provide wide bandwidth [9]-[11]. Limiting amplifier chains are connected to both inputs of the phase detectors.

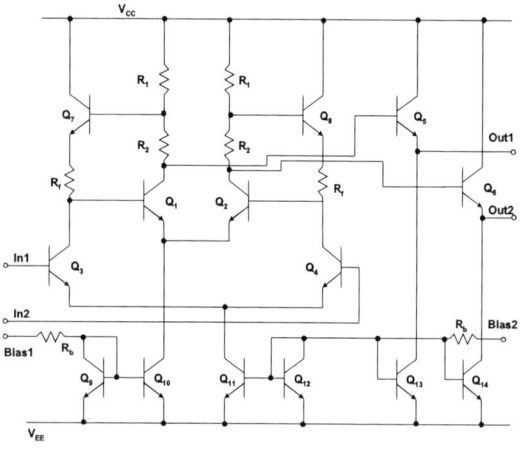

Figure. 4. Cherry Hooper limiting amplifier stage

As shown in Fig. 1, peak detectors are used to detect the signal amplitude. The peak detector circuit [12] has relatively high dynamic range compared with the phase detector.

IV. SIMULATION RESULTS

The 10 GHz DPDT switch, Cherry-Hooper amplifier, phase detector and peak detector circuits were simulated using Cadence SpectreS. As shown in Fig. 5, the simulated insertion loss of the RF switch at 10 GHz is 1.9 dB. The simulated isolation is -30.1 dB.

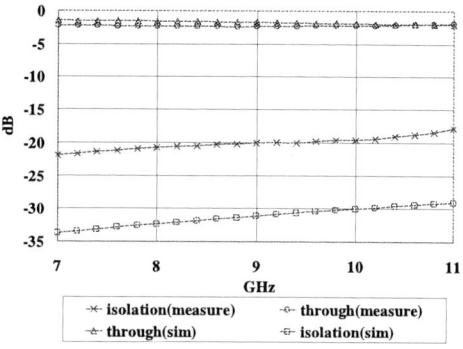

Fig. 5.   Simulation and measurement results of DPDT switch.

Simulations of the S-R flip-flop type phase detector, here called Type I, are shown in Fig. 6. When 0 dBm input signals are applied, the differential output varies from -91 mV to 91 mV as the delay varies from 36 to 324 degrees. However, the output is highly amplitude-dependent.

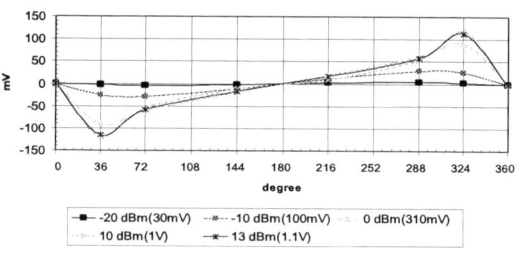

Fig. 6. Simulation results of phase detector type I at 10 GHz.

The simulation results for the Gilbert cell phase detector with Cherry-Hooper limiting amplifiers (Type 2) are shown in Fig. 7. The differential output varies from -836 mV to +836 mV as the phase varies from 144 to 324 degrees. The amplitude dependence is greatly reduced.

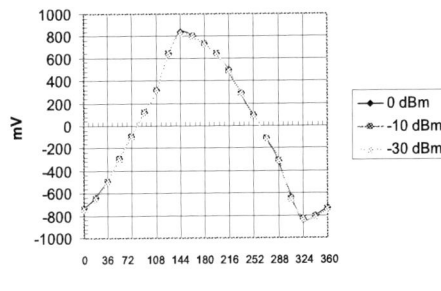

Fig. 7. Simulation results of phase detector type II at 10 GHz.

Simulations of the peak detector are shown in Fig. 8. The DC output changes from 1 mV to 935 mV as the input is swept from -29 dBm to 11 dBm.

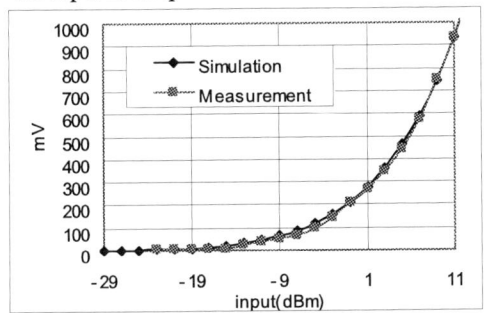

Fig. 8.   Measured and simulated results for the peak detector at 10 GHz .

V. MEASUREMENT RESULTS

The DPDT switch and a Type I phase detector were fabricated with IBM 0.18 micron SiGe BiCMOS-7WL technology through the MOSIS fabrication service. A photomicrograph is shown in Fig. 9. The total chip size is about 1.2 mm^2 (1.0 mm x 1.2 mm).

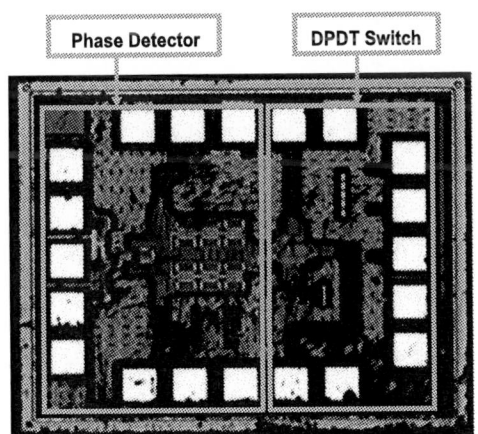

Fig. 9. Chip micrograph (1.2 mm X 1.0 mm) of phase detector and DPDT switch

Measured results for the DPDT switch are shown in Fig. 5. The insertion loss is -2.2 dB and the isolation is -19 dB at 10 GHz, agreeing well with simulations.

Measurement results for the Type I phase detector (without limiting amplifiers) are shown in Fig. 10. With 0 dBm input, the DC output varies from -128 mV to 88 mV as phase varies from 36 to 324 degrees. In agreement with the simulations, the DC output varies significantly with input amplitude.

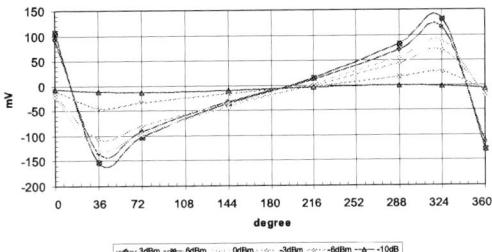

Fig. 10. Measurement results of phase detector type I at 10 GHz.

The peak detector and the Type II phase detector with 4-stage limiting amplifiers were fabricated on a second chip using the same technology. A photomicrograph is shown in Fig. 11. The total size of this chip is about 2.64 mm^2 (1.2 mm x 2.2 mm). Measured results for the Type II phase detector are shown in Fig. 12. The results are very similar to the simulations in Fig. 8. DC output varies from -645 mV to 798 mV with 0 dBm input as phase varies from about 144 to 324 degrees. Also consistent with simulations is the reduced amplitude dependence.

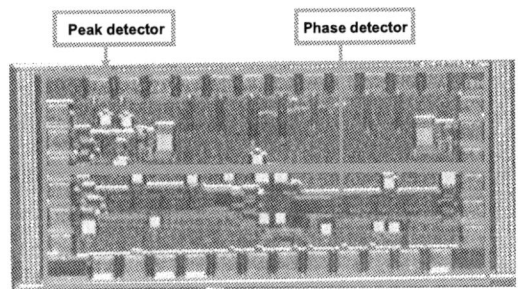

Fig. 11.   Chip micrograph (1.2 mm X 2.2 mm) of phase detector and peak detector.

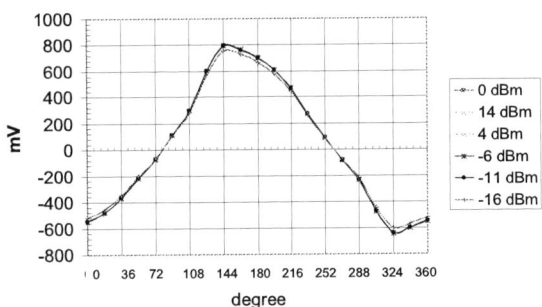

Fig. 12.   Measurement results of phase detector type II at 10GHz.

217

Measured results for the peak detector are shown in Fig. 8. The DC output varies from 1 mV to 935 mV. The measured and simulation results agree well.

VI. S-PARAMETER APPLICATION

A series of simulations demonstrate how the proposed circuits can be used in measuring s-parameters. A DUT circuit was created using passive elements from the IBM design library; the DUT's simulated S_{21} was (-0.54 dB, -90.0°). As shown in Fig. 13, the s-parameter setup uses two dividers, a 90° phase delay circuit, a peak detector and a phase detector. In an initial calibration step, the DUT and the 90° delay are both bypassed. The dc peak detector output at P2 is found to be 275 mV, corresponding to 0.96 dBm according to Fig. 12. The phase detector output at P3 is -367 mV, corresponding to -291° from Fig. 7. To measure the DUT's s-parameters, Bypass-Line1 is removed and the 90° phase delay is left bypassed. The peak detector output becomes 257 mV, corresponding to 0.44 dBm, and the phase detector output becomes 620 mV, corresponding to -20.7° or -305°. To resolve the ambiguity in the phase, Bypass-Line2 is removed and the phase is measured again. The phase detector output with 90° phase delay becomes 325 mV, corresponding to -108.7° or -230.5°. Only the first set of phase values is self-consistent, so the phase measurement with the DUT but without the 90° phase delay is interpreted as -20.7°. By subtracting the calibration values, the magnitude of the DUT's S_{21} is found to be -0.52 dB, and the phase is -89.3°, consistent with expectation.

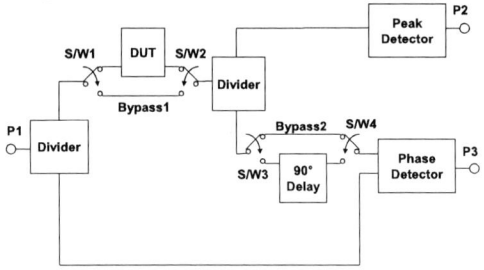

Fig. 13. Block diagram for verification of s-parameter measure.

VII. SUMMARY

This paper presents the design, simulation and measurements of integrated BiCMOS s-parameter measurement circuits for low cost IC test. The functionality of the circuits has been verified and the use of the circuits blocks in s-parameter extraction has been demonstrated through simulations. Measurements suggest that with calibration, s-parameter magnitudes can be measured on-chip with tolerances below 0.5 dB. We believe that calibrated phase measurement precision is better than 3°.

ACKNOWLEDGEMENT

The authors wish to acknowledge the assistance and support of the C2S2 FRCP Marco Focus Center and the MOSIS IBM MEP donation program.

REFERENCES

[1] J.-S. Yoon and W. R. Eisenstadt, "Embedded loopback test for RF ICs," *IEEE Trans. on Inst. And Meas.*, Special Issue on VLSI Test, Vol. 54, No. 5, pp. 1715-1720, October 2005.

[2] Q. Yin, W. R. Eisenstadt, R. M. Fox, and T. Zhang, "A translinear-based RF RMS detector for embedded test of RF ICs," *IEEE Trans. on Inst. And Meas.*, Special Issue on VLSI Test, Vol. 54, No. 5, pp. 1708-1714, October 2005.

[3] T. Zhang, W. R. Eisenstadt, and R. M. Fox, "20 GHz bipolar RF RMS power detectors," *IEEE JSSC*, 2006.

[4] S. Puligundla, R. M. Fox and W. R. Eisenstadt, "An accurate high-dynamic range 1.8V RF CMOS signal strength detector," WTW2005, 4th Workshop on Test of Wireless Circuits and Systems, Cannes, France, June, 2005, 6 pp.

[5] J. Yoon, and W. R. Eisenstadt, "Lumped passive circuits for 5GHz embedded test of RF SoCs," *ISCAS 2004*, Vancouver, Canada, May 23-26, 2004, Vol. 1. pp I-241 to I-244.

[6] F.-J. Huang and K. O, "A 900-MHz T/R switch with a 0.8-dB insertion loss implemented in a 0.5-_m CMOS process,". *IEEE 2000 CICC*, pp. 341–344, 2000.

[7] William F. Egan, *Frequency Synthesis by Phase Lock*. New York, NY, John Wiley & Sons, Inc., 1981.

[8] Paul R. Gray, Robert G. Meyer, *Analysis and Design of Analog Integrated Circuits*. New York, NY, John Wiley & Sons, Inc., 2001.

[9] Wolfgang Pöohlmann "A silicon-bipolar amplifier for 10 Gbit/s with 45 dB Gain", *IEEE JSSC*, pp 551-556, May 1994.

[10] H.M. Rein "Design considerations for very-high-speed Si-Bipolar IC's operating up to 50 Gb/s", *IEEE JSSC*, pp 1076-1090, Aug. 1996.

[11] Chris D. Holdenried, James W. Haslett, and Michael W. Lynch, "Analysis and design of HBT Cherry-Hooper amplifier with emitter-follower feedback for optical communications" *IEEE JSSC*, pp 1959-1967, 2004

[12] T. Zhang, W. R. Eisenstadt and R. M. Fox, "A novel 5GHz RF power detector", in *Proc. ISCAS*, May 2004, vol. 1, pp. 897-900

2006 Bipolar/BoCMOS Circuits and Technology Meeting Proceedings

RF Switch on Standard SiGe-CMOS Technology for System-on-a-chip Radio Transceivers

Domenico Zito and Bruno Neri

Radio-Frequency & Microwave Integrated Circuits Laboratory (RFLab), Dipartimento di Ingegneria dell'Informazione (DIIEIT), University of Pisa, I-56122 Pisa, Italy

Abstract — **The design of a fully integrated RF switch on standard SiGe-CMOS technology is presented and discussed. A case study for 5-GHz indoor WLAN applications is reported. It allows on-chip T/R antenna switching operations providing better performance (typical total insertion losses are 0.2 and 0.9 dB, in R and T modes respectively) with respect to those obtained by using other solid-state devices as switching element.**

Index Terms — **Bi-CMOS, receiver, RF switch, transmitter.**

I. INTRODUCTION

Nevertheless fully integrated transmitters and receivers for Wireless Local Area Networks (WLANs) are even more widespread in literature [1, 2], the T/R (Transmission/Reception) antenna switch (AS) is still representing the last obstacle to the complete transceiver integration on standard silicon technologies. The main limit for RF switching operations on silicon substrate is represented by the low quality factor of the integrated LC filters. Particularly, the quality factor is drastically reduced by loss mechanisms into the inductors. Thus, the most common commercial solutions consist of using bulky and expensive off-chip components, such as MEMS [3] or GaAs [4] devices. Only recently, some active circuits have been realized as switching element on silicon technologies [5]. However, they are characterized by a not negligible insertion loss (IL) that, in the best case, is around 1.5 dB.

In this paper, a fully integrated RF switch based on the active circuit (namely 'Boot-strapped Inductor') is presented. Particularly, the design presented hereinafter relates the case of the T/R antenna switching operations. In Section II, the operating principle and the circuit design are shown. The case study for 5-GHz WLAN is dealt with in Section III. Finally, in Section IV, the conclusions are drawn.

II. OPERATING PRINCIPLE AND CIRCUIT DESCRIPTION

The RF switch herein presented focuses the case of T/R antenna switching operations, which allows the proper connection between the antenna and the RF front-end. The scheme of principle of the RF front-end which includes the antenna switch (AS), the low noise amplifier (LNA) and the power amplifier (PA),

is reported in Fig. 1. The antenna switch (AS) is realized by using the Boot-strapped Inductor (BSI) circuit, which provides a behaviour equivalent to that of a high quality factor inductor (Q greater than 50) [6]. The equivalent inductance L_{eq} obtained by the BSI circuit can be varied by means of a control voltage by even a factor of 10.

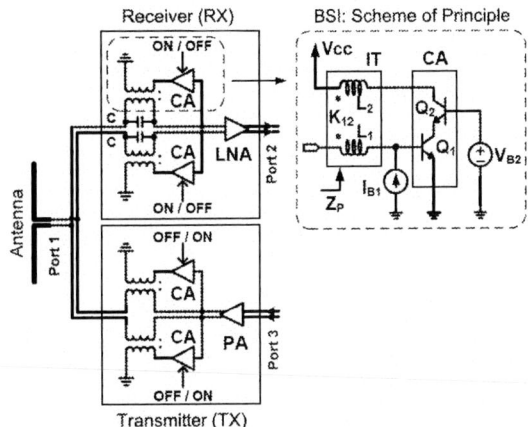

Fig. 1. Fully integrated RF transceiver front-end (LNA, PA and BSI-based AS): scheme of principle.

The BSI consists of an integrated transformer (IT) and a current amplifier (CA), as shown in Fig. 1. The operating principle can be basically explained by a simplified small signal analysis [6]. The impedance seen from the primary (Z_P) of the transformer is:

$$Z_P = [R_1 + r_\pi - \omega M_{12} g_m^* r_\pi \sin\varphi] + j\omega[L_1 + M_{12} g_m^* r_\pi \cos\varphi]$$

where, g_m^* is the equivalent transconductance gain of the CA which takes into account the capacitive effects of parasitic capacitances (mainly C_π), φ is the phase difference between the currents into the two spirals of the IT (L_1, L_2), M_{12} is their mutual inductance, whereas R_1 is the parasitic resistance of the primary. If φ is close to zero (which can be achieved by introducing an additional inductor [6]), then Z_P exhibits a high-Q equivalent inductance behaviour (L_{eq}), which can be tuned acting on the quiescent points of Q_1 and Q_2 (in principle, by means of I_{B1} and V_{B2}) [6].

However, in the case of the T/R antenna switch

1-4244-0458-4/06/$20.00 ©2006 IEEE

(patented) [7], the BSI circuit is on/off driven into two operating conditions: the former, in which the CA is switched-on and then the increase of the equivalent inductance seen from the primary (boot-strap effect) is obtained, realizing the blocking impedance for signal; the latter, in which the CA is switched-off and then the aforementioned effect is missing, realizing a low impedance state.

Thus, with reference to the scheme of Fig. 1, in T mode, the BSI at the input of the low noise amplifier (which is now switched-off) is in on state, providing a high impedance path (L_{eq} is resonant with C) with respect to that of the antenna (typically, 50 Ohm). In R mode, the LNA is obviously switched-on and the CA of the receiver (RX) is in off state; therefore, the value of L_{eq} is switched to the low value. During this time interval, the BSI circuit at the output of the transmitter (TX) is activated in order to provide a high impedance path for the signal which could flow from the antenna toward the power amplifier.

The different approaches (resonance for the RX and high impedance for the TX) are related to their different linearity constrains. Actually, the resonant approach has shown higher blocking impedance (obtained as ratio between the first harmonics of voltage and current) and wider linear behavior of the BSI (RX) circuit itself for a larger signal available from the source. It is clear that the critical case is represented by the larger output signal available from the TX, so that the asymmetric approach has been adopted.

The proposed solution can be implemented in different technologies and processes, in a single-ended or in a fully differential architecture (like that drawn in Fig. 1) and can be adapted for wide range of RF frequencies. As matter of fact, as far as the highest frequency of application is concerned, the success for the mentioned approach depends on the cut-off frequency of the available active devices (MOSFETs or BJTs) and self-resonance frequency of the integrated spirals inductors and transformers. Thus, if we consider the cut-off frequencies (beyond 150 GHz) of the modern Bi-CMOS/CMOS processes and the self-resonance frequencies obtained by using theirs copper metal layers (up to 50 GHz [8]), then such an approach can be reasonable implemented up to millimiter-waves.

The key factor for a successful design consists of properly sizing the CA and IT, in order to alternatively (T/R) obtain a nearly perfect impedance matching between the antenna, PA and LNA, respectively. As shown, the operating principle of the BSI does not directly depend on the characteristics of the secondary spiral of the integrated transformer, so that a stacked structure is typically preferred. Moreover, it is worth of mentioning that the primary spiral of the integrated transformers (with the thickest top metal layer) of the BSI circuits allow the effective

accomplishment of the impedance integrated matching between the antenna, the receiver [9] and the transmitter. Thus, the primary spirals of the two transformers can be not considered as additional components but only as those which have to be typically inserted in a traditional design. For these reasons, an optimal design of such a switch cannot be de-embedded from the low noise amplifier and power amplifier designs.

III. CIRCUIT DESIGN AND PERFORMANCE

As case study, the RF switch has been designed for T/R antenna switching operations for low power (50 mW) 5.15-5.35 GHz WLAN systems, according to the scheme of Fig. 1 and relatively to a fully differential topology with a 50 Ω antenna.

The circuits (T/R switch, LNA and PA) have been realized by using a standard SiGe-CMOS 0.35 μm technology (by AMS) with a top metal thickness of 2.5 μm and a maximum cut-off frequency (f_{Tmax}) equal to 75 GHz.

The equivalent single-ended scheme of the overall RF front-end and the summary of the RX and TX performance are reported in Fig. 2 and Table I, respectively.

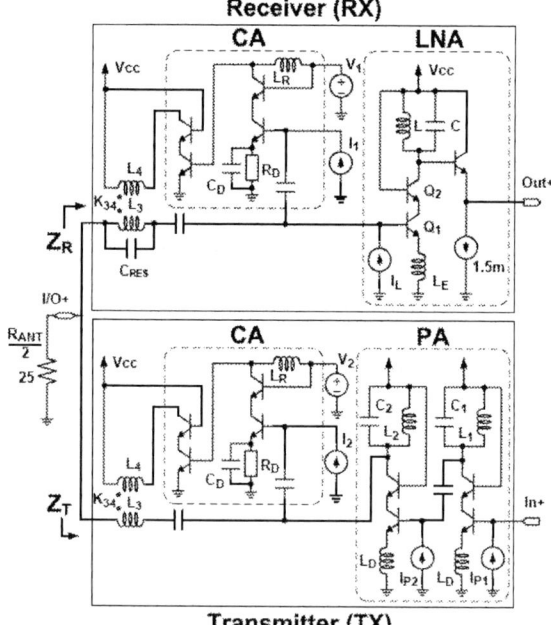

Fig. 2. RF front-end for 5-GHz WLAN: simplified scheme.

The LNA has been implemented by a cascode stage with inductive emitter degeneration (L_E) for an integrated matching to a 50 Ω antenna [9] and a subsequent emitter follower buffer stage.

TABLE I
SUMMARY THE RX AND TX PERFORMANCE

RX	G_T* [dB]	NF* [dB]	iCP_{1dB} [dBm]	P_C (CA) [mW]	P_C (LNA) [mW]	S11* [dB]
Simulated	20.3	3.74	-21	26	27	-25
Measured	20.04	3.8	-20.5	25.3	26.6	-13.3

TX	G_P* [dB]	P_{SAT} [dBm]	oCP_{1dB} [dBm]	P_C (CA) [mW]	P_C (PA) [mW]
Simulated	22.6	12	10	12.9	144
Measured	20.4	7	3	12.7	112

(*) at 5.25 GHz

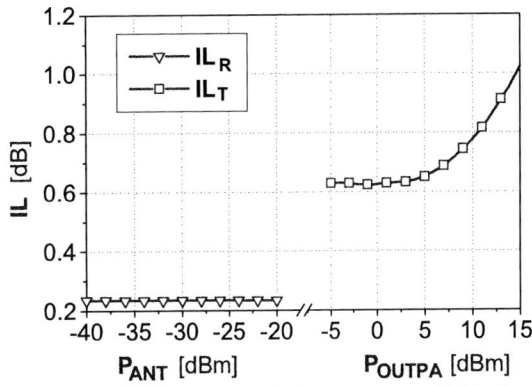

Fig. 3. IL (in T and R modes) vs. power available from sources (antenna and output of the PA, respectively): simulation results obtained by Harmonic Balance.

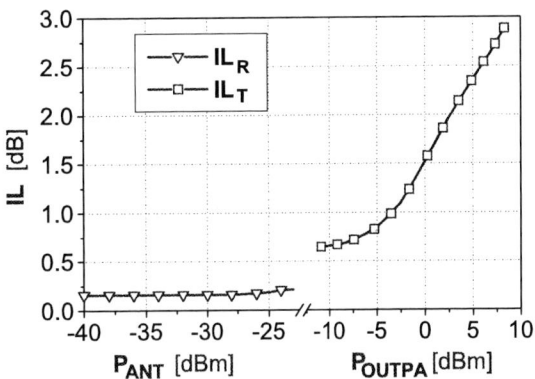

Fig. 4. IL (in T and R modes) vs. power available from sources (antenna and output of the PA, respectively): experimental results.

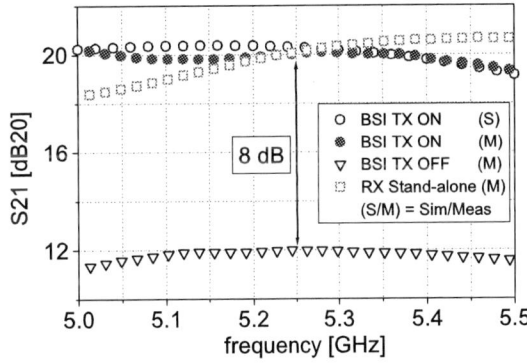

Fig. 5. S21 (in R mode) vs. frequency, obtained with -27 dBm of power available from the antenna (P_{ANT}). To be noted the effect obtained when the CA of the BSI (TX) is switched-off.

A basic design of a linear power amplifier has been realized only in order to attest the effectiveness of the antenna switch design based on the BSI circuit herein presented. The PA has been implemented with a cascade of two cascode emitter stages with inductive degeneration and LC resonant load filters (L_1C_1, L_2C_2) which have been realized by means of on-chip integrated inductors.

As for the layout of the test-chip (see Fig. 7), the G-S-G-S-G pad structures has allowed us to carry-out both for on-wafer and on-board tests.

The signal pads have been realized by using the upper metal levels (3 and 4) and without any ESD protections, in order to reduce the parasitic capacitance toward the substrate.

Circuit and EM (2D½ and 3D) simulations have been performed by means of SpectreRF™ and ADS, Momentum™ and HFSS™, respectively.

As for the AS performances, in R mode, the input impedance of the RX is very close to 50 Ω, whereas the output impedance of the TX ($|Z_T|$, see Fig. 2) is about 840 Ω. In T mode, Z_T is very close to 50 Ω, whereas $|Z_R|$ becomes approximately equal to 1 KΩ. The relevant IL, defined as the ratio between the power available from source and that delivered to the load, is 0.235 dB and 0.585 dB (the maximum IL_T is nearly equal to 1.1 dB at 50 mW) for R and T modes, respectively. The insertion losses versus power available from the sources (antenna and output of the PA in T mode and R mode, respectively) are reported in Fig. 3.

Post-layout simulations have been widely confirmed by measurements on five prototypes which have been wire bonded on microstrip test boards (Roger 4003, 1/2 oz, 10 mils) with ε_r equal to 3.38.

The measured IL (both for T and R modes) versus power available from the source are reported in Fig. 4. To be noted the excellent agreement with the simulation results, except for higher values of the power available from the output of the PA, which is due to a current mismatch of its bias circuit (see Table I).

Performance in terms of power delivered to the load are reported in Figg.5 and 6.

To be noted that, when the BSI is switched-on, the results are very close to those measured for RX and TX as stand-alone circuits. Particularly, in the frequency band of interest, the typical IL_R and IL_T are 0.2 and 0.9 dB (mismatches included), respectively.

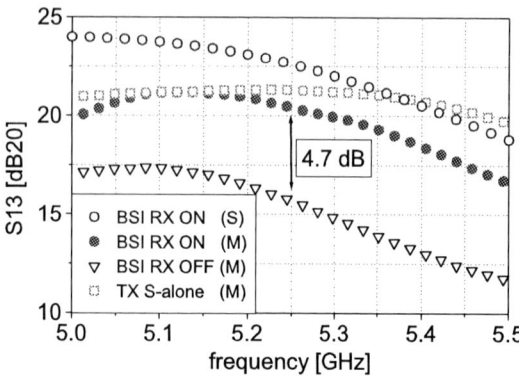

Fig. 6. S13 (in T mode) vs. frequency, obtained with -27 dBm of power available from the source at the input of the PA. To be noted the effect obtained when the CA of BSI (RX) is switched-off.

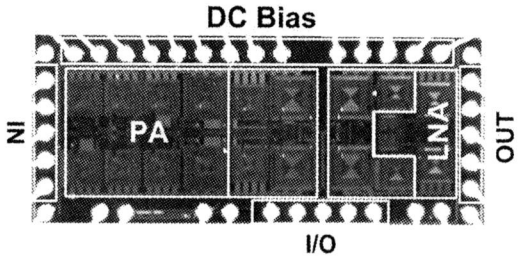

Fig. 7. Chip micrograph. The total are occupancy on die is 2.5 mm x 1 mm (pad included).

IV. CONCLUSIONS

A novel solution for integrated RF switch has been presented and demonstrated by measurements for 5-GHz T/R antenna switching operations.

The design criteria and the limits of applications has been discussed considering the main circuit and technological aspects. From basic considerations on the present state-of-the art, it results that the approach based on the Boot-Strapped Inductor circuit can be exploited up to millimeter-waves.

The case study has been implemented in a 0.35 µm SiGe-CMOS standard and low cost process (S35 by Austriamicrosystems).

The performances exhibited at 5-GHz are very promising if compared with those typically obtained by using external switches or duplexer which not only have to be mounted as external components to the transceiver causing additional losses due to the impedance mismatches but, like those on silicon

recently presented [5], introduce insertion losses (IL) of 1-2 dB at least.

Thus, the herein presented RF switch for 5-GHz WLAN represents a significant new step toward the realization of single-chip transceivers on standard silicon Bi-CMOS/CMOS technologies for next generation system-on-a-chip wireless interfaces.

ACKNOWLEDGEMENT

The authors wish to acknowledge the financial support of the Italian Ministry of Education, University and Research (namely, Ministero Italiano dell'Istruzione, dell'Università e della Ricerca – MIUR), and "Fondazione Cassa di Risparmio di Pisa".

REFERENCES

[1] H. Samavati, T. H. Lee, R. H. Rategh, "5-GHz CMOS Wireless LAN" IEEE Transaction on Microwave Theory and Techniques (MTT), vol.50, pp.268-280, Jan 2002;

[2] M. Zargari, J. Leung, et al., "A 5-GHz CMOS Transceiver for IEEE 802.11a Wireless LAN Systems" IEEE Journal of Solid-State Circuits (JSSC), vol.37, pp.1688-1692, Dec 2002;

[3] C. Goldsmith, J. Kleber and B. Pillans, "RF MEMS: benefits & challenges of an evolving RF switch technology", Tech. Digest of IEEE GaAs IC Symp. 2001, pp.147-148, Oct 2001;

[4] C.-H. Lee, S. Chakraborty, et al., "Broadband highly integrated LTCC front-end module for IEEE 802.11a WLAN applications", IEEE Proc. of MTT Symp. 2002, vol.2, pp.1045-1048, Jun 2002;

[5] N. A. Talwalkar et al.,"Integrated CMOS Transmit-Receive Switch using LC-Tuning Substrate Bias for 2.4-GHz and 5.2 GHz Applications", IEEE JSSC, vol.39, pp.863-870, Jun 2004;

[6] G. D'Angelo, A. Monterastelli, B. Neri et al., "High-quality active inductors", IEE Electronics Letters, vol.35, 30 Sep 1999;

[7] L. Fanucci, A. Hopper, B. Neri, D. Zito, "A Novel Fully Integrated Antenna Switch for Wireless Systems", IEEE Proc. of ESSDERC 2003, pp.553-556, Sep 2003;

[8] T. O. Dickson, M.-A. LaCroix, S. Boret, D. Gloria, R. Berkens, S.P. Voinigescu, "30-100-GHz Inductors and Transformers for Millimiter-Wave (Bi)CMOS Integrated Circuits", IEEE Transactions on Microwave Theory and Techniques (MTT), vol.53, n.1, Jan 2005;

[9] G. Palmisano, G. Girlando, "Noise Figure and Input Matching in RF Cascode Amplifiers", IEEE Trans. on Circuits and Systems II: ADSP,vol.46, pp.1388-1396, Nov 1999.

2006 Bipolar/BiCMOS Circuits and Technology Meeting Proceedings

SiGe BiCMOS for Analog, High-Speed Digital and Millimetre-Wave Applications Beyond 50 GHz

S.P. Voinigescu, T. Chalvatzis, K.H.K. Yau, A. Hazneci, A. Garg, S. Shahramian, T. Yao,
M. Gordon, T.O. Dickson, E. Laskin, S.T. Nicolson, A.C. Carusone, L. Tchoketch-Kebir,
O. Yuryevich, G. Ng, B. Lai, and P. Liu

ECE Dept., University of Toronto, 10 King's College Rd., Toronto, ON, M5S 3G4, Canada

Abstract — This paper explores the application of SiGe BiCMOS technology to mm-wave transceivers with analog and digital signal processing. A review of 10-80Gb/s SERDES performance across 3 SiGe BiCMOS and CMOS technology nodes reveals remarkable similarities with digital CMOS IC scaling and points to the benefits of a SiGe BiCMOS roadmap. Examples of 40-Gb/s equalizers, track-and-hold amplifiers and ADCs with mm-wave sampling clocks are provided, along with GHz-range opamp filters and 65-GHz wireless transceivers. Automotive radar and imaging applications in the 80-100 GHz range are also briefly discussed.

I. INTRODUCTION

Historically, due to larger-scale economies, wireless and (briefly) fibre-optic applications have driven SiGe BiCMOS process development [1]-[3]. While SiGe HBT performance has steadily improved over the last few years [4]-[8], the frequency of most applications has remained in the 2-10 GHz range. This has allowed CMOS technology to catch up in some of the larger volume markets, such as WLAN and 10-Gb/s datacom, where SiGe BiCMOS technology had carved a niche. The shift away from continual speed enhancement and towards higher digital signal processing content will place greater demand in the future for ultra-fast ADC techniques in applications such as 40-Gb/s fibre-optic, backplane and chip-to-chip communication, instrumentation, and digital cinema. In these applications, SiGe HBTs must be paired with nanoscale MOSFETs to deliver the required performance at acceptable power dissipation levels. Ultra high f_T and f_{MAX} values, simultaneously exceeding 250 GHz, as well as low phase noise are, however, required for the potentially lucrative 77/94GHz automotive, mm-wave imaging, and 100-Gb/s Ethernet applications. This paper addresses transistor performance scaling and building blocks for 40-80Gb/s transceivers with analog and digital signal processing and for mm-wave radio and radar.

II. SiGe HBT AND SERDES PERFORMANCE SCALING

Benefiting from the clear guidelines set forth by the International Roadmap for Semiconductors (ITRS),

CMOS technology scaling has continued unabated to nanometre dimensions. Cut-off and maximum oscillation frequencies in general purpose (GP) 65-nm CMOS technologies with physical gate lengths below 45 nm now exceed 200 GHz at 1V supply [9]. Power dissipation, noise figure, and phase noise performance in CMOS RF and fibre-optic ICs improve with scaling. At the same time, as the data compiled in Fig. 1 illustrate, SiGe BiCMOS technology evolution has been rather irregular, with missed nodes and at least 3 foundries jockeying up for the leading position at one time or another. Interestingly, a clear scaling law exists, as indicated by the dashed trend line, which allows SiGe BiCMOS to retain a 2-generation advantage over CMOS in terms of f_T and f_{MAX}.

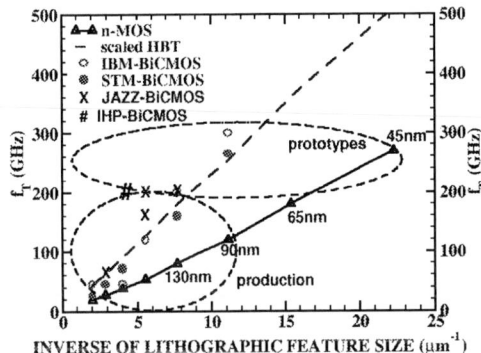

Fig. 1 f_T scaling in CMOS and SiGe BiCMOS technology nodes

Fig. 2 shows the measured f_T of production and of prototype SiGe HBTs as a function of collector current density, along with the f_T of a scaled device simulated using Silvaco's T-CAD platform. The scaled transistor has 100-nm emitter width, a 5-nm thick base doped 8×10^{19} cm^{-3}, a 25-nm thick collector, and features a trapezoidal Ge profile varying from 20% to 40%. Its NF_{MIN} is 1.1 dB at 65 GHz while f_T and f_{MAX} simultaneously exceed 500 GHz at a current density of 30 mA/μm^2 with a peak of 560 GHz at 60 mA/μm^2. The simulator was calibrated on experimental 160-GHz SiGe HBT characteristics from production SiGe BiCMOS processes [4], [6]. For comparison, an InP HBT with 12.5-nm base width and 710/340 GHz f_T/f_{MAX} at 20 mA/μm^2 was

1-4244-0458-4/06/$20.00 ©2006 IEEE

recently fabricated [10]. These results indicate that a SiGe HBT structure, much like the ones prototyping today, is capable of maintaining the 2× advantage over 45-nm MOSFETs. However, the significant increase in current density with scaling, and slightly higher NF_{MIN} than MOSFETs at comparable f_T, remain weak points of SiGe HBTs. Indeed, experimental device [11] and 60-GHz circuit data [12] indicate comparable or lower NF_{MIN} for 90/65-nm MOSFETs than for the best SiGe HBTs.

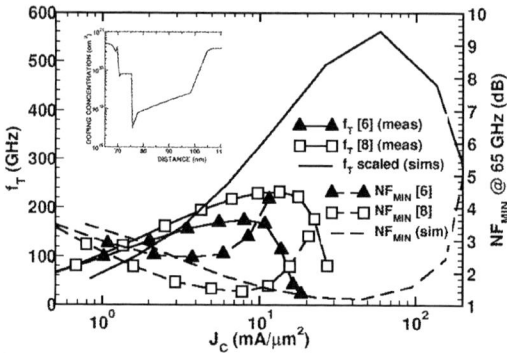

Fig. 2 f_T and NF_{MIN} at 65 GHz vs J_c in production [6], prototype [8], and in simulated scaled SiGe HBTs. The simulated structure is shown in the inset.

A fully integrated 5-GHz wireless transceiver and a 10-Gb/s SONET receiver were first reported at MRS in 1998 [13]. These circuits were implemented in an early version of a 0.5μm SiGe BiCMOS process with 45/60-GHz f_T/f_{MAX} [1]. The SONET receiver employed E²CL logic and operated from a -5.2V supply. The first single-chip 10-Gb/s SONET SERDES [14] and a 10-Gb/s Ethernet (10GE) SERDES (Fig. 3) fabricated in 0.35μm [2] and 0.25μm [3] SiGe BiCMOS processes were introduced in December 2000 and the summer of 2001, respectively. Both these ICs rely on simpler ECL families to reduce the supply voltage of the 10-Gb/s circuitry to 3.3V.

The 0.35μm SiGe HBT has an f_T of 45 GHz and allows for a 10-GHz latch to be realized with 2-mA tail current and no inductive peaking. The total power dissipation of the SONET SERDES is 2.2 W and includes a VCSEL driver with output swing and pre-emphasis control, laser bias and monitoring circuitry. The output swing and VCSEL modulating current are adjustable over a 3:1 range up to 2×750mV$_{pp}$ and 15mA, respectively [14]. The 10GE SERDES dissipates 3.2 W from 3.3-V, 2.5-V and 1.8-V supplies. It incorporates over 200,000 digital CMOS gates embedded between the analog 10.3-Gb/s and 3.125-Gb/s interfaces (Fig.3). Taking advantage of both SiGe HBT and CMOS scaling across technology nodes, and reducing the number of vertically-stacked transistors, two new generations of this chip have since been ported, first to 180-nm SiGe BiCMOS [4] dissipating about 2 W from 3.3-V and 1.8-V supplies,

and next to 130-nm CMOS, dissipating approximately 1 W from a 1.2-V supply.

The most interesting aspect of this SERDES family is that it demonstrates the benefits of scaling for mixed-signal SiGe BiCMOS products. Scaling leads to an almost linear improvement in jitter, rise/fall times, die area, supply voltage and power dissipation [15]. It also confirms the prediction made in [14] that only at the 130-nm node would CMOS be able to compete with SiGe BiCMOS technology for 10-Gb/s SERDES applications. Indeed, several 10-Gb/s SERDES chips in 130-nm CMOS, including a VCSEL driver [16], have been reported during the last 2 years while 180-nm CMOS solutions could not compete [17].

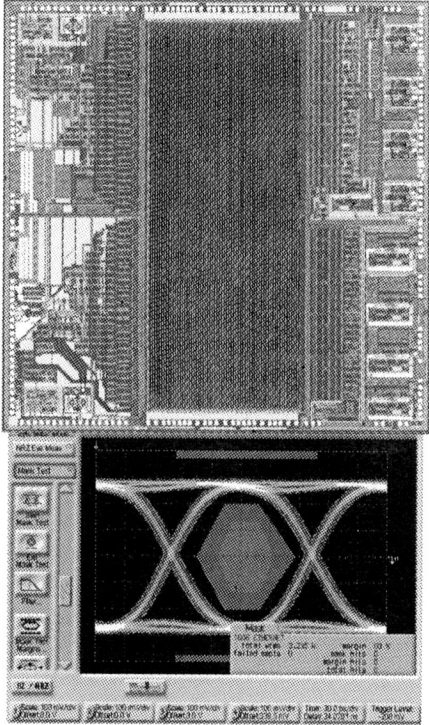

Fig. 3 Single-chip 10GE transceiver in 0.25μm SiGe BiCMOS: a) die photograph and b) 10.3-Gb/s output eye diagram (Quake Technologies 2001)

Although a 40-GHz flip-flop [18] and a 60-Gb/s 2:1 MUX [19] have recently been realized in 90-nm CMOS, a 45-GHz flip-flop, as needed for a full-rate 40-Gb/s SONET SERDES with FEC, has only been demonstrated in InP [20] and SiGe BiCMOS technologies [21]-[23]. Scaled 40/80-Gb/s SERDES can now be fabricated in 130-nm SiGe BiCMOS technology [6], operating from 2.5-V supply with similar power dissipation and performance margin (Fig.4) as the first-generation 10GE chip (Fig. 3). This (4-8)× speed increase [23] at comparable power dissipation and cost has been made possible by the 4× improvement in transistor f_T/f_{MAX} and by relying on a MOS-SiGe HBT cascode topology and inductive

224

peaking [23]. The combination of MOSFETs and HBTs on the high-speed path capitalizes on the low V_T of nanoscale MOSFETs, the cancellation of Miller capacitance, and on the low gate resistance to improve switching speed over HBT-only logic while reducing the power supply voltage to 2.5 V and 1.8 V [11, 24].

Fig. 5 illustrates the scaling of MOS-HBT cascode inverter delay between production 180-nm and 130-nm SiGe BiCMOS nodes with comparable SiGe HBT performance. The benefits of future scaling to 90-nm BiCMOS are also clearly apparent where the delay reduction from 5.1 ps to 3.3 ps is solely due to replacing 130-nm with 90-nm MOSFETs. The inset shows how emitter-followers (EF) can be employed for level shifting and further increase speed. Note that the optimal delay is achieved when the tail current corresponds to 0.3mA/μm of gate width for the MOSFETs that make up the differential pair [27].

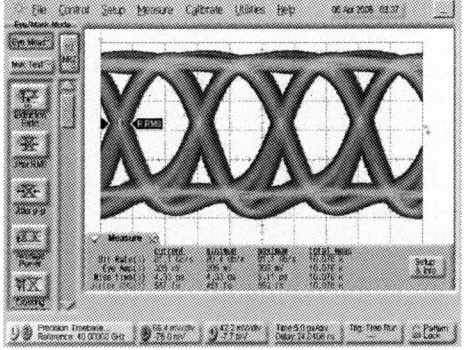

Fig. 4   2×300mV$_{pp}$, 80-Gb/s output eye diagram (running for 1 hour) of a 2.5-V, 1.4W transmitter [23] implemented in a 130-nm SiGe BiCMOS process [6].

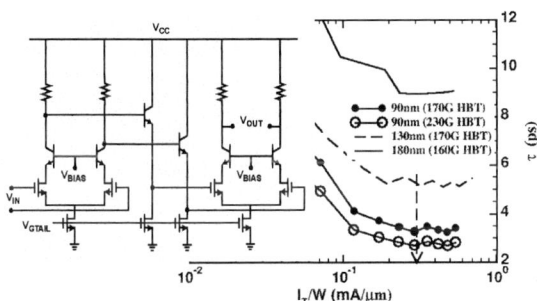

Fig. 5   Scaling of BiCMOS cascode inverter delay across nodes obtained from measured DC and high-frequency characteristics. All inverters have a gain of 1.5 and a fanout of 1. The impact of the EF stage is not included.

III. ANALOG CIRCUITS

Operational amplifiers (opamps) can be used in GHz-range filters and delta-sigma modulators if their unity gain bandwidth exceeds 10 GHz [25], [26]. The schematic of a fully differential opamp employing a MOS-HBT cascode with cascode p-MOSFET load [25] is shown in Fig. 6. The circuit has an EF output for level-shifting purposes and to reduce the output

impedance. It was implemented in 130-nm and 180-nm generations of SiGe BiCMOS technologies with similar HBT f_T/f_{MAX} [25],[26]. The MOS-HBT cascode has speed, linearity and noise advantages over the HBT-HBT cascode. This is the result of the lower input time constant $R_G \times (C_{gs}+C_{gd})$, better phase margin and concomitant low-noise and high-linearity bias in MOSFETs, as opposed to HBTs. To maximize the unity gain bandwidth, all MOSFETs and HBTs are biased at the peak f_{MAX} current density which, for n-MOSFETs and p-MOSFETs, is approximately 0.2 mA/μm and 0.08 mA/μm, respectively, across technology nodes and foundries [27]. A record unity gain bandwidth of 37 GHz was measured in the 130-nm half-circuit with 10-mA current [25]. Opamp biquad filters (Fig. 7) were fabricated and their performance was compared with that of 2-stage g_m-LC bandpass filters based on the same MOS-HBT cascode topology (Fig. 8). The measured transfer characteristics, NF and linearity of both filters are shown in Figs. 9-11. With comparable linearity and power dissipation, the opamp filter occupies 10 times smaller area than the g_m-LC filter.

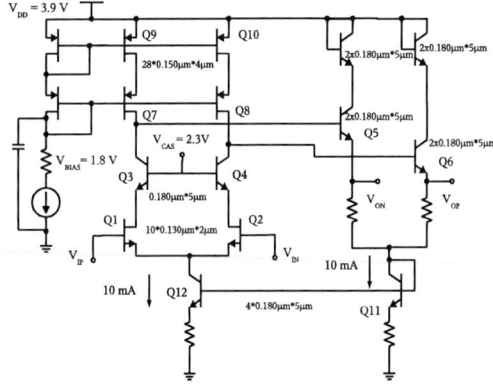

Fig. 6   130-nm SiGe BiCMOS opamp schematics.

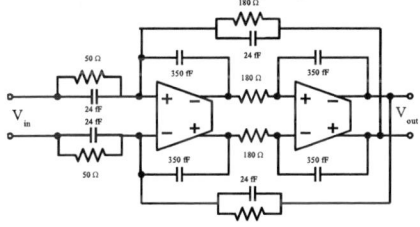

Fig. 7   130-nm SiGe BiCMOS biquad filter diagram.

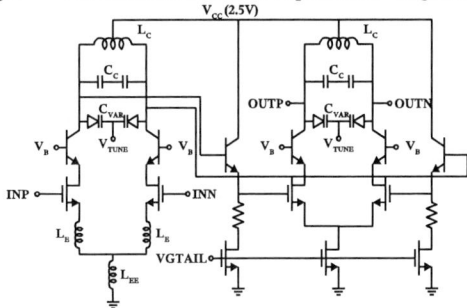

Fig. 8   Block diagram of 130-nm SiGe BiCMOS g_m-LC filter based on a BiCMOS cascode.

IV. 40-Gb/s Equalizers

Production SiGe BiCMOS technology is well-poised to address 45-Gb/s transceivers with equalization. A typical block diagram of an equalizer with analog DSP is shown in Fig. 12. While the FFE block could be realized in 90-nm CMOS, the DFE requires a full-rate 45-GHz flip-flop which is unlikely to be implemented in CMOS with adequate margin before the 65-nm node.

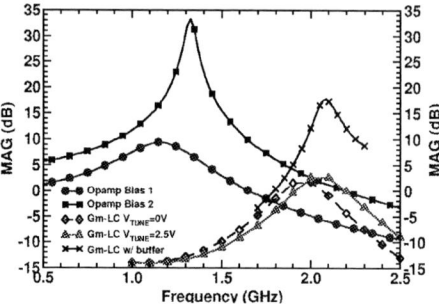

Fig. 9 Measured biquad and gm-LC filter characteristics for different bias conditions. Measurements of the g_m-LC filters with and without an output buffer are provided.

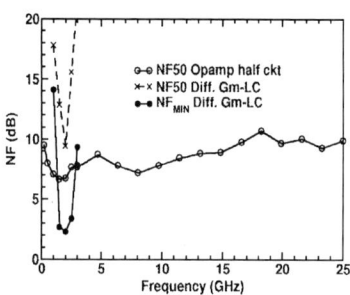

Fig. 10 Measured 50-Ω NF of opamp half-circuit and of differential gm-LC filter.

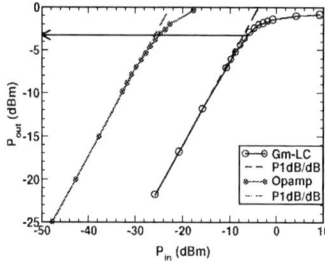

Fig. 11 Comparison of the measured linearity of two-stage opamp and g_m-LC filters operating at 1.2 GHz and 1.9 GHz, respectively.

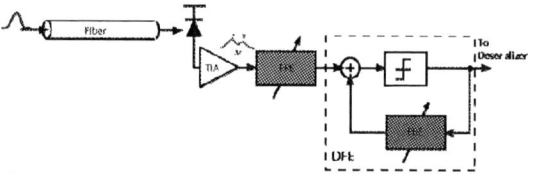

Fig. 12 Block diagram of electrical equalizer for polarization mode dispersion in optical fibers.

A. Feed Forward Equalizer

Fig. 13 describes the block diagram of a 7-tap 40-Gb/s FFE [28] implemented in a 180-nm SiGe BiCMOS process [4]. A differential, distributed topology is employed which is realized with M6-over-M2 microstrip lines and HBT-cascode taps with gain and sign control. The measured and modelled characteristics of the 9-ft SMA cable used in system-level simulations and experiments are illustrated in Figs. 14-16. The circuit is capable of compensating for more than 20dB loss at 20 GHz. Error-free operation was verified from 5 Gb/s to 40 Gb/s, and equalized eyes were obtained up to 49 Gb/s [28].

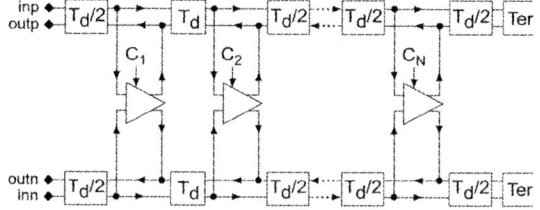

Fig. 13 Schematics of 7-tap FFE implemented in 180-nm SiGe BiCMOS.

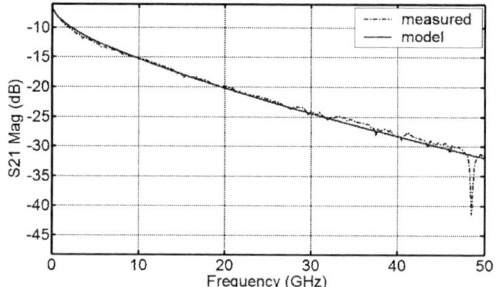

Fig. 14 Modeled vs. measured transmission loss of 9-ft SMA cable.

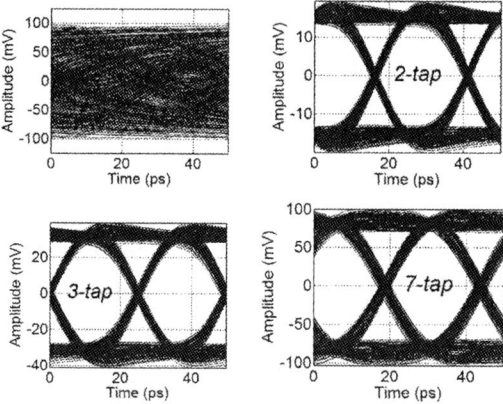

Fig. 15 Modelled 40-Gb/s equalized eye diagrams as a function of the number of equalizer taps. a) input signal after passing through the SMA cable and output of b) 2-tap, c) 3-tap, d) 7-tap equalizers.

B. 40 Gb/s DFE

The DFE whose block diagram is shown in Fig. 17, was implemented in the same technology [4] with 3.3-V ECL and verified to correctly equalize a PRBS

226

passing through a 9-ft SMA cable at data rates from 5 Gb/s to 39.5 Gb/s, Fig.18 [29].

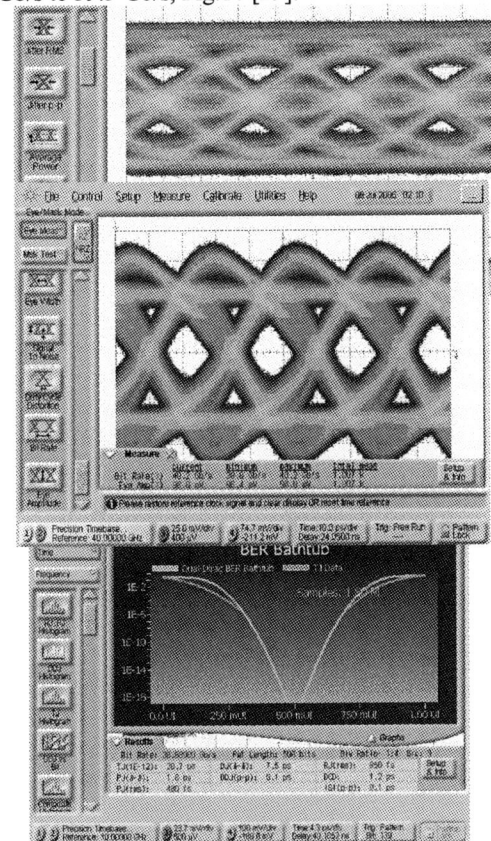

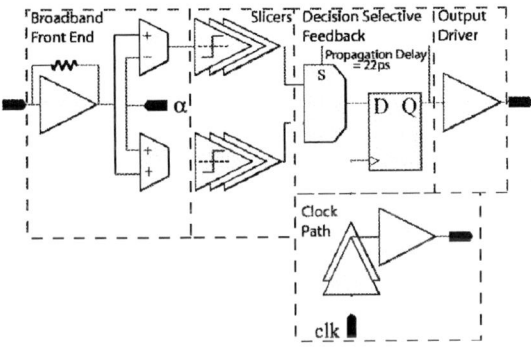

Fig. 16   FFE input/output 40-Gb/s eye diagrams and bathtub.

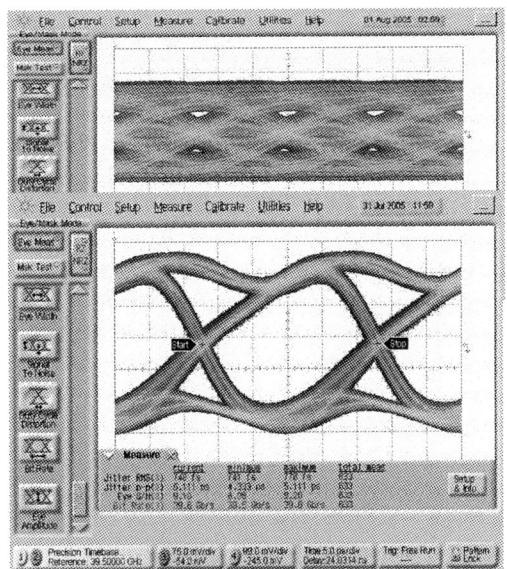

Fig. 18   39.5-Gb/s DFE input and output eye diagrams after passing through a 9-ft long SMA cable.

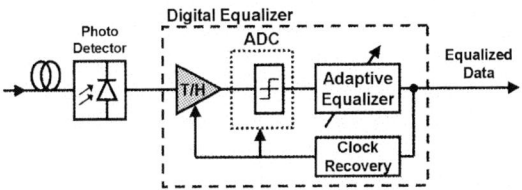

Fig. 19 Block diagram of a DSP based fiber optic equalizer.

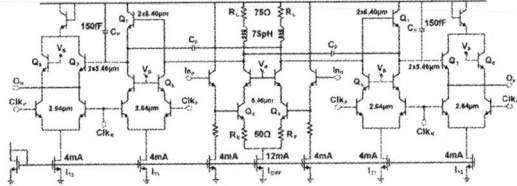

Fig. 20   Circuit diagram of the track & hold block in 180-nm SiGe BiCMOS [4]. Signals In_N and In_P are provided by a low-noise, broadband transimpedance amplifier [30].

Fig. 17 1-tap look-ahead DFE architecture.

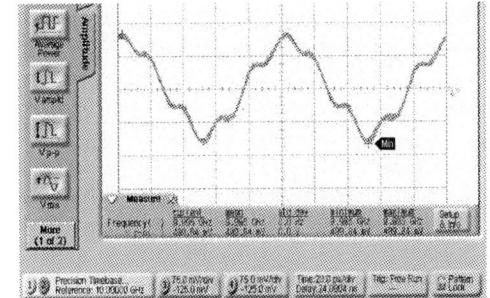

Fig. 21   Differential output of the T&H amplifier for an 10-GHz sinusoid sampled at 40GHz (70 mV/div) [30].

C. 40 Gb/s Digital Equalizer Blocks

When the technology has adequate speed, a digital DSP solution (Fig. 19) is preferred over an analog one due to its increased flexibility. In this architecture, the most critical block is the T&H amplifier (Fig. 20), which must simultaneously satisfy broad band, low noise and high linearity requirements. A T&H amplifier with over 40-GHz bandwidth (Fig. 21) and 4.5-bit linearity was reported in [30].

V. ADCs

The high intrinsic speed of both SiGe HBTs [1]-[8] and CMOS [11], [12] transistors provides an excellent incentive for the introduction of digital signal processing techniques relying on high oversampling ratios and mm-wave clock frequencies. Such techniques typically require only simple circuit topologies, similar to those used in a fiberoptic SERDES, that can be designed to be robust to process variations, transistor leakage and non-linearity. The 2-GHz bandpass $\Delta\Sigma$ design described in [31] employs the g_m-LC transconductor from Fig. 8 and the BiCMOS cascode MSM flip-flop [22] as the 40-GSamples/sec quantizer. The IIP3 test in Fig. 23 shows an SFDR of 61 dB with an ENOB of 8.5 over a 120 MHz band centred at 2 GHz. The 40-Gb/s output stream, reproduced in Fig. 24, can be decimated with the same 2.5-V CML BiCMOS logic family [22]-[24].

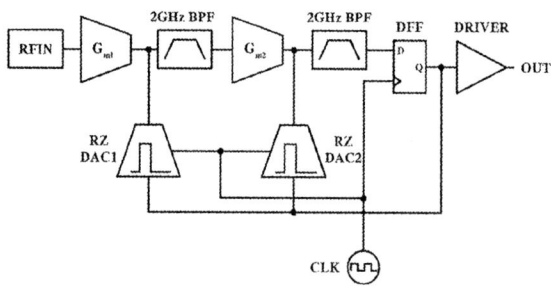

Fig. 22   System level $\Delta\Sigma$ ADC diagram [31].

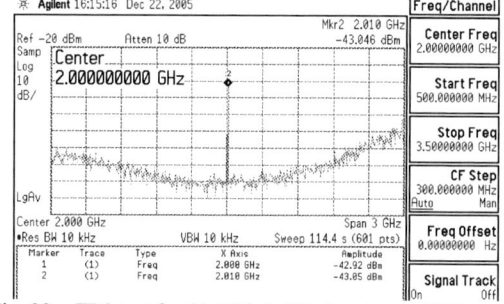

Fig. 23   IIP3 test for 40-GS/s 2-GHz bandpass ADC.

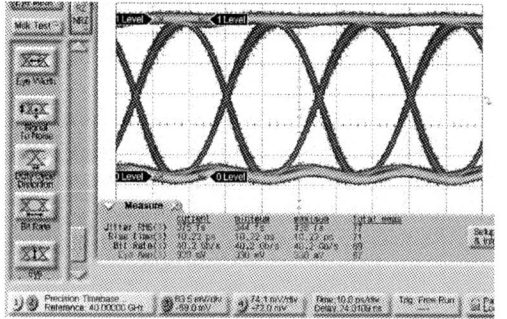

Fig. 24 $\Delta\Sigma$ bandpass ADC output eye diagram in open loop at 40Gb/s with a 2-GHz sinusoidal input.

VI. MILLIMETRE WAVE TRANSCEIVERS

Figs. 25 and 26 reproduce the schematics and die photos of a 60-GHz WLAN transceiver chip set and of a 65-GHz Doppler sensor with on-die patch antenna [32], [33] implemented in a production 180-nm SiGe BiCMOS process with f_T/f_{MAX} of 160 GHz

[4]. The transmitter has an image-reject architecture with stagger-tuned, two-stage, lumped poly-phase filters at 5 GHz and at 65 GHz. All three ICs include for the first time a low-phase-noise 60-GHz VCO with 10% tuning range [34] and rely on inductors and transformers to minimize die area. The VCO, downconvert mixer, LNA, and power amplifier employ SiGe HBT cascodes with inductive degeneration to improve isolation and simplify inter-stage matching. The entire design process was conducted using hand analysis and Spectre, as in the 2-10 GHz range [35], while inductors and transformers were designed using ASITIC.

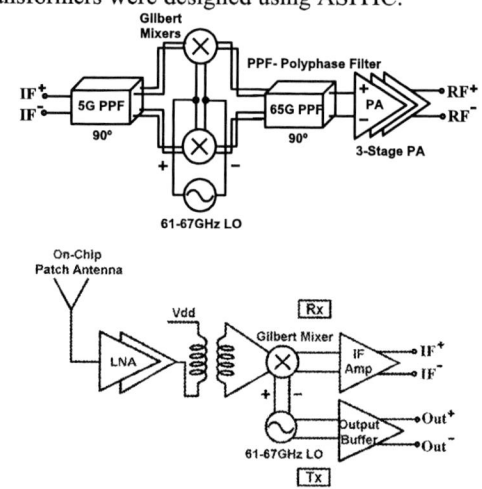

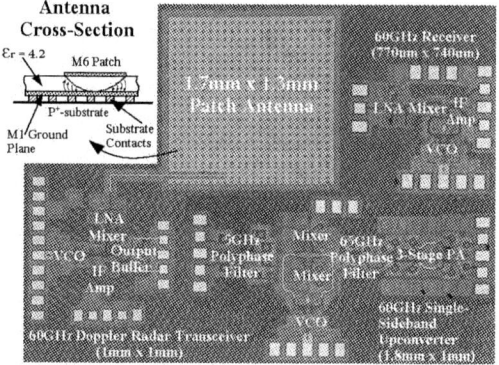

Fig. 25   Block diagrams of a) 60-GHz radio transmitter, b) Doppler radar implemented in 180-nm SiGe BiCMOS.

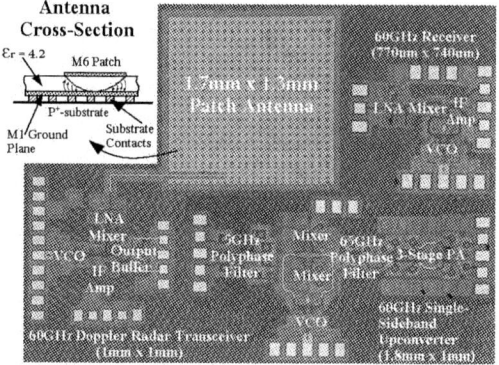

Fig. 26   Die photos of Doppler radar with patch antenna (left) WLAN transmitter (bottom) and receiver (top right).

Fig. 27 shows the measured conversion gain of the WLAN receiver measured on wafer, and of the Doppler sensor with and without the on-die antenna. The Doppler sensor with antenna was tested either by placing a horn antenna, or a GGB probe at different elevations above the on-die antenna. The double-sideband (DSB) noise figure, typically 12 dB, and the corresponding single-ended down-conversion gain of the WLAN receiver, 15 dB at 2.5 V, and 20 dB at 3.3 V supply, are compiled in Fig. 28. More than 45 dB of image rejection has been measured in the WLAN transmitter, Fig. 29.

228

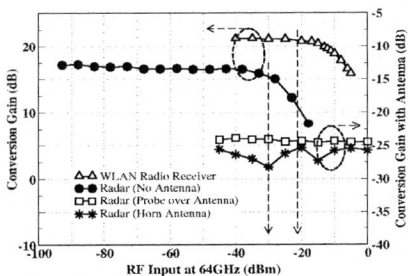

Fig. 27 Single-ended conversion gain for radar and WLAN receiver with radio IF at 1 GHz and radar IF at 730 MHz.

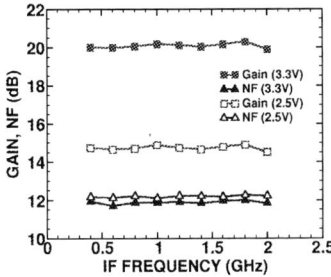

Fig. 28   Measured WLAN receiver gain and noise figure.

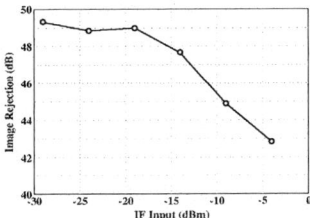

Fig. 29   SSB WLAN radio transmitter image rejection vs. IF input power with 61-GHz LO and 5-GHz IF.

Even though prototype SiGe HBTs and production 65-nm MOSFETS [12] now have adequate performance for automotive radar and imaging at 94 GHz [36],[37], on-chip isolation remains a major concern and will influence the choice of system architecture. For example, beyond 100 GHz, antennas are small enough for phase arrays to be integrated in silicon but crosstalk will limit their applicability. Alternatively, rectangular waveguide arrays are reasonably small, have excellent isolation and, in conjunction with finlines, provide more efficient radiation. For mm-wave medical imaging, where arrays of tens of transceivers are required, *low (phase) noise* [36], *isolation* and *power dissipation* are the most important design goals. Fig. 30 reproduces the measured NF_{MIN} at 5 GHz and at 65 GHz in HBTs with f_{MAX} of 200 GHz and 300 GHz, respectively. NF_{MIN} was obtained from Y-parameter measurements as in [35],[39]. The impact of transistor vertical profile scaling on NF_{MIN} is negligible at 5 GHz but becomes significant at 65 GHz. Furthermore, since the transit time through the collector space charge region is significantly larger than the base transit time, correlation between the collector and base shot noise currents increases, pushing the optimum noise current

density at mm-waves close to the peak f_T/f_{MAX} current density and changing the value of the optimum noise impedance. The latter has important implications for LNA and VCO design, and allowed us to achieve a record phase noise of -101.3 dBc/Hz at 1 MHz from the 103.6 GHz carrier, as shown in Fig. 31 [36].

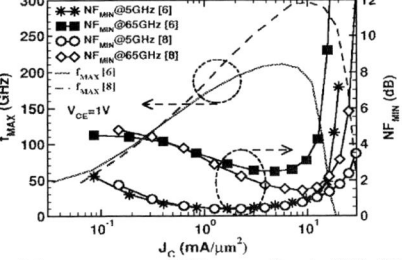

Fig. 30   Measurements of NF_{MIN} scaling in SiGe HBTs.

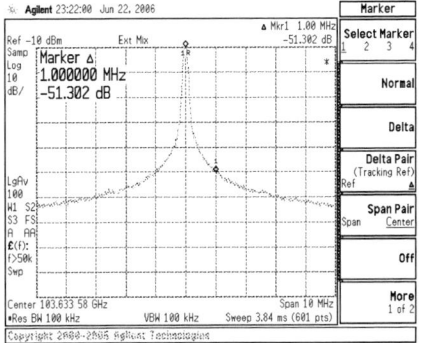

Fig. 31 Measured spectrum, phase noise of 103-GHz VCO.

VII. CONCLUSIONS

Low-frequency analog figures of merit such as g_m/I, g_m/g_o, lower $1/f$ noise and the larger $f_T \times BV_{CEO}$ product remain the main strengths of SiGe HBTs well into the mm-wave regime. However, historical data, moderate volume markets and relentless pressure from more aggressively scaled CMOS point to the importance of pairing next generation 250/500GHz SiGe HBTs with 90-nm CMOS as a viable and economical way to capture all applications demanding fast HBTs and at least some of the 90/65-nm CMOS market. In that scenario the discontinued supply-voltage scaling of 65-nm CMOS will help release most of the pressure off SiGe. Stand-alone SiGe HBT technology, stripped of its CMOS advantage, will have a difficult time competing with InP or CMOS.

ACKNOWLEDGEMENTS

We thank NSERC, Micronet, CITO, Gennum, NORTEL and STM for funding, and STM and Jazz Semiconductor for fabrication. Equipment grants from CFI and OIT, and CAD tools from CMC are also acknowledged. We are grateful to Dan Trepanier of Quake for the 10GE SERDES data, and to Pascal Chevalier and Alain Chantre of STM for 230-GHz HBT discussions. S. Voinigescu would like to thank

Rudy Beerkens, Bernard Sautreuil, Paul Kempf and Marco Racanelli for their support over many years, and his former colleagues at NORTEL and Quake Technologies in Ottawa, Canada.

REFERENCES

[1] D.L. Harame et al., "Si/SiGe Epitaxial-Base Transistors –Part II: Process Integration and Analog Applications," *IEEE Trans. ED*, Vol.42, pp. 459-482, 1995.

[2] A. Monroy, et al., "A high performance 0.35μm SiGe BiCMOS technology for wireless applications," *Proc. IEEE BCTM*, pp.121-124, Sept. 1999.

[3] H. Baudry et al., "High performance 0.25μm SiGe and SiGe:C HBTs using non-selective epitaxy," *Proc. IEEE BCTM*, pp.52-55, Sept. 2001.

[4] M. Racanelli, et al., "Ultra High Speed SiGe NPN for advanced BiCMOS technology," *IEEE IEDM Techn. Digest*, pp. 336-339, Dec. 2001.; M. Racanell and P. Kempf, "SiGe BiCMOS Technology for RF Circuit Applications," *IEEE Trans. ED*, Vol.52, pp. 1259-1270, 2005.

[5] B. Jaganathan, et al. "Self-aligned SiGe NPN transistors with 285 GHz f_{MAX} and 207 GHz f_T in a manufacturable technology," *IEEE Electron Device Lett.*, Vol.23, No.5, pp.258-260, March. 2002.

[6] M. Laurens et al, "A 150 GHz f_T/f_{MAX} 0.13 μm SiGe:C BiCMOS technology," *Proc. IEEE BCTM*, pp.199-202, Sept. 2003.

[7] T.F. Meister et al., "SiGe Bipolar Technology with 3.9ps gate delay,"*IEEE BCTM*, pp.103-106, Sept. 2003.

[8] P. Chevalier et al, "300-GHz f_{MAX} self-aligned SiGeC HBT optimized towards CMOS compatibility," *Proc. IEEE BCTM*, pp.120-123, Oct. 2005.

[9] B. Jagannathan, et al, "RF CMOS for Microwave and MM-wave Applications," *IEEE SiRF, Techn. Digest.* pp.259-264, Jan. 2006.

[10] W. Hafez et al., "12.5 nm base pseudomorphic heterojunction bipolar transistors achieving f_T=710 GHz and f_{MAX} = 340 GHz," *Appl.Phys.Lett.*, Vol.87, pp.252109-1/3, Dec. 2005

[11] P. Chevalier et al., "Advanced SiGe BiCMOS and CMOS Platforms for Optical and mm-wave ICs," *IEEE CSICS*, San Antonio, Nov. 2006.
D. Grindberg, "Silicon Heterostructure Handbook," pp.4.5-439-457 Taylor & Francis, 2006.

[12] T. Yao, et al., "60-GHz PA and LNA in 90-nm RF-CMOS," *IEEE RFIC Symposium Digest*, pp. 147-150, June 2006.
D. Alldred et al. "A 1.2V, 60GHz radio receiver with on-chip transformers and inductors in 90nm CMOS," *IEEE CSICS*, San Antonio, Nov. 2006.

[13] R. Hadaway, et al., "Application of SiGe HBTs to Datacom and Wireless," MRS, April, 1998.

[14] S.P. Voinigescu et al., "Circuits and Technologies for Highly Integrated Optical Networking ICs at 10Gb/s to 40Gb/s," *Proc. IEEE CICC*, pp.331-338, May 2001.

[15] D. Trepanier, *IEEE CSICS Panel* Nov. 2005.

[16] S. Rabii, et al., "An Integrated VCSEL Driver for 10 Gb/s Ethernet in 0.13μm CMOS," *IEEE ISSCC Digest*, pp.246-247, Feb. 2006.

[17] J. Cao et al, "OC-192 Receiver in Standard 0.18μm CMOS," *ISSCC Digest*, pp.250-251, Feb. 2002.

[18] T. Chalvatzis, et al., "A 40Gb/s Decision Circuit in 90-nm CMOS," *ESSCIRC*, Montreux, Sept. 2006.

[19] D. Kehrer et al., "A 60Gb/s 2:1 Selector in 90nm CMOS," *IEEE CSICS Digest*, pp.105-108, Oct. 2004.

[20] A. Hendarman, et al., "STS-768 Multiplexer With Full-Rate Output Data Retimer in InP HBT," *IEEE J.Solid-State Circuits*, Vol.38, pp.1497-1503, Sept. 2003.

[21] M. Meghelli, "A 108Gb/s 4:1 Multiplexer in 0.13 μm SiGe-Bipolar Technology," *ISSCC Digest*, pp.236-237, Feb. 2004.

[22] T.O. Dickson, et al., "A 2.5-V, 45-Gb/s Decision Circuit Using SiGe BiCMOS Logic," *IEEE J. Solid-State Circuits*, Vol.40, No.4, pp.994-1003, 2005.

[23] T.O. Dickson and S. P. Voinigescu., "Low-power circuits for a 10.7-to-86 Gb/s 80-Gb/s serial transmitter in 130-nm SiGe BiCMOS," *IEEE CSICS*, Nov. 2006.

[24] E. Laskin and S.P. Voinigescu, "A 60 mW per Lane, 4 × 23-Gb/s 2^7-1 PRBS Generator," *IEEE CSICS, Techn. Digest*, pp.192-195, Nov. 2005.

[25] S.P. Voinigescu, et al., "Design Methodology and Applications of SiGe BiCMOS Cascode Opamps with up to 37-GHz Unity Gain Bandwidth," *IEEE CSICS, Techn. Digest*, pp.283-286, Nov. 2005.

[26] G. Ng, et al., "1 GHz Opamp-Based Bandpass Filter," *IEEE SiRF, Techn. Digest*. pp.369-372, Jan. 2006.

[27] T.O. Dickson et al., "The Invariance of Characteristic Current Densities in Nanoscale MOSFETs and its Impact on Algorithmic Design Methodologies and Design Porting of Si(Ge) (Bi)CMOS High-Speed Building Blocks," *IEEE J. S-St. Circuits*, Aug. 2006.

[28] A. Hazneci and S. P. Voinigescu, "49-Gb/s, 7-Tap Transversal Filter in 0.18μm SiGe BiCMOS for Backplane Equalization," *IEEE CSICS*, pp.101-104, Oct.2004.

[29] A. Garg, et al., "A 1-Tap 40-Gbs Lookahead Decision Feedback Equalizer in 0.18μm SiGe BiCMOS Technology," *IEEE CSICS*, pp.37-41, Nov.2005.

[30] S. Shahramian, et al., "A 40-GSamples/Sec Track & Hold Amplifier in 0.18μm SiGe BiCMOS Technology," *IEEE CSICS*, pp.101-104, Nov. 2005.

[31] T. Chalvatzis and S. P. Voinigescu, "A Low-Noise 40-GS/s Continuous-Time Bandpass ΔΣ ADC Centered at 2GHz," *IEEE RFIC Symposium Digest*, pp. 323-326, June 2006.

[32] M. Gordon, et al.,"65-GHz Receiver in SiGe BiCMOS Using Monolithic Inductors and Transformers" *IEEE SiRF Techn. Digest*. pp.265-268, Jan. 2006.

[33] T. Yao, et al., "65GHz Doppler Radar Transceiver with On-Chip Antenna in 0.18μm SiGe BiCMOS," *IEEE IMS Digest*, pp.1493-1496, June 2006.

[34] C. Lee, et al., "SiGe BiCMOS 65-GHz BPSK Transmitter and 30 to 122 GHz LC-Varactor VCOs with up to 21% Tuning Range," *IEEE CSICS, Technical Digest*, pp.179-182, Oct. 2004.

[35] S.P. Voinigescu et al, "A scalable high-frequency noise model for bipolar transistors with application to optimal transistor sizing for low-noise amplifier design," *IEEE J. Solid-State Circuits*, vol. 32, pp. 1430-1438, Sept. 1997.

[36] S.T. Nicolson, et al, "Design and Scaling of SiGe BiCMOS VCOs Operating near 100 GHz," *IEEE BCTM 2006*.

[37] E. Laskin et al. "Low-Power, Low-Phase Noise SiGe HBT Static Frequency Divider Topologies up to 100 GHz," *IEEE BCTM 2006*.

[38] T.O. Dickson, S.P. Voinigescu, "SiGe BiCMOS Topologies for Low-Voltage Mm-Wave Voltage-Controlled Oscillators and Frequency Dividers," *IEEE SiRF, Techn. Digest*. pp.273-276, Jan. 2006.

[39] K.H.K. Yau and S.P. Voinigescu, "Modeling and Extraction of SiGe HBT Noise Parameters from Measured Y-Parameters and Accounting for Noise Correlation,"*IEEE SiRF Digest*. pp.226-229, Jan. 2006.

A 64-bit High-Speed Read-Write Look-Up Table Memory Implemented in GaAs HBT

Andre G. Metzger and Peter M. Asbeck

University of California, San Diego, California, USA

Abstract — This work describes a novel low-density bipolar memory circuit which functions near the maximum clock rate of the technology. An InGaP/GaAs HBT implementation of a 64-bit programmable Look Up Table memory is demonstrated. By simulation, the read function shows a 23ps delay through the memory array. Measurements of the circuit in package validates successful read operation at 5GHz.

Index Terms — heterojunction bipolar transistors, switching circuits, emitter coupled logic, read only memories, flip-flop memories

I. INTRODUCTION

High speed data processing frequently requires read/write memory cells that can be driven at the maximum clock rate. These functions are typically fulfilled by single-bit latching flip-flop circuits. When more than a few bits are required, the power demands of multiple flip-flops are very substantial, and one typically would resort to an arrayed memory design. The design of typical static random access memories, however, does not permit very high clock rates to be maintained. In this paper, a novel low-density bipolar memory circuit topology is presented which allows read and write functions near the maximum clock rate of the technology.

Where down-conversion and parallel processing is not possible, high-speed look up table (LUT) memories enable architectures that require processing data at the incoming bit rate. Applications of this type include nonlinear fiber-optic equalization [1,2]. Certain equalizer topologies and mixed signal conversion circuits require a low-density memory circuits that can be driven at the incoming bit rate. If the required memory size is sufficiently large, it is evident that a large bank of high speed latching flip-flop circuits cannot be used for this application. High-speed flip-flop cells, such as described in [3], are extremely fast but also very area and power hungry.

In this work, an InGaP/GaAs Heterojunction Bipolar Transistor (HBT) integrated circuit implementing a 64-bit programmable LUT memory was designed, fabricated, and evaluated in package. A read function fast enough for 10GHz clock rate was demonstrated by simulation and a 5GHz clock rate was confirmed experimentally.

In [4], a fiber-optic equalizer integrated circuit employing this high-speed 64-bit HBT LUT is used to associate received signal vectors, after conversion from analog to digital domain, with an estimate of the transmitted pattern.

II. INTEGRATED CIRCUIT TECHNOLOGY

CMOS is by far the most dominant technology for memory applications; it features high density integration, extremely low power dissipation, and low cost. But FET technologies have inherent series resistance in the channel that can slow down static memory circuits. To decrease channel on-resistance and increase transconductance, large FET widths are required. Although CMOS technologies have now exceeded 200GHz f_t and 100GHz f_{MAX}, large gate width NMOS structures can be significantly slower in high-speed switching applications than modern microwave bipolar transistors. Classical figures of merit such as f_t and f_{MAX} do not adequately represent expected circuit performance.

High transconductance and low output resistance make bipolar technologies intrinsically well suited for applications where high switching speeds and long line drivers are required. The biggest advantage over CMOS is the high current handling capability. This advantage is exploited to reduce RC time constants when driving resistive loads with rapidly switching currents.

To demonstrate the high-speed memory application, a mature and readily available InGaP/GaAs HBT technology was selected. Simulations show that the CML architecture allows the 50GHz f_t / 50GHz f_{MAX} technology to be sufficiently fast to demonstrate an adequately large memory circuit clocked at 10GHz. Characteristics of this technology include a minimum emitter size of 1.4x3.0mm, a double non-self aligned base contact, a beta of about 100, three layers of metal, and integrated schottky diodes and MIM capacitors [5]. High-speed LSI circuits were previously demonstrated in this technology [6].

III. INTEGRATED CIRCUIT DESIGN

Figure 1(a) shows a typical NMOS static random access memory (SRAM) unit cell while figure 1(b) shows a conventional bipolar variant [7]. Conventional bipolar SRAM takes advantage of a special dual-emitter transistor structure and the sharp exponential turn-on current vs. voltage characteristics of the transistors. Current is steered between the two emitters depending on the base-emitter voltage. The bipolar memory cell is "enabled" when the ROWU and ROWL lines are raised to logic high levels.

The conventional bipolar memory cell and possible high-speed variants thereof [8,9] were determined to be too slow for the target fiber-optic equalizer application. Standby current is minimized if ROWU is allowed to sit at a reduced logical low voltage (as close as possible to ROWL while maintaining the memory state); but the delay time to power-up any given ROW line is directly correlated to charge time. Standby current and ROW rise-times are improved when the load resistors of the unit cell are made larger. Unfortunately, the time to sense a voltage change on the COL output lines (such as would occur when the memory state is being read through a sense amplifier) is compromised by the choice of larger load resistors. A 64-bit bipolar SRAM design will result in load resistors in the k-ohm range to optimize standby currents and ROWU charge time.

To achieve a sub-50ps read-out time with eight memory cells per row and eight memory cells per column, load resistances must be reduced by at least one order of magnitude. Moreover, reduction of RC charge and discharge times must be decoupled from an excessive increase in standby current; the average power density in the array must be limited to avoid high temperature performance degradation and reliability concerns. A new memory cell design is required that can allow the designer to independently control standby current and load resistor value.

Figure 2(a) shows the new unit cell developed for high-speed low-density memory applications. For the fastest switch times, the memory cell and supporting circuitry is based exclusively on Current Mode Logic (CML). Standby current is set by the current source qcs and reference voltage VCS. Larger resistors r12 and r22 are cross-connected to define a positive feedback condition for the differential pair formed by q12 and q22. This maintains a bistable differential voltage across nodes A and B, which will assign a digital high or low to the memory cell. Line driver transistors q11 and q21 pull current from load resistors r11 and r21 to drive sense lines COL and COL-B. When ROW is brought to a logical high, the voltages at nodes A, B, C, and D are raised and all of the eight memory cells on the selected ROW are enabled for a read or write operation through COL.

The CML force-sense amplifier topology is shown in simplified form in figure 2(b). Current source qcs1 (enabled by column selection circuitry omitted from this figure) steers current between qse1 and the selected memory cell's q11 COL driver. When REF is set to the exact mid-level of the memory logic, a differential reading of the memory state will be produced at OUT. When REF is set to a differential

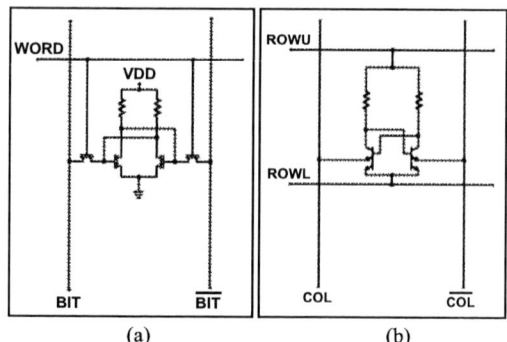

Fig. 1. Comparison of high-speed memory cells (a) NMOS SRAM, (b) bipolar SRAM

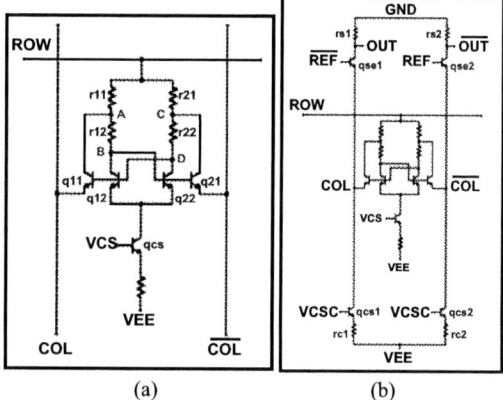

Fig. 2. (a) High-speed memory cell of this work (b) memory cell with CML sense amplifier

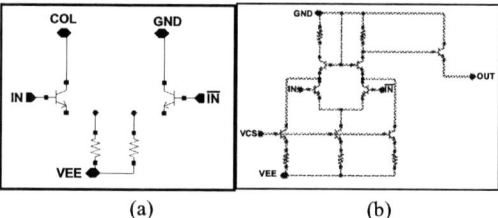

Fig. 3. (a) Row-select and (b) column select circuit

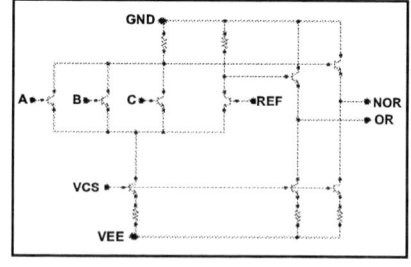

Fig. 4. 3-Bit X (ROW) and Y (COL) decode circuit

logic level, the memory state of the selected unit cell can be overwritten. Details of the row and column drivers are shown in figure 3.

232

An individual memory cell is selected within the 64-bit memory array with a 6-bit address vector. The leading 3-bits are used to decode the X-axis coordinates (ROW) and the trailing 3-bits are used to decode the Y-axis (COL). The CML decode circuit is shown in figure 4. Final bit selection of the entire array comes from one of the eight read-write amplifiers at the top of each column: qse1 and qse2 from figure 2(b). All eight read-write amplifiers have collectors tied to a common load resistor set: rs1 and rs2 (not specifically shown). As only one of the eight columns will be activated at any one time, the output at rs1 and rs2 is the output of the high-speed programmable LUT.

A block diagram of the memory circuit in its entirety is shown in figure 5. True to the layout style, columns and rows are placed staggered on both sides of the memory array to keep interconnect lines as short as possible and evenly spread line drivers across the die. An additional clock delay is inserted before a final retiming flip-flop.

IV. SIMULATION

Figure 6 shows the output eye diagram of the memory as simulated at 5GHz using Cadence's SPECTRE tools and including estimated parasitic capacitances. The circuit model is driven by a pseudo-random source which determines a repeating random address set. The memory is synchronously toggled between read and write modes at various times during the transient simulation. Vertical markers on the eye diagram are used to measure the worst-case time delay from the rowdriver's output to the sense amplifier's output; the most critical delay path on the chip. The latency during a read cycle is indicated with vertical markers: 22ps for the worst-case read cycle, 52ps for the worst case write cycle. This result suggests that the circuit will be able to run in read mode at the target clock frequency of 10GHz. The current design, as laid out, will not allow for a simple method to experimentally validate high-speed write performance.

V. LAYOUT AND PACKAGING

The GaAs HBT integrated circuit was packaged in an 32-pin multi-lead frame package. The package is mounted on a custom PCB with ample heat-sinking and with an interface for a computer controller. The die floor-plan was optimized to the package such that bondwires could be kept short and connect directly to 50-ohm board traces

The 64-bit LUT occupies 2.0mm x 2.0mm of chip area (including output buffers but not bondpads), contains approximately 500 transistors, and dissipates approximately 6 Watts at –5.2V. A die photo of the assembled unit is shown in figure 7.

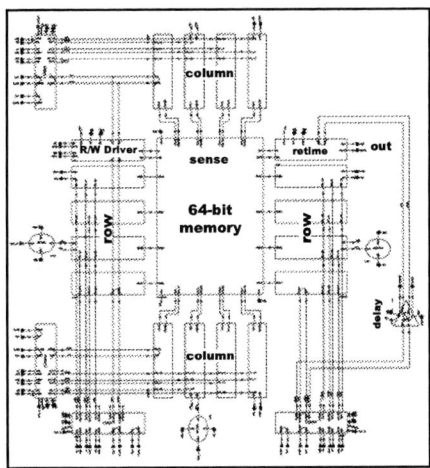

Fig. 5. Memory cell block diagram indicating staggered placement strategy for row and column drivers

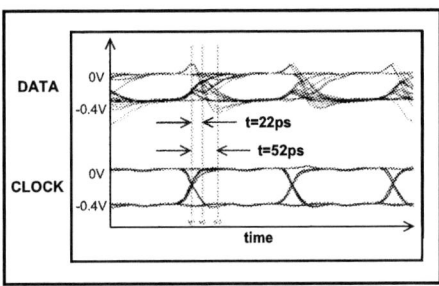

Fig. 6. Simulated read/write timing at 5GHz

Fig. 7. GaAs die in package, as wirebonded for test

V. TESTS AND MEASUREMENTS

Functionality is demonstrated as shown in figure 8 with a digital test sequence that validates all memory states using an Agilent 16500B mainframe pattern generator and logic analyzer up to 100MHz. A pattern representative of a 2^7 PRBS sequence is first written to the memory, and then this same sequence is read out of the memory. Data in the memory was successfully retained over a 30 minute period.

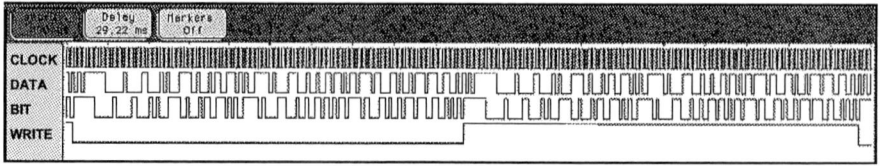

Fig. 8. Measured low-speed functionality read/write test

High speed tests made use of an Anritsu 12.5Gbps pseudo random pattern generator source. After low-speed programming, a sequence of 1's and 0's was used to toggle the LUT between two memory cells holding different bit values. The results indicated successful read operation up to 5GHz, and the associated output eye diagram is shown in figure 9. Above 5GHz, there was some indication of bit errors. It is unclear whether the bit errors were a result of limitations of front-end circuit performance (other components driving the memory), packaging issues, or difficulty in getting the heat out of the package.

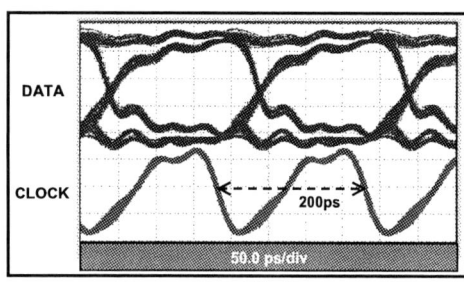

Fig. 9. Data and clock output as measured at 5GHz with GaAs die bonded and packaged

VII. CONCLUSION

An InGaP/GaAs HBT implementation of a 64-bit programmable LUT is demonstrated by simulation up to 10GHz and validated experimentally up to 5GHz. This circuit validates a novel low-density high-speed programmable LUT topology which can enable data processing at close to the maximum bit rate possible.

ACKNOWLEDGEMENT

The authors wish to acknowledge Skyworks Solutions Inc for the fabrication of the GaAs IC and assistance in prototyping. Additional thanks to Peter Zampardi, Charles Chang, Ravi Ramanathan, Theresa Leyva, and Ken Weller.

REFERENCES

[1] A. Garg, A. C. Carusone, and S. P. Voinigescu, "A 1-Tap 40-Gbps Lookahead Decision Feedback Equalizer in 0.18μm SiGe BiCMOS Technology", *27th IEEE Compound Semiconductor IC (CSIC) Symposium*, Session B.4, Palm Springs, CA, Nov 2nd, 2005.

[2] H-M. Bae, J. Ashbrook, J. Park, N. Shanbhag, A. Singer, S. Chopra, "An MLSE Receiver for Electronic-Dispersion Compensation of OC-192 Fiber Links", *2006 IEEE International Solid-State Circuits Conference*, Session 13.2, San Francisco, CA, Feb 5-9, 2006.

[3] R.B. Nubling, N.H. Sheng, K.C. Wang, M.F. Chang, W.J. Ho, G.J. Sullivan, C.W. Farley, and P.M. Asbeck, "25 GHz HBT Frequency Dividers", *IEEE Gallium Arsenide Integrated Circuit (GaAs IC) Symposium*, 22-25 Oct. 1989, p: 125 - 128.

[4] A.G. Metzger, P.M. Asbeck, "A Nonlinear Electronic Equalizer implemented in InGaP/GaAs HBT for Dispersion Compensation of Gigabit Optical Fiber Links", *Accepted paper: 2006 IEEE Compound Semiconductor Integrated Circuit Symposium (CSICS)*, San Antonio, Texas, Nov 12-15, 2006.

[5] R.T. Huang, D. Nelson, S. Mony, R. Tang, R. Pierson, J. Penney, R. Sahai, "Manufacturing AlGaAs/GaAs HBTs on 100 mm wafers", *Gallium Arsenide Integrated Circuit (GaAs IC) Symposium*, Oct 10-13, 1993, p. 345 - 348.

[6] Metzger, A.G.; Chang, C.E.; Pedrotti, K.D.; Beccue, S.M.; Keh-Chung Wang; Asbeck, P.M, "A 10-Gb/s high-isolation, 16×16 crosspoint switch implemented with AlGaAs/GaAs HBT's", *IEEE Journal of Solid-State Circuits*, Volume 35, Issue 4, April 2000, p. 593 – 600.

[7] D. Hodges, H. Jackson, Analysis and Design of Digital Integrated Circuits, McGraw-Hill Inc., New York, New York, 2nd edition, p. 342-386, 1998

[8] Heald, R.; Herndon, W.; In-Nan Wu; Si-Yu Chen, "A 15ns 64K bipolar SRAM", *IEEE Solid-State Circuits Conference Digest of Technical Papers*, Feb 13, 1985, p. 50 – 51.

[9] P.J. Zampardi, S.M. Beccue, R.L. Pierson, W.J. Ho, J. Yu, M.F. Chang, A. Sailor, K.C. Wang, "Threshold Tunable MESFET and Ultra-Fast Static RAM", International Semiconductor Device Research Symposium, San Diego, 18-22 September 1994, p. 657-662.

2006 Bipolar/BoCMOS Circuits and Technology Meeting Proceedings

Low-Power, Low-Phase Noise SiGe HBT Static Frequency Divider Topologies up to 100 GHz

E. Laskin[1], S. T. Nicolson[1], P. Chevalier[2], A. Chantre[2], B. Sautreuil[2], S. P. Voinigescu[1]

1) Edward S. Rogers Sr. Dept. of ECE, University of Toronto, Toronto, ON M5S 3G4, Canada
2) STMicroelectronics, 850 rue Jean Monnet, F-38926 Crolles, France

Abstract — Static 2:1 frequency dividers with different latch configurations were designed and fabricated in two SiGe HBT technologies. The self-oscillation and maximum operation frequency are found to be correlated with the f_{MAX} but not with the f_T of the technology. A self-oscillation frequency of 77 GHz, the highest among static dividers in SiGe HBT technology, is achieved with the lowest power consumption of 122 mW from 3.3 V supply. Phase noise measurements of the 100-GHz input and 50-GHz output signals indicate ideal behavior with no measurable noise contribution from the divider. All fabricated dividers feature an integrated single-ended-to-differential transformer for ease of testing, yet broadband 20 GHz-100 GHz operation is maintained.

Index Terms — Static frequency dividers, SiGe HBT.

I. CIRCUIT DESIGN

A study of static divider performance in two SiGe technologies was performed. For a fair comparison, the same divider topology, shown in Fig. 1, was maintained and the same supply voltage was used. The single-ended input is converted to a differential signal through an on-chip transformer. A 50-Ω, 250-mVpp swing buffer is used at the output. Two different latch configurations are implemented in the core of the divider for comparison.

A. Integrated Transformer

A 1:1 transformer similar to that in [1] was scaled to 100 GHz and realized with vertically-stacked, symmetrical square coils in the top 2 metal layers of the technology. The transformer diameter is 30 μm, with 2-μm stripe width and 1-μm spacing. The coupling coefficient between the two coils is 0.855. The role of the transformer is solely to facilitate divider measurements and to avoid the use of expensive and bulky test setups. A matching network consisting of an inductor and a capacitor was added to match the divider input to 50 Ω. Simulation results of the transformer with matching network are

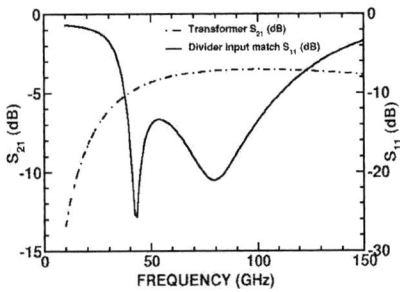

Fig. 2.   Simulated on-chip transformer performance.

reproduced in Fig. 2. S_{21} peaks at 100 GHz, with a maximum value of -3.5 dB, while S_{11} remains below -10 dB from 30 GHz to 110 GHz. This is the first demonstration of an integrated transformer in a circuit operating at 100 GHz.

B. Latch Design

Two different ECL latches, with and without emitter followers (EF) on the clock input path, shown in Fig. 3a and 3b, respectively, were designed. Notice that, apart from peaking inductors operating at 100 GHz, no bandwidth improvement techniques, such as split-resistor load [1] or double emitter-followers in the feedback [2], are employed. The use of inductors is critical to reduce power dissipation while still achieving high frequency operation. The first version, with double EF on the clock input path, operates from 3.3 V and consumes 19.5 mA. The second latch consumes only 11.5 mA from 3.3 V.

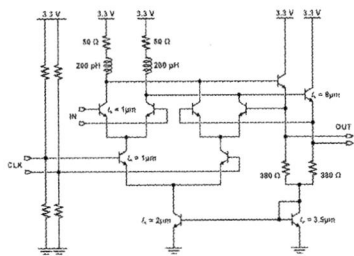

Fig. 3a.   Schematic of the latch with emitter followers.

Fig. 3b.   Schematic of the latch without emitter followers.

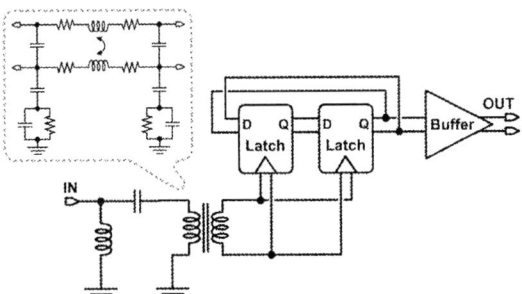

Fig. 1.   Static 2:1 frequency divider topology.

II. FABRICATION

Die photos of the two dividers are shown in Fig 4. Both dividers were fabricated in two SiGe HBT technologies with exactly the same circuit layout and without redesign between the two technologies. The first technology is a 0.13 μm SiGe BiCMOS process (referred to as BiCMOS9 [3]). Only the BiCMOS9 HBTs, with an f_T of 170 GHz and f_{MAX} of 200 GHz, were utilized. The second technology (referred to as BipX [4]) features HBTs with an f_T of 230 GHz and f_{MAX} of 300 GHz. Measured f_T and f_{MAX} curves for the two technologies are plotted versus collector current density in Fig. 5. Dividers were also fabricated and tested in two other BipX process splits with f_T/f_{MAX} in the 220 GHz to 270 GHz range.

III. TEST SETUP

The test setup used for divider characterization is shown in Fig 6. All measurements were performed directly on wafer with 110-GHz GGB probes. An HP 83752A 0.01-20-GHz signal source and Millitech AMC 15 00000 ×4 and AMC 10 00000 ×6 frequency multipliers were employed to generate the input signal in the 50-75 GHz and 75-100 GHz frequency bands, respectively. Below 50 GHz, measurements were carried out with the Agilent E8257D PSG analog 250 kHz-67 GHz signal generator as the signal source. A 50-GHz Agilent E4448A PSA with an Agilent 11970W 75-110 GHz waveguide mixer was used to observe the spectrum and phase noise of the input and output signals.

IV. MEASUREMENT RESULTS

All fabricated dividers operate from 3.3 V and consume 122 mW without clock EF, and 145 mW with clock EF. Their sensitivity curves are plotted in

 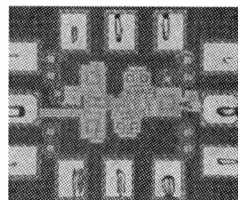

Fig. 4.   Die photos of the divider a) with (left - 515μm × 473μm) and b) without EF (right - 502μm × 360μm).

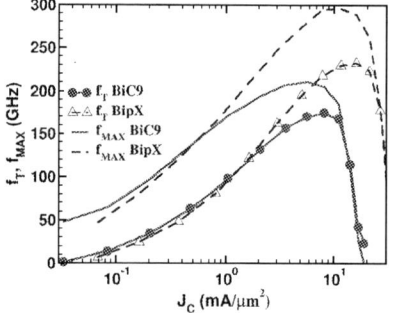

Fig. 5.   Measured f_T and f_{MAX} at V_{CE}=1V on transistors with 3 emitter stripes, 6 base, and 4 collector contacts in the two technologies (BiCMOS9 and BipX).

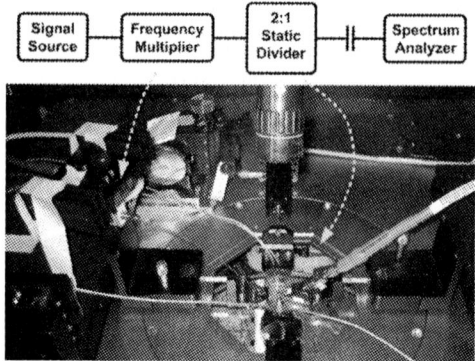

Fig. 6.   Divider test setup.

Fig. 7. In both technologies, the elimination of the EF stages from the clock input path results in a 20% increase in the self-oscillation frequency (SOF). This is in agreement with simulations and with the reduced voltage gain of the EF at high frequency [5]. Fig. 7 also shows that the SOF is correlated with the speed of the technology. By moving to a faster technology, the transistor f_{MAX} is increased 1.5 times from 200 GHz to 300 GHz, and the SOF of the dividers increases from 54 GHz to 77 GHz. The results on different BipX splits, with f_T of 230 GHz and 265 GHz, reveal lower divider SOF for the 265-GHz f_T circuits, in agreement with the HBT power gain at 65 GHz, rather than its f_T. Measurements also indicate that the maximum divider frequency tracks its SOF. Therefore, we recommend that the self-oscillation frequency be used to more fairly compare the performance of static dividers because it depends solely on the circuit parameters. On the other hand, the maximum operating frequency of the divider depends to a large extent on the test setup and on the power available from the signal source.

The fastest divider has a SOF of 77 GHz, the highest reported to date in any SiGe HBT/BiCMOS technology. It divides correctly up to at least 100 GHz. The spectrum of the 50-GHz divided-by-two output signal is shown in Fig. 8.

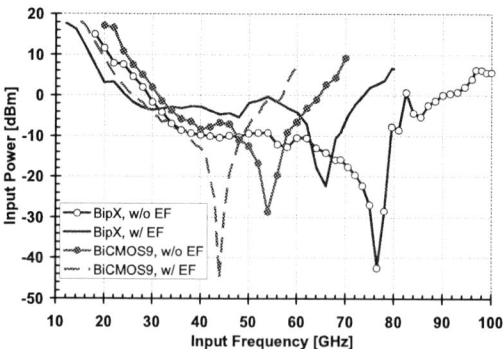

Fig. 7.   Measured sensitivity curves of the 2 static dividers in the BiCMOS9 and BipX technologies.

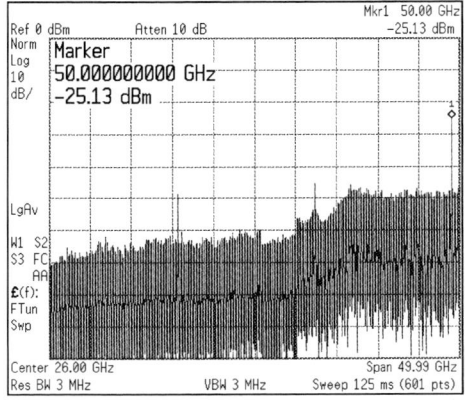

Fig. 8. Spectrum of the divide-by-two 100-GHz signal obtained from the BipX divider without EF.

In order to assess the noise added by the divider, the phase noise of the 100 GHz multiplier output signal was measured as a function of the multiplier input power at 16.667 GHz (Fig. 9). The measurement accuracy is limited by the mixer noise floor for low-power inputs, and by the mixer non-linearity for high-power inputs. Fig. 10 shows a phase noise of -96.4 dBc/Hz at 100 kHz offset from the 50-GHz divider output. For comparison, the measured phase noise at a 100 kHz offset from the 100-GHz divider input signal is -90.4 dBc/Hz (with 0.3 dBc/Hz measurement error). This proves that the divider noise contribution is negligible since the phase noise is improved by an ideal 6 dB after the divide-by-two operation. This is an important result for PLL applications, where the phase noise of the VCO must not be degraded by the divider. Lack of a signal source with sufficient output power prevented divider testing above 100 GHz. The same divider was tested over temperature and found to divide up to 97 GHz at 50°C and up to 91 GHz at 100°C, Fig. 11. To explore the effect of process variations on the performance of the divider without EF, a wafer map of its SOF in the BipX process is presented in Fig. 12. Sensitivity curves collected from 12 dice on 4 wafers are shown in Fig. 13 for the BiCMOS9 version.

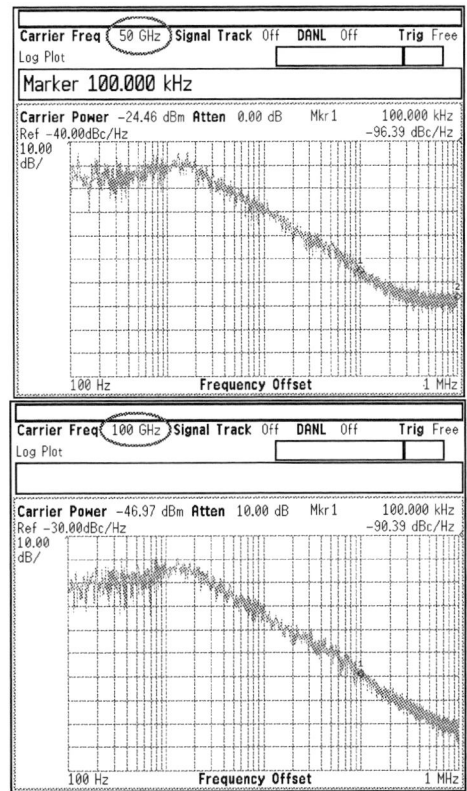

Fig. 10. Phase noise of the divider without emitter followers fabricated in BipX (top) with 100-GHz input signal. Phase noise of the 100-GHz input signal (bottom).

V. COMPARISON WITH PREVIOUS DIVIDERS

The two dividers fabricated with 170-GHz HBTs, are compared in Fig. 14 with a previously published divider fabricated in the same technology and which is based on a MOS-HBT (BiCMOS) cascode topology [1]. In the BiCMOS cascode latch, the clock HBT pair is replaced by an NMOS pair [1]. Since the BiCMOS latch is 15% faster, it is expected that the self-oscillation frequency of the 230-GHz HBT divider can further be improved by employing a BiCMOS cascode when 90-nm MOSFETs are available.

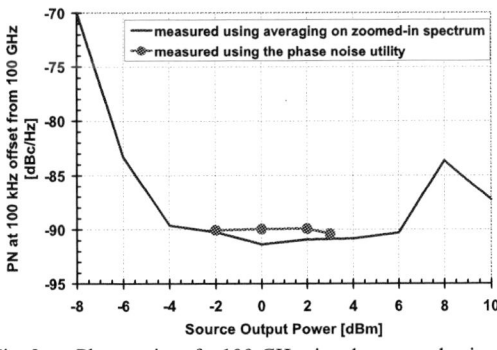

Fig. 9. Phase noise of a 100-GHz signal measured using two methods. This 100-GHz signal was used as the divider input.

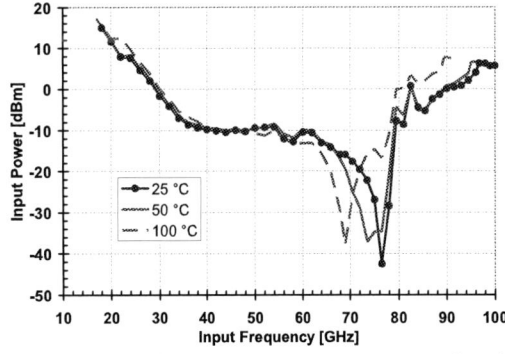

Fig. 11. Sensitivity curves across temperature for the divider without EF fabricated in the BipX technology.

237

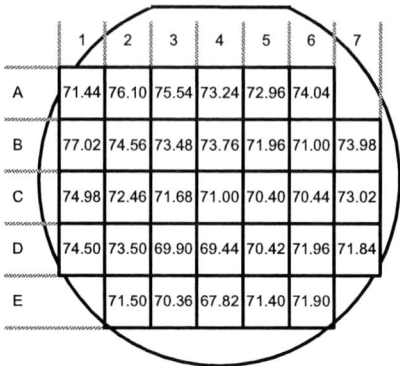

Fig. 12. Wafer map of the self-oscillation frequency (in GHz) for the BipX dividers without EF.

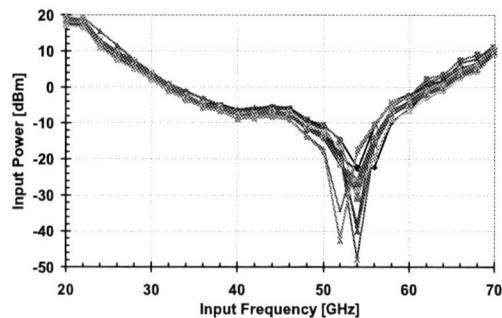

Fig. 13. Effect of process variations on the divider without EF fabricated in the BiCMOS9 process.

Finally, dividers presented in this work are compared to other state-of-the-art dividers in Table 1. With a SOF of 77 GHz and a power dissipation of 122 mW, this design has the lowest power and the highest self-oscillation frequency among all SiGe HBT dividers.

VI. CONCLUSION

Static frequency dividers with two different topologies were designed and fabricated in 170-GHz and 230-GHz SiGe HBT technologies. It was observed from measurements and simulations that dividers operate faster when EF are removed from the clock path. The maximum operation frequency and the SOF were found to be correlated with the transistor f_{MAX} but less so with f_T. A 1:1 transformer was employed to provide single-ended-to-differential conversion of the input signal, yet the circuit operates from 20 GHz to 100 GHz. A record low power of 122 mW and operation up to at least 100 GHz (limited by the measurement setup) was demonstrated with no measurable phase noise contribution from the divider.

REFERENCES

[1] T. O. Dickson and S. P. Voinigescu," SiGe BiCMOS Topologies for Low-Voltage Millimeter-Wave Voltage-Controlled oscillators and Frequency Dividers" *SiRF 2006 Techn. Digest*, pp. 273–276.

[2] A. Rylyakov, and T. Zwick, "96-GHz Static Frequency Divider in SiGe Bipolar Technology," *JSSC*, vol. 39, no. 10, pp. 1712–1715, Oct. 2004.

[3] M. Laurens, *et al.* "A 150GHz f_T /f_{MAX} 0.13μm SiGe:C BiCMOS technology," *BCTM 2003 Proceedings*, pp. 199–202.

[4] P. Chevalier, *et al.* "300 GHz f_{max} self-aligned SiGeC HBT optimized towards CMOS compatibility" *BCTM 2005 Proceedings*, pp. 120-123.

[5] M. Mokhtari, *at al.* "100+ GHz Static Divide-by-2 Circuit in InP-DHBT Technology," *JSSC*, vol. 38, no. 9, pp. 1540–1544, Sept. 2003.

[6] M. Rodwell, *et al.* "Transistor and Circuit Design for 100-200 GHz ICs," *JSSC*, vol. 40, no. 10, pp. 2061–2069, Oct. 2005.

[7] J. O Plouchart, *et al.* "Performance Variations of a 66GHz Static CML Divider in 90nm CMOS," *ISSCC 2006 Digest*, pp. 526–527.

[8] S. Trotta, *et al.* "110-GHz Static Frequency Divider in SiGe Bipolar Technology," *CSICS 2005 Digest*, pp. 291–294.

[9] D. A. Hitko, *et al.* "A Low Power (45mW/Latch) Static 150 GHz CML Divider," *CSICS 2004 Digest*, pp. 167–170.

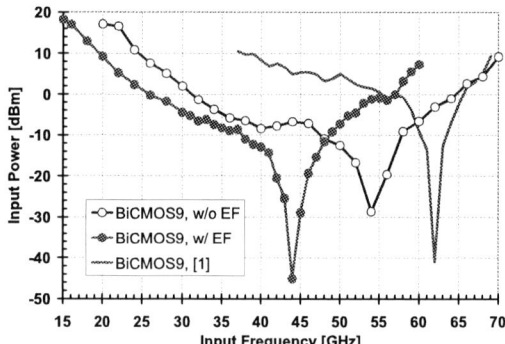

Fig. 14. Comparison of the two HBT-only dividers to a MOS-HBT static divider from [1]. All dividers were fabricated in the same BiCMOS9 process and were biased from a 3.3V supply.

Table. 1. Comparison to Previously Published Static Frequency Dividers.

Reference	Self-Oscillation Freq.	Max. Divider Freq.	Power Consumption	Technology
[6]	86 GHz	150 GHz	?	450-GHz f_T InP
[5]	33 GHz	100 GHz	750 mW	135-GHz f_T InP
[7]	48 GHz	66 GHz	80 mW (1.8V)	90nm CMOS
[8]	65 GHz	110 GHz	1.35 W (-5.2V)	225-GHz f_T SiGe:C HBT
[9]	95 GHz	143.6 GHz	90 mW	400-GHz f_T InP
[2]	71 GHz	96 GHz	770 mW (-5.0V)	210-GHz f_T SiGe HBT
This Work	77 GHz	>100 GHz	122 mW (3.3-3.6V)	230-GHz f_T SiGe HBT
This Work	54 GHz	70 GHz	145 mW (3.3-3.6V)	170-GHz f_T SiGe HBT

2006 Bipolar/BoCMOS Circuits and Technology Meeting Proceedings

A Low Power 12.5Gb/s SiGe Limiting Amplifier Using a Feed-forward Adjustable Threshold Loss-Of-Signal Detector

A. Maxim, D. Smith

Maxim Inc., 4201 Monterey Oaks Blvd., Austin TX 78749, Email: adrianmaxim@ieee.org

Abstract — A 12.5Gb/s limiting amplifier was realized in a 0.18μm 80GHz f_T SiGe BiCMOS process. The wide bandwidth signal path consists of a cascade of emitter followers and cascoded differential stages, resulting in a significant power consumption reduction in comparison with standard Cherry-Hooper implementations. The output offset voltage was reduced to fractions of mV by using a dedicated active offset cancellation loop for each gain stage. The compensation capacitors were integrated on-chip by using a Miller multiplication effect. A loss-of-signal detector that can sense signals down to mV level was achieved using a feed-forward architecture with data dependent comparison thresholds. The main IC specifications include: 60dB gain, 13GHz bandwidth, 1mV sensitivity, 0.2mV output offset voltage, 30mA current consumption from a 3.3V supply and 1.5x1.5mm² die area.

Index terms — limiting amplifier, SiGe, SONET.

I. INTRODUCTION

Most existing 10Gb/s limiting amplifiers use the Cherry-Hooper (CH) gain stage due to its high bandwidth efficiency, at the expense of larger power dissipation and higher circuit complexity [1-6]. This architecture is optimal for 10Gb/s ICs realized in processes with relatively low transition frequency (f_T=30 to 50GHz). The modern SiGe processes having transition frequencies in excess of 80 to 100GHz allow the use of standard differential pairs as gain stages, achieving a lower power consumption in comparison with standard CH implementations [7]. This paper proposes an alternative way of realizing a wide bandwidth signal path using cascoded differential stages that present a lower input capacitance, which further reduces the power consumption in the driving stages.

DC offset cancellation loops are usually used to minimize the output offset voltage. Passive [1-3] and active [4-7] offset cancellation loops were implemented in the past, their low cut-off frequency requiring off-chip compensation capacitors. Recently, a dual offset cancellation loop architecture was introduced to minimize the output offset voltage, while the Miller multiplication was used to integrate on-chip the compensation capacitors [7]. This design proposes a multi-loop architecture that uses a dedicated DC offset cancellation loop for each gain stage, achieving output offset voltages down to fractions of mV and allowing the on-chip integration of the offset loops compensation capacitors. The nowadays trend towards compact optical receiver assemblies requires on one hand a low power and thus a low supply voltage limiting amplifier and on the other hand a minimization of the off-chip components. Reducing the die area (cost) places an additional challenge on the DC offset cancellation loop design, since their compensation capacitors take a large area. In wide bandwidth signal paths a significant power dissipation is brought by the inter-stage isolation emitter followers. Using cascoded differential gain stages lead to a significant power consumption reduction due to their lower input capacitance.

II. LIMITING AMPLIFIER CIRCUIT DESCRIPTION

The top level diagram of the proposed 12.5Gb/s limiting amplifier is presented in Fig.1. The high gain (60dB) wide bandwidth signal path consists of six cascode differential gain stages (10dB each) separated with emitter followers that isolate the large collector load resistors from the relatively large input capacitance of the following gain stage. Implementing a single offset cancellation loop around a high gain signal path requires a very large compensation capacitance value that cannot be integrated on-chip. The solution for a fully-integrated limiting amplifier is to use individual offset cancellation loops around each signal path gain stage. This reduces the overall size of the compensation capacitors size to a value that can be implemented on-chip and also provides a very low residual output offset voltage (fractions of mV). Each offset cancellation loop consists of a differential input amplifier (A_{f1}), a single-ended Miller capacitance multiplication stage (A_{miller}) and an output attenuation stage (A_{f2}) that injects the correction current into the signal path. To avoid instability in the high gain signal path due to parasitic coupling through supply lines, the front-end, the core amplifier and the back-end stages use separate supply lines: V_{CCi}/V_{EEi}, V_{CCm}/V_{EEm}

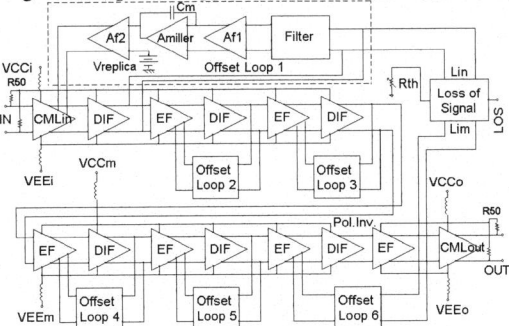

Fig.1 12.5Gb/s limiting amplifier top level diagram

1-4244-0458-4/06/$20.00 ©2006 IEEE   239

and V_{CCo}/V_{EEo} Multiple bondwires were used for the high current supply pins to reduce the on-chip supply voltage bouncing due to the L·di/dt voltage drops. The IC includes also a received-signal-strength-indicator (RSSI) circuit and a loss-of-signal (LOS) detector.

Fig.2 presents the input emitter followers Q_{20},Q_{21} and the first gain stage Q_{22}-Q_{25}. All the other gain stages in the signal path have a similar architecture. Most existing designs inject the offset loop output current into the 50Ω input termination resistors to create the offset correction voltage [4-6]. This technique results in a large bandwidth reduction at the IC input due to the high impedance at the correction node and the relatively large parasitic capacitance of the offset loop output devices. Emitter resistors were also used to create the offset correction voltage, providing a lower signal path bandwidth degradation [7]. This design uses emitter resistors R_{off} in conjunction with two differential pairs Q_{26},Q_{27} and Q_{28},Q_{29} that steer the I_{tail} current either directly to the Q_{20},Q_{21} emitter followers, or through the R_{off} resistors. The main advantage of this architecture is keeping a constant current in the emitter follower pair and therefore achieving a lower required offset cancellation voltage range. This results in smaller current consumption in the offset loop, which in many designs constitute a significant percentage of the overall IC power consumption [1-5]. Two feed-forward capacitors C_{ff} were used to add a zero in the signal path, which cancels the pole introduced by the R_{off} resistors. The Q_{22},Q_{23} gain stage uses a cascode configuration Q_{24},Q_{25} to minimize the Miller gain of its C_{bc} parasitic capacitances. The output pole given by R_c and the total output capacitance is cancelled with a zero created by the emitter peaking capacitor C_e. A good tracking over process and temperature between the pole and the zero was ensured by using two back-to-back shorted base-emitter devices Q_{ce1}, Q_{ce2} to create the C_e capacitance, such that it matches the C_{bc} capacitances of the output devices.

Fig.3 presents the detailed schematic of the offset cancellation loop used for each signal path gain stage. The high frequency component of the gain stage output is eliminated with a low corner frequency R_f, C_f filter. This filter also isolates the large input capacitance of the offset loop amplifier from the high speed main signal path. In the selected BiCMOS

process the best matching is achieved by the low f_T lateral PNP transistors. Their low operating frequency is not an issue for the offset cancellation loop, since it has a near DC operation. The first gain stage Q_{30},Q_{31} uses lateral PNP devices with large emitter degeneration R_{deg1} to achieve a low offset voltage. The Q_{32},Q_{33} diode connected devices perform a signal level shifting which allows the use of a PNP gain stage. The Q_{34},Q_{35} load current mirrors use large degeneration resistors R_{deg2} that ensure a high output impedance. The dominant offset loop pole is created at this high impedance node by the C_m capacitance, which is Miller multiplied at the input of the M_{36},Q_{38} stage. The M_{36} follower stage preserves the high impedance at the dominant pole node, while Q_{38} provides a moderate voltage gain ($A_v\approx30$), which results in an enough low value for the on-chip C_m capacitor. A single-ended Miller amplifier was used to minimize the size of the on-chip compensation capacitance, while the conversion back to differential mode was ensured by a replica bias stage. The DC potential at the dummy node (D) in the Q_{30},Q_{31} stage is provided by a matched DC local feedback loop M_{37},Q_{39} that ensures the circuit symmetry.

The output CML stage and its driving emitter followers are presented in Fig.4. The Q_{44},Q_{45} output differential switch uses a large emitter degeneration (R_{eo}) to set its gain to unity ($A_v\approx1$). Therefore, even if left outside the offset cancellation loop it has a negligible impact on the output offset voltage. The R_{eo} resistors also help reducing the C_{be} capacitive loading on the pre-driver emitter followers Q_{40},Q_{41}. The Miller gain seen by C_{bc} of Q_{44}, Q_{45} was minimized using a cascode switch architecture Q_{46},Q_{47}. This further reduces the capacitive load presented to the driving stage and therefore allows the reduction of the main signal path power consumption. Small value I_{on} current sources keep the cascode devices active all the time, eliminating the waveform distortion due to their turn-on delay. A significant power reduction was achieved in the past using push-pull emitter followers that have an asymmetric switched tail current, which boosts the drive strength of the leg that controls the turning-off output switch [8]. This design proposes a dynamic drive strength boosting that further reduces the power consumption in the Q_{40},Q_{41} emitter followers, without using additional DC power. It uses

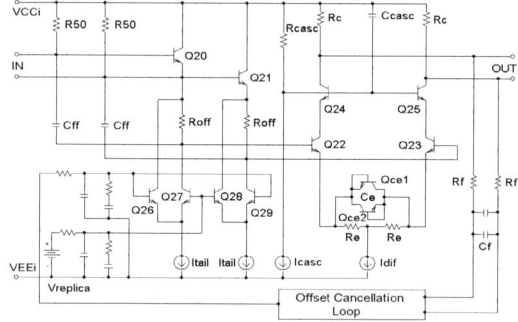

Fig.2 Input CML stage and first gain amplifier

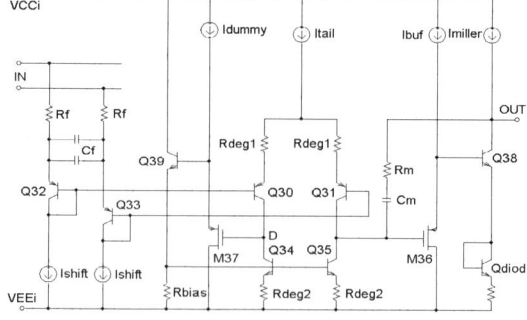

Fig.3 DC offset cancellation loop amplifier

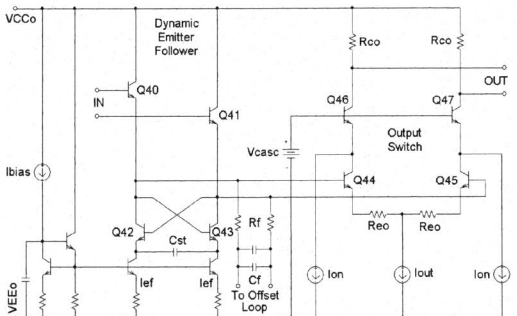

Fig.4 Output CML driver with dynamic emitter follower

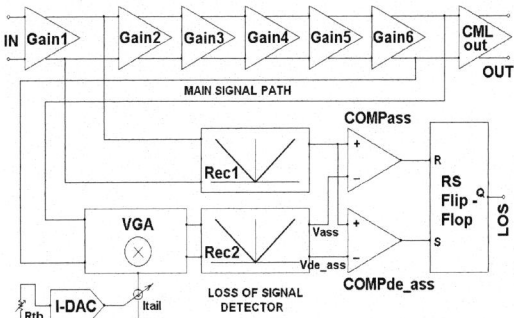

Fig.5 Feed-forward loss-of-signal detector circuit

constant tail currents I_{ef} that has a cross-coupled Q_{42},Q_{43} stage. The local positive feedback provides additional AC drive strength for the output switch leg that is turning-off. The charge required to discharge the base of the output switch devices Q_{44},Q_{45} when turning-off is provided by the C_{st} storage capacitor.

Fig.5 presents the block level diagram of the loss-of-signal (LOS) detector. The few existing limiting amplifiers that have an LOS capability use feedback detector architectures that are sensitive to the data pattern, resulting in an increased minimal detectable signal level. Furthermore, they are usually connected at the IC input, making the first stages offset voltage the limiting factor for the lowest detectable signal.

This design proposes a feed-forward architecture that is connected after the first gain stage (G_{ain1}) of the main signal path, reducing thus the impact of the input offset voltage of the LOS detector on the minimum detectable signal. The signal level uncertainty was avoided by using in the main signal path resistive degenerated gain stages that have their amplification set by a ratio of well matched resistors. This results in an accurate and reproducible relationship between the input signal power and the power after the first gain stage. The signal level amplitude provided by the first gain stage (which is always operating in linear regime) is measured by rectifying the signal (R_{ec1}) and then detecting its peak with a high frequency peak detector. Two data dependent threshold levels for assert and de-assert states are generated by rectifying with a second rectification block R_{ec2} the limited signal from the input of the output CML stage that is used to switch a variable gain amplifier (VGA), which has its tail current derived from a high precision off-chip resistor R_{TH}. This resistor sets the V_{ass} and V_{de_ass} threshold levels. Two comparators (COMP$_{ass}$, COMP$_{de\_ass}$) compare the detected signal level with the assert and de-assert threshold levels and drive an RS flip-flop that generates the output loss-of-signal alarm (LOS).

III. EXPERIMENTAL RESULTS

Fig.6 shows the die photo of the proposed 12.5GHz limiting amplifier that was fabricated in a 0.18μm 80GHz f_T SiGe BiCMOS process. A "butterfly" style symmetric layout was used to

minimize the metal interconnect parasitic inductances and capacitances, resulting in a lower signal path peaking and therefore a lower deterministic jitter in the band-limited data path. Fig.7.a presents the output eye diagram in the worst-case corner: minimum supply voltage (3V), maximum die temperature (125°C), and minimum input signal level (1mV). An 18ps$_{pp}$ DJ was measured (15ps$_{pp}$ after subtracting the measurement setup contribution), while the 20-80% rise/fall times are <25ps. Fig.7.b illustrates the output eye diagram for typical operating conditions (15mV input signal, 25°C, 3.3V supply) showing a much lower deterministic jitter (7.5ps$_{p\_p}$) and also faster rise/fall times (20ps). Fig.8 shows the output data eye deterministic jitter dependence on the input signal amplitude. At signal levels larger than 10mV the DJ decreases to few ps$_{pp}$. Fig.9 presents the dependence of the assert and de-assert levels on the precision off-chip resistor value. The proposed LOS detector is capable of sensing signals as low as 3 mV.

Table.1 presents a performance comparison between existing 10Gb/s limiting amplifiers and present design. Using individual offset cancellation loops around each gain stage resulted in a best in class output offset voltage (0.2mV), while allowing in the meantime the on-chip integration of the compensation capacitors. This brought a more compact design with a lower number of off-chip components. Using cascoded differential pairs to reduce the capacitive loading on the signal path and dynamic emitter followers to drive the high current output CML stage lead to a three-fold reduction of the power consumption in comparison with existing SiGe

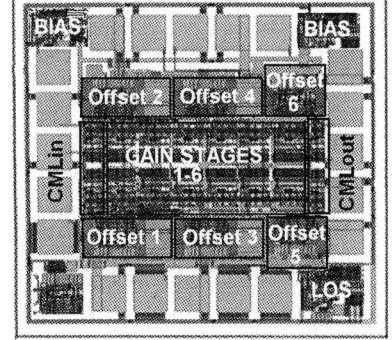

Fig.6 12.5Gb/s limiting amplifier IC die photo

241

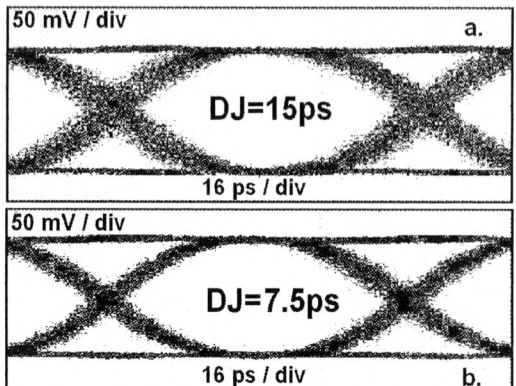

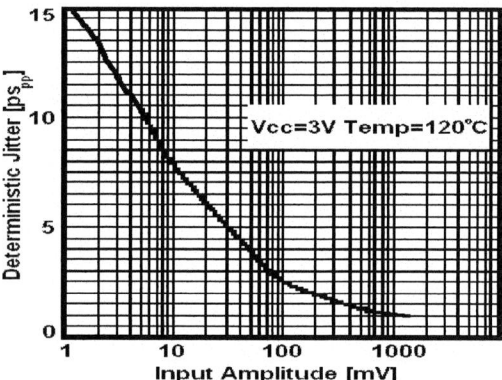

Fig.7.Output eye diagrams a. 1mV b. 15mV input level

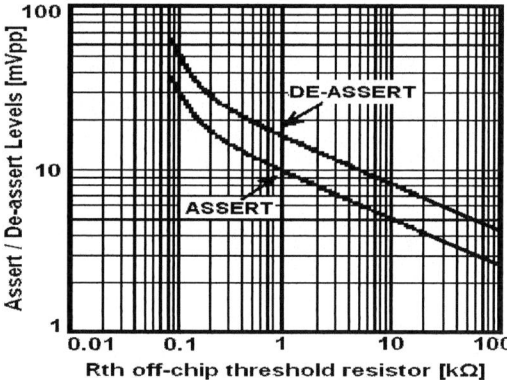

Fig.8 Deterministic jitter versus input amplitude

Table.1. 10Gb/s LA performance comparison

Reference	[3]	[5]	This work
Process	SiGe 47GHz	0.18µ CMOS	SiGe 80GHz
Gain	60dB	50dB	60dB
Bandwidth	15GHz	9.4GHz	13GHz
Sensitivity	3.5mV	4.6mV	1mV
Det. Jitter	$20ps_{p\ p}$	$13ps_{p\ p}$	$10/15ps_{p\ p}$
Rise/Fall times	35ps	30ps	20/25ps
Output offset	4.5mV	10mV	0.2mV
Supply voltage	3.5V	1.8V	3.3V
Supply current	100mA	75mA	30mA
Die area [mm²]	1.2x2.6	0.5x1.5	1.5x1.5

the output leg that is turning-off is provided by a local positive feedback loop that uses the charge stored on a differential capacitor to discharge the base of the turning-off output device. The output offset voltage was reduced to fractions of mV by using a multi offset cancellation loop architecture, which corrects individually the offset of each gain stage in the signal path. A single-ended Miller multiplied compensation capacitor has reduced to half the required die area in comparison with a fully differential DC offset cancellation loop, having a negligible detrimental impact on the offset voltage performance. The output offset voltage was further reduced by using a differential current steering circuit to generate the offset correction voltage, while keeping constant the current level in the signal path devices. A minimum detectable signal of 3 mV was achieved by the loss-of-signal circuit by using a feed-forward architecture with data dependent assert and de-assert levels.

Fig.9 Assert / De-assert levels versus R_{th} resistor

limiting amplifiers [3]. The low power consumption achieved by this SiGe design is comparable with the one of CMOS implementations operating at a lower supply voltage [5]. However, this SiGe IC achieves a much better input sensitivity and lower offset voltage.

IV. CONCLUSION

Replacing the traditional Cherry-Hooper wide bandwidth stages with cascoded differential stages lead to a two-fold power consumption reduction in the main signal path. A further power reduction was achieved by using a dynamic emitter follower to drive the last CML stage. The additional drive strength for

REFERENCES

[1] M. Moller, H.M. Rein, "15Gbit/s high gain limiting amplifier fabricated using Si-bipolar technology", Electron Lett., vol.30, no.18, pp.1519-1521, Sep. 1994.

[2] M.Lang, et al,"20-40Gb/s 0.2µm GaAs HEMT chip set for optical data receiver", IEEE JSSC, vol.32, no.9, pp.1384-1393, Sep. 1997.

[3] Y. Greshishchev, P. Schran, "A 60dB gain, 55dB dynamic range 10Gb/s broad-band SiGe HBT limiting amplifier", IEEE JSSC, vol.34, no.12, pp.1914-1920, Dec. 1999.

[4] E. Sackinger, W. Fischer, "A 3GHz 32dB CMOS limiting amplifier for SONET OC48 receivers", IEEE JSSC, vol.35, no.12, pp.1884-1888, Dec. 2000.

[5] S. Galal, B. Razavi, "10Gb/s limiting amplifier and laser/modulator driver in 0.18µm CMOS technology", IEEE JSSC, vol.38, no.12, pp.2138-2146, Dec. 2003.

[6] H. Tran, et al, "6KΩ 43Gb/s differential transimpedance limiting asmplifier with auto-zero feedback", IEEE JSSC, vol.39, no.10, pp.1680-1689, Oct. 2004.

[7] A. Maxim, et al., "A 12.5GHz SiGe BICMOS Limiting Amplifier Using Dual Offset Cancellation Loop", Proc. of ESSCIRC 2005 Conference, pp.97-100, Sep. 2005.

[8] A. Maxim, "A 10Gb/s SiGe compact laser diode driver using push-pull emitter followers and Miller compensated output switch", Proc. of ESSCIRC 2003 Conference, pp.557-560, Sep. 2003.

2006 Bipolar/BoCMOS Circuits and Technology Meeting Proceedings

250-GHz self-aligned Si/SiGeC HBT featuring an all-implanted collector

P. Chevalier, C. Raya, B. Geynet, F. Pourchon, F. Judong, F. Saguin, T. Schwartzmann,
R. Pantel, B. Vandelle, L. Rubaldo, G. Avenier, B. Barbalat, and A. Chantre

STMicroelectronics, 850 rue Jean Monnet, F-38926 Crolles Cedex, France
tel.: (+33) 476.926.730, fax: (+33) 476.926.814, e-mail: pascal.chevalier@st.com

Abstract — **This paper presents investigations led to simplify the collector module of SiGeC HBTs in order to reduce technology cost. Outcome of this work is an HBT featuring an all-implanted collector with record f_T and f_{max} (>250 GHz).**

Index Terms — **BiCMOS integrated circuits technology, Heterojunction bipolar transistors, Ion implantation, Modeling, Silicon Germanium.**

I. INTRODUCTION

Si/SiGeC HBTs have now reached performances enabling applications such as high-speed optical communications or millimeter-wave radars. Devices built on 0.13-μm core CMOS offer both f_T and f_{max} > 250 GHz that will only be available in 45 nm CMOS node, but with lower voltage capabilities [1]. Part of the performance of state-of-the-art HBTs can however be sacrificed to reduce technology cost. In this paper we report the work that has been carried out to divide by two the number of additional masks needed to offer an HBT into a 0.13-μm CMOS technology while 250 GHz f_{max} is still demonstrated. First, we review the investigations done on the collector module that led to the choice of an all-implanted collector technology. Second, we present this 4-mask HBT technology and its performances. Then the limitations of this technology are explored thanks to a modeling comparison between this all-implanted collector HBT and a conventional epitaxial collector HBT featuring exactly the same emitter–base technology. Finally, we conclude on the capabilities of this low-cost technology and its application scope.

II. FROM A CONVENTIONAL EPITAXIAL COLLECTOR TO AN ALL-IMPLANTED COLLECTOR

Different approaches can be considered to build a low-cost derivative of a SiGe BiCMOS technology, by simplifying either the emitter–base architecture [2] or the collector module [3]–[4]. We have shown that the self-alignment of the emitter–base system is a key advantage to reach state-of-the-art performances [5]–[6]. Therefore, we have focused our attention on the collector module since it uses half of the total number of additional masks needed to build the HBT described in [5]: Zero Level, Buried Layer, Deep

Trenches Isolation (DTI) and Sinker. The first point was to investigate the influences of the collector contact implantation and the collector-to-emitter distance L_{E-C}. The conventional collector construction is shown in Fig. 1.a.

A first attempt was to replace the dedicated collector sinker implant by the N+ S/D implant coming from NMOS devices (Fig. 1.b). Results of this comparison are presented in Table I for a standard CBEBC layout with L_{E-C} of 1.04 μm. f_T drops by more than 20 GHz, keeping a reasonable value of 237 GHz, while f_{max} drops only by approximately 10 GHz to 267 GHz. This is the consequence of the $R_C - C_{BC}$ trade-off: R_C increases but C_{BC} decreases (lower lateral capacitance). Shorter L_{E-C} distances (down to ~0.5 μm) have been investigated by removing the shallow trenches between emitter–base and collector areas. Base contacts are then moved in another plane, leading to the $C_B E^B C$ cellular layout described in [7]. Evolution of both f_T and f_{max} with L_{E-C} are compared in Table I for the two collector contact implantations presented before i.e. the sinker implant (Fig 2.a) and the N+ S/D implant (Fig 2.b).

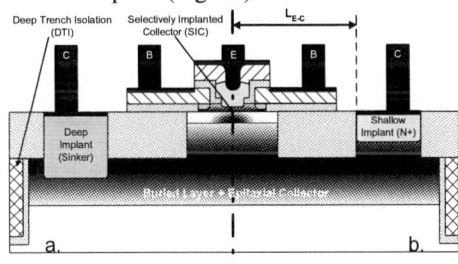

Fig. 1. CBEBC transistor with a (DTI + buried layer + epitaxial collector + SIC) collector module connected either by a deep (a) or a shallow (b) implantation.

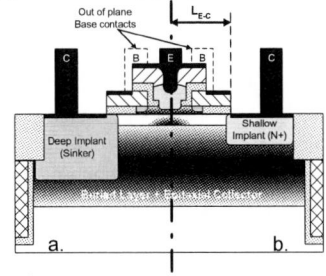

Fig. 2. $C_B E^B C$ transistor with a (DTI + buried layer + epitaxial collector + SIC) collector module connected either by a deep (a) or a shallow (b) implantation.

1-4244-0458-4/06/$20.00 ©2006 IEEE   243

TABLE I.

f_T AND f_{MAX} OBTAINED FOR DIFFERENT TRANSISTOR LAYOUTS (CBEBC VS. C_BE^BC) AND COLLECTOR CONTACT IMPLANTATIONS (SINKER VS. S/D IMPLANT).

Structure	Development (µm²)	L_{E-C} (µm)	Sinker Implant		S/D implant	
			f_T (GHz)	f_{max} (GHz)	f_T (GHz)	f_{max} (GHz)
CBEBC	0.14×5.70	1.04	261	279	237	267
C_BE^BC	5×0.14×1.20	0.68	272	226	228	240
C_BE^BC	5×0.14×1.20	0.63	282	214	228	241
C_BE^BC	5×0.14×1.20	0.58	293	196	232	235
C_BE^BC	5×0.14×1.20	0.53	301	184	233	229

It is clear that the diffusion of the heavily phosphorous doped sinker implant promotes the collector resistance decrease (f_T reaches 300 GHz) but at the expense of a high C_{BC} value, which strongly penalizes f_{max} (~185 GHz). More balanced results are obtained when sinker implantation is replaced by N+ S/D implantation. Compared to the CBEBC layout, f_T's of C_BE^BC structures are hardly impacted (−4/−9 GHz) while f_{max}'s decrease (−26/−38 GHz) but remain equal to or higher than f_T's. Absence of STI between emitter–base and collector areas indeed penalizes C_{BC} (cf. section IV).

A second attempt was to remove the DTI module. We have shown in [8] that removing deep trenches has a negligible impact on HF performance but strongly degrades the collector–substrate capacitance C_{CS}. While C_{CS} increase may be penalizing for digital applications (such as ring oscillators), an efficient way to reduce this capacitance is to reduce the depth of the buried collector layer. In fact simulations have shown that a thick buried layer is useful to collect electrons when an STI is present between emitter–base and collector areas but is of no need when STI is removed since electron transport occurs in the top part of the buried layer (meaning that the link between the buried layer sheet resistance and the collector resistance is not so direct...). As a consequence we have considered replacing the buried layer – epitaxial collector module by a shallow implant, as reported in the next section.

III. ALL-IMPLANTED COLLECTOR TECHNOLOGY

The transistor emitter–base architecture is identical to the mother technology i.e. fully self-aligned thanks to the selective epitaxial growth of the SiGeC base and based on 0.13-µm core CMOS. The conventional collector module (buried layer, collector epitaxy, deep trenches isolation and collector sinker implantation) has been replaced by an implanted shallow collector, done after STI formation, contacted by N+ S/D implantation (Fig. 3). The shallow collector requests the use of the C_BE^BC layout, resulting in a one active area HBT, since the STI between emitter and collector contacts would cut the shallow collector layer. It is important to highlight that in spite of an identical one

active area HBT layout, this technology is significantly different from the one presented in [2]. Contrary to [3], our technology features: 1) a true all-implanted collector since no Si buffer layer is grown after collector implantation, 2) a fully self-aligned emitter–base architecture and 3) only 4 dedicated masking steps (vs. 5). These steps are indeed: 1) Collector implantation, 2) Polybase patterning, 3) Emitter window patterning, and 4) Polyemitter patterning.

The shallow arsenic implantation allows the realization of a retrograde collector profile at base–collector junction, which is comparable to what is obtained with a SIC. Then, addition of a SIC aligned on the emitter window is not mandatory. The main advantage of using an additional SIC is to allow reducing the dose of the collector implantation to reach the same cut-off frequencies [7]. Indeed, the implantation conditions (high-energy / high-dose) of such a collector lead to silicon amorphisation. Recrystallisation occurs during the high temperature anneal performed right after the collector implantation. After recrystallisation of the amorphous Si layer, a deep band of end-of-range (EOR) defects (stacking faults and dislocation loops) remains. These EOR defects, that can not be avoided, do not generate any electrical issue as long as they are outside the base–collector depletion region (Fig. 4.a). A worst case, obtained when implantation conditions are not correctly set, is when Si layer is not completely amorphised and then 2 bands of EOR defects appear after recrystallisation process (Fig. 4.b). The upper band leads to collector leakages.

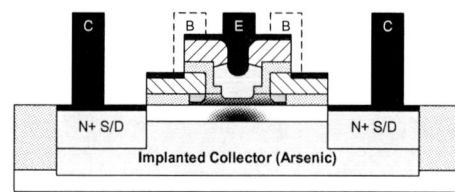

Fig. 3. C_BE^BC transistor with an implanted arsenic collector and N+ S/D collector contact (DTI module is omitted).

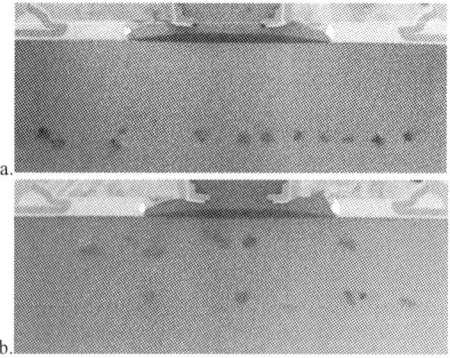

Fig. 4. TEM pictures of the collector regions of Si/SiGeC HBTs featuring an arsenic all-implanted collector generating 1 band (a) or 2 bands (b) of EOR defects.

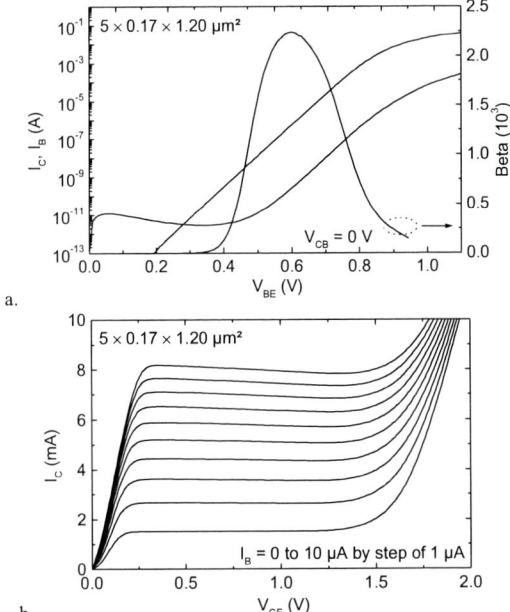

a.

b.

Fig. 5. Gummel (a) and I – V (b) characteristics of a $5\times0.17\times1.20$ μm² low-cost bulk HBT ($L_{E\text{-}C}$ = 0.53 μm).

f_T

246.1	242.0	238.5	241.8	248.3	
245.0	235.6	235.7	239.4	239.6	240.9	255.6
238.7	235.5	250.5	259.4	249.3	236.2	242.9
235.6	234.8	242.7	249.0	238.0	236.8	235.8
	238.8	236.3	234.2	235.2	235.0	

Min	Max	Mean	Stdev
234.2	259.4	241.1	6.50

f_{max}

244.3	245.7	242.8	245.4	241.5	
244.1	245.9	245.8	248.6	233.8	246.7	246.3
246.9	243.6	250.3	255.6	250.4	246.5	248.8
244.4	245.7	247.9	244.6	243.2	245.4	243.3
	244.5	245.5	244.9	243.7	240.5	

Min	Max	Mean	Stdev
233.8	255.6	245.4	3.60

Fig. 6. f_T and f_{max} mappings [GHz] of a $5\times0.17\times1.20$ μm² transistor at V_{CB} = 0.5 V and V_{BE} = 0.92 V ($L_{E\text{-}C}$ = 0.53 μm).

Devices corresponding to Fig. 4.a show good DC characteristics, as presented in Fig. 5. The maximum current gain β_{max} is about 2200 at V_{BE} ~ 0.7 V. The high doping levels at emitter–base junction lead however to a base band-to-band tunneling current [9]. The Early voltage extracted from output characteristics of Fig. 5.b at medium collector currents (where there is no self-heating) is close to 100 V thanks to a low pinched base resistance value of 2.2 kΩ/sq. The collector–emitter breakdown voltage BV_{CEO}, derived from base current reversal measurements, is 1.5 V at medium injection (V_{BE} = 0.7 V). Open emitter collector–base breakdown voltage BV_{CBO} is measured around 5.1 V. Record HF characteristics are demonstrated: mappings presented in Fig. 6 exhibit average values higher than 240 GHz for both f_T and f_{max} with 259 GHz f_T and 256 GHz f_{max} as best couple of values. The low standard deviations

(2.7 % for f_T and 1.5 % for f_{max}) demonstrate the good on-wafer uniformity of this technology. These results are to the authors' knowledge the best reported results for Si/SiGeC HBT requesting only 4 additional masks to CMOS.

IV. IMPACT OF THE COLLECTOR MODULE ON THE DEVICE PERFORMANCE

In spite of the exceptional performances reported above, the all-implanted collector HBT features lower f_{max} than conventional epitaxial collector HBT [6]. Then we have extracted the SPICE model cards of two different HBTs featuring exactly the same emitter–base technology and identical emitter area but different collector technologies (conventional epitaxial vs. all-implanted collectors) to identify the differences. Results, obtained for an all-implanted collector with a lower doping at collector–base junction than the one used for the record performances reported in previous section, are summarized in Table II.

TABLE II.

COMPARISON OF THE MAIN ELECTRICAL CHARACTERISTICS BETWEEN A CONVENTIONAL EPITAXIAL COLLECTOR HBT AND AN ALL-IMPLANTED COLLECTOR HBT FEATURING EXACTLY THE SAME EMITTER–BASE TECHNOLOGY.

Structure Device (μm²) Layout	Conv. 0.17×5.60 CBEBC	Low-Cost 5×0.17×1.20 C_BE^BC
f_T (GHz)	252	225
f_{max} (GHz)	276	245
V_{BE} @ peak f_T (V)	0.88	0.91
β @ V_{BE}=0.75V (-)	1990	2030
BV_{CBO} (V)	5.5	5.3
BV_{CEO} (V)	1.5	1.5
R_E (Ω)	2.4	3.3
R_{Bi} (kΩ/sq)	2.3	2.1
R_{Bx} (Ω)	8.6	8.3
R_{Ci} (Ω)	9.0	11.9
R_{Cx} (Ω)	4.7	6.7
C_{BE} (fF)	15.4	16.9
C_{BC} (fF)	9.6	15.8
C_{CS} (fF)	8.8	21.0
R_{TH} (K/W)	2900	1200

I–V characteristics and f_T–/f_{max}–I_C characteristics are compared in Fig. 7.a and Fig. 7.b respectively. First, it is important to notice that even if BV_{CBO} value is indeed higher than the one reported previously (5.3 V vs. 5.1 V) it remains lower than the one of the conventional technology (5.5 V) showing a larger doping at collector–base junction.

f_T of low-cost technology is however lower than the one of the conventional technology. It is penalized by both a higher resistance of the implanted collector (not fully compensated by the shorter $L_{E\text{-}C}$) and higher base–emitter C_{BE} and base–collector C_{BC} capacitances due to the emitter fragmentation essential for the one active area layout. The slightly larger emitter resistance could be explained by the device layout too.

245

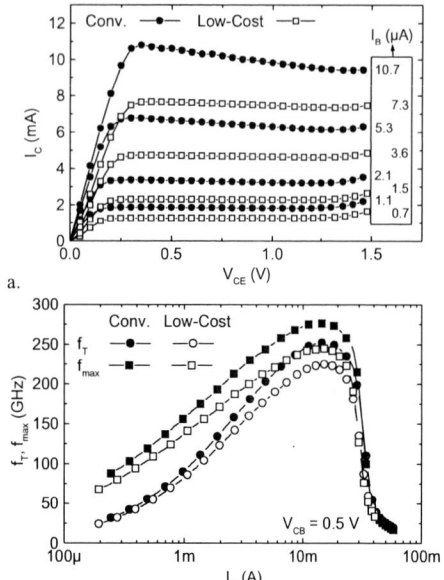

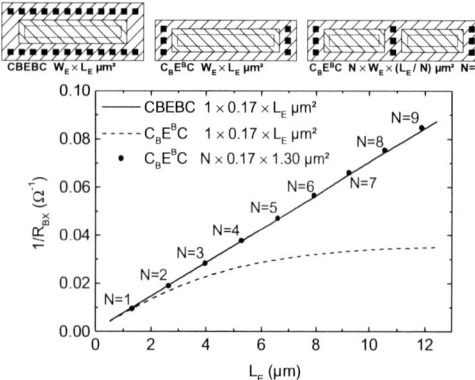

Fig. 7. Comparison of $I-V$ (a) and f_T / f_{max} $-$ I_C (b) characteristics of a 0.17×5.60 µm² conventional HBT vs. a 5×0.17×1.20 µm² low-cost HBT.

Fig. 8. Evolutions of the inverse of the extrinsic base resistance vs. emitter length for CBEBC and C_BE^BC (single stripe vs. fragmented emitters) layouts (calculations).

Fig. 8 demonstrates that the emitter fragmentation, leading indeed to a higher emitter perimeter-to-area ratio (P_E / A_E) [7], is mandatory to reduce the extrinsic base resistance R_{BX}. Calculations exhibit an R_{BX} degradation > 10 % for C_BE^BC layout compared to CBEBC layout for emitter length L_E larger than 2 µm. R_{BX} of fragmented emitter C_BE^BC layouts is comparable to the one of CBEBC layouts and even lower for large L_E due to the larger P_E / A_E. The larger P_E / A_E of low-cost HBTs contributes also to reduce the thermal resistance R_{TH} of the device, even if the main part of the reduction noticeable in Table II must be attributed to DTI removal [8]. It reduces device self-heating, as can be seen in Fig. 7.a, which is beneficial to f_T. Choice of the emitter fragmentation (N vs. L_E) is therefore the result of a trade-off between capacitances and base resistance.

f_{max} of the low-cost HBT suffers of course from the lower f_T compared to the conventional structure. There is however no limitation due to base resistance since both intrinsic and extrinsic base resistances of the low-cost HBT are lower than the ones of the conventional architecture (consistent with Fig. 8). The main burden to f_{max} is in fact the collector–base capacitance, whose large increase is due to STI removal between emitter–base and collector areas. Finally, absence of deep trenches isolation leads to a twofold increase of C_{CS}, which may be an issue for digital applications. Restoring the DTI module may be requested for such applications.

V. CONCLUSION

We have demonstrated that impressive performances can be obtained with low-cost HBTs: f_T and f_{max} higher than 250 GHz are to the authors' knowledge the best reported results for SiGeC HBT requesting only 4 additional masks to CMOS. Many applications can be satisfied with this HBT, except those requiring a very large $f_{max} \times BV_{CBO}$ product (>1.5 THz.V). 300 GHz f_{max} seems indeed out of reach of this low-cost architecture since R_C and C_{BC} can not be tuned independently. Investigations on collector module go on to overcome this limitation.

ACKNOWLEDGEMENT

The authors thank the staffs of the 200 mm Si plants in STMicroelectronics Crolles and CEA-LETI Grenoble involved in this work and more especially D. Lagarde for the technical contribution, A. Halimaoui for fruitful discussions on implantation defects and M. Roche for the managerial support.

REFERENCES

[1] P. Chevalier et al, "Advanced SiGe BiCMOS and CMOS platforms for Optical and Millimeter-Wave Integrated Circuits", in CSICS Tech. Dig., 2006, to be published.
[2] A. T. Tilke et al, "A Low-Cost Fully Self-Aligned SiGe BiCMOS Technology Using Selective Epitaxy and a Lateral Quasi-Single-Poly Integration Concept", IEEE Trans. Electron Devices, vol. 51, no. 7, pp 1101-1107, Jul. 2004.
[3] B. Heinemann et al "Novel collector design for high speed SiGeC HBTs", in IEDM Tech. Dig., 2002, pp. 775-778.
[4] L. Lanzerotti et al "A Low Complexity 0.13 µm SiGe BiCMOS Technology for Wireless and Mixed Signal Applications", in Proc. BCTM, 2004, pp. 237-240.
[5] P. Chevalier et al, "230 GHz Self-Aligned SiGeC HBT for Optical and Millimeter-Wave Applications", IEEE J. Solid-State Circuits, vol. 40, no. 10, pp. 2025-2034, Oct. 2005.
[6] P. Chevalier et al, "300 GHz f_{MAX} self-aligned SiGeC HBT optimized towards CMOS compatibility", in Proc. BCTM, 2005, pp. 120-123.
[7] P. Chevalier et al, "Low-Cost Self-Aligned SiGeC HBT Module for High-Performance Bulk and SOI RFCMOS Platforms", in IEDM Tech. Dig., 2005, pp. 983-986.
[8] B. Barbalat et al, "Deep Trench Isolation Effect on Self-Heating and RF Performances of SiGeC HBTs", in Proc. ESSDERC, 2005, pp. 129-132.
[9] D. Lagarde et al, "Band-to-band Tunneling in Vertically Scaled SiGe:C HBTs", IEEE Electron Device Lett., vol. 27, no. 4, April 2006, pp. 275-277.

2006 Bipolar/BiCMOS Circuits and Technology Meeting Proceedings

Development of a Cost-Effective, Selective-Epi, SiGe:C HBT Module for 77GHz Automotive Radar

Jay P. John, Jim Kirchgessner, Matt Menner, Hernan Rueda, Francis Chai, Dave Morgan, Jill Hildreth, Morgan Dawdy, Ralf Reuter, and Hao Li

Microwave and Mixed Signal Technology Lab, Freescale Semiconductor, Tempe, AZ, 2100 E. Elliot Road, MD: EL741, Tempe, AZ, 85284.

Abstract — The development of a selective-epi, SiGe:C HBT module for 77GHz automotive radar applications is described. A cutoff frequency (f_T) of 185GHz, in conjunction with a maximum oscillation frequency of 260GHz has been achieved through the implementation of a self-aligned selective-epi base structure and a simple, cost-effective collector construction without buried layer or deep trench isolation.

Index Terms — Silicon bipolar/BiCMOS process technology, Bipolar transistors, Millimeter wave technology, SiGe, SiGe:C.

I. INTRODUCTION

Silicon germanium carbon (SiGe:C) HBT technologies have become the standard for many wireless applications as consumer demand for low power portable products continues to increase [1] - [3]. More recently, SiGe:C HBT technologies have gained increasing interest for emerging high frequency (>60GHz) markets, such as automotive radar (77GHz) [4] – [6], as f_T and f_{MAX} of the HBT devices has exceeded 200GHz.

In this paper, we describe the development of a selective-epi, SiGe:C HBT technology, leveraging the simple collector construction of Freescale's established 0.18μm BiCMOS technology. The goal of this effort was to achieve a device with a peak f_T/f_{MAX} of >180/250GHz to enable 77GHz auto-radar circuits. Aggressive scaling of the base profile, combined with significant reduction in extrinsic base resistance was required. This paper discusses the opportunities and trade-offs associated with various approaches to optimize key HBT performance characteristics.

II. BASELINE TECHNOLOGIES

The starting point for the technology was an established 0.18μm BiCMOS platform, designed for wireless RF applications [1]. Although this technology has been described elsewhere, the key features will be reviewed briefly. Fig. 1 illustrates the basic HBT structure for this technology.

The HBT emitter/base structure is a simple quasi-self-aligned structure, with the extrinsic base to emitter dimension set by lithography alignment tolerances. The 0.18μm process utilizes only shallow trench isolation and a high-energy implanted collector well to minimize overall process complexity. In addition, an enhanced version of the original SiGe:C HBT exists, with a 2nd generation intrinsic profile [2]. This version also includes a collector enhancement – a novel, low resistivity sub-isolation buried layer ("SIBL") under selected regions of the shallow trench isolation. This gives the device significantly improved collector resistance, while avoiding the more traditional (and expensive) buried layer/epi plus deep trench approach.

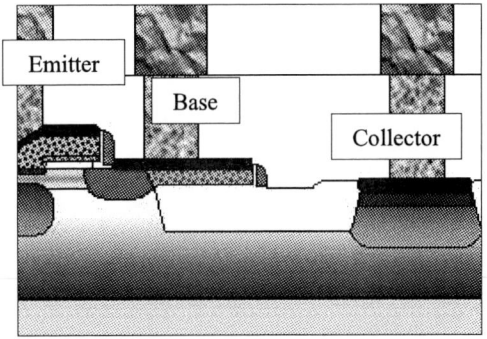

Fig. 1. Illustration of the original HBT device [1].

Fig. 2 shows the evolution to the "xHBT" structure, which combines the SIBL layer from the enhanced device, with a self-aligned, selective-epi base top structure.

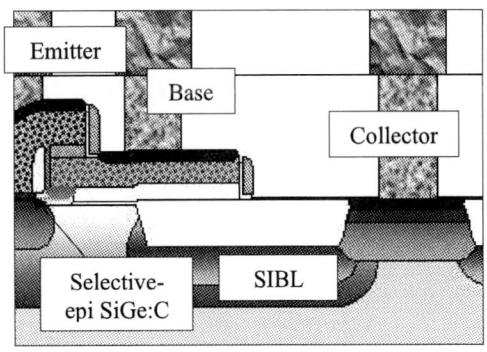

Fig. 2. Illustration of the xHBT device.

1-4244-0458-4/06/$20.00 ©2006 IEEE   247

Table 1 gives a brief comparison of some of the key device parameters of the xHBT compared to the earlier HBT versions.

In this paper we will describe some of the work done to develop an HBT device suitable for 77GHz radar applications. We include both measured data as well as simulated data using a complete 2-D structure which accounts for both intrinsic and extrinsic effects. Some of the obvious opportunities for performance improvement include: 1) intrinsic emitter/base profile scaling, 2) extrinsic base resistance reduction, 3) device layout optimization.

III. INTRINSIC PROFILE IMPROVEMENTS

In order to push f_{MAX} to the required performance level (>250GHz), a high f_T (in the range of 180-200GHz) is also required. The intrinsic profile from the 0.18µm enhanced HBT (f_T~120GHz) was used as the "starting point" for process/device simulations. In order to achieve greater than 50% improvement in f_T, a significant reduction in the intrinsic device transit time is required. A simplistic view of the HBT transit time equations is given below for conceptualization:

$$\tau_{EC} = \tau_E + \tau_B + \tau_C, \quad (1)$$
$$\tau_E = (V_T/I_C)*(C_{JE} + C_N) \quad (2)$$
$$\tau_B = W_B^2/2*D_n \quad (3)$$
$$\tau_C = W_C/(2*v_{sat}) + C_{JC}*(R_C + R_E) + C_{JC}* V_T/I_C \quad (4)$$

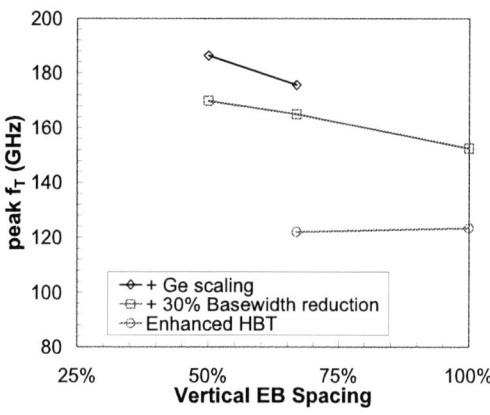

Fig. 3. Simulated results showing f_T improvement as a function of E/B spacing, basewidth, and Ge scaling.

The enhanced HBT utilized limited τ_B reduction (reduced basewidth), τ_E reduction (emitter/base scaling), as well as significant τ_C reduction (increased collector doping), and the reduction of the collector resistance (addition of the SIBL module). However, to meet the more stringent performance requirements for 77GHz applications, further scaling in the τ_E and τ_B components was required. Fig. 3 shows simulated peak f_T as a function of vertical emitter-base spacing, coupled with reduced base width and increased germanium grading across the base (to reduce τ_B by enhanced drift field). These changes reduce τ_E, (by reducing the delay associated with the emitter-base junction) and reduce τ_B (due to the implied base width reduction and enhanced drift field). These simulations utilized 1-D doping profiles obtained from the existing 0.18um HBT devices, in order to project the changes required to obtain the necessary cut-off frequency enhancements. From this plot, a 50% emitter-base spacing reduction and a 30% basewidth reduction, in conjunction with germanium scaling, were required to push f_T above 180GHz.

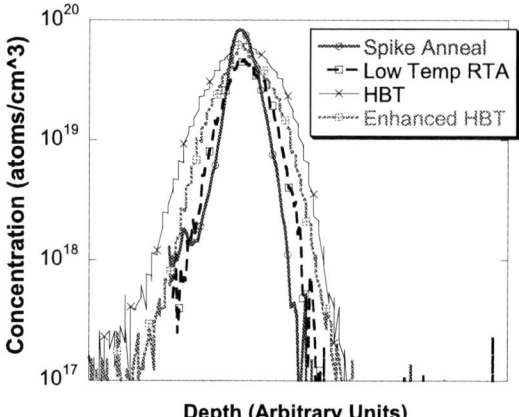

Fig. 4. Comparison of various boron base profiles, showing profile scaling with technology.

While scaling the vertical emitter-base spacing is straightforward through a simple reduction in the epi Si cap thickness, scaling the basewidth is more challenging. Clearly, the overall thermal budget must be reduced. Fig. 4 compares the boron base profiles

TABLE I
BASIC SIGE:C HBT PARAMETER COMPARISON WITH PREVIOUS 0.18µM HBT'S

	Units	0.18µm HBT [1]	Enhanced HBT [2]	xHBT
Emitter Size	um^2	0.25x10	0.25x10	0.16x10
Peak Current Gain	--	120	360	425
BV$_{CEO}$	V	3.3	1.8	2.0
BV$_{CBO}$	V	12.3	6.2	6.2
Peak f_T (Vce=2V)	GHz	50	120	185 [1]
Peak f_{MAX} (Vce=2V)	GHz	110	120	260 [1]

[1] V_{CE}=1.5V

from the standard and enhanced HBT from Table 1, with two alternate, reduced thermal cycle final anneals – a low temperature RTA and a high temperature Spike anneal. While both of these final anneals result in a large reduction in final basewidth, the Spike anneal gives the narrowest base, while maintaining high activation that is required for low base resistance and CMOS device integration [6].

Measured xHBT f_T vs. I_C results are shown in Fig. 5. This plot reflects the following improvements: 1) basewidth scaling (low temperature RTA and high temperature Spike anneal), 2) vertical E/B scaling (Si cap), 3) Ge grading scaling.

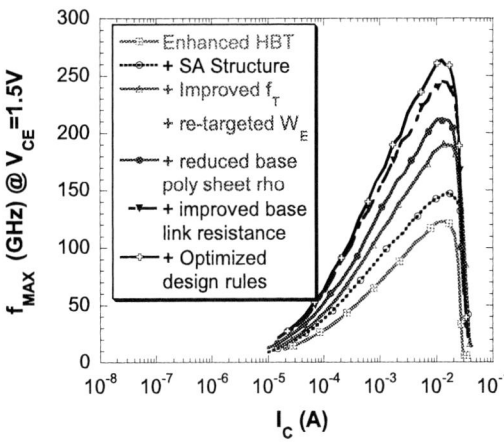

Fig. 6. f_{MAX} optimization (A_E ~0.16x10um^2).

However, even for a self-aligned structure, achieving low R_B and high f_{MAX} can still present an engineering challenge, as shown in Fig. 6. Preliminary results for the xHBT, while superior to the enhanced HBT, still fell short of the technology targets. Numerous process improvements have been implemented to improve base resistance and increase f_{MAX}. The effect of these improvements is highlighted in Fig. 6. In particular, these efforts have been focused on several areas: 1) improved f_T, 2) reduced W_E, 3) optimized base poly sheet rho, 4) improved link base resistance, and 5) design rule optimization. The combination of these improvements has resulted in a ~2x improvement in f_{MAX} over the enhanced HBT.

The f_T and f_{MAX} results for the new device are shown in Fig. 7, with a peak f_T/f_{MAX} of 185/260GHz. This f_{MAX} is the highest known value for a SiGe:C HBT not utilizing buried layer/epi and deep trench modules [10], [11]. The Gummel characteristics of this device are shown in Fig. 8.

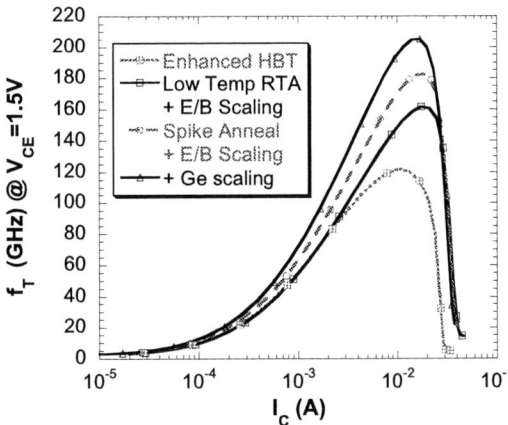

Fig. 5.   Measured f_T improvements for intrinsic profile optimization (A_E ~0.16x10um^2).

IV. EXTRINSIC DEVICE STRUCTURE IMPROVEMENTS

The main objective of moving from a quasi- to a fully-self-aligned emitter-base structure is to reduce the extrinsic base resistance by reducing the spacing between the highly-doped extrinsic base poly and the intrinsic device [7], as well as to reduce the total collector-base capacitance (which also benefits from the reduced dimensions). Without an effective emitter-base self-alignment scheme, $f_{MAX} > 250$GHz has not been consistently achieved on SiGe:C HBT's. Peak f_{MAX} performance for conventional quasi-self aligned SiGe HBT's is limited to below ~200GHz [8], [9]. Although higher f_{MAX} values might be feasible with very aggressive scaling of key dimensions, there would likely be some compromise in device yield and manufacturability. The fully self-aligned structure schematically shown in Fig. 2 allows this spacing to be reduced by almost ~4x, compared to the quasi-self aligned structure in Fig. 1. The E/B spacer width reduction also enables a reduction in the final emitter width, which we have scaled from ~0.25μm to ~0.16μm in this work.

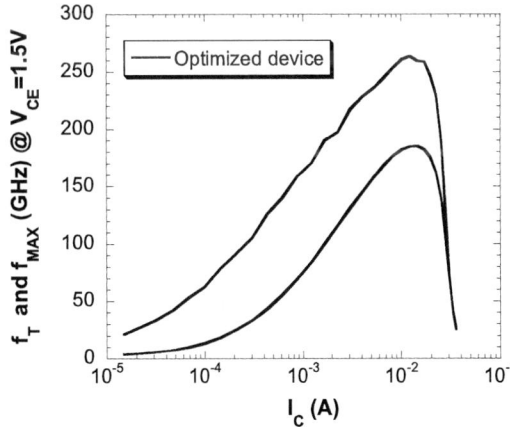

Fig. 7. f_T and f_{MAX} for the optimized xHBT device (A_E ~0.16x10um^2).

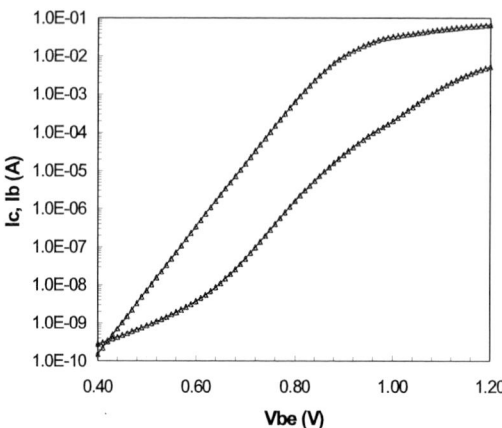

Fig. 8. Gummel characteristics for the xHBT device (A_E ~0.16x10um^2).

V. CIRCUIT PERFORMANCE

Noise parameter measurements were carried out on xHBT devices at 77GHz. The measured S_{OPT} agrees with transistor models, improving confidence in the design and optimization of LNA noise matching. NF_{MIN} of 5.5dB has been measured on non-optimized xHBT devices.

A number of 77GHz circuits have been fabricated in this process and measured on silicon, including a W-band low-noise amplifier, which demonstrates a 6.2dB noise figure at 77GHz, with adjustable gain from 0 to 33dB at 77GHz [12] and a 77GHz VCO, which demonstrates ~14dBm single-ended output power.

VI. CONCLUSION

The development of a self-aligned, selective-epi SiGe:C HBT module compatible with an established 0.18µm technology has been described. Utilization of a Spike anneal for the final emitter anneal, in conjunction with aggressive vertical emitter/base scaling and germanium profile optimization has resulted in a peak f_T of >180GHz. Efforts to reduce extrinsic base resistance have been discussed, which have enabled a boost in f_{MAX} to >260GHz. This excellent performance is achieved with a low-complexity technology that does not require the integration of buried layer/epi and deep trench isolation modules.

ACKNOWLEDGEMENT

The authors wish to acknowledge the ATMC, Oak Hill Fab, PALAZ, and PALTX organizations for outstanding wafer processing and analytical support.

REFERENCES

[1] J. Kirchgessner, et al, "A 0.18 µm SiGe:C RFBiCMOS technology for wireless and gigabit optical communication applications", *Proc. BCTM*, pp. 151-154, 2001.

[2] J. P. John, et al, "Optimization of a SiGe:C HBT in a BiCMOS technology for low power wireless applications," *Proc. BCTM*, pp. 193-196, 2002.

[3] A. Joseph, et al, "A 0.18um BiCMOS Technology Featuring 120/100 GHz (f_T/f_{MAX}) HBT and ASIC-Compatible CMOS Using Copper Interconnect," *Proc. BCTM*, pp. 143-146, 2001.

[4] B.A. Orner, et al, "A 0.13um BiCMOS Technology featuring a 200/280GHz (f_T/f_{MAX}) SiGe HBT," *Proc. BCTM*, pp. 203-206, 2003.

[5] J. Bock, et al, "3.3ps SiGe Bipolar Technology," *IEDM Tech. Dig.*, pp 255-258, 2004.

[6] P. Chevalier, et al, "300 GHz f_{MAX} self-aligned SiGeC HBT optimized towards CMOS compabitility," *Proc. BCTM*, pp. 120-123, 2005.

[7] M. Racanelli and P. Kempf, "SiGe BiCMOS Technology for RF Circuit Applications," *IEEE Trans. Electron Devices*, Vol. 52, No. 7, pp. 1259-1270, 2005.

[8] M. Laurens, et al, "A 150GHz f_T/f_{MAX} 0.13um SiGe:C BiCMOS Technology," *Proc. BCTM*, pp. 199-202, 2003.

[9] S.V. Huylenbroeck, et al, "Lateral and Vertical Scaling of a QSA HBT for a 0.13um 200GHz SiGe:C BiCMOS Technology," *Proc. BCTM*, pp. 229-232, 2004.

[10] P. Chevalier, et al, "Low-Cost Self-Aligned SiGeC HBT Module for High-Performance Bulk and SOI RFCMOS Platforms", *IEDM Tech. Dig.*, pp 983-986, 2005.

[11] B. Heinemann, et al, "A Low-Parasitic Collector Construction for High-Speed SiGe:C HBT's," *IEDM Tech. Dig.*, pp 251-254, 2004.

[12] R. Reuter and Yi Yin, "W-band Low-Noise SiGe-Amplifier with State-of the Art Noise Figure and Adjustable High Gain", *submitted to BCTM*, 2006.

2006 Bipolar/BoCMOS Circuits and Technology Meeting Proceedings

Experimental Study of Metallic Emitter SiGeC HBTs

B. Barbalat[1,2], F. Judong[1], L. Rubaldo[1], P. Chevalier[1], M. Proust[1], C. Richard[1], G. Borot[1],
B. Vandelle[1], F. Saguin[1], D. Dutartre[1], N. Zerounian[2], F. Aniel[2] and A. Chantre[1]

[1]STMicroelectronics, 850 rue Jean Monnet, 38926 Crolles Cedex, France
[2]Institut d'Electronique Fondamentale, Bât 220, Université Paris-Sud 11, 91405 Orsay, France

Abstract — **This paper deals with the integration of a metallic emitter in a high-speed SiGe HBT technology. An innovative integration process called PRETCH is detailed, along with static and high-frequency electrical results. Hall Effect measurements down to 20 K on As-doped mono-emitter layers are also reported, which help understand the physics of hole injection in highly doped emitters. The optimization of the process is detailed, and a significant increase of the $f_T \times BV_{CEO}$ product in metallic emitter SiGe HBTs is finally demonstrated.**

Index Terms — **BiCMOS, Hall effect, HBT, high-speed, hole diffusion, metallic emitter, SiGeC.**

I. INTRODUCTION

SiGe HBTs exhibit nowadays cut-off frequencies f_T and f_{MAX} higher than 250 GHz [1]. Such advances are made possible by the increase of collector doping, that delays the onset of the Kirk effect, and enables high collector currents. As a drawback, the current gain increases, with detrimental effect on collector-to-emitter breakdown voltage BV_{CEO}. This paper explores the use of a metallic emitter to increase the base current I_B without degrading the collector current I_C and then to improve the BV_{CEO} [2].

This effect has been experimentally reported in [3] with a complete silicidation of a thick emitter, whereas the metallic emitter process presented here uses a thin poly-emitter approach thanks to an innovative integration flow. Next, we discuss both static and RF electrical results, and Hall Effect measurements used to understand the electrical behavior of the mono-emitter. We finally conclude on the possibility to create a metallic emitter with the thin poly-emitter approach.

II. METALLIC EMITTER CONCEPT

In the NPN bipolar transistor, I_B is proportional to the gradient of the hole concentration at the entrance of the emitter [4]. This gradient depends on the recombination rate in the emitter, which is a function of the emitter doping (Auger or SRH recombination) when the emitter is thick. If a very shallow emitter is used, the neutral emitter width W_E becomes smaller than the hole diffusion length, and I_B is controlled by the recombination at the emitter contact, the metal / silicon interface acting as a fast recombination surface (Fig. 1) [4]. Hence, a "metallic emitter" is a thin and well controlled emitter that leads to a large I_B [2].

The I_B increase has a beneficial impact on BV_{CEO} due to the reduction of current gain β. As the collector current I_C only depends on the base profile and width, it is theoretically independent of the emitter thickness and doping. The metallic emitter then offers a way to increase the base current solely.

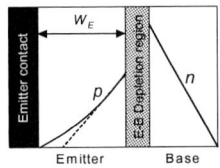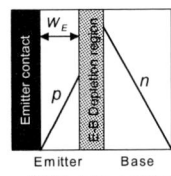

(a) Classic poly-Si emitter   (b) Metallic emitter

Fig. 1. Hole concentrations in the case of (a) classic poly-emitter and (b) metallic emitter. When W_E is reduced, the hole gradient is increased, leading to a higher base current.

III. DEVICE FABRICATION

A. Standard Process

Devices studied in this work are fabricated using the technology detailed in [5]. The architecture of the HBT is a Fully Self-Aligned (FSA) double-polysilicon structure using Selective Epitaxial Growth (SEG) of the SiGeC base. The SiGe growth is terminated by an undoped silicon capping layer (Si-cap) separating the base from the emitter. The Si-cap is necessary to account for the diffusion of the doping species from the emitter. The arsenic In-Situ Doped Poly-emitter (ISDP) is deposited in a single wafer RTCVD reactor, also used for the SiGeC base growth. The emitter is partially mono-crystalline, leading to a mono-emitter in the intrinsic region. Different doping levels can be obtained by changing the arsine (AsH_3) flow and the emitter growth conditions.

The HBT front-end fabrication finishes with a spike activation annealing, cobalt silicidation and contacts formation. The silicide thickness is ~ 20 nm, so W_E is somewhat thinner than the deposited poly-emitter. The contact is composed of successive layers of Ti, TiN and W. A TEM view of the standard structure can be seen in Fig. 2a.

B. Thin Silicided Polyemitter (TSP)

The first investigated approach to realize a metallic emitter was to reduce the emitter thickness from 150

1-4244-0458-4/06/$20.00 ©2006 IEEE

to 40 nm, while keeping the rest of the process unchanged, as can be seen in Fig. 2b. Particularly, the silicidation module was not modified in order to keep CMOS compatibility, whereas in [2] the silicide thickness was tuned to create a thin neutral emitter with a thick polysilicon layer.

Different ISDP thicknesses were experimented, with some impact on base current. However, the silicide / silicon interface is not smooth enough to allow good control of the emitter thickness. Also, depending on thermal budget variations, the silicide thickness and roughness may vary, leading to non-ideal base currents [6]. This is detrimental to a proper use of the metallic emitter effect. We have therefore developed an alternative process allowing thinner emitters and smoother metal / silicon interfaces.

C. PRETCH Process

The approach retained for the metallic emitter study is based on an innovative process used to integrate metal gates or high-K dielectrics in advanced MOSFETs. The process is called PRETCH, for Poly REplacement Through Contact Holes [7].

The realization of the HBT with PRETCH architecture is close to the standard one. As a main difference, after the thin ISDP deposition a thin oxide layer is deposited, which will act as an etch stop. Then, we deposit a 100 nm sacrificial poly-silicon, overlapped by a nitride layer which is necessary to prevent the top of the emitter from silicidation. The whole stack ISDP – oxide – poly-Si – nitride is then patterned. The fabrication is continued by borderless nitride deposition, and the first Inter-Metal Dielectric (IMD) level. The contact etching through the IMD stops on the silicide (base and collector contacts) and in the sacrificial poly-Si above the emitter (Fig. 3a). The sacrificial poly-Si is etched out thanks to a SF_6 based plasma, very selective to all dielectrics and to silicide (Fig. 3b). The thin oxide layer is then wet etched and the metal of the contacts is deposited by sputtering (Ti) and CVD (TiN and W) (Fig. 3c). The emitter cavity and the contacts are filled at the same time. This process was made possible by the use of centered stripe emitter contacts, having the same length as the emitter window [1]. TEM view of the achieved PRETCH emitter device is shown in Fig. 2c.

IV. ELECTRICAL RESULTS

A. Highly Doped Emitters

Static electrical parameters are reported in Table I for $0.15 \times 3.6\ \mu m^2$ HBTs with Si-cap thickness fixed to 18 nm and emitters doped using standard arsine flow F. Metallic emitters using 80, 60 and 40 nm ISDP are compared to the standard 150 nm thick non-metallic emitter. Collector current density is quite constant, and maximum current gain is divided by a factor of 2 when the emitter thickness is reduced from

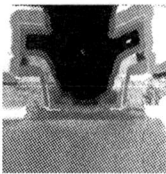

(a) Std Emitter   (b) TSP   (c) PRETCH

Fig. 2.   TEM views of finished transistors: (a) Standard mono-emitter; (b) Thin Silicided Poly-emitter, 20 nm thick; (c) Emitter made with the PRETCH process, 40 nm thick.

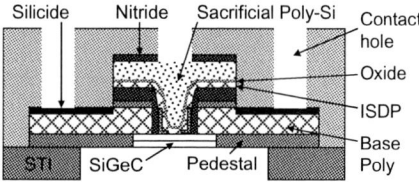

(a) Contact etching, stopping on the silicide (base and collector) and in the sacrificial poly-Si (emitter).

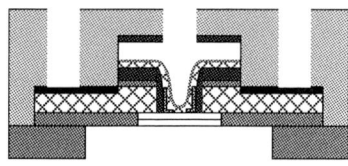

(b) Sacrificial poly-Si isotropic etching. Etching is selective towards thin oxide and silicide.

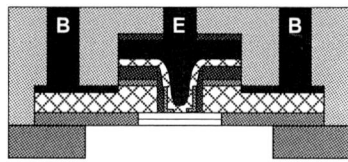

(c) Thin oxide removal and metal filling through contact holes.

Fig. 3.   PRETCH process successive steps (collector contacts are not shown).

150 nm to 40 nm, thanks to a twofold increase of J_B. Gain reduction is however not sufficient to significantly increase BV_{CEO} (+ 0.1 V). Besides, emitter resistivity increases from 3.2 $\Omega.\mu m^2$ for the standard poly-emitter to 8.3 $\Omega.\mu m^2$ for the thinnest emitter (40 nm).

Fig. 4 compares the Gummel plots and gain of standard emitter and 40 nm metallic emitter HBTs, illustrating the gain decrease due to emitter thickness reduction. The atypical behavior of I_B below 0.4 V is due to a band-to-band tunneling effect reported in [8]. This tunnel current is reduced when the emitter is thinner, which means that the emitter-base junction is less abrupt and the arsenic-to-boron distance is increased. SIMS analyses proved indeed that arsenic diffusion depends on the whole doping atoms

TABLE I
HIGH EMITTER DOPING – $0.15 \times 3.6\ \mu m^2$ HBTs

Emitter thickness [nm]	150 (ref)	80	60	40
J_C (0.7 V) [$\mu A / \mu m^2$]	22.8	21.4	22.2	21.5
J_B (0.7 V) [$nA / \mu m^2$]	9.1	10.5	12.1	18.5
β (0.7 V)	2500	2040	1830	1160
BV_{CEO} [V]	1.44	1.46	1.48	1.54
ρ_E [$\Omega.\mu m^2$]	3.2	3.4	5.0	8.3

252

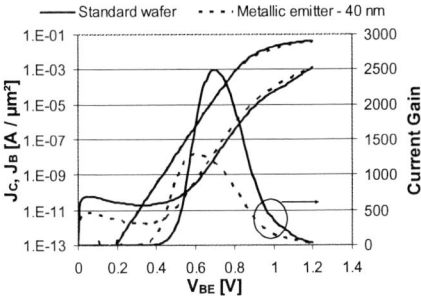

Fig. 4. Gummel plots comparison of standard and metallic emitter 0.15 × 3.6 μm² HBTs. The arsine flow is set to the standard flow F, and the Si-cap is 18 nm thick.

reservoir, explaining why arsenic diffuses less when the emitter is thinner. As reported in [1], one of the main contributions to the emitter resistance is the low doped tail region of the emitter profile, which depends on the final arsenic-to-boron distance. Then, when thin poly-emitters are used, Si-cap thickness needs to be reduced to maintain low R_E.

The absence of significant metallic emitter effect was corroborated by Hall Effect measurements down to 20 K. Fig. 5 shows the effective carrier concentration in 40 nm thick As-doped mono-emitter layers, as a function of arsine flow (F, F/3 and F/6) and temperature. There is no degradation of the carrier concentration at low temperature, which means that the highly-doped emitter is degenerated, and behaves not as a semiconductor, but as a metal. It should be pointed out that the hole diffusion lengths that can be derived from these experiments are significantly smaller than anticipated from literature [4]. As schematically shown in Fig. 6a, these results explain the small magnitude of the metallic emitter effect seen with these highly doped emitter layers.

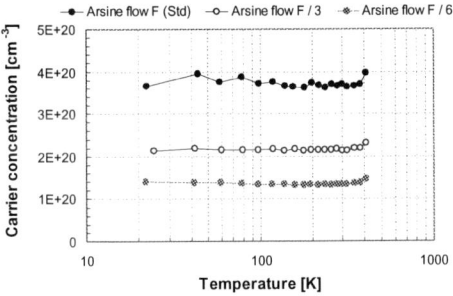

Fig. 5. Carrier concentration in mono-Si with arsine flow F, F/3 and F/6 measured by Hall Effect. The absence of carrier freeze-out is typical of a degenerated semiconductor.

B. Lowly Doped Emitters

In order to increase the base current by metallic emitter effect, emitter doping concentration has to be reduced. This increases the hole concentration at the entrance of the emitter. The other advantage of reducing the emitter doping is to decrease volume

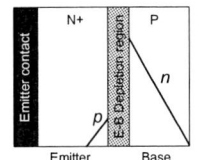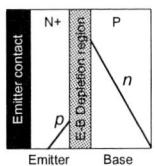

(a) Highly-doped poly-emitter: There is almost no effect of emitter thickness reduction because of the small hole injection level and small hole diffusion length.

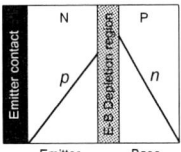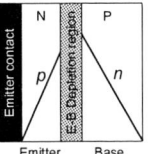

(b) Lowly-doped poly-emitter: The injected hole concentration is large, and the diffusion length is increased, so the emitter thickness has an effect on the hole gradient, then on the base current.

Fig. 6. Influence of emitter doping and thickness on hole injection and recombination.

recombination and increase hole diffusion length. Hence, the hole concentration gradient becomes controllable by the emitter thickness (Fig. 6b). To balance the emitter resistance increase due to doping reduction, Si-cap thickness has to be adapted to the new doping reservoir.

Devices were fabricated by reducing the arsine flow up to 3 times and decreasing Si-cap thickness. Full epitaxial growth of the emitter was also tested, leading to emitter doping up to one decade lower than the initial poly-emitter one. Fig. 7 plots the Gummel curves of 0.15 × 3.6 μm² HBTs with reduced emitter doping, having 40 nm and 20 nm thick emitters and 6 nm thick Si-cap. Metallic emitter behavior is evidenced through the significant increase of the base current, while collector current is left unchanged. Current gain is divided by 10 due to base current increase, with significant increase of BV_{CEO} from 1.44 V to 1.92 V.

Table II summarizes the static and high-frequency parameters of metallic emitter HBTs having different Si-cap thickness, emitter thickness and doping level. A comparison with non-metallic reference process is

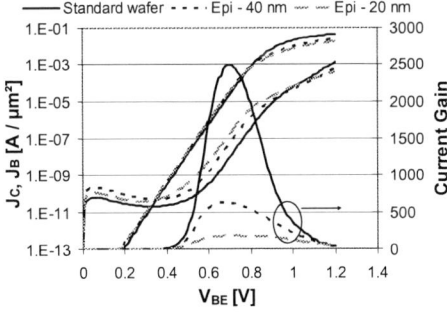

Fig. 7. Gummel plots comparison of 0.15 × 3.6 μm² HBTs with different emitters: Reference device uses 18 nm Si-cap, 150 nm emitter and arsine flow F; Metal emitter devices have varying emitter thickness, 6 nm Si-cap and use epitaxial growth conditions for the emitter.

provided. The low values of f_T and f_{MAX} observed with low Si-cap thicknesses are largely due to an unwanted increase of the base access resistance in these non-optimized devices.

Indeed, the thickness of the pedestal oxide, in which the base cavity is formed in this selective epitaxy technology, has not been decreased when reducing the Si-cap thickness. This is known to degrade the resistance of the link region between intrinsic and extrinsic bases. Fig. 8 illustrates this issue in the extreme case of an ultra-thin Si-cap (2 nm) where there is no base connection at all between the poly-base and the epitaxial base, leading to an infinite value of R_B. When the Si-cap is 6 nm thick, the case must be intermediate, leading to a high – but finite – base resistance. On the other hand, since R_E extraction is dependent on R_B [9], the R_E increase seen in Table II when R_B is large is mainly due to the degradation of the base link.

Meanwhile, the observed increase of BV_{CEO} enables to push the $f_T \times BV_{CEO}$ product from 375 GHz.V for a standard transistor up to 412 GHz.V with a metallic emitter.

V. DISCUSSION AND CONCLUSION

We have demonstrated an innovative CMOS-compatible integration scheme to create a high-speed SiGeC HBT with metallic emitter. Thanks to the robustness of PRETCH process, the emitter thickness is well controlled, leading to a smooth metal / silicon interface.

We have pointed out the unexpectedly short hole diffusion lengths observed in our heavily doped emitters, which initially prevented metallic emitter behavior. With a strong reduction of the emitter doping (more than an order of magnitude), the hole diffusion length becomes large enough to authorize a control of the base current by the emitter thickness. The base current was then increased by 10, improving significantly the breakdown voltage BV_{CEO}.

This doping reduction led to the increase of series resistance with detrimental effect on cut-off frequencies f_T and f_{MAX}: R_E increase was solved by reducing the Si-cap, which in turn strongly degraded the base resistance. To that effect, devices with thin Si-cap and adapted pedestal base oxide should demonstrate metallic emitter effect with no degradation of the cut-off frequencies. Predicted performances for optimized devices are 260 GHz f_T with 1.9 V BV_{CEO}, corresponding to a $f_T \times BV_{CEO}$ product around 500 GHz.V.

Alternative approaches involving materials studies must be considered. For instance, a better trade-off between emitter resistance and hole diffusion length is anticipated for Phosphorous doped emitters. Other solutions are also interesting to increase the $f_T \times BV_{CEO}$ product, such as devices with enhanced neutral base recombination [10].

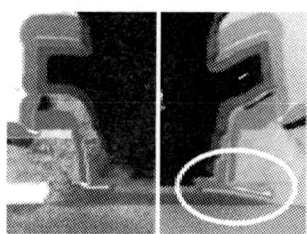

Fig 8. TEM view illustrating a bad base connection. Left: Correct link; Right: No base link in case of very thin Si-cap.

TABLE II
STATIC AND RF PARAMETERS –
REDUCED EMITTER DOPING – $0.15 \times 3.6 \ \mu m^2$ HBTs

Emitter type & doping	Std	F / 3	F / 3	Epi	Epi
Emitter thickness[nm]	150	40	20	40	20
Si-Cap thickness [nm]	18	10	8	6	6
J_C 0.7 V [$\mu A / \mu m^2$]	22.8	27.5	28.9	32.3	31.6
J_B 0.7 V [$nA / \mu m^2$]	9.1	27.7	55.9	51.5	172.7
β 0.7 V	2500	996	516	627	183
BV_{CEO} [V]	1.44	1.61	1.75	1.69	1.92
ρ_E [$\Omega.\mu m^2$]	3.2	6.8	13.5	19.3*	23.7*
R_B [Ω]	28	28	33	184	138
f_T [GHz]	260	256	234	198	190
f_{MAX} [GHz]	282	284	247	158	138
$f_T \times BV_{CEO}$ [GHz.V]	375	412	410	335	363

(*) Erroneous values due to high R_B values.

ACKNOWLEDGEMENTS

The authors wish to thank the many actors of ST Crolles involved in the fabrication, simulation and electrical characterization of the HBT. They also thank Xavier Mescot at IMEP for Hall Effect measurements.

REFERENCES

[1] P. Chevalier et al., "300 GHz f_{max} Self-Aligned SiGeC HBT optimized towards CMOS compatibility", *Proc. BCTM*, 2005, pp. 120-123.

[2] European Patent Office, Serial No. EP 1 517 377 A1, filed on September 17, 2003 (A. Chantre et al., ST).

[3] J.J.T.M. Donkers et al., "Metal Emitter SiGe:C HBTs", *IEDM Tech. Dig.*, 2004, pp. 243-246.

[4] P. Ashburn, "Design and Realization of Bipolar Transistors", J. Wiley & Sons, 1988.

[5] P. Chevalier et al., "230 GHz Self-Aligned SiGeC HBT for Optical and Millimeter-Wave Applications", *IEEE Journal of Solid State Circuits*, Vol. 40, No 10, October 2005, pp. 2025-2034.

[6] N. Wils et al., "Identification and analysis of a new BJT parametric mismatch phenomenon", *Proc. BCTM*, 2005, pp. 224-227.

[7] S. Harrison et al., "Poly-gate Replacement Through Contact Hole (PRETCH): A new method for high-K / metal gate and multi-oxide implementation on chip", *IEDM Tech. Dig.*, 2004, pp. 291-294.

[8] D. Lagarde et al, "Band-to-band Tunneling in Vertically Scaled SiGe:C HBTs", *IEEE Elec. Dev. Letters*, Vol. 27, No. 4, April 2006, pp. 275-277.

[9] D. Berger et al., "HiCUM Parameter Extraction Methodoloy for a Single Transistor Geometry", *Proc. BCTM*, 2002, pp. 116-119.

[10] B. Barbalat et al., "Carbon effect on neutral base recombination in high-speed SiGeC HBTs", *ISTDM Conf. Dig.*, 2006, pp. 238-239.

2006 Bipolar/BoCMOS Circuits and Technology Meeting Proceedings

Schottky Barrier Diodes for Millimeter Wave SiGe BiCMOS Applications

R.M. Rassel, J.B. Johnson, B.A. Orner, S.K. Reynolds, M.E. Dahlström, J.S. Rascoe, A.J. Joseph, B.P. Gaucher, J.S. Dunn, S.A. St. Onge

IBM, Systems & Technology Group, 1000 River Road, Essex Junction, VT 05452 USA

Abstract — **For the first time, a high performance, low leakage Schottky barrier diode (SBD) with cutoff frequency above 1.0 THz in a 130nm SiGe BiCMOS technology for millimeter-wave application is described. Device optimization has been evaluated by varying critical process and layout parameters such as, anode size, cathode depth, cathode resistivity, junction tailoring, and guardring optimization is investigated.**

Index Terms — **BiCMOS process technology, cutoff frequency, millimeter wave diodes, millimeter wave technology, Schottky diodes, terahertz (THz).**

I. INTRODUCTION

There is an increasing demand for integrated circuits to operate at millimeter wave (mmWave) frequencies. SiGe BiCMOS technology is a natural choice for such applications due to its high performance bipolar transistor integrated with high quality passives and CMOS devices [1]. SiGeHBTs have shown steady improvement with a recent report of 300/350 GHz fT/fmax [2], which enables various mmWave applications. However, for a technology to be millimeter wave capable one must look beyond the high speed transistors and evaluate the passive components such as Schottky Barrier Diodes (SBD), PIN diodes [3], and varactor diodes [4].

An SBD with a cutoff frequency (Fc) above a terahertz enables designs such as mixers, high speed sample and hold circuits at mmWave frequencies [5]. In other applications, the SBD is used as a rectifier in power detectors for power amplification circuits. Low leakages in the off state translate to superior hold times in sample & hold circuits. For a power detector low leakage is necessary due to noise constraints.

Presented in this paper is a low leakage (<0.004 A/cm^2) Schottky barrier diode for use as an element in a mixer circuit. The performance goals of a cutoff frequency greater than 1 THz and a quality factor greater than 10 at 55 GHz were achieved. In addition, the technology goal is to have the SBD diode integrated modularly into a high-performance SiGe BiCMOS technology [6].

II. DEVICE DESCRIPTION

A cross-sectional sketch of the integrated SBD is shown in Figure 1. The device is a vertical Schottky diode with a buried n+ layer naturally arising from a SiGe BiCMOS HBT process. The n+ cathode layer is formed by high dose implant into a p-type substrate and then buried by a nominally intrinsic epitaxial Si growth. An n-type junction tailor implant is added between the buried n+ layer and the overlaying Schottky junction. The p-type guardring is typically formed from the standard PFET source/drain ion implant and defines the Schottky junction area. The Schottky barrier is formed by cobalt salicidation of n-silicon. Effectively the p-guardring and the Schottky junction are shorted together by silicide. The buried layer is contacted with an n+ reach-through implant, and deep trench isolation provides perimeter isolation from the substrate.

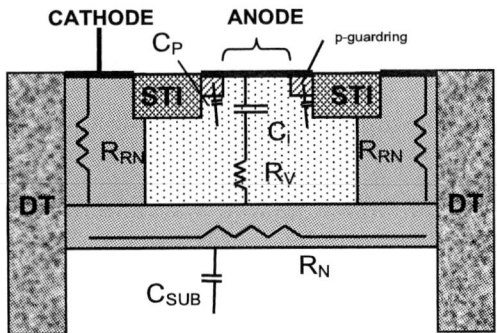

Fig. 1.   SBD Cross Section.

Parasitic resistances, which decrease Fc, include the down resistance in the intrinsic region, buried layer lateral resistance, reach through resistance, and contact and wiring resistances. The intrinsic region resistance dominates the resistive parasitic. Parasitic capacitance occurs between the device and substrate, p-guardring and buried layer, and between anode and cathode. Anode-to-cathode capacitance sources include the capacitance down to the buried layer, the capacitance between the anode and the reach through implants, and capacitances between the anode and cathode wiring. Capacitance minimization improves Fc.

Both layout and process optimization have been studied. Layout optimization includes variations in the anode as well as anode-to-reachthrough distance and cathode width. Process optimization focused on epitaxial thickness (i.e. cathode distance), junction

---

1-4244-0458-4/06/$20.00 ©2006 IEEE   255

tailor ion implant, buried layer resistance, and guardring depth and concentration.

A high performance HBT is usually fabricated with an n+ buried subcollector and an intrinsic epitaxial layer, and this subcollector is often used as the SBD cathode. The trend has been to decrease the epitaxial layer thickness to increase the HBT speed, and this negatively impacts SBD reverse bias leakage as well as the junction capacitance. In order to tune this parameter independent of the HBT, we have introduced a dedicated low resistance n+ layer that provides a deeper cathode that can be tuned independently for the performance of mmWave devices.

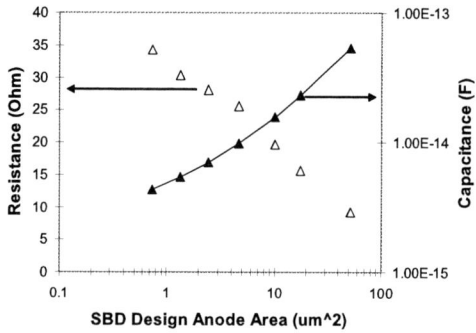

Fig. 2. Median SBD resistance & capacitance extracted from 2-port s-paramerters taken from 10 to 100 Ghz for various anode sizes.

III. SBD DEVICE CALCULATIONS

Two port s-parameters were used to analyze the Schottky barrier diode. One port is connected to the anode and the second port is connected to the cathode while substrate contacts wired to a ground plane were placed in close proximity of the SBD. The SBD capacitance is given as:

$$C = Imaginary(1/Z_{22}4\pi\omega) \qquad (1)$$

and the on resistance is extracted from s-parameters as:

$$R = Real(1/-Y_{12}). \qquad (2)$$

S-parameter data is taken at the zero volt bias condition. The resistance is constant in frequency, but typically a significant amount of measurement noise is encountered. To reduce this noise, the median value over a given frequency range has been used by other authors [7]. We have adopted this technique and use the median C and R over a wide frequency range of 10 to 100 GHz. The extracted capacitance was observed to have some frequency dependence; the median capacitance value was effectively the capacitance at approximately 55 GHz. Fc was calculated using the above R and C as:

$$Fc = 1/(2\pi RC). \qquad (3)$$

The quality factor for the SBD is given as:

$$Q = imag(-Y_{11}) / real(Y_{11}). \qquad (4)$$

IV. SBD DEVICE RESULTS

Figure 2 shows the effects of decreasing the SBD anode size. As the anode size decreases, the resistance increases while the capacitance decreases. As shown in Figure 3 the Fc increases as anode sizes decreases with a cutoff frequency reaching 1.1 THz for a moderately sized dimension.

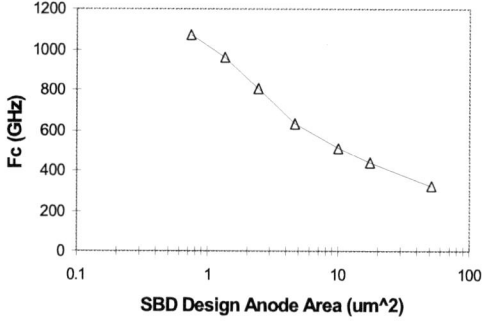

Fig. 3. SBD cutoff frequency for various anode sizes.

Similarly the quality (Q) factors across frequencies from 10 to 100GHz for various device sizes are shown in Figure 4.

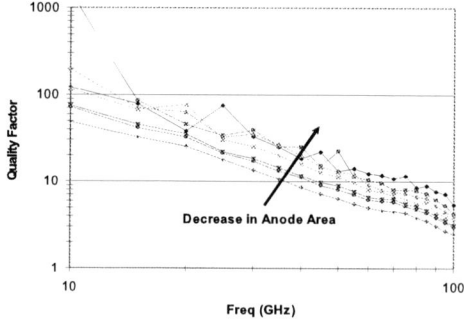

Fig. 4. Quality factor for various SBD anode sizes.

The decrease in device size increases the Q to greater than 100 at 10GHz and to about 11 at 70Ghz. Higher Q and Fc seem achievable by simply scaling the device size more aggressively. Our data is limited to anode areas of 0.74 μm^2, but anode size as small as 0.10 μm^2 that are achievable using the advance lithography is expected to improve the device performance further.

The anode to cathode spacing was varied to reduce parasitic capacitance between the anode and reachthrough regions as well as contact/wiring capacitances. Likewise, increasing the cathode width in order to decrease the down resistance of the reachthrough region was investigated. Neither approaches showed capacitance improvement (Figure 5) suggesting that the Fc was limited by other factors.

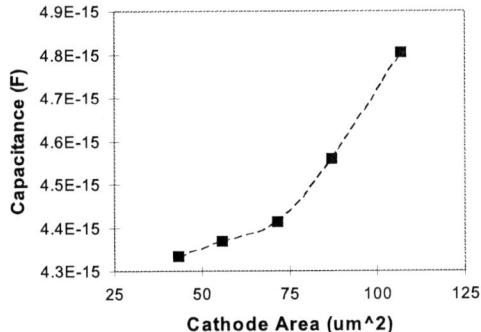

Fig. 5. Capacitance for various cathode areas resulting from increases in anode to cathode space as well as cathode width. Schottky anode = 0.74μm².

The buried n+ layer distance from the surface is found to be a strong lever for performance optimization (Figure 6). An optimal n-epi thickness similar to the PIN diode in this technology [3] is observed for various device sizes.

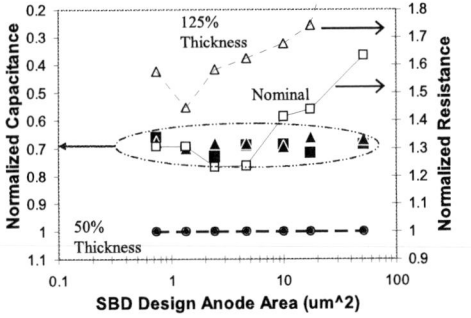

Fig. 6. Normalized resistance & capacitance data to highlight the effects of increased n-epi thickness. The capacitance data is shown with solid symbols & is highlighted with blue circle.

The capacitance from the 50% to 100% nominal n-epi thickness shows a dramatic drop nearly proportional to the resistance increase for the smaller size devices. The increase in resistance out weighs the capacitance drop from the 50% to the 125% nominal n-epi thickness split. By optimizing the n-epi thickness, both mmWave capable SBD and PIN diode are attainable [3] in this SiGe BiCMOS technology.

An additional n-type implant is required to lower the resistance from the junction down to the buried n+ layer since the PIN diode requires an intrinsic region. This n-type implant to tailor the dopant profile between the Schottky junction and the buried layer was investigated in great detail. Figure 7 shows the Fc enhancement by doubling the n-type dopant concentration in this tailor implant as well as pushing it farther from Si surface. This Fc is observed to be a combination of resistance & capacitance reduction.

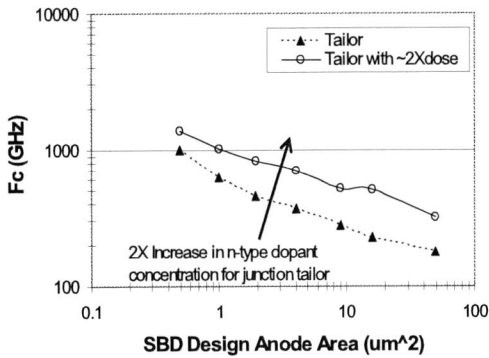

Fig. 7. Fc across various anode sizes for two different junction tailor dopant profiles. The solid line is 2 times the dopant concentration as the dashed line.

The buried n-layer resistivity is analyzed to give a better understanding of all parasitic components in this SBD (Figure 8). The cathode with the 10x higher sheet resistance shows around 5 to 7ohms higher SBD device resistance with no significant scaling on device anode area.

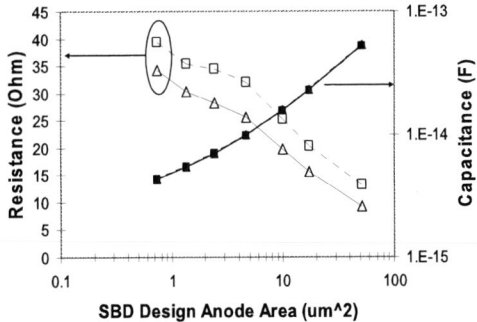

Fig. 8. Resistance and capacitance across numerous anode areas for nominal and 10 times greater cathode resistivity. The triangles with solid line represent the nominal cathode resistivity and squares with dashed line show the data for cathode resistivity ten times greater than nominal.

Therefore the smallest tested device shows >20% reduction in resistance while the largest area device shows >40% reduction for the lower cathode resistivity. The SBD capacitance delta between the two different cathodes is insignificant in respect to the percent delta observed with the resistance. However, the SBD does show a small systematic percent decrease in capacitance of around 0% to 2.5% as the anode area is increased from 0.74 to 51 μm², respectfully. This is due to the fact the SBD designs have a fixed p+ guardring width for all anode sizes. As the active anode area increases, the cathode back-plate area also increases which contributes a parasitic capacitance to the p- substrate. The data suggests that cathode to substrate parasitic becomes a larger contributor to the total device capacitance with larger anode areas, but is still a very small contributor in comparison to the p-type guardring.

The above analysis suggests that p+ guardring optimization is the strongest lever for decreasing the capacitance of the smallest device sizes and therefore increasing Fc. Previous publications [7] suggest that this p+ guardring could simply be removed, but at the cost of increased leakage. The absence of the guardring is also a reliability concern. To investigate this important aspect of the device, we built Schottky barrier diodes with and without the guardring and also varied the guardring ion implant conditions.

The reverse bias leakage for different SBD structures is analyzed. A SBD built from the PFET N-well implant & base CMOS processing [7] as well as our SiGe BiCMOS described above was investigated. The reverse bias leakage at 1 V is observed to be around two orders of magnitude larger (>0.1 A/cm^2) for the CMOS SBD structures without a p-type guardring [7] than our SiGe BiCMOS SBD structure with a guardring (<0.004 A/cm^2).

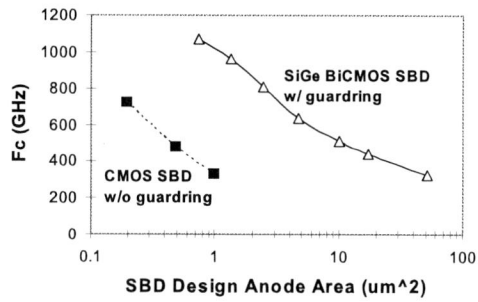

Fig. 9. Cutoff frequency across various anode sizes. A CMOS SBD without a p-type guardring is shown with squares/dashed line and a SiGe BiCMOS SBD with a p-type guardring is shown with triangles/solid line.

Figure 9 shows the Fc for a standard CMOS SBD without a p-type guardring [7] and our SiGe BiCMOS SBD with a p-type guardring. The cutoff frequency for our SiGe BiCMOS SBD is ~200% greater than the CMOS SBD with equivalent anode sizes. The addition of the buried n+ layer and the n+ reachthrough contact reduces the resistance beyond what seems to be achievable from utilizing standard CMOS processing. The SiGe BiCMOS SBD is far superior to the SBD built from standard CMOS processing.

The significant increase in reverse bias leakage suggests that a complete removal of the guardring is not possible. Figure 10 shows the capacitance decreasing proportionally to the guardring ion implant energy and ion implant concentration while maintaining the DC characteristics. This suggests that the Fc could be increased furthermore by

guardring optimization without any impact of reverse bias leakage or DC characteristics.

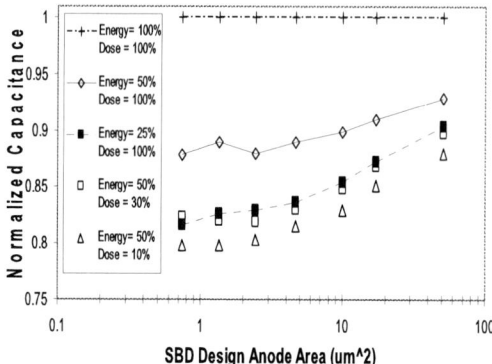

Fig. 10. The normalized capacitance for numerous p-type guardring ion implant conditions over various anode sizes.

V. SUMMARY

The device design for a high performance, low leakage Schottky barrier diode in a 130nm SiGe BiCMOS technology has been discussed. Device optimization included cathode depth, cathode resistivity, junction tailoring, and guardring optimization reduction. The optimal device design described has a cutoff frequency above 1.0 THz and a Q greater than 10 at 55GHz. This device combines with other critical passive devices [3] [4], and the high-performance active devices available in 130nm SiGe BiCMOS technologies to enable mmWave circuit design in silicon manufacturing technology. This work is supported in part by funding from the Defense Advanced Research Projects Agency (DARPA) under contract number N-66001-05-C-8013.

REFERENCES

[1] B. Gaucher, et al., "Silicon Germanium Based Millimeter-Wave IC's for Gbps Wireless Communications and Radar Systems", *GOMAtech Conference* March 2006.

[2] M. Khater, et al., *Technical Digest of the IEEE International Electron Devices Meeting*, 247 – 250, (2004).

[3] B. A. Orner, et al., "p-i-n Diodes for Monolithic Millimeter Wave BiCMOS Applications", *ISTDM* May 2006H.

[4] D. Coolbaugh, et al., *Radio Frequency Integrated Circuits (RFIC) Symposium 2002 IEEE*, 341-344 (2002).

[5] B.Gaucher, et al., "Fully Integrated SiGe mmWave Transmitter and Receiver ICs", ISTDM May 2006.

[6] B. A. Orner, et al., *Proceedings of the 2003 Bipolar/BiCMOS circuits and Technology Meeting*, 203-204 (2003).

[7] S. Sankaran, et al, *IEEE Electron Device Letters*, 26 (7), 492 – 494 (2005).

2006 Bipolar/BoCMOS Circuits and Technology Meeting Proceedings

High Performances 3D Damascene MIM Capacitors Integrated in Copper Back-End Technologies

S. Crémer [1], C. Richard [1], D. Benoit [1], C. Besset [1], J.-P. Manceau [1,2], A. Farcy [1],
C. Perrot [1], N. Segura [1], M. Marin [1], S. Bécu [1,3], S. Boret [1], M. Thomas [1,4],
S. Guillaumet [1], A. Bonnard [1], P. Delpech [1] and S. Bruyère [1] – sebastien.cremer@st.com

(1) ST Microelectronics, 850 rue Jean Monnet, 38926 Crolles Cedex, France
(2) LEMD, UMR CNRS 5517, 25 rue des martyrs, BP 166, 38042 Grenoble Cedex 9, France
(3) L2MP, UMR CNRS 6137, 49 rue Joliot-Curie, BP 146, 13384 Marseille Cedex 13, France
(4) LMGP, UMR CNRS 5628, ENSPG BP46, 38402 St Martin d'Hères, France

Abstract — RF and analog designs require high performances MIM capacitors. In order to continue downscaling of MIM devices, we have developed and integrated a 3D damascene MIM capacitor in the copper back-end of a 0.13 μm BICMOS technology. Si_3N_4 has been used as dielectric to reach excellent performances for a 5 fF/μm² capacitance density in term of leakage current, voltage linearity, dielectric relaxation, and reliability. This 3D architecture is a very promising candidate to carry on capacitance density increase.

Index Terms — Dielectric material, Leakage currents, MIM Devices, Reliability

I. INTRODUCTION

As all the other devices, the occupied area of Metal-Insulator-Metal (MIM) capacitors integrated in CMOS and BiCMOS technology back-end has to be continuously reduced. Contrarily to MOS transistor, the required capacitance density increase has been achieved by introducing for each MIM generation a new dielectric material, maintaining almost a constant dielectric thickness by taking advantage of a higher dielectric constant. This choice has been mainly driven by mixed-signal technologies. Indeed, they require high performances for MIM devices like low leakage current and good voltage linearity associated with the constraint of 5 V supply voltage for each technology node (from 0.35 μm to 0.13 μm), while digital and radio frequency circuits generally use reduced voltage at each generation. Thus, 350 Å SiO_2, 320 Å Si_3N_4 and recently 450 Å Ta_2O_5 have enabled to reach capacitance density equal respectively to 1 fF/μm², 2 fF/μm² and 5 fF/μm² for planar MIM capacitors. These devices are today in production in aluminum BEOL using one additional mask and RIE process to define top and bottom electrodes [1]. All these values correspond to maximum density achievable with such dielectrics based on reliability criteria as discussed in [2]. Nevertheless this density increase leads also to leakage current and voltage linearity degradation as depicted in Fig. 1 and Fig. 2. Using 2D architecture and industrial dielectric,

5fF/μm² is close to the maximum capacitance density to guarantee C_2 coefficient below 100 ppm/V² and reliability for 5V applications. Other solutions like stacking of various materials [3-4] are a good way to reduce leakage and/or voltage linearity. Nevertheless, $Ta_2O_5/HfO_2/Ta_2O_5$ stack does not enable to solve high-K material weaknesses like dielectric relaxation [5]. Moreover, SiO_2/High-K stacks, particularly interesting to reduce voltage linearity, are suspected to be limited by reliability aspects to weak voltage supply, due to the thinning of the dielectric thickness requires to reach high capacitance density.

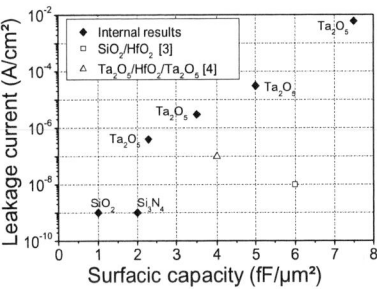

Fig. 1. Leakage current at 3.3 V and 125°C for SiO_2, Si_3N_4 and Ta_2O_5 dielectrics and for dielectric stacks

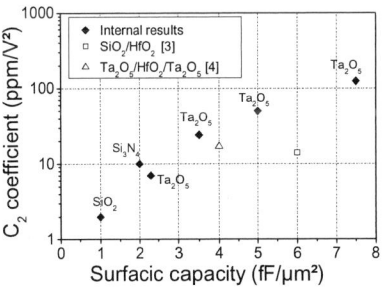

Fig. 2. Quadratic voltage linearity coefficient (C_2) for SiO_2, Si_3N_4 and Ta_2O_5 dielectrics and for dielectric stacks

In this context, achievement of 3D MIM capacitors is now required to carry out capacitance density increase, while maintaining high electrical

1-4244-0458-4/06/$20.00 ©2006 IEEE

performances (low leakage current and good voltage linearity ($C_2 < 100$ ppm/V²)) to satisfy requests for analog circuits and good reliability.

This paper presents the realization and full electrical characterization of 3D damascene MIM capacitor integrated in copper back-end technologies.

II. PROCESS INTEGRATION

This 3D damascene MIM capacitor is integrated between the two last metal levels of a 0.13 µm BICMOS technology having six interconnection levels. After via4/metal 5 process by a classical dual damascene process, capacitor trenches are etched in IMD after photolithography step: it is the only additional mask used for this 3D MIM integration. The MIM stack is then deposited. Top and bottom electrodes are composed of either TaN or TiN materials deposited respectively by PVD and CVD process, while Si_3N_4 dielectric is deposited by PECVD method. Copper deposition allows filling of the trenches. The following CMP step enables to isolate the capacitor trenches. It is the most critical step of the integration since electrical parameters like capacitance density and leakage current depend on this process step. Then wet cleaning is performed: its interest will be demonstrated in the next part based on leakage current results. Finally, MIM electrodes are connected thanks to via5/metal6 level as represented in Fig. 3.

HRSEM cross-section for a pitch equal to 0.7 µm is shown in Fig. 4 demonstrating low roughness of metal 5 bottom electrode and good planarization of the MIM capacitor.

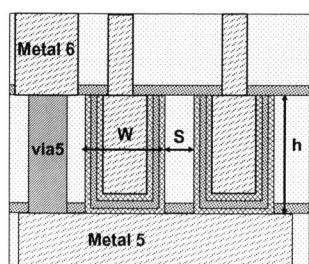

Fig. 3.   Schematic cross-section of 3D damascene MIM

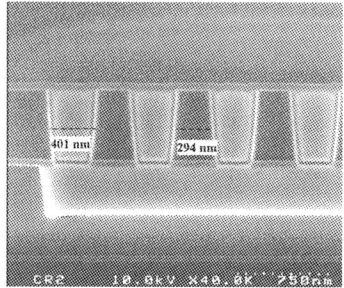

Fig. 4.   HRSEM cross-section of 3D damascene MIM capacitor (via 5 and metal 6 are not visible here)

III. ELECTRICAL RESULTS AND DISCUSSION

MIM capacitor key parameters are capacitance density, leakage current, breakdown voltage and dielectric lifetime. For analog applications, voltage linearity, dielectric relaxation and matching have also to be considered. Finally, for RF circuits, Q factor is crucial. All of these parameters will be presented and discussed in this paragraph.

A. Capacitance density

We define C_{3D} (also called capacitance density) as the total capacitance divided by the projected area.

C_{3D} can be modeled by the following equation:

$$C_{3D} = \frac{\varepsilon_r \varepsilon_0}{t} \times \left(\frac{W.L + 2h.L + 2h.W}{(W+S).L} \right) \quad (1)$$

where t is dielectric thickness, W is trench width, S is the space between trenches, h is the trench height equal to 0.65 µm and L is the trench length in the direction perpendicular to the cross-section (Fig. 3). Since C_{3D} depends of trench height it is important to have a good uniformity of CMP process.

Variation of C_{3D} with W and S are represented and compared to (1) in Fig. 5 and Fig. 6 assuming a constant dielectric thickness of 300 Å whatever W.

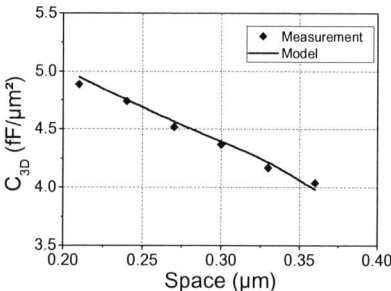

Fig. 5. C_{3D} variation with S for W = 0.4 µm

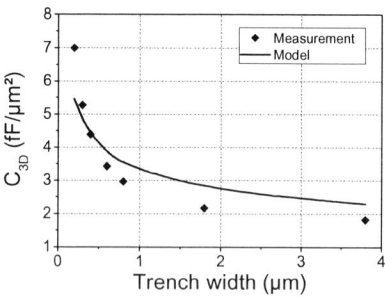

Fig. 6. C_{3D} variation with W for S = 0.3 µm

While C_{3D} variation with S is consistent with (1), a clear mismatch is observed considering W variation. This behavior is well explained by the non ideal conformity of Si_3N_4 deposed by PECVD amplified with trench width reduction. Note that even if non still ideal, the results shown correspond to an optimized

Si_3N_4 recipe using mix-frequency deposition. Using aggressive pitch of 0.43 μm (W= 0.2 μm and S = 0.23 μm), high capacitance density up to 8 fF/μm² have been successfully reached.

B. Leakage Current

Since integration of this 3D MIM capacitor is based on CMP process, we have measured leakage current for various trench widths in order to compare test structures having different perimeter/area ratio. Fig. 7 depicts median value of leakage current density measured at 5 V as a function of W with and without wet cleaning process step.

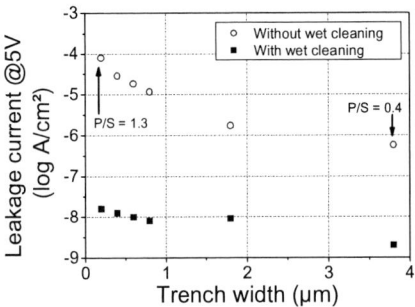

Fig. 7.   Median value of leakage current density at 5 V

Adding a wet cleaning process step allows to reduce leakage current by at least three decades for small W. This is explained by an important contribution of peripheral leakage current to the total one, that is suppressed by wet cleaning process, which removes residues due to CMP process. Thus, for optimized process, leakage current at 5 V as low as 10^{-10} A/cm² and breakdown voltage higher than 20 V are obtained as shown on I-V characteristics performed at 25 °C and 125 °C (Fig. 8).

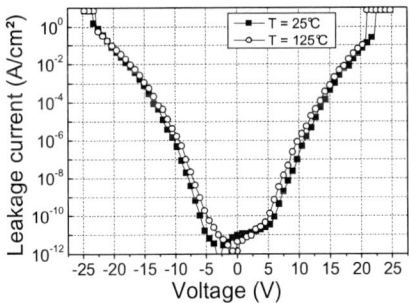

Fig. 8.   I-V characteristics for W = 0.4 μm

C. Breakdown Voltage

In order to compare structures having different trench width, we introduce C_{2D} parameter equal to the total capacitance divided by the developed area. This parameter is only linked to the dielectric thickness in the trenches, while C_{3D} is also structure geometry dependent like W and S as shown in Fig. 5 and Fig. 6.

In Fig. 9, we represent $BV.C_{2D}$ parameter, which is proportional to an average electrical breakdown field. Whatever the trench width, $BV.C_{2D}$ is quite constant indicating that dielectric quality is preserved whatever dielectric thickness in the trenches. 46 fF/μm².V is obtained: it corresponds to breakdown voltage of 23 V for a planar 2 fF/μm² MIM capacitor, which demonstrates an excellent dielectric quality in this 3D architecture. Since breakdown voltage decrease with W reduction, we have evaluated on next section dielectric reliability for W = 0.4 μm to maintain breakdown voltage above 20 V.

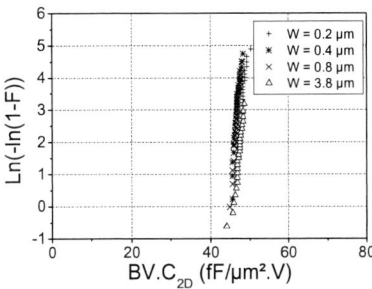

Fig. 9.   Weibull distribution of $BV.C_{2D}$ for W ranging from 0.2 μm to 3.8 μm

D. Reliability Evaluation

In this section, time-to-breakdown experiments performed at 125 °C on 3D MIM capacitors will be developed. Several test structures have been investigated with a constant trench width equals to 0.4 μm and different trench length from 10 to 500 μm, offering a developed dielectric area in a range of 7600 μm² to 0.35 mm². The samples have been stressed at two voltages at wafer level: the corresponding time-to-breakdown Weibull distributions scaled to a 100 000 μm² developed area are depicted in Fig. 10.

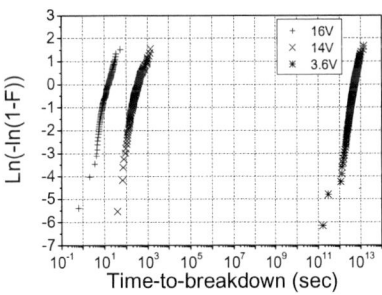

Fig. 10.   Time-to-breakdown Weibull distributions for both accelerated stress and nominal condition

As demonstrated in [9], the √E model can well describe the dielectric reliability of PECVD nitride MIM capacitor, based in particular on a constant charge to breakdown approach. It has been checked that in our experimental voltage range the charge to breakdown can be considered as constant. Therefore

$\sqrt{E}$ model has been applied to extrapolate TDDB lifetime on 3D MIM capacitor: classical reliability criterion of 10 years at 3.6 V for 1 mm² projected area at 0.01 % failure is widely reached, demonstrating good reliability of this device.

E. Voltage Linearity

First results after optimization of top electrode allowed to reach a quadratic coefficient of 25 ppm/V² for a 5 fF/µm² MIM capacitor. This result is comparable to stack solutions [3-4].

F. Dielectric Relaxation

Dielectric Relaxation (DR) consists in the decrease of the capacitance with frequency [6]. The amplitude of the variation observed for 5 fF/µm² Ta_2O_5 MIM capacitor at high temperature (Fig.11) is known to degrade some particularly sensitive applications like A/D converters and VCO [7]. Considering the 3D MIM, its capacitance variation with frequency is much lower.

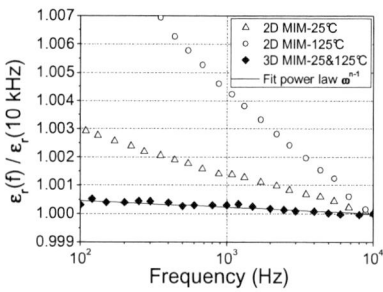

Fig. 11. Comparison of dielectric relaxation between 3D MIM with Si_3N_4 and 2D MIM with Ta_2O_5 for 5 fF/µm²

To quantify this variation, the permittivity evolution has been fitted with a frequency power law: ω^{n-1} [6] with $1-n = 10^{-4}$. This value represents a very low dispersion of the dielectric permittivity. Moreover, the power law shows that the DR of our device is due to the Si_3N_4 quality [6]. During the development of the 3D MIM device, it has been identified that several process steps, such as top electrode process, can degrade the Si_3N_4 quality leading up to a n decrease (a higher DR).

G. Matching Results

For analog applications, the matching of 3D MIM devices was evaluated using a set of capacitor pair with minimal spacing and symmetrical routing to pads. For devices smaller than 500 µm², the experimental mismatch follows $\sigma(\Delta C/C) = 1\% \cdot \mu m/(WL)^{1/2}$. This gives a similar level to the one reported for usual MIM capacitor [8].

H. High Frequency Characterization

Finally, S parameters have been measured up to 50 GHz for capacitance between 100 fF and 10 pF. For all capacitors cut-off frequency is above 25 GHz. We obtained very good Q factor of 100 @2 GHz for a 2 pF capacitor allowing integration of such device in RF designs like VCO.

IV. CONCLUSIONS AND PERSPECTIVES

For the first time, the integration of 3D damascene MIM capacitor using Si_3N_4 dielectric deposited by PECVD has been successfully achieved. Note that only one additional mask has been required. High electrical performances have been measured: capacitance density of 5fF/µm² with a good voltage linearity, leakage current below 10^{-10} A/cm² at 5 V and 125 °C, breakdown voltage above 20 V, very low dielectric relaxation and very good reliability. This architecture will allow continuing MIM roadmap either by increasing trench height and using ALD technique for electrodes and dielectric deposition and/or by the integration of high-K materials.

REFERENCES

[1] M. Gros-Jean, S. Crémer, C. Besset and O. Salicio, "High linear Ta_2O_5 MIM capacitors embedded in Al-interconnected BICMOS technology" , *WODIM 2002*, pp. 73-76.

[2] K.-H. Allers, P. Brenner and M. Schrenk, "Dielectric reliability and material properties of Al_2O_3 in Metal Insulator Metal capacitors (MIMCAP) for RF bipolar technologies in comparison to SiO_2, SiN and Ta_2O_5" , *BCTM 2003*, pp. 35-38.

[3] S.J. Kim, B. J. Cho, M.-F. Li, S.-J. Ding, M.B. Yu, C. Zhu, A. Chin and D.-L. Kwong, "Engineering of voltage vonlinearity in high-K MIM capacitor for cnalog/mixed-signal ICs" , VLSI *2004*, pp. 218-219.

[4] Y-K Jeong, S-J Won, D-J Kwon, M-W Song, W-H Kim, M-H Park, J-H Jeong, H-S Oh, H-K Kang, K-P Suh , "High quality high-K MIM capacitor by $Ta_2O_5/HfO_2/Ta_2O_5$ multi-layered dielectric and NH_3 plasma interface treatments for mixed-signal/RF applications" , *VLSI 2004*, pp. 222-223.

[5] J.R. Jameson, P. B. Griffin, A.Agah, J. D. Plummer, H-S. Kim, D. V. Taylor, P.C. McIntyre and W.A. Harrison, "Problems with metal-oxide high-K dielectrics due to 1/t dielectric relaxation current in amorphous materials" , *IEDM 2003*, p. 91-94.

[6] A. K. Jonscher, "Dielectric relaxation in solids", *London, Chelsea Dielectric* (1983).

[7] J.P. Manceau, S. Bruyère, E. Picollet, M. Minondo, C. Grundrich, D. Cottin, and M. Bely, "Dielectric relaxation characterization and modeling in large frequency and temperature domain. Application to 5fF/µm² Ta_2O_5 MIM capacitor", *IEEE International Conference on Microelectronics Test Structures* (ICMTS2006)

[8] C. H. Ng, C.-S. Ho, S.-F. Chu and S.-C. Sun, "MIM Capacitor Integration for mixed-signal/RF applications", *IEEE Trans. on Electron Devices*, vol. 52, N°7, 2005, pp. 1399- 1408.

[9] K.H Allers, "Prediction of dielectric reliability from I-V characteristics: Poole-Frenkel conduction mechanism leading to $\sqrt{E}$ model for silicon nitride MIM capacitor", *Microelectronics Reliability*, vol. 44 (2004), pp. 411-423

2006 Bipolar/BoCMOS Circuits and Technology Meeting Proceedings

5.8 GHz RF Transceiver LSI Including On-Chip Matching Circuits

Minoru Nagata, Hideaki Masuoka, Shin-ichi Fukase, Makoto Kikuta[*1], Makoto Morita[*1], Nobuyuki Itoh,
Semiconductor Company, Toshiba Corporation,
[*1]Toshiba Microelectronics Corporation,
2-5-1 Kasama Sakae-ku,
Yokohama 247-8585, Japan

Abstract—**A fully integrated 5.8 GHz band transceiver LSI for electronic toll collection systems has been developed. The transceiver consists of LNA, down-converter, ASK detector, ASK modulator, voltage controlled oscillator, and ΔΣ-fractional-N PLL. Since internal RF matching circuitry for input terminal of LNA and output terminal of ASK modulator were also integrated, almost no external matching devices are necessary with VSWR<1.25. SBM for ASK modulator and VGA for LO amplitude control were adopted to avoid local leakage and to keep modulation linearity, result, obtained modulation index were over 95%. The ∞-shape resonant coil was implemented for VCO to diminish magnetic coupling. Manufacturing process was SiGe BiCMOS process with 47 GHz cut-off frequency.**

I. INTRODUCTION

Recently dedicated short range communication (DSRC) systems have been gradually entering service in Japan and similar IEEE 802.11p radio systems are expected to be widely implemented worldwide in the near future. Among the various DSRC systems, electronic toll collection (ETC) systems, using 5.8GHz DSRC band and ASK modulation, have been widely introduced, since electronic account settlement using ETC systems is available for many toll roads, many cars are being equipped with ETC terminals. Given that downsizing and cost-effectiveness are important requirements for ETC terminals, a fully integrated RF transceiver LSI is required in order to minimizing external components and ensures ease of use. It is in this context that RF LSI is presented in this paper.

The block diagram of the developed LSI is shown in Fig.1. The reception part of the LSI consists of down-conversion mixer, and ASK detector. Input 5.8 GHz signal is amplified by LNA and down-converted to 40 MHz IF signal by the following down-conversion mixer, and then the down-converted signal is demodulated to 1024 kbps bit stream signal by ASK demodulator. The transmitter part consists of roll-off filter, ASK modulator and driver amplifier. The input bit stream data from base-band LSI is directly modulated to 5.8 GHz RF signal by ASK modulator. The local part consists of fully integrated VCO, frequency doubler and ΔΣ-Fractional-N PLL to

generate 5.8 GHz local signal. To avoid injection pulling, VCO oscillates at half of carrier frequency, 2.9 GHz, and 5.8 GHz local signal is generated by frequency doubler.

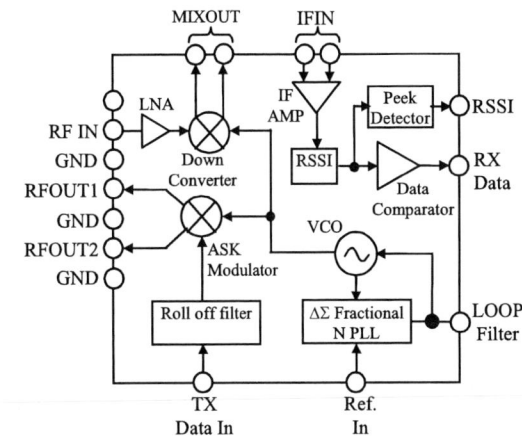

Fig.1 Block diagram of 5.8 GHz band RF transceiver LSI for ETC terminal

II. BUILDING BLOCKS

A. Integrated matching circuitry

In many case, integrated RF LSIs are mounted on general purpose packages. However, in the case of LSIs operating in the frequency range up to 5.8 GHz band, it is difficult to adjust on-board matching conditions of RF terminals because the effects of parasitic capacitances and inductances become significant. LNA and its matching circuits are shown in Fig. 2(left hand). In the case of off-chip matching, the base terminal and the emitter terminal of the LNA are connected directly to the terminal of the LSI chip. Hence, these terminals are very sensitive to the effect of parasitic elements. Fig. 2(right hand) shows simulation results for the effect of the parasitic capacitance (C_p) connected to base terminal of Q1 and parasitic inductance (L_p) of emitter terminal of Q1. Input impedance matching is strongly influenced by values of parasitic components.

1-4244-0458-4/06/$20.00 ©2006 IEEE   263

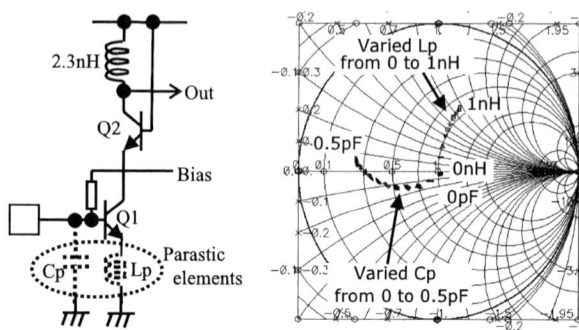

Fig. 2 Equivalent circuit of LNA input in the conventional way (left hand) and simulated input matching influenced by C_p and L_p (right hand)

To avoid these difficulties we integrated the on-chip matching circuits in order to dispense with matching devices on the RF module board. On-chip matching equivalent circuit of RX input terminal and its layout are shown in Fig.3.

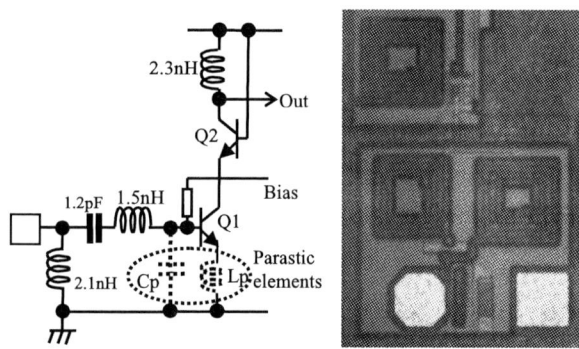

Fig. 3 On-chip equivalent circuit and its layout

Although the internal matching seems complete using this matching methodology, it is imperfect because bonding wires and lead frame of LSI package are between the LSI chip's RF terminal (matching to 50 Ω) and on-board transmission line (also matching to 50 Ω). Ideally the impedances of the bonding wires part and lead frame part would be 50 Ω, but this is impossible using general-purpose packages or the conventional mounting method. GSG (ground-signal-ground) structure of the RF signal transmutation line shown in Fig. 4 was adopted so that the line impedance of connecting parts would approach 50 Ω. Existing unclear impedance parts were bond wire and frame of package. Estimated impedance of bond wire was approximately 200 Ω, and that of lead frame was approximately 50 to 70 Ω. But their electronic length was about 3 mm (about λ/16 at 5.8 GHz). So, the effect of those parts is slight in terms of overall matching conditions. Therefore, stable matching characteristics will be obtained and external matching elements can be eliminated.

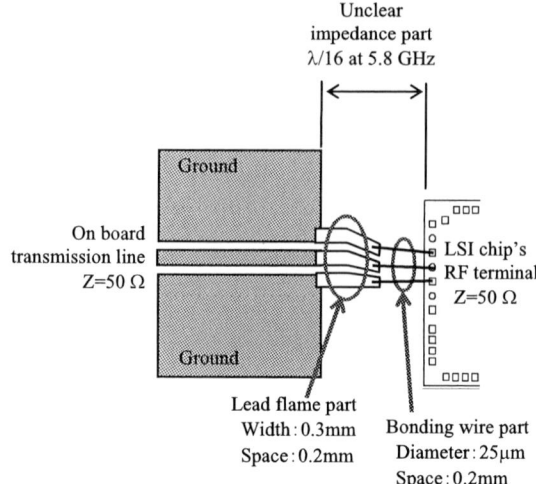

Fig. 4 RF signal Input/Output Interface between board and LSI chip

B. ASK modulator

For the transmitter part, we selected a direct modulation system, that modulates using 5.8 GHz local signals directly. In the case of direct modulator, the frequencies of local and transmit signal are identical and sometimes local leakage is a problem at higher operating frequency such as 5.8 GHz.

There are two candidates for modulator topology, double balanced mixer (DBM) and single balanced mixer (SBM). Equivalent circuit of DBM and simulated results of ASK modulation are shown in Fig. 5. From the result in Fig. 5, it is clear that when RF frequency is lower, for instance 100 MHz, modulated RF output signal is translated exactly. On the other hand, when RF frequency is 5.8 GHz, local leakage to modulated signal output is remarkable. As a result of this local leak, modulation index is significantly lower.

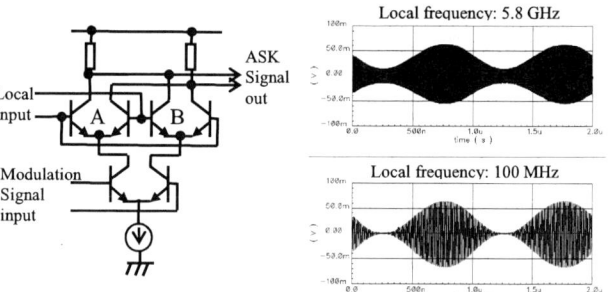

Fig.5 Equivalent circuit of double balanced mixer for ASK modulator (left hand) and the simulated output wave form of double balanced mixer (right hand)

Since output level is suspended using DBM, RF signal should be canceled between the left differential pair (A) and right differential pair (B). Therefore, differential pair balance between A and B is very important for minimize local leakage; however, it is rather difficult to maintain a balance for 5.8 GHz signal.

We chose a modulator structure based on a single balanced mixer (SBM). As output level is suspended, SBM differential pair is also suspended but SBM ensures no balance problem arises. Hence, it may be possible to reduce local leakage by using SBM.

In the case of using SBM topology for ASK modulator, RF output level is proportional to collector current, Ic. In other words, it is proportional to equivalent input capacitances of base terminal of differential pair of the local signal input, and C_π is expressed in equation (1).

$$C_\pi = \tau_F \frac{qIc}{kT} + Cje \approx \tau_F \frac{qIc}{kT} \qquad (1)$$

In the modulation scheme, large amplitude of modulation signal induced large collector current, resulting in large output amplitude according to simple estimation. Therefore, as the amplitude modulation signal is large despite collector current being set small, local signal level becomes lower and the efficiency of the modulator decreases. As a result, modulation linearity becomes worse. On the other hand, as the modulation signal is small despite collector current being set large, input capacitance C_π decreases and the impedance of local input terminal becomes large. So local signal level becomes too large and the local leak level increases. This consideration indicates the balance between incoming signal and local signal is very important for realizing fine ASK modulator. To keep modulation linearity and decrease local leak level, we adopted variable gain amplifier for the previous stage of the modulator's local input terminal as shown in Fig. 6, and its gain is controlled by the modulation data.

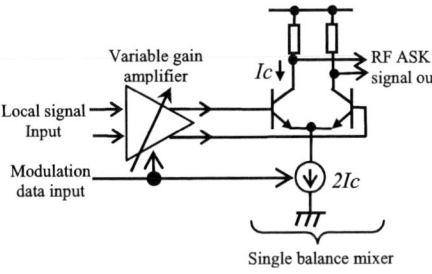

Fig. 6 ASK Modulator's block diagram

C. ∞-shape VCO and ΔΣ-PLL

For generating RF local signals for ASK modulator and for down-converter, integrated 5.8 GHz band Voltage Controlled Oscillator (VCO) is also integrated. VCO oscillates at half of carrier frequency, 2.9 GHz, and 5.8 GHz local signal is generated by frequency doubler.

As VCO block is very sensitive to the outer noises, power supply and ground terminals of VCO were completely separated from those terminals of other circuit blocks and oscillator's core parts were completely enclosed by guard ring. Of course they were effective for reducing electrical and capacitive coupling to other circuits but ineffective in terms of reducing the influence of magnetic coupling on the resonant coil. In particular, doing transmission of ASK signal, output stages of transmit

blocks were switched on and off by modulation signal. So, matching coils placed at the transmit output section scatter the magnetic fields. Also, if resonant coil of VCO were affected by this magnetic field, self-modulation would occur, causing spur characteristics of transmit signal to deteriorate. In view of this background, we selected the ∞-shape resonant coil to diminish magnetic coupling between VCO resonant coil and others. VCO structure and its layout are shown in Fig. 7. The magnetic field generated by the outer noise sources passing through all loops of the ∞-shape coil in the same direction. Voltages at loops caused by these magnetic fields are generated in opposite directions in the ∞-shape coil, thereby canceling each other. Therefore, use of the ∞-shape coil reduces the influence of magnetic noise.

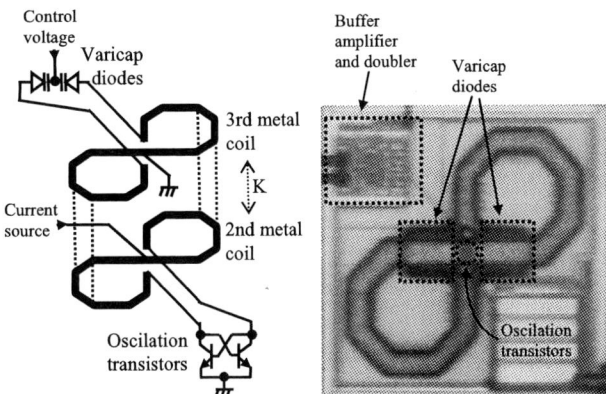

Fig. 7 VCO Structure and its layout

In the Japanese DSRC band standard used by ETC systems frequencies of RF channels are set at integer times multiple as large as 5 MHz. On the other hand, since symbol rate of modulation signal is set to 1.024 Mbps, the clock frequency of almost all base band processors is set to the integer times multiple as large as 1.024 MHz, for instance 16.384 MHz. In order to operate the ETC terminal with only one reference frequency, this reference clock must be used as a reference frequency of PLL. However, to generate RF carrier whose frequency is integer times multiple as large as 5 MHz, reference frequency of PLL is reduced to their greatest common divisor, such as 8 kHz. Also, RF channel switching time of less than 100 μ sec is required. We selected ΔΣ fractional N PLL to achieve both requirements.

III. MEASUREMENT DATA

SiGe-BiCMOS process was used with SiGe npn transistor, two resistors, MIM capacitor, and 3-μm-thick top metal layer for high-Q on-chip spiral inductor. f_{Tmax} of SiGe npn transistor was 45 GHz.

The photograph of RF transceiver LSI on the evaluation board is shown in Fig. 8(left hand). Signal line of board was coplanar line as shown in Fig. 4. There are no matching devices on the board.

Measured reflection coefficient of the LNA input terminal on this evaluation board is shown in Fig. 8(right hand). The voltage standing wave ratio (VSWR) is below 1.25 at 5.8 GHz band without any external matching elements. This result indicates on-chip matching circuitry works correctly.

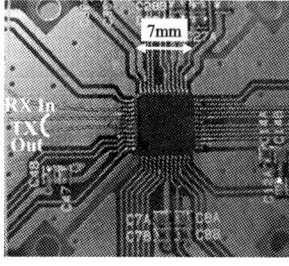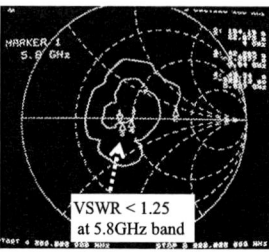

Fig. 8 LSI on the evaluation board (left hand) and reflection coefficient of LNA input terminal (right hand).

Fig. 9(left hand) shows the measured eye diagram of ASK modulated RF output's signal level. Observed local leak level is very small and modulation index is over 95%. This result is a very high index value for an on-chip modulator.

The CW spectrum of VCO controlled by $\Delta\Sigma$ fractional-N PLL is shown in Fig. 9(right hand). No conspicuous fractional noise is observed and no spurious is observed, either.

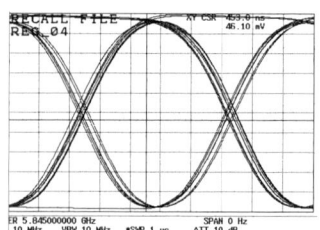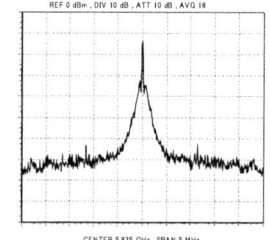

Fig. 9 Measured eye-pattern of modulated TX output (left hand) and measured spectrum of VCO and $\Delta\Sigma$-PLL (right hand)

Typical electronic charactaristics of the developed LSI are shown in Table 1. RX sensitivity is -80 dBm with BER of less than 0.001 % and current consumption of TX and RX are 40 mA under 3.0 V power supply.

The microphotograph of the RF transceiver LSI chip is shown in Fig. 10.

Table 1 Electronic characteristic

RF frequency	5.775 ~ 5.845 GHz
IF frequency	40 MHz
TX output level	-5 dBm Max
Modulation	2-level ASK 1.024 Mbps
Modulation index	> 95 %
99 % band width	< 4.4 MHz
ACPR	40 dBc (5 MHz spacing)
RX sensitivity	-80 dBm (BER <10^{-5})
Operating voltage	2.7 ~ 3.3 V
Current consumption	40mA(TX/RX)
Operating temperature	-40 ~ 85 centigrade
Package type	QFP 48pin

Fig. 10 Micro-photograph of LSI chip

IV. SUMMARY

Fully integrated 5.8 GHz band transceiver LSI for electronic toll collection system has been developed. Internal RF matching circuitry for input terminal of LNA and output terminal of ASK modulator were integrated, and almost all external matching devices are not necessary.

REFERENCES

[1] T. Matsuda, K. Ohhata, N. Shiramizu, S. Hanazawa, M. Kudoh, Y. Tanha, Y. Takeuchi, H. Shimamoto, T. Nagashima, K. Washio, "Single-chip 5.8GHz ETC transceiver IC with PLL and demodulation circuits using SiGe HBT/CMOS," Digest of Technical Papers of ISSCC, pp. 96-97, February, 2002.

[2] Japanese Ministray of Land, Infrastructure, and Transport, Road Burreau. Introduction about Japanese elecrtonic toll collection system. Online at: http://www.mlit.go.jp/road/ITS/

2006 Bipolar/BiCMOS Circuits and Technology Meeting Proceedings

5-GHz WLAN Standards Compliant Image Reject Radio Receiver on Low-cost SiGe-CMOS Technology

Domenico Zito and Bruno Neri

Radio-Frequency & Microwave Integrated Circuits Laboratory (RFLab), Dipartimento di Ingegneria dell'Informazione (DIIEIT), University of Pisa, I-56122 Pisa, Italy

Abstract — **A fully integrated heterodyne receiver with 1.1 GHz of intermediate frequency is presented. The circuits has been designed to be compliant with the multi-standard (IEEE 802.11a and HiPerLAN/2) indoor 5-6 GHz WLAN applications. It consists of a selective low noise amplifier with a passive notch filter centred at the image frequency and a image reject mixer realized according to Hartley's scheme. The circuit has been implemented in a standard SiGe-CMOS process and exhibits 54 dB of total image rejection, 4.7 dB of noise figure, 68 dB of LO-to-RF leakage attenuation and a input-referred 1–dB compression point equal to -20dBm.**

Index Terms — **Bi-CMOS technology, image reject mixer, low noise amplifier,.RF receiver.**

I. INTRODUCTION

The even more growing interest for Wireless Local Area Network (WLAN) mass-market interfaces has contributed to renew the research aimed to low cost and fully integrated radio transceivers design.

The traditional obstacle toward the realization of fully integrated solutions is represented by the loss effects on silicon at RF frequency. Thus, high quality factor LC integrated filters for undesired tones suppression can not be realized together with the rest of the receiver. This is the case of the heterodyne architecture, in which the image signal suppression is typically obtained by means of highly selective external and expensive component, as SAW filter. Even if other solutions are available (such as Homodyne and Low-IF), the heterodyne architecture offers best performance especially in terms of sensitivity and isolation between LO and RF ports.

Therefore, the common approach to accomplish the image rejection (IR) requirements with fully integrated solutions consists of using image rejection mixers (IRMs). Unfortunately, in practical cases the effectiveness of this solution is limited by the process tolerances, which around 5 GHz roughly reduce the maximum obtainable rejection to 25-35 dB. This value is not adequate for some applications, so that, special processes with extremely reduced tolerances and high costs should be used [1].

In this paper a low cost, fully integrated down-converter from 5-6 GHz to 1.1 GHz of intermediate frequency is presented. The image reject mixer (IRM) is preceded by a selective Low Noise Amplifier (LNA) with a passive notch filter centred at the image

signal frequency band. Thus, thanks to the frequencies planning, a large image signal suppression can obtained by summing up the effects of the selective LNA and image reject mixer. This approach has been successfully adopted to realize the first down-conversion in a single-chip RF transceiver front-end for 5-6 GHz Wireless Local Area Network (WLAN) multi-standard (IEEE 802.11a and HiPerLAN/2) applications [1].

This paper is organized as follows. In Section II, the circuit (selective LNA and IRM) description and design are presented. In Section III, the performance are reported and compared with those of the most representative solutions previously presented in literature. Finally, in the Section IV, the conclusions are drawn.

II. CIRCUIT DESCRIPTION AND DESIGN

The simplified scheme of the proposed image reject receiver is reported in Fig.1.

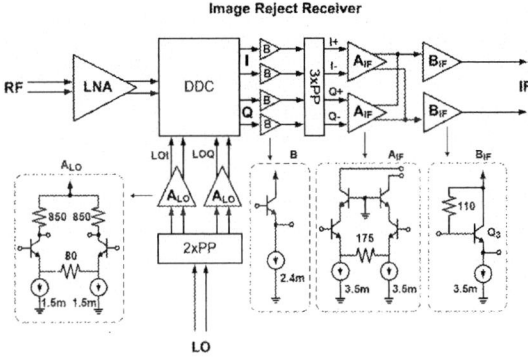

Fig.1. Simplified scheme of the proposed single-chip heterodyne receiver (LNA with notch filter and Image Reject Mixer). DDC states for double-balanced down-converter. For sake of clarity, the details about bias circuitry have been omitted.

The total image rejection (IR) is obtained summing up the attenuation provided by a selective LNA (which includes a passive notch filter centered at the image frequency) together with the image rejection provided by the IRM realized according to Hartley's scheme.

1-4244-0458-4/06/$20.00 ©2006 IEEE   267

The basic scheme of the LNA is shown in Fig.2.

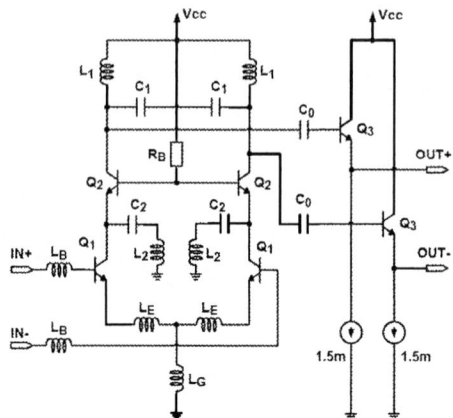

Fig. 2. Simplified schematic of the LNA with passive notch filter.

The LNA has been implemented by a cascode differential topology, in which the two sides are coupled each other through the bases of Q_2. With respect to the common emitter differential pair, this topology, named 'Base Coupled Differential' [2], provides a better common mode rejection ratio and input linearity range. The inductance L_G (0.9 nH) takes into account the parasitic effect due to the bonding wire toward the ground plane underlying the silicon die (down-bonding). The inductors L_B (0.9 nH) have been realized by means of bonding wires which, unlike integrated spiral inductors, do not require any area on die and provide a lower contribution to the overall noise figure (NF). The inductive emitter degeneration (L_E of 0.26 nH, realized by integrated spiral inductors) together with the transistors area sizing and L_0 allow the realization of the input integrated matching [3] to a 50 Ohm antenna. The selective frequency response is obtained by means of two couples of passive LC integrated filters (L_1C_1). The capacitors C_1 have been opportunely sized in order that the total equivalent capacitance at the collector node of Q_2 resonates with the inductance provided by L_1 (1.8 nH, with an associated quality factor of 11) at 5.35 GHz (the upper band limit for the indoor applications). The L_2C_2 series resonant filters realize the notch at the image frequency f_{IM} (3.15 GHz). The subsequent buffer stage drives the RF inputs of the IRM for a better performance in terms of noise figure of the overall receiver. The power consumption (P_C) is 27 mW (including 9 mW for the buffer) with 3V of power supply.

The IMR (shown in Fig.1) has been implemented by using a double-balanced architecture which shows a better immunity to large LO feedthroughs toward RF and IF ports. Particularly, the LO stage consists of a two-stages poly-phase filter (2xPP) and two amplifiers (A_{LO}). Each A_{LO} has been implemented by

a common emitter differential pair with resistive degeneration, as shown in Fig.1. The A_{LO} drains a total current of 3 mA and provides the proper I/Q signals at $f_{LO} = 4.25$ GHz. The 2xPP reduces the mismatch sensitivity and increases the quadrature accuracy within the 4.05-4.25 GHz frequency range.

The double-balanced down converter (DDC) is shown in Fig.3.

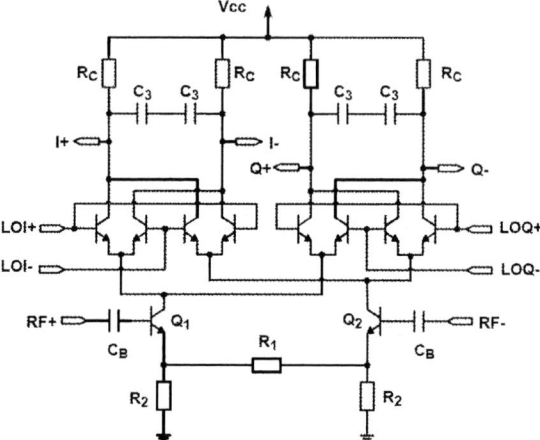

Fig. 3. Schematic of the Double-balanced Down Converter (DDC).

It has been realized by means of a typical double switching quad Gilbert's cell, which reduces the gain mismatch between I and Q signal paths. The resistors R_1 (200 Ω) and R_2 (100 Ω) allow the achievement of the linearity requirements without significantly impairing the NF performance. The C_3 (240 fF) capacitors attenuate the component at sum frequency ($f_{RF} + f_{LO}$). The DDC drains a bias current of 9.6 mA.

The IF stage of the IRM consists of the buffers B, a three-stages PP filter (3xPP), the amplifiers A_{IF} and the final buffers B_{IF}. Each A_{IF} has been implemented by a cascode stage which provides a high isolation between the outputs of 3xPP and the respective current sum nodes. The buffer stages B_{IF} allow the proper driving of different load impedances relative to the different test configurations (directly to the subsequent mixer for the second down-conversion or toward the input of the vector network and spectrum analyzers).

The design has been implemented by using a low-cost SiGe-CMOS 0.35 μm process (by AMS) with four levels of metal and a maximum cut-off frequency (f_{Tmax}) of 75 GHz.

III. CIRCUIT PERFORMANCE

Post-layout circuit and 2D½ EM simulations have been performed by means of SpectreRF™, Advanced Design System™ and Momentum™, respectively. All the integrated inductors have been designed as square spirals by using the top metal layer with a thickness of 2.5 um. The underlying patterned ground shield for a better substrate effects decoupling has been realized using the polysilicon layer.

The LNA frequency response in terms of S21 parameter is reported in the following Fig. 4.

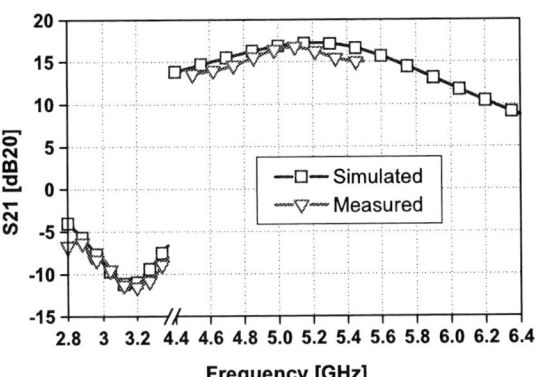

Fig. 4. $|S21|_{dB}$ of the LNA vs. frequency

To be noted that $|S21|_{dB}$ amounts to 17.8 dB at 5.35 GHz (signal frequency band), whereas it is equal to -11.5 dB at 3.15 GHz (image frequency band). Therefore, the LNA provides an image attenuation approximately equal to 28 dB. A NF of 2.6 dB and a Return Loss (20 log $|S11|$) of -18 dB have been obtained. As far as the linearity performance is concerned, the input-referred 1-dB Compression Point (CP_{1dB}) is -15.8 dBm; the input-referred 3rd-order Intercept Point (iIP3) is equal to -8.2 dBm (obtained considering two input tones at frequencies $f_1 = 5.25$ GHz and $f_2 = 5.3$ GHz). The reported performance have been obtained with a 100 Ohm load impedance.

As for the IRM performance, the conversion gain (G_C) is 6.84 dB and the NF is 23.4 dB at 1.1 GHz; the iCP_{1dB} amounts to -3.63 dBm. The Monte Carlo Analysis (for a maximum spreading of the resistance and MIM-capacitance values equal to 10%) has shown that the 3xPP allows us to achieve a image rejection greater than 30 dB in 200 MHz around 1.1 GHz, in 90% of cases. All the IRM performance are relative to the load and source impedances of 100 Ohm (differential).

Post-layout simulations have been widely confirmed by measurements on four prototypes wire bonded on microstrip test boards (Roger 4003, 1/2 oz, 10 mils) with ε_r equal to 3.38. A test-chip is shown in Fig. 5.

To be noted that LNA and IMR have been integrated as stand alone and after cascaded by interstage wire bonding, in order that they could be separately tested as well. As far as the test-chip layout is concerned (see Fig. 5), the G-S-G-S-G pad structure has allowed to carry-out both for on-wafer (for a preliminary characterization) and on-board tests.

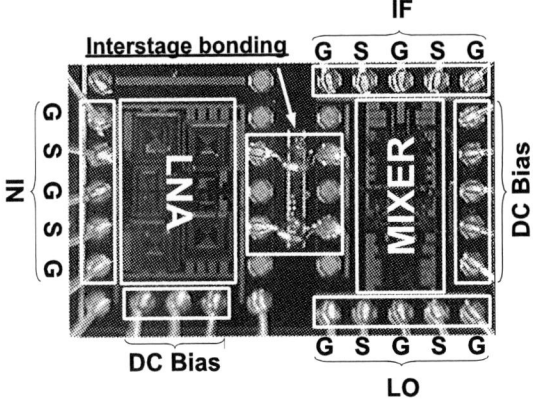

Fig. 5. Chip picture. The outputs of the LNA have been wire bonded to the input of the IMR (overall receiver front-end).

The signal pads have been realized by using the upper metal levels (3 and 4) and without any ESD protections in order to reduce the parasitic capacitance toward the substrate.

The wire bonding effects have been evaluated by using HFSS™ 3D EM simulator. The measurements on the LNA, IMR and IF receiver (LNA + IMR) have provided results in excellent agreement with those obtained by simulations, as shown in Fig.4 and in the following Fig 6.

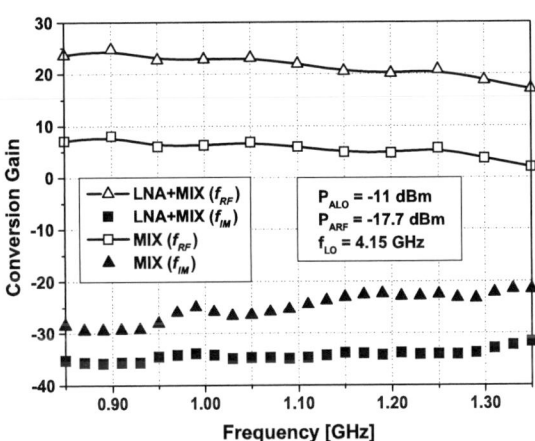

Fig. 6. Measured Conversion Gain (G_C) vs. frequency (f_{IF}) of the overall receiver (LNA + IMR). The term in round brackets is referred to the frequency of the input signal.

Moreover, a LO-to-RF leakage attenuation greater than 68 dB has been measured in all the operating conditions.

In the Table I, the performance of the overall receiver are summarized and compared with obtained with the most relevant solutions (in terms of IR) presented in literature [1, 4, 5].

269

TABLE I
SUMMARY AND COMPARISON OF THE OVERALL PERFORMANCE

Work	Tech	NF [dB]	IR [dB]	G_G [dB]	iCP$_{1dB}$ [dBm]	P$_C$ [mW]	Area [mm^2]
[1]	CMOS	7.35	50	26 (voltage)	-18	57.9	-
[4]	SiGe	5.9	40	19 (power)	-22	157.5	-
[5]	GaAs	3.8	35	26 (power)	-18	228	3.12
Our	SiGe	4.7	54	20 (power)	-20	152.6	1.37*

(*) LNA stand-alone + IMR stand-alone (pad included)

To be noted that our work reaches the best image rejection performance.

IV. CONCLUSIONS

A fully integrated heterodyne radio receiver on standard silicon technology has been presented. The proposed solution sums up the effects of a selective low noise amplifier with a passive notch filter and a image reject mixer in order to overcome the traditional limits concerning the achievement of high-image rejection on standard process.

The experimental results have shown that the proposed radio receiver is compliant with the most stringent requirements (iCP$_{1dB}$ ≥ -21 dBm, NF ≤ 10 dB, IR ≥ 35 dB and LO-to-RF leakage ≤ -57 dBm) of both standards (HiPerLAN/2 and IEEE 802.11a) for 5-6 GHz WLAN systems [1].

Moreover, the obtained results have been favorably compared with those of the previously presented solutions, particularly, as far as 54 dB of image rejection (IR) is concerned. This result has been confirmed by the tests on four manufactured prototypes and represents the best image rejection ever reached before at 5-GHz, without any additional circuitry for the tracking notch filter [1, 4, 5] and using a low cost standard SiGe-CMOS process.

ACKNOWLEDGEMENT

The authors wish to acknowledge the financial support of the Italian Ministry of Education, University and Research (namely, Ministero Italiano dell'Istruzione, dell'Università e della Ricerca – MIUR), and "Fondazione Cassa di Risparmio di Pisa".

REFERENCES

[1] H.Samavati, T. H.Lee , H.R.Rategh "5-GHz CMOS Wireless LAN" IEEE Microwave Theory and Tech. (MTT), vol. 50, pp. 268-280, Jan 2002;

[2] S. Di Pascoli, L. Fanucci, B. Neri and D. Zito, "Base coupled differential amplifier: a new topology for RF integrated LNA", Int. Journal of Circuit Theory and Applications, Wiley, vol.31, is.4, pp.351-360, Aug 2003;

[3] S.P. Voinigescu, et al., "A scalable high-frequency noise model for bipolar transistors with applications to optimal transistors sizing for Low Noise Amplifier design", IEEE JSCC, vol.32 is.9, pp. 1430-1439, Sep 1997;

[4] M. Copeland, S.P. Voinigescu, et al., "5-GHz SiGe HBT monolithic radio transceiver with tunable filtering", IEEE Trans. on MTT, vol.48 is.2, pp. 170, Feb 2000;

[5] T. Brauner, R. Vogt, et al. "5-6 GHz monolithically integrated calibratable Low-Noise downconverter for Smart Antenna Arrays" IEEE RFIC Symposium, pp. 435-438, Jun 2003.

2006 Bipolar/BoCMOS Circuits and Technology Meeting Proceedings

A SiGe BiCMOS 9.75/10.6GHz Frequency Synthesizer for DBS Satellite LNB Down-Converters Using Half-Rate Oscillators

A. Maxim, M. Gheorghe, C. Turinici

Integrated Products, 8713 Cobblestone, Austin TX 78735, Email: adrianmaxim@ieee.org

Abstract — **A fully-integrated frequency synthesizer for DBS satellite front-ends was realized in a low cost 50GHz f_T SiGe BiCMOS process. Two half-rate VCOs followed by Gilbert frequency doublers generate the 9.75/10.6 GHz LO signals with lower phase noise than a full-rate oscillator. The loop filter was integrated on-chip by using a passive feed-forward architecture that provides a noiseless resistor multiplication. A high PSRR regulator with a V_{GS}/R low noise reference was used to minimize the supply pushing impact on VCO phase noise and spur performance. The synthesizer performance includes: -106dBc/Hz phase noise at 100KHz offset, $<0.4°_{rms}$ integrated phase noise, 1x1.9mm^2 die area and drawing 35mA from a 3.3V supply voltage.**

Index terms — **LNB, PLL, satellite receiver, VCO.**

I. INTRODUCTION

Most existing DBS satellite low-noise-blocks (LNB) which down-convert the TV signal from the Ku-band (10.7-12.75GHz) to the L-band (0.95-2.15GHz) are using discrete implementations with open-loop dielectric resonator oscillators (DRO) that require a manual alignment [1]. The temperature drift of the oscillator frequency is usually compensated with a second down-conversion implemented in the digital demodulator. Recently, integrated closed-loop frequency synthesizers were developed for the LNB applications [2-6]. In [3] a dual-conversion architecture was implemented using a 5GHz oscillator. Its main drawback is the poor image rejection due to the oscillator's second harmonic component, limited by device matching. A single conversion receiver was proposed in [4] using a half-rate oscillator and a high frequency rectifier followed by a high speed comparator that generates the full-rate 10GHz LO signal. Its downside is the large phase

noise degradation brought by the rectifier-comparator frequency doubler implementation. Full-rate oscillators were also implemented for single conversion LNB receivers, but they require a higher f_T process, resulting in a higher cost [6]. This paper proposes an alternative way of realizing a half-rate oscillator based LNB frequency synthesizer, which achieves a much lower phase noise degradation, by implementing the frequency doubler with a poly-phase filter followed by a Gilbert multiplier and a tuned filter. A low cost 50GHz f_T SiGe process was used. The VCO phase noise was reduced by using a tail resistor bias, controlled by a discrete AAC loop. The loop filter was implemented on-chip by using a passive feed-forward resistor multiplication architecture.

II. LNB SYNTHESIZER CIRCUIT DESCRIPTION

Fig.1 presents the top level diagram of the single conversion LNB receiver. The signal path consists of a front-end low noise amplifier (LNA) built with an inductive degenerated cascode differential stage, a double-balanced Gilbert mixer using linearity boosting and an RF output buffer. The fully-integrated synthesizer uses a 50MHz crystal reference oscillator to reduce the PLL gain and thus minimize the magnification of the PLL front-end noise. A passive feed-forward loop filter (LF) was realized using two charge-pumps (an integral (CP$_i$) and a proportional (CP$_p$) one) and a standard RC filter. Two constant frequency LO paths were implemented for the low and high Ku-bands. Each of them consists of a half-rate oscillator (VCO$_L$/VCO$_H$ operated at 4.875/ 5.3GHz), followed by a clock buffer (BUF) and a Gilbert frequency doubler (fx2). The two paths are multiplexed (MUX) and buffered by Clk-BUF. The feedback divider uses a dual-modulus architecture and provides two fixed division ratios: 195 and 212.

Low noise PLLs require a low value loop filter resistance. Achieving a high bandwidth requires a very large loop filter capacitance (nF), which cannot be implemented on-chip [2-6]. Furthermore, bringing the sensitive VCO control line off-chip may result in noise and spur coupling that degrades PLL's phase noise performance. This paper proposes a passive feed-forward loop filter presented in Fig.2, which achieves a noiseless resistor multiplication, such that the loop filter capacitance was reduced to hundreds of pF and thus can be integrated on-chip. By pumping the proportional charge-pump current into the series resistance R$_z$ and summing the resulting voltage with

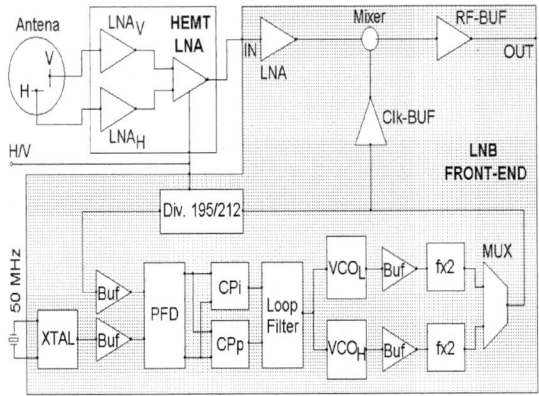

Fig.1 LNB block diagram with synthesizer architecture

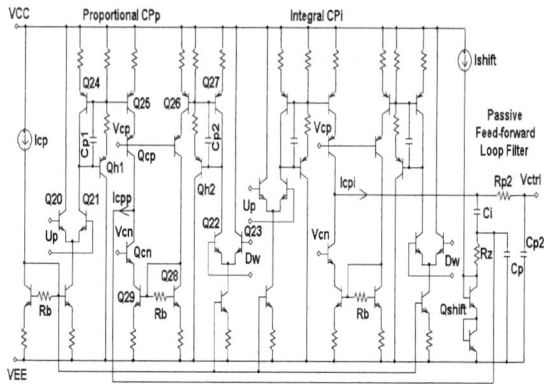

Fig. 2 Charge-pump and passive feed-forward loop filter

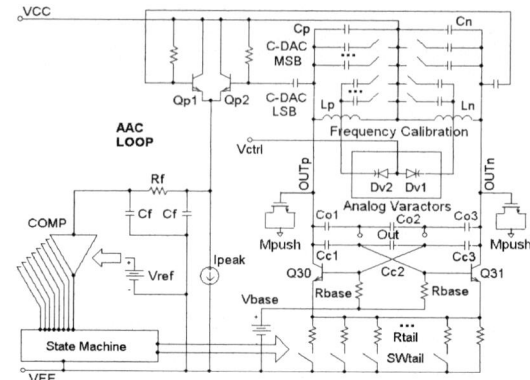

Fig.3 Half-rate VCO with frequency calibration network

the integral control stored on the C_i capacitance, the equivalent loop filter resistance becomes $R_{eq}=R_z\cdot(1+I_{cpp}/I_{cpi})$. Selecting a large charge-pump current ratio (e.g. 20 to 30) the loop filter capacitance can be reduced by the same factor. The Q_{shift} diode connected devices biased from the I_{shift} current avoid the saturation of the proportional charge-pump at the beginning of the pump period. The C_p capacitance introduces a first ripple pole, while R_{p2}, C_{p2} add a higher frequency pole that further reduces the reference spurs and high frequency phase noise.

The low speed available PNPs prevent the realization of a complementary NPN-PNP charge-pump. An all-NPN differential current steering charge-pump was built with Q_{20}, Q_{21} and Q_{22}, Q_{23} pump-up and pump-down switches (see Fig.2). To preserve symmetry both up and down currents are passed through low speed PNP current mirrors (Q_{24}, Q_{25} and Q_{26}, Q_{27}), that add a fourth pole to the loop. To minimize current mirroring errors, base current helpers (Q_{h1}, Q_{h2}) were used. The C_{p1} and C_{p2} compensation capacitances extend the PNP mirrors bandwidth (capacitive peaking). The pump-down current is passed through a second fast NPN current mirror (Q_{28}, Q_{29}) that uses the R_b resistor to reduce its base current error. Cascode N and P-side mirrors (Q_{cn}, Q_{cp}) further improve the up/down current matching over the entire output voltage range.

Fig.3 presents the differential negative-g_m LC oscillator. A capacitive cross-coupled C_{c1}-C_{c3} differential stage Q_{30}, Q_{31} was used as amplifier, achieving a 4V peak differential amplitude that significantly reduces the phase noise level. A further reduction in phase noise was achieved by using a 1/f noise-free tail resistor R_{tail} and a low noise bias voltage V_{base} to set the tail current value. The digital automatic amplitude control loop (AAC) was implemented with a high frequency peak detector Q_{p1}, Q_{p2}, a flash ADC using an array of comparators and a state machine that controls the oscillator's tail current by adjusting the tail resistance value with the SW_{tail} switches. The L_p-L_n inductor uses a three turn symmetric layout having a 2-to-1 tap point to connect the low quality factor varactor decreasing thus its phase noise impact. The oscillator gain was reduced

by using a frequency calibration circuit that compensates for the process variation. A ten bit calibration capacitor DAC ensures less than 0.5% residual frequency error. A 2% tuning range varactor compensates for the temperature drift and any residual process or aging frequency variation.

In oscillators using a discrete AAC loop that is open for long time intervals, the amplitude modulation (AM) to phase modulation (PM) conversion may dominate the VCO phase noise performance. The supply pushing due to the nonlinear parasitic capacitances shown to the LC tank by the energy-restoring amplifier need to be minimized. The nonlinear capacitance at the VCO output is dominated by the $Q_{30,31}$ C_{bc} base-collector capacitance that has a negative voltage coefficient. A partial supply pushing cancellation was achieved by adding MOSFET gate capacitors M_{push} at the two output nodes that have a positive voltage coefficient. A perfect supply pushing cancellation can be achieved only for a single design corner. However, this pushing neutralization circuit gives a pushing gain lower than 100KHz/V over all corners, making negligible its contribution to the VCO phase noise performance.

Fig.4 presents the detailed schematic of the Gilbert type frequency doubler that generates the 9.75/10.6GHz clock starting from the 4.875/5.3GHz outputs of the two VCOs. Two pairs of emitter followers Q_{46}-Q_{49} were used to drive the upper and respectively lower ports of a balanced Gilbert mixer (Q_{40}-Q_{45}). Quadrature phases of the 5GHz clock are obtained with a first order poly-phase filter (R_1, C_1 and R_2, C_2) that embeds the parasitic base resistance and base-emitter capacitance of the lower differential pair of the mixer. The two upper differential pairs drive a low Q L_{10g},C_{10g} load tuned at 9.75/10.6GHz. The R_{damp} resistors limit the quality factor around 5 to 6. A "blind" trimming using the MSBs of the 5GHz VCO center frequency calibration is implemented to reduce the process variation of the 10GHz center frequency. The frequency doubler devices are biased at peak f_T current density for fast switching and thus lower noise contribution.

The LNB supply is provided by a DC-DC converter that contaminates it with low frequency

272

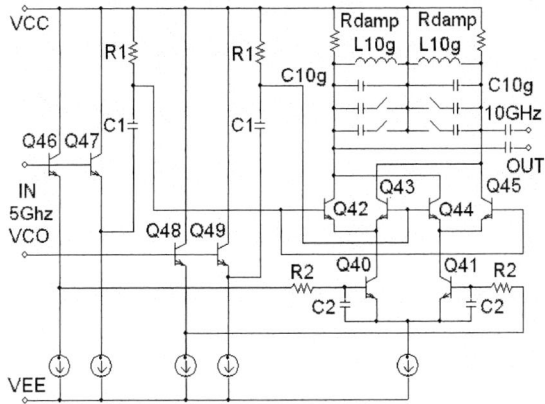

Fig.4 Frequency doubler with 1st order poly-phase filter

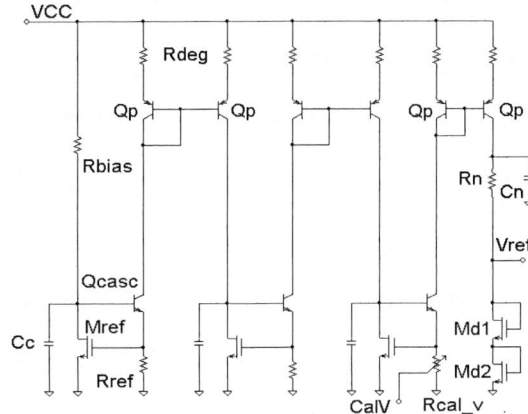

Fig.6 V_{GS}/R low noise reference voltage generator

spurious tones that may jeopardize the PLL spurious performance through the VCO supply pushing effect. Fig.5 presents the high PSRR regulator used to bias the oscillator. Optimizing the VCO phase noise requires a high value local supply voltage that results in only a 0.4V to 0.5V voltage drop on the regulator. A PFET M_{pout} series regulator was used due to its minimum required V_{DS} voltage and reasonable current capability in comparison with lateral PNP devices. The noise of the regulator (both 1/f and thermal) can degrade the VCO phase noise through the supply pushing mechanism. A low noise regulator ($<10nV/\sqrt{Hz}$) was achieved by using a V_{GS}/R reference presented in Fig.6. To reduce the noise contribution of the regulator operational amplifier, a single ended OpAmp was implemented using the M_{amp} input comparison device and the Q_{fold} folded cascode stage. A folding resistor R_{fold} was used instead of a current source in order to minimize amplifier's noise. The VCO supply voltage is set by $2V_{GS}(NFET)+2V_{SG}(PFET)$ given by M_{d1}, M_{d2}, M_{amp} and M_{amp*}. The opposite temperature coefficients of the NFET and PFET threshold voltages helped achieving a precise temperature compensation of the oscillator supply voltage. Large area FETs are required to minimize their 1/f noise component.

Achieving a low phase noise VCO bias regulator requires a very low noise reference voltage generator. Bandgap references offer a good process and temperature stability, but they are notorious for

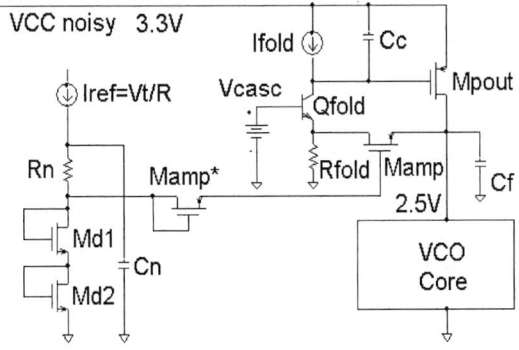

Fig.5 Low noise high PSRR VCO regulator

their large noise that may dominate the VCO phase noise performance. This paper uses a V_{GS}/R generator (see Fig.6) to create a low noise reference. V_{BE}/R references can also offer a low noise, but they come with a large temperature variation that is hard to eliminate through calibration. The main advantage of using an N-MOSFET reference is that its threshold voltage V_T has a negative temperature coefficient, while the $V_{on}=V_{GS}-V_T$ has a positive temperature coefficient. The current density in the N-FETs was selected such that a precise cancellation of the two opposite temperature coefficients was achieved. An overall temperature variation less than 25ppm/C was achieved for the VCO supply voltage over the entire temperature range.

The reference voltage is provided by the M_{ref} 0.35μm NFET available in the selected low cost BiCMOS SiGe process, while the voltage-to-current conversion is realized by the R_{ref} resistor. The Q_{casc} device increases the output impedance of the current mirror, improving the supply rejection. The turn-around PNP mirror use large resistive degeneration R_{deg} to minimize its noise contribution. The input current is set by the R_{bias} resistor, resulting in a low noise, but having a strong V_{CC} supply dependence. Three V_{GS}/R cells were connected in cascade to achieve a PSRR in excess of -70dB. The output voltage is generated by injecting the low noise V_{GS}/R current into diode connected NFETs (M_{d1}, M_{d2}). The V_T threshold voltage has a large process variation (±100mV). The process stability of the reference voltage was achieved by calibrating at power-up the V_{GS}/R based reference to a bandgap voltage by tuning the $R_{ref}=R_{cal-v}$ resistor of the last V_{GS}/R cell.

III. EXPERIMENTAL RESULTS

Fig.7.a presents the open-loop phase noise characteristic of the LO signal after the frequency doubling circuit. It achieves –106dBc/Hz at 100KHz offset. Fig.7.b gives the closed-loop PLL phase noise plot, resulting in a $0.4°_{rms}$ total double-sided integrated phase noise from 1KHz to 10MHz. The phase noise is

273

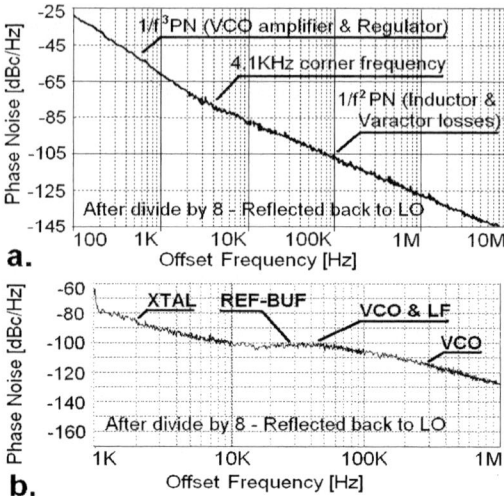

Fig. 7. Open-loop VCO (a) and closed-loop PLL (b) phase noise measurements

Table.1 LNB front-ends performance comparison

Ref.	[4]	[5]	[6]	This work
Process	BJT 46GHz	BJT 46GHz	SiGe 70GHz	SiGe 50GHz
Architect.	Dual Conv.	Single Conv.	Single Conv.	Single Conv
VCO freq.	4.87/5.3 GHz	4.87/5.3 GHz	9.75/10.6 GHz	4.87/5.3 GHz
VCO gain	360 MHz/V	360 MHz/V	640 MHz/V	150 MHz/V
PN @ 100K	-101 dBc/Hz	-101 dBc/Hz	-103 dBc/Hz	-106 dBc/Hz
Integ. PN	-	$0.8°_{rms}$	$1.2°_{rms}$	$0.4°_{rms}$
Sup.Volt.	3.3V	3.3V	3.3V	3.3V
Sup.Curr.	80mA	100mA	45mA	35mA
XTALfreq	25MHz	25MHz	50MHz	50MHz
Inductor Q	17 @5GHz	-	26 @10GHz	25 @5GHz
Loop filter	Off-chip	Off-chip	Off-chip	On-chip

low-end f_T technologies. The poly-phase filter and Gilbert mixer frequency doubling results in a nearly ideal frequency doubler characteristic. The noiseless resistor multiplication loop filter ensures a negligible noise contribution, while allowing the on-chip integration of the filter capacitors. Avoiding the off-chip exposure of the sensitive VCO control node reduces the sensitivity to board level coupled noise and helps improving the synthesizer spurious tones performance. A low noise V_{GS}/R reference and a single-ended high PSRR series regulator ensure low supply noise and spur injection.

Fig. 8. LNB front-end die-photo; detail of the half-rate frequency synthesizer

dominated by the on-chip planar inductor losses and the relatively modest diode varactor quality factor. Fig.8 presents a detail of the LNB down-converter die photo showing the LNB synthesizer. Table.1 provides a performance comparison between existing LNB PLLs and the proposed half-rate VCO single-conversion architecture. Present design offers a fully-integrated frequency synthesizer with on-chip loop filter, which achieves the lowest power consumption and the best-in-class phase noise performance.

The half-rate oscillators allow the use of a low cost 50GHz f_T SiGe BiCMOS process, while achieving a low phase noise performance. The Gilbert frequency multiplier with 1st order poly-phase filter introduces only 1dB phase noise degradation from the ideal 6dB change, providing a lower phase noise LO signal in comparison with a full-rate 10GHz oscillator built in the same low cost process.

IV. CONCLUSION

Using half-rate oscillators to generate the fixed frequency high and low LO signals brings both a lower cost and a better phase noise performance in

REFERENCES

[1] P. Wallace, et al, "A low cost high performance MMIC low noise down converter for direct broadcasct satellite reception", Proceedings of Microwave and Milimiter Wave Monolithic Circuits Symposium 1990, pp. 7-10.

[2] N. Shiga, et al, "MMIC family for DBS downconverters with pulse doped GaAs MESFET", Proceedings of GaAs Symposium 1991, pp.139-142

[3] T.Copani, et al, "A single chip receiver for multi-user low noise block down converter", ISSCC 2005 Digest of technical papers, pp.438-439.

[4] C. Copani, et al., "A 12 GHz silicon bipolar dual-conversion receiver for digital satellite applications", IEEE Journal of Solid State Circuits, vol.40, no.6, pp.1278-1287, June 2005.

[5] G. Girlando, et al., "A monolithic 12 GHz heterodyne receiver for DVB-S applications in silicon bipolar technology", IEEE Trans. On MTT, vol.53, no.3, pp.952-959, March 2005.

[6] C. Vaucher, et al, "Silicon-Germanium ICs for satellite microwave front-ends", Proceedings of BCTM 2005 Conference, pp.196-203.

2006 Bipolar/BoCMOS Circuits and Technology Meeting Proceedings

Ruggedness Improvement by Protection

André van Bezooijen[1], Anton de Graauw[1], Lennart Ruijs[1], Skule Pramm[1], Christophe Chanlo[1], Henk Jan ten Dolle[1], Freek van Straten[1], Reza Mahmoudi[2], and Arthur H.M. van Roermund[2]

[1] Philips Semiconductors, Gerstweg 2, 6534 AE Nijmegen, The Netherlands
[2] Eindhoven University of Technology, Den Dolech 2, 5600 MB Eindhoven, The Netherlands

Abstract — Cellular phone power amplifier transistors have to withstand extreme voltages, temperatures and currents. Requirements on IC and packaging technology are relaxed by using over-voltage and over-temperature protection. To avoid breakdown, protection circuits are used that detect the collector peak voltage and die temperature to limit the output power once a threshold level is crossed. For a supply voltage of 5 V and a nominal output power of 2 W, no breakdown is observed for a VSWR of 10 over all phases. For a VSWR of 4 and worst case mismatch phase the maximum die temperature is reduced from 143°C to 112°C when the output power is adaptively reduced from 32.1 dBm to 27.7 dBm.

Index Terms — Adaptive control, avalanche breakdown, over-voltage protection, power amplifiers, temperature.

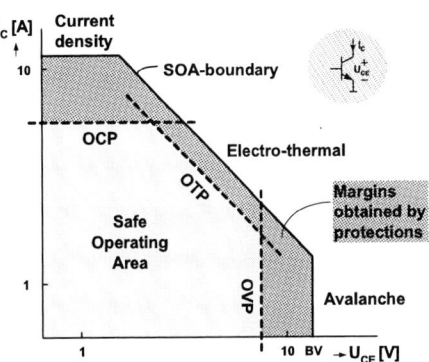

Fig. 1. Requirements on power transistor safe operating area, bounded by avalanche breakdown, electro-thermal instability, and current limited blow-out, can be relaxed by over-voltage (OVP), over-temperature (OTP), and over-current protection (OCP).

I. INTRODUCTION

Cellular phone power amplifiers (PAs) have to withstand extreme environmental operating conditions. Destructive breakdown of the power transistor has to be prevented, even when simultaneously a large antenna mismatch, high ambient temperature, and a fully charged battery is present, while the phone is at maximum output power.

For a bipolar power transistor three different causes of break-down can be distinguished: (1) avalanche break-down of the collector-base junction, (2) thermal run-away due to electro-thermal instability, and (3) blow-out of interconnect due to local dissipation.

To prevent avalanche breakdown, under such extreme conditions, power transistors are commonly implemented in an IC-technology with high breakdown voltage. This is achieved by using a relatively thick epi-layer of a silicon transistor (or mesa-layer thickness of a GaAs HBT device) and by optimizing the collector doping profile. However, such a ruggedness optimization inevitably results in a reduction of the f_T, and consequently, in a lower gain-bandwidth product of the power transistor and thus in a lower PA efficiency [1].

Over-heating of a bipolar power transistor causes destructive breakdown when thermal instability of the device occurs. And, over-heating reduces the product life-time, due to acceleration of failure mechanisms. Most often, ballasting resistors, heat spreading interconnect, exposed heat sinks, and copper filled laminate vias are used to guarantee electro-thermal stability.

In this paper we present protection circuits [2,3,4] that prevent over-voltage, and over-temperature conditions of the power transistor. An integrated voltage detector and temperature detector are used to monitor the collector peak voltage, and die temperature. Once a detected parameter value crosses a certain threshold level the effective power control voltage is reduced to limit the amplifier output power.

By applying protection circuits, the requirements on power amplifier IC- and packaging technology can be relaxed, which provides headroom for further cost and size reduction. Consequently, low cost mainstream silicon IC-technology (with relatively low breakdown voltages) can be applied to integrate the power amplifier together with biasing circuits, power control functions and control logic. Moreover, PA modules with smaller form factors (thus larger thermal package resistance) become feasible when extreme temperatures are avoided by over-temperature protection.

II. PROTECTION CIRCUITS

A. Collector Peak Voltage Detector

In Figure 2 the circuit diagram of a power transistor *T3* and its biasing circuit are shown, as well as the collector peak voltage detector and its succeeding threshold comparator. A capacitive voltage divider with shunt resistors is used to sense both the DC and RF part of the collector voltage. Consequently, the detector output voltage U_{DET} is a function of the battery supply voltage V_{SUPPLY} and of the RF voltage

1-4244-0458-4/06/$20.00 ©2006 IEEE   275

wave amplitude. The collector peak voltage is distributed over the capacitors keeping the maximum voltage across each capacitor well below their breakdown voltage of approx. 10 V. To avoid DC current discharging the battery, in PA OFF-mode, a LPNP transistor $T4$ switch is placed in series with the shunt resisters to interrupt the DC-path when the PA is OFF. A NPN collector-base junction diode $T5$, with high reverse breakdown voltage, is used as detector, charging the filter capacitor during the peaks of the RF voltage. Once the detected voltage U_{DET} crosses a threshold level U_{REF}, that is applied externally, the comparator output transistor $T10$ sinks a current I_{PROT} to actuate the PA power control loop.

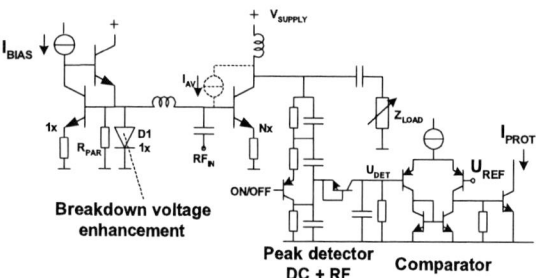

Fig. 2. Power transistor with biasing circuit, collector peak voltage detector, and threshold comparator.

The biasing circuit, with diode $D1$ to ground, provides, under avalanche conditions, a low output impedance of the biasing circuit, which enhances the effective breakdown voltage of the power transistor. The avalanche current is sunken by the biasing circuit rather than amplified, times β, by the power transistor itself.

B. Avalanche Breakdown Based Peak Voltage Detector

In modern Si and SiGe:C processes a Selective Implant in the Collector (SIC) is often used to optimize, in combination with the epi-thickness, the transistor breakdown voltage and f_T. Blocking this SIC results in a transistor with higher breakdown voltage (HV-NPN) than with SIC (LV-NPN), see Figure 3.

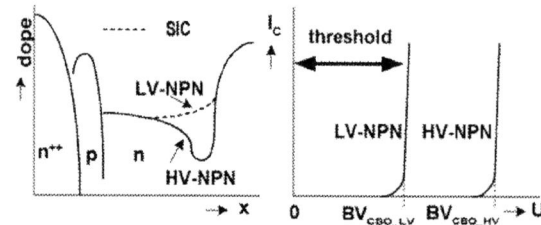

Fig. 3. Doping profiles of a low-voltage and high-voltage NPN and their corresponding breakdown voltages BV_{CBO_LV} and BV_{CBO_HV}.

In Figure 4 an alternative implementation of an over-voltage detector is shown that makes use of

avalanche breakdown of the collector-base junction to activate the protection loop. Once the collector peak voltage exceeds the BV_{CBO} of the LV-NPN, a small current (limited by a 2 kΩ series resistor to protect the detection diode against destructive breakdown) flows into the NPN and PNP currents mirrors, that are scaled to give additional loop gain. An RC-filter rejects the RF carrier. The LV-NPN breakdown voltage defines the threshold value at which protection occurs. It results in a very simple detector circuit topology.

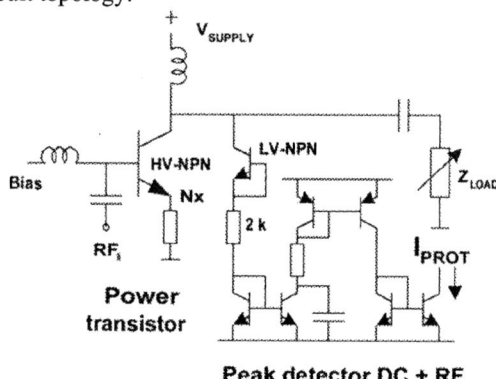

Fig. 4. Circuit diagram of the HV-NPN power transistor and a LV-NPN avalanche breakdown based over-voltage detector.

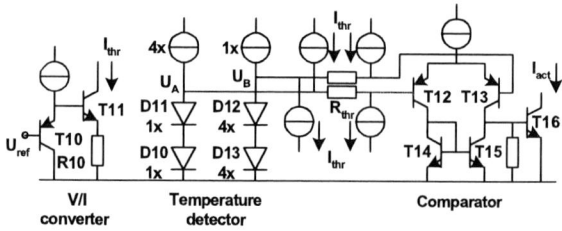

Fig. 5. Circuit diagram of the temperature detector and comparator with adjustable threshold.

C. Temperature Detector

A differential temperature detector circuit topology was chosen to minimize its susceptibility to RF interference, and to obtain an accurate reading of the absolute die temperature. The detector, shown in Figure 5, consists of the four diodes $D10 \ldots D13$, scaled in area by a factor four, and biased by currents, also scaled by a factor four, but in opposite order. It results in a differential detector output voltage U_A-U_B proportional to temperature. This detected voltage is independent of bias currents and process parameters and, consequently, calibration of the detector is not needed. This detected voltage is fed to a comparator circuit $T12 \ldots T15$ that activates its output buffer transistor $T16$ when the comparator input voltage crosses zero. The detector threshold level is set by the differential offset voltage generated across the two threshold resistors R_{THR} and threshold currents I_{THR}. The reference voltage U_{REF} is applied externally and controls a simple V/I-converter (*T10, T11, and R10*) to generate this current.

III. MEASUREMENT RESULTS

A. Safe Operating Area

Investigations have been performed on the safe operating area (SOA) of a 4 W power transistor and its breakdown enhancing biasing circuit. The transistor output characteristic I_C vs. U_{CE} is measured for 4 different bias levels labeled as VI-1, 2, 3, and 4 in Figure 6. The boundary of avalanche instability, labeled *Av*, is given by the points of infinite slope and is measured at approx. 8.5 V. During these experiments no snap-back nor destructive breakdown is observed.

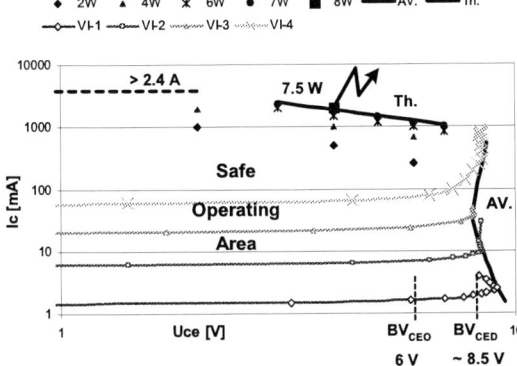

Fig. 6. Measured safe operating area of a 4 W power transistor determined by the boundaries of avalanche instability and thermal instability. $T_{MB} = 25°$ C.

In another set of experiments the boundary of thermal stability is determined at 7.5 W. In this case the collector emitter voltage is varied from 2 to 7 V and the bias current is gradually increased up to the point of thermal run-away, which often results in a burned out device. This thermal boundary, labeled *Th*, turns out to be very distinct once the mounting base temperature is kept constant by using water cooling.

B. Over-voltage Protection I

The ruggedness of the over-voltage protected 4 W device is verified with a continuous wave at 900 MHz. A three stage power amplifier has been used as driver for the power transistor. The over-voltage detector output current I_{PROT} is externally fed back via a resistor to adapt the power control voltage U^*_{DAC}. Initially, the control voltage U_{DAC} is set to obtain 33dBm output power for a nominal 50 Ω load and 3.5 V supply.

Then, the supply voltage is set at 5 V and a large VSWR of 10 is applied at the output and varied over all phases. The results in Figure 7 show that actuation occurs over a wide range of mismatch phases, when the reference is set at a relatively low value of 0.8 V, for which a maximum reduction in output power of

2.7 dB occurs. Under these conditions no breakdown is observed. It is worthwhile noting that even without correction (U_{REF} = 1.4 V) some of the devices passed these extreme tests, but many of them did not.

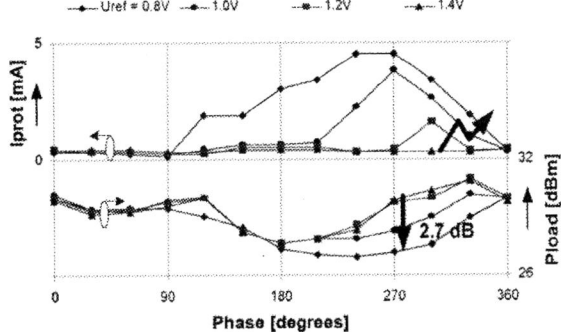

Fig. 7. Measured actuation current I_{PROT} and output power P_{LOAD} as a function of the phase of mismatch for reference voltages of 0.8, 1.0, 1.2, and 1.4 V. VSWR = 10, U_{SUPPLY} = 5 V.

C. Over-voltage Protection II

Similar ruggedness measurements have been performed on a power transistor protected by the avalanche breakdown based over-voltage detector. Figure 8 depicts the protection current and adapted power control voltage as function of the mismatch phase for open and closed loop conditions, while VSWR = 4 and U_{SUPPLY} = 3.5 V. Note that the difference in mismatch phases at which actuation occurs between Figure 7 and 8 is due to changes in the measurement set-up.

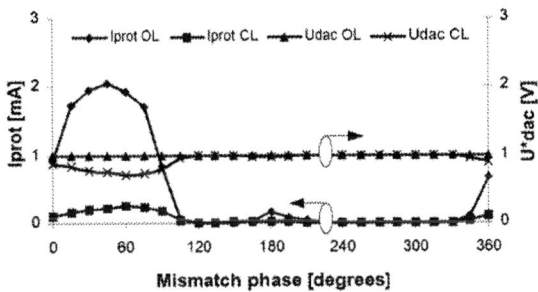

Fig. 8. Measured actuation current I_{PROT} as function of the mismatch phase of the avalanche breakdown based over-voltage detector circuit in open-loop (OP) and closed-loop (CL) condition. VSWR = 4 and U_{SUPPLY} = 3.5V.

D. Over-temperature Protection

In a similar manner the over-temperature protection circuit is tested. Figure 9 shows, for a VSWR of 4, the protection current I_{PROT} for three different over-temperature reference voltages. Relatively strong actuation occurs over a wide range of phases for U_{REF} = 1.1 V, whereas no actuation takes place for mismatch phases between 180 and 270 degrees. At U_{REF} = 1.4 V the loop does not activate at all.

For these conditions a maximum die temperature of 143°C is found at a mismatch phase of 30 degrees, which is reduced to 112 and 98°C when U_{REF} is set at 1.2 and 1.1 V respectively, as shown in Figure 9. The corresponding output power P_{LOAD} reduces from 32.1 dBm to 27.7 and 25.8 dBm. It is worthwhile noting that a very good correlation is found between dissipated power and adaptive output power correction.

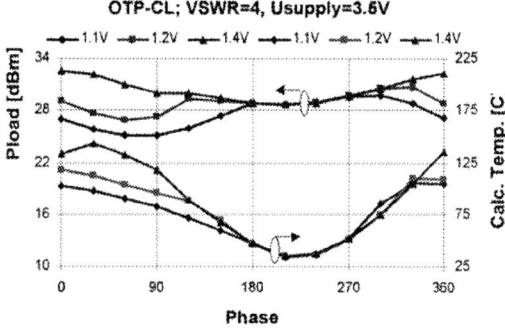

Fig. 9. Adaptively controlled output power P_{LOAD} and calculated die temperature as a function of the phase of mismatch while the over-temperature protection loop is active by setting U_{REF} at 1.1, 1.2, and 1.4V.

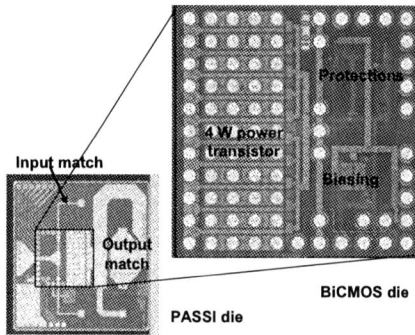

Fig. 10. Photographs of a power transistor die, with integrated biasing and protection circuits, flip-chipped on a passive-silicon die, to be placed on top of, and to be wire-bonded to, a laminate.

IV. TECHNOLOGIES

The active circuits are processed in a 0.25 μm BiCMOS technology [5]. This process offers a high voltage NPN (HV-NPN) with a collector-emitter breakdown voltage BV_{CEO} of 6 V, a collector-base breakdown voltage BV_{CBO} of 18 V, and has a top f_T of approx. 26 GHz. This active die is flip-chipped on a passive silicon (PASSI) die with a high-ohmic (5 kΩ.cm) low loss substrate. It provides high-Q Metal-Insulator-Metal (MIM) capacitors with a high breakdown voltage (200 V) for the implementation of rugged, low loss, matching networks, and a high-density capacitor (25 nF/mm^2) for supply decoupling. Plated through-wafer vias are used to connect top metal layers with the back-side metal for proper RF and DC-grounding of the power transistor, supply

decoupling capacitors and matching capacitors. Its 5 μm thick top metal layer is well suited for implementation of low loss inductors and power routing. The good thermal conductance of the relatively large, heat spreading, PASSI die ensures a low thermal resistance. See Figure 10.

V. CONCLUSIONS

Under extreme operating conditions, power transistors of cellular phone power amplifiers suffer from destructive breakdown due to avalanche and thermal breakdown.

We have successfully demonstrated that breakdown can be avoided by detection of collector peak voltage and die temperature to limit the output power adaptively once a threshold level is crossed.

The safe operating area of the tested 4 W silicon power transistor is bounded by avalanche instability at 8.5 V and by electro-thermal instability at 7.5 W.

For a supply voltage of 5 V and nominal output power of 2 W, no breakdown is observed for a VSWR of 10 over all phases when output power is adaptively reduced by 2.7 dB at maximum.

At 3.5 V and a VSWR of 4 a maximum die temperature of 143°C is found, which is adaptively reduced to 112°C and 98°C when the output power is adaptively reduced from 32.1 dBm to 27.7 and 25.8 dBm by setting the reference voltage at 1.2 and 1.1 V respectively.

REFERENCES

[1] D. L. Harama, "Current status and future trends of SiGe BiCMOS technology," *IEEE Transactions on Electron Devices*, vol. 48, no 11, Nov. 2001, pp. 2575-2594.

[2] A. van Bezooijen, F. van Straten, R. Mahmoudi, and A.H.M. van Roermund, "Avalanche breakdown protection by adaptive output power control," *IEEE RSW Proceedings*, pp. 519-522, Jan. 2006.

[3] K. Yamamoto et al., "A 3.2V Operation single-chip dual band AlGaAs/GaAs HBT MMIC power amplifier with active feedback circuit technique," *IEEE J. Solid-State Circuits*, vol. 35, no 8, August 2000, pp. 1109-1120.

[4] A. Scuderi, F. Carrara, A. Castorina, and G. Palmisano, "A high performance RF power amplifier with protection against load mismatches," *IEEE MTT-S Digest*, 2003, pp. 699-702.

[5] P. Deixler et al., "QUBiC4plus: A cost-effective BiCMOS manufacturing technology with elite passive enhancements optimized for silicon-based RF-system-in-package environment," *Proc. BCTM*, 2005, pp. 272-275.

2006 Bipolar/BiCMOS Circuits and Technology Meeting Proceedings

Compact Modeling of High Frequency Correlated Noise in HBTs

P.Sakalas[1,2], J.Herricht[1], A.Chakravorty[1], M.Schroter[1,3]

[1]CEDIC, Dresden University of Technology, Mommsenstrasse 13, 01062 Dresden, Germany, [2]Semiconductor Physics Institute, Vilnius, Lithuania, sakalas@iee.et.tu-dresden.de, [3]ECE Dept., University of California, San Diego, USA.

Abstract ---A compact model solution, consistent with the system theory for correlated base and collector shot noise sources, is derived and implemented in the bipolar transistor model HICUM using Verilog-A. Compiled (with Tiburon) Verilog-A model is simulated using ADS 2004A and the results are tested against measured noise parameters for high-frequency (f_T at 150 GHz) SiGe HBTs. Very good agreement between simulated and measured data is obtained.

Index Terms --- Noise, SiGe Heterojunction bipolar transistors, HICUM, correlation, shot noise, noise modeling, Verilog-A.

I. INTRODUCTION

For bipolar transistor, noise behaviour at high frequency substantially differs from that at the low-frequency region due to the effect of correlation between the base and collector (i.e., the input and output) current shot noise sources [1][2][3][4][5][6][7][8][9]. The correlation between the base and collector shot noise sources in HBTs is well described analytically in [2]. For modern HBTs (e.g., SiGe-HBTs), the noise correlation becomes important at frequencies beyond ~1/10 of peak f_T [8]. Unfortunately, implementation of correlated noise in a compact model is not straight-forward, and from a system theoretical approach a consistent implementation has neither been investigated yet, nor been done for bipolar transistors. Conventional noise computation, implemented in all simulators, can only handle uncorrelated noise sources. Even through the adjoined network concept [10], noise calculation does not handle correlation terms. From the point of view of SPICE implementation, a possible Verilog-A solution of correlated noise for MOS transistors has recently been given in [11]. However, a complete and systematic analysis of the bipolar transistor noise from the perspective of SPICE-like implementation is still missing. In this paper, we deal with a complete theoretical analysis of noise behaviour in bipolar transistors. Our solution handles the correlation terms in such a manner that the existing SPICE-like simulators can additionally compute the correlated noise in a similar way they do for the uncorrelated noise sources. The problem was addressed almost two decades ago through the noise calculation of complex active filter circuits [12] and the corresponding solution provides a useful background for our investigation.

II. THEORY AND VERILOG-A IMPLEMENTATION

Transfer of the noise signal in the linear network in frequency domain is described as follows[13]:

$$\mathbf{S}_{YY}(j\omega) = \mathbf{G}(j\omega) \cdot \mathbf{S}_{XX}(j\omega) \cdot \mathbf{G}^+(j\omega) \quad (1)$$

where $\mathbf{G}(j\omega)$ is transfer function matrix of any noise source to a

free of choice output in the circuit and $\mathbf{G}^+(j\omega)$ is the adjoint matrix of the transfer function. The $\mathbf{S}_{XX}(j\omega)$ is the power spectral density (PSD) matrix of noise sources and is defined as:

$$\mathbf{S}_{XX}(j\omega) = \begin{pmatrix} S_{i_{nb}} & \underline{S}_{i_{nb}i_{nc}} \\ \underline{S}_{i_{nc}i_{nb}} & S_{i_{nc}} \end{pmatrix} \quad (2)$$

Off-diagonal elements of the matrix correspond to the cross-PSDs, which are not taken into account by the conventioanl SPICE-like circuit simulators while simulating noise behaviour. Consequently, the correlation between noise sources is omitted in the noise computation. For the bipolar transistors, correlation between base and collector shot noise plays a significant role at high frequencies and, therefore, should be accounted for.

In order to force circuit simulators to compute correlation terms additionally, transformation of the input matrix into a diagonal matrix is performed. Here, the input matrix is expressed through a diagonal $\mathbf{D}_X(j\omega)$ and a transformation matrix $\mathbf{T}(j\omega)$:

$$\mathbf{D}_X(j\omega) = \begin{bmatrix} \mathbf{D}_1 & 0 \\ 0 & \mathbf{D}_2 \end{bmatrix}, \quad \mathbf{T}(j\omega) = \begin{bmatrix} 1 & 0 \\ t_{21} & 1 \end{bmatrix} \quad (3)$$

$$\mathbf{S}_{XX}(j\omega) = \mathbf{T}(j\omega) \cdot \mathbf{D}_X(j\omega) \cdot \mathbf{T}^+(j\omega) \quad (4)$$

where $\mathbf{T}^+(j\omega)$ is the corresponding adjoint matrix. Now the modified input matrix becomes:

$$\mathbf{S}_{XX}(j\omega) = \begin{bmatrix} \mathbf{D}_1 & \mathbf{D}_1 t_{21}^* \\ \mathbf{D}_1 t_{21} & \mathbf{D}_1 |t_{21}|^2 + \mathbf{D}_2 \end{bmatrix} \quad (5)$$

Comparing matrix elements of eq. (5) and eq. (2), one obtains: $\mathbf{D}_1 = S_{i_{nb}}$, $\mathbf{D}_2 = S_{i_{nc}} - S_{i_{nb}} |t_{21}|^2$. The cross-PSD is:

$$\underline{S}_{i_{nc}i_{nb}} = S_{i_{nb}} t_{21} \quad (6)$$

Expression for the control factor t_{21} can be found later and associated with a controlled source in the equivalent circuit (EC).

Interpretation of the above matrix manipulation into a noise EC model can provide a better understanding for further implementation. After transforming $\mathbf{S}_{XX}(j\omega)$ into a diagonal matrix, the practically correlated noise sources i_{nb} and i_{nc} (see Fig. 1a) take the form of three uncorrelated ones $\overline{i_{nb}}$, $\overline{i_{nc}}$ and $t_{21}i$ (see Fig.1b). The transformation yields an additional noise source, which carry the correlated terms. This is understood as the inclusion of the additional controlled source during the transfomation through the control factor t_{21}. The control factor of this source contains the correlation between the noise sources i_{nb} and i_{nc}. This additional controlled current source is tagged in parallel to the output noise source (see Fig.1b) keeping consistency with the system theory [13].

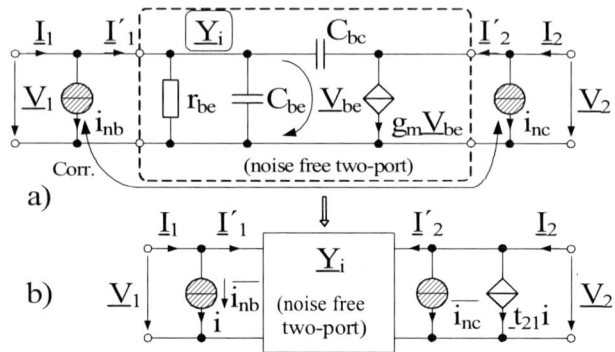

Fig. 1. a) Noise free two-port with small signal equivalent circuit (SSEC) of the internal BJT in the dashed box and correlated noise sources i_nb and i_nc, at the input and output respectively b) SSEC with three modified uncorrelated noise sources. Note that $i \equiv \overline{i_{nb}}$.

Now the input voltage noise source in Fig.2.b will be expressed as a function $v_{n\_ers} = f(\overline{i_{nb}}, \overline{i_{nc}}, t_{21}i)$ of uncorrelated noise sources in Fig.2.a. This further analysis is carried out in connection with the noise computational methods adopted in conventional SPICE-like circuit simulators. Two-port Y-parameter representation for circuits in Fig.2a/b can be given by,

$$I_1 = \underline{Y}_{11}\underline{V}_1 + \underline{Y}_{12}\underline{V}_2 + \overline{i_{nb}} \quad , \quad \underline{V}_1 = v_{nRs} - R_S I_1 \quad \text{(Fig. 2a)} \quad (7)$$
$$I_2 = \underline{Y}_{21}\underline{V}_1 + \underline{Y}_{22}\underline{V}_2 + \overline{i_{nc}} + t_{21}i$$

$$I_1 = \underline{Y}_{11}\underline{V}_1 + \underline{Y}_{12}\underline{V}_2 \quad , \quad \underline{V}_1 = v_{n\_ers} - R_S I_1 \quad \text{(Fig. 2b)} \quad (8)$$
$$I_2 = \underline{Y}_{21}\underline{V}_1 + \underline{Y}_{22}\underline{V}_2$$

Setting $\underline{V}_2 = 0$ (output shorted) in both eq. (7) and eq. (8) and resolving expression for $\underline{I}_2$, one obtains,

$$v_{n\_ers} = \underbrace{\frac{1}{G1}v_{nRs}}_{} + \left(t_{21}\underbrace{\frac{1+R_S\underline{Y}_{11}}{\underline{Y}_{21}}}_{G3} + \underbrace{\left(\frac{-R_S}{}\right)}_{G2}\right)\overline{i_{nb}} + \underbrace{\frac{1+R_S\underline{Y}_{11}}{\underline{Y}_{21}}}_{G3}\cdot\overline{i_{nc}} \quad (9)$$

Expressing eq. (9) in terms of power spectral densities, i.e., in terms of eq. (1), one gets the final expression for PSD as,

$$S_{v_{n\_ers}} = |G_1|^2 S_{v_{nRs}} + |t_{21}G_3 + G_2|^2 S_{\overline{i_{nb}}} + |G_3|^2 S_{\overline{i_{nc}}} \quad (10)$$

The expressions for PSDs in eq. (10) are:

$$S_{v_{nRs}} = 4kTR_S, \quad S_{\overline{i_{nb}}} = 2qI_B \quad (11)$$

Note that $i \equiv \overline{i_{nb}}$ and R_S is the source resistance.

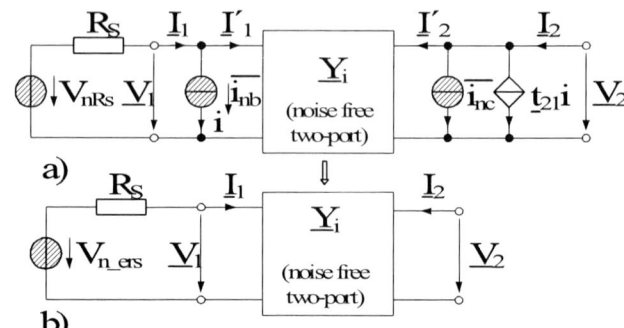

a)

b)

Fig. 2. a) modified SSEC (cf. framed part in Fig. 1 a)) with uncorrelated noise sources $\overline{i}_{nb}$, $\overline{i}_{nc}$ and $t_{2l}i$, b) SSEC with equivalent noise voltage source v_{n_ers} at the input (usually the input port source).

$$S_{\overline{i_{nc}}} = S_{i_{nc}} - |t_{21}|^2 S_{i_{nb}} = S_{i_{nc}}\left(1 - B_f\left(\omega\frac{\tau_{Bf}}{3}\right)^2\right) \quad (12)$$

with

$$S_{i_{nb}} = 2qI_B, \quad S_{i_{nc}} = 2qI_C \quad (13)$$

where B_f is dc gain and τ_{Bf} is base transit time. Control factor t_{21} for the eq. (10), following from eq. (6) is defined as,

$$t_{21} = \frac{S_{i_{nc}i_{nb}}}{S_{i_{nb}}} = -j\omega\frac{\tau_{Bf}}{3}\frac{I_C}{I_B} \text{ with } S_{i_{nc}i_{nb}} \approx -2qI_C\left(j\omega\frac{\tau_{Bf}}{3}\right) \quad (14)$$

where I_B and I_C are the direct base (intrinsic base emitter diode) and direct collector (forward transfer) currents, respectively. The drivation of the cross-PSD $S_{i_{nc}i_{nb}}$ (eq. (14)) may be found in [1][2].

Now one may try to implement the above mentioned method of correlated noise computation through compact model and thereafter verify the implementation with a circuit simulator. Emergence of Verilog-A as a preferred language for writing compact models [14][15], enables the model developer to implement and test any new concept in a short period of time. For the noise simulation in Verilog-A, only a limited number of functions are supported, among which "white_noise ( )" and "flicker_noise ( )" are useful. The language reference manual (LRM)[15] has an example of correlated noise implementation with a real valued correlation coefficient. However, one can use the "ddt ( )" operator with a concept of capacitive coupling to implement any imaginary correlation [11]. Inside "white_noise ( )", it is not obviously permitted to specify any argument that depends upon frequency, also the argument should not be negative in any case. As per our theoretical discussion, one requires to implement three controlled sources, one at the input and the other two at the output. The source at the input side is the same as the base current uncorrelated noise source. At the output side, one controlled source can be implemented by tagging a capacitor (with capacitance value of noise transit time) from the input noise source maintaining a proper sign of the control factor t_{21}. However, the main problem, according to eq. (12), is to implement the remaining noise source in the output with the proper PSD, part of which depends upon the square of the frequency preceded by a negative sign!

280

According to [11], we understand that to obtain a "negative sign" inside the noise PSD is not possible from simple addition and subtraction of controlled sources and one can not multiply controlled sources in any SPICE-like implementation of noise. Therefore, it is not possible to straightforwardly realize the new PSD $S_{i_{nc}}$ in Verilog-A, since it is clear that any noise PSD corresponds to a squared quantity cancelling any negative sign. However for this case, we can use an approximation as in eq. (15), which can imitate the theoretical prediction:

$$S_{\overline{i_{nc}}} \approx S_{i_{nc}}\left(1 - \frac{B_f}{2}\left(\omega\frac{\tau_{Bf}}{3}\right)^2\right)^2 = S_{i_{nc}} - |t_{21}|^2 S_{i_{nb}} + \underbrace{\frac{|t_{21}|^4}{4B_f}S_{i_{nb}}}_{\text{error term}} \quad (15)$$

It is found that an error (last term in eq. (15)) does not exceed a 10% limit up to one third of the peak transit frequency (for 150 GHz SiGe process HBTs). It is also worth to mention that this PSD is part of the total noise correlation, which is again part of the total noise of the transistor. In the calculation of the minimum noise figure (NF$_{min}$), ultimately the error remains very small up to the peak value of f$_T$. Putting eq. (15) into eq. (10), and re-arranging the terms, we obtain:

$$S_{v_{n\_ers}} \approx |G_1|^2 A + |G_2|^2 B + |G_3|^2 C + 2Re\left\{G_3 G_2^* t_{21}\right\}B + K \quad (16)$$

$$A = S_{v_{nRs}}, \quad B = S_{i_{nb}}, \quad C = S_{i_{nc}}, \quad K = |G_3|^2\frac{|t_{21}|^4}{B_f}B$$

In eq. (12), factor 1/3 is relevant for pure diffusion transistors. In modern transistor models like HICUM, a bias dependent total transit time is formulated including the contributions from the emitter and the collector regions [16]. Therefore the factor 1/3 may be found a little smaller than the one actually required from device physics, if the total transit time is used in the implementation. To maintain consistency and generality for all other processes including Si-based ones, in our implementation we used the bias dependent total transit time and a parameter instead of the factor 1/3. The VCCSs, shown in Fig. 3, are dependent on the voltages V(b_n1) and V(b_n2):

$$I(bi, ei) = gV(b\_n1), \quad V(b\_n1) = \frac{1}{g}\sqrt{2 \cdot q \cdot i_{bei}},$$

$$I(ci, ei)_1 = \left(1 - \frac{B_f}{2}(alit \cdot Tf \cdot \omega)^2\right) \cdot g \cdot V(b\_n2), \quad (17)$$

$$V(b\_n2) = \frac{1}{g}\sqrt{2 \cdot q \cdot i_t},$$

$$I(ci, ei)_2 = -j\omega \cdot B_f \cdot alit \cdot Tf \cdot g \cdot V(b\_n1)$$

Fig. 3 shows a SPICE-like implementation of transistor noise (including correlation). where "g" corresponds to a uniform conductance of 1S, "alit" is a parameter dependent on the process technology (e.g. 1/3 for diffusion transistor), "Tf" is the bias dependent total transit time that takes into account the total delay for the carrier in emitte base, and collector regions, "q" is the elementary charge and "B$_f$" is equivalent to dc gain, "i$_{bei}$" is internal base and "i$_t$" is transfer currents, as defined in HICUM. Now we can get the PSDs as:

$$S_{I(bi, ei)} = 2qi_{bei} = S_{\overline{i_{nb}}} = S_{i_{nb}} \quad (18)$$

$$S_{I(ci, ei)_1} = 2qi_t \cdot \left(1 - \frac{B_f}{2}(alit \cdot Tf \cdot \omega)^2\right)^2 \approx S_{\overline{i_{nc}}}$$

$$S_{I(ci, ei)_2} = (2qi_{bei}(B_f \cdot alit \cdot Tf \cdot \omega)^2) = |t_{21}|^2 S_{\overline{i_{nb}}} = |t_{21}|^2 S_{i_{nb}}$$

,

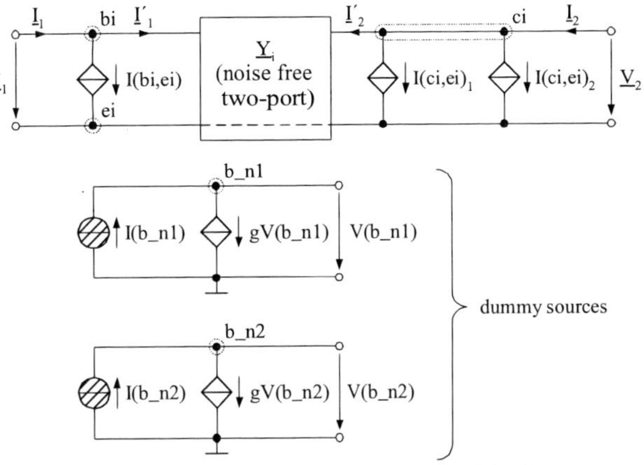

Fig. 3. Realization of correlation in compact models from a system theory perspective.

A Verilog-A implementation of corresponding correlated noise in bipolar transistors is given in Fig. 4. Note that for modular representation, p1 and p2 are used to indicate the input and the output node of the two-port network, that is to say, the branch b_p1 is equivalent to branch (bi,ei) and b_p2 to (ci,ei) of Fig. 3. The variable betadc is precalculated from the direct internal base current (ibei) and the transfer current (it) of the transistor. Two dummy noise current sources are created with spectral densities 2*Q*ibei and 2*Q*it at (n1,0) and (n2,0) branches. Connecting with 1Ω resistor, it is ensured that the noise currents are same as the noise voltages at the respective branches. In the input node, only the base current noise contribution is directly tagged. The output node is tagged with two separate noise contributions, one from the base current noise associated with frequency dependent control factor and other

```
inout  p1, p2;
branch (p1)    b_p1;
branch (p2)    b_p2;
branch (n1)    b_n1;
branch (n2)    b_n2
parameter real alit = 0.333 from [0:1];
I(b_n1) <+ white_noise(2*'P_Q*ibei, "shot");
I(b_n1) <+ V(b_n1);
I(b_n2) <+ white_noise(2*'P_Q*it, "shot");
I(b_n2) <+ V(b_n2);
I(b_p1) <+ V(b_n1);
I(b_p2) <+ V(b_n2);
k1 = alit*Tf;
I(b_p2) <+ 0.5*betadc*k1*k1*ddt(ddt(V(b_n2)));
I(b_p2) <+ betadc*ddt(-k1*V(b_n1));
```

Fig. 4. Verilog-A implementation of HBT transistor correlated noise.

from the transfer current (or collector current) noise. Part of this second term is associated with a squared-frequency dependency preceded by a negative sign as shown in eq. (18).

III. MODEL VERIFICATION

Noise parameters were measured for 0.18 μm SiGe-BiCMOS HBTs [17], with a peak transit frequency of f_T=150 GHz at V_{CE}=1.5V. Measured transistors have a CBEBC contact configuration with emitter window areas $A_{E0}[\mu m^2]$=0.2*4.52, 0.2*10.16. Measurements were performed in the 2-26 GHz frequency band. Noise parameters were de-embedded with the correlation matrix technique. Simulations were performed with ADS 2004A using compiled (with Tiburon) Verilog-A code. This piece (Fig.4) of Verilog-A code is introduced into the existing Verilog-A HICUM Level 2 model to test against the actual measured noise data. Using simulation results from both, the compiled Verilog-A HICUM as well as the simulator built-in HICUM level 2 model with the same set of model parameters, perfect agreement was obtained against the measured dc and high-frequency data. Since Verilog-A based model is very flexible, it can be easily modified and compiled over again, after any change. This opens a possibility to seek impact to noise parameters the effect of correlation between the base and the collector current shot noise sources by setting model parameter to desired value. For example at ("*alit*"=0) HICUM noise model does not account correlation effect. Measured and simulated results are presented in the Figures 5, 6. At microwave low frequency (2 GHz) no impact of correlation is observed, where both simulated curves (with and without correlation) coincide with

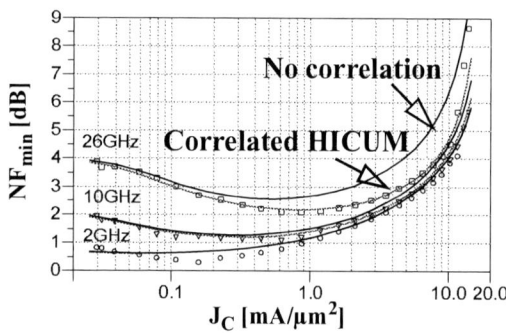

Fig. 5. NF_{min} vs. J_c for SiGe HBTs, symbols: meas., lines: sim.

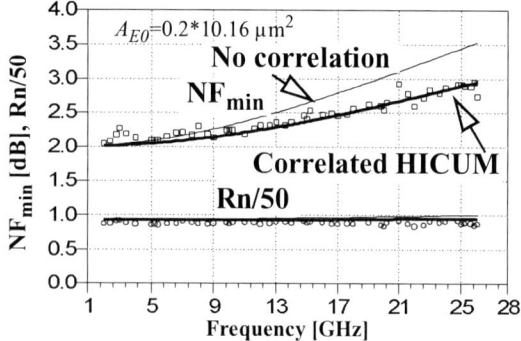

Fig. 6. NF_{min} and R_n/50 vs frequency at J_c=2.74mA/μm^2, V_{CE}=1.5 V, meas., lines: sim.

measured data. Beyond ~10 GHz, simulation results without the effect of correlation deviate from the measured data significantly, whereas those including the effect of correlation are found to be in a perfect agreement up to 26 GHz.

IV. CONCLUSIONS

Based on system theory, bipolar transistor noise is formulated. Developed equations include noise correlation terms in such a way that the approach can be implemented into a compact model to be used with any conventional SPICE-like circuit simulator. The concept is successfully realized in existing Verilog-A code for HICUM. Implementation is verified against measured high-frequency (f_T=150 GHz) SiGe-HBTs noise data. The corresponding results are found to be in perfect agreement.

ACKNOWLEDGEMENTS

The authors are thankful to Jazz Semiconductors Inc. for providing wafers. Fruitful discussions with C.McAndrew from Freescale Semiconductors, G.Coram from Analog Devices, Marek Mierzinski from Tiburon Inc. are acknowledged. We are indebted to Falk Korndoerfer from IHP Frankfurt(Oder) for the noise equipment and to German Research Society (DFG) for the financial support.

REFERENCES

[1] M. J. Buckingham, "Noise in electronic devices and systems", Ellis Horwood Lim., Halsted Press, NY, ISBN 0-470-27467-0, 1983.

[2] A. Blum, "Elektronisches Rauschen", B.G.Teubner Stuttgart, printed: Präzis-Druck GmbH, ISBN 3-519-06183-X, 1996.

[3] L. Escotte et alii, "Noise Modeling of Microwave Heterojunction Bipolar Transistors", *IEEE TED*, v.42, pp. 883-888, 1995.

[4] M. Rudolph et alii, "An HBT Noise model Valid up to Transit Frequency", *IEEE EDL*, v.20, No.1, pp. 24-26, 1999.

[5] J. Gao et alii, "Microwave Noise Modeling for InP-InGAs HBTs", *IEEE Trans. MTT*, v.52, No.4, pp. 1264-1272, 2004.

[6] J. Möller et alii, "An Improved Model for High-Frequency Noise in BJTs and HBTs Interpolating between the Quasi-Thermal Approach and the Correlated-Shot-Noise Model", Dig. *IEEE BCTM*, pp. 228-231, 2002.

[7] J. Herricht, P.Sakalas et alii, "Verification of π-Equivalent Circuit based Microwave Noise Model on $A_{III}B_V$ HBTs with Emphasis on HICUM", Dig.-*CDR, IEEE MTT-S IMS*, Long Beach, CA, 2005.

[8] P. Sakalas et alii, "Microwave noise in III-V and SiGe based HBTs, comparison, trends, numbers", Proc.of *Spie Conf: Noise in Devices and Circuits II*, Vol 5470, pp. 151-163, Gran Canaria, 2004.

[9] G. Niu et alii, "A Unified Approach to RF and Microwave Noise Parameter Modeling in Bipolar Transistors", *IEEE TED*, v.48, No.11, pp. 2568-2574, 2001.

[10] R. Rohrer, "Computationally Efficient Electronic-circuit Noise Calculations", JSSC 6, No.4, pp. 204-213, 1971.

[11] C. C. McAndrew et alii, "Correlated Noise Modeling and Simulation", *WCM*, 2005.

[12] G. Elst et alii,"Rauschanalyse linearer Netzwerke mit korrelierten rauschquellen", Nachrichtentech., Elektron., v.34, Nr.11, pp. 428-430, 1984.

[13] G. Wunsch, H.Schreiber, Stochastische Systeme, Springer-Verlag Berlin Heidelberg, ISBN 3-540-54313-9, 1992.

[14] G. J. Coram, "How to (and how not to) write a compact model in Verilog-A", Proc. *IEEE* Intern. Behavioral Modeling and Simulation Conference, BMAS 2004, pp. 97-106, 2004.

[15] Verilog-AMS Language Reference Manual, Ver. 2.2, http://www.eda.org/verilog-ams.

[16] M. Schroter et alii,"Physics-Based Minority Charge and Transit Time Modeling for Bipolar Transistors", *IEEE TED*, v.46, No.2, pp. 288-300, 1999.

[17] M. Racanelli et alii, "Ultra High Speed SiGe NPN for advanced BiCMOS Technology", *IEDM Tech*. Digest, pp. 336-339, 2001.

2006 Bipolar/BoCMOS Circuits and Technology Meeting Proceedings

Modeling of Intrinsic Base Majority Carrier Thermal Noise for SiGe HBTs Including Fringe BE Junction Effect

Kejun Xia and Guofu Niu

Electrical and Computer Engineering Department
200 Broun Hall, Auburn University, Auburn, AL 36849, USA
Tel: 334 844-8263 / Fax: 334 844-1888 / E-mail: xiakeju@auburn.edu

Abstract — **This paper investigates the impact of fringe BE junction on base majority carrier RF noise in SiGe HBTs. Due to fringe effect, the base hole noise should be modeled by correlated noise voltage source and noise current source in hybrid representation. The noise voltage source can be modeled by a weakly bias dependent noise resistance that is different from the intrinsic base resistance. The correlation between the two noise sources is important for biases around peak f_T.**

I. BACKGROUND

Base hole noise is a major noise source for SiGe HBTs. Traditionally this noise is modeled by the thermal noise of r_{bx} and r_{bi}, the small signal base resistance for the extrinsic and intrinsic region respectively. r_{bx} is the resistance of a true resistor whose noise can be well modeled with $4kTr_{bx}$. However, r_{bi} is a lumped resistance. There are two kinds of r_{bi}, depending on whether quasi-static (QS) equivalent circuit or non-quasi-static (NQS) equivalent circuit is used. As studied in [1], $r_{bi,nqs}$ is more physical and also smaller than $r_{bi,qs}$. We found problems of using $4kTr_{bi}$ for noise modeling based on either QS or NQS lumped equivalent circuit. Firstly, S_{ic} extracted for the intrinsic transistor is larger than $2qI_C$ [2] when QS equivalent circuit is used. If we brutally choose a larger $r_{bi,qs}$ to make $S_{ic} = 2qI_C$, S_{ib} becomes negative at low frequencies [2]. The result remains the same if we use NQS equivalent circuit. Another problem is that the absolute value of the imaginary part of noise parameter Y_{opt}, i.e. $|\Im(Y_{opt})|$, is overestimated by van Vliet model [3] based on NQS equivalent circuit. The deviation cannot be eliminated by choosing appropriate $r_{bi,nqs}$. This work aims to solve these two problems by modeling the distributive effect of base hole noise.

The distributive effect is a significant feature of intrinsic base hole noise [4]. The best way to examine this effect is through microscopic noise simulation. Fig. 1 shows DESSIS simulated 2D spatial contribution of base hole noise for S_{vb}, the base terminal noise voltage in Z-representation. Double base contact is used. Observe that only the base region beneath the edge of BE junction contributes to base hole noise significantly. There exist two kinds of distributive effect, the fringe effect associated with the edge transistor and the crowding effect associated with the intrinsic transistor. To account for these effects, we divide the BE/BC junction into four segments A_{1-4}, leading to five equivalent base resistances of three types as shown in Fig. 2. Further analysis shows that at least four segments (five resistors) are needed. Type I resistances are for the edge transistors. Type III resistances are for the main intrinsic transistor. Type II resistances are a combination of resistances from the main

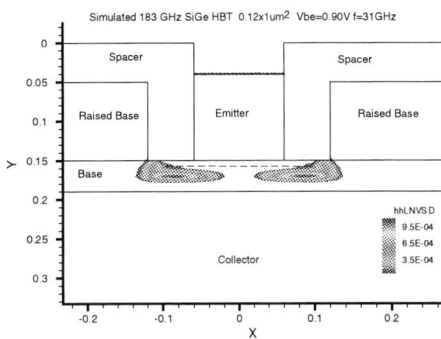

Fig. 1. DESSIS simulated spatial contribution of base hole noise for S_{vb} in a 183 GHz SiGe HBT at V_{BE}=0.90V, f=31 GHz.

and edge transistors and contribute much more noise than type III as shown by Fig. 1 due to AC crowding effect. Because of the narrow emitter width and high base doping,

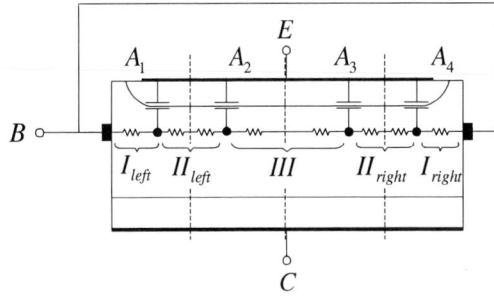

Fig. 2. Illustration of base distribution effect by dividing the base resistances into five segments of three types. Double base contact is used.

DC crowding effect is negligible in practice. Hence the traditional $4kTr_{bi}$ description is theoretically true only for the main intrinsic transistor *without* the fringe region [5]. In lumped equivalent circuit based modeling, the fringe region or edge transistor is not explicitly separated from the main intrinsic region [2][6][7]. However it is unknown how the fringe effect affects base hole noise and how important the effect is.

This paper shows that the base hole noise should be modeled by a noise voltage source at the input and a correlated noise current source at the output due to the fringe effect. It is the correlation of the two noise sources that cause the S_{ic} and $|\Im(Y_{opt})|$ problems described above. DESSIS device simulation is used as guidance, as base hole and electron noises can be separated in simulation. Experimental data are used to verify the new model.

II. PHYSICAL CONSIDERATIONS

As discussed in Section I, the five resistance model has captured both the fringe effect and crowding effect of base hole noise in a lumped fashion. Fig. 3 shows the small signal equivalent circuit that corresponds to Fig. 2. The five

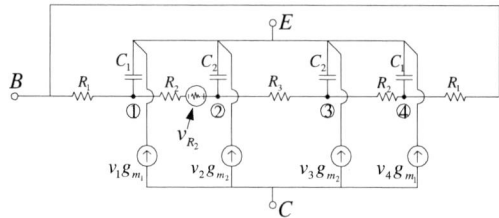

Fig. 3. Small signal equivalent circuit of five segments model. Only the noise voltage source of left R_2 is shown. g_{be} is neglected. Four nodes are labelled.

resistances correspond to those five segments. The four capacitors and transconductances correspond to segment A_{1-4}. Note that $g_{m2} >> g_{m1}, C_2 >> C_1$. g_{be} is neglected in Fig. 3 which is only used for base hole noise derivation. The g_{be} in the small signal equivalent circuit of SiGe HBT is not neglected. All the small signal components are connected through four inner nodes. The resulting equivalent circuit is symmetric.

Although the DC base-emitter bias is the same for A_{1-4} segments, the local f_T varies along the emitter junction. A_2 and A_3 have the same f_T. A_1 and A_4, however, have lower f_T because of wider base of the edge transistor, meaning that $g_{m1}/C_1 < g_{m2}/C_2$. The smaller local f_T does not affect the transistor f_T much because of the small area of A_1 and A_4 compared to A_2 and A_3. However we will show that just because of the non-uniformity of f_T, base hole will produce noise current at the collector. This result cannot be obtained in [5] where uniform f_T is assumed. Although chain representation of noise is directly related to noise parameters [8], for the directness of physics we will model the base hole noise using hybrid representation as shown in Fig. 4. v_h and i_h are the noise source for the hybrid representation, while v_a and i_a are the noise source for the chain representation. The

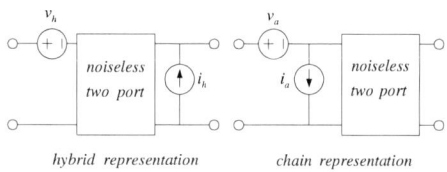

hybrid representation   chain representation

Fig. 4. Hybrid representation and chain representation.

hybrid representation noise is then transformed into chain representation by

$$S_{va} = S_{vh} + \frac{S_{ih}}{|Y_{21}|^2} + 2\Re\left[\frac{S_{ihvh^*}}{Y_{21}}\right], S_{ia} = S_{ih}\left|\frac{Y_{11}}{Y_{21}}\right|^2,$$

$$S_{iava^*} = S_{ihvh^*}\frac{Y_{11}}{Y_{21}} + S_{ih}\frac{Y_{11}}{|Y_{21}|^2}. \tag{1}$$

where the Y-parameters are for the intrinsic transistor *including* r_{bi}.

III. MODEL EQUATION DERIVATION

We first use the five resistance model to derive model equations for the base hole noise. The equations include three model parameters R_{bn}, K_1 and K_2. We then examine the bias dependence of these model parameters using device simulation.

A. v_h and i_h

Because of the symmetry of the circuit in Fig. 3, the noise of R_3 does not contribute to either v_h or i_h. Each R_1 gives $4kTR_1/4$ noise for v_h. The two R_1 totally contribute $4KTR_1/2$ noise to v_h. Again because of symmetry, R_1 does not contribute to i_h. The two R_2 resistors contribute to both v_h and i_h.

Now consider the left R_2. We insert a test noise voltage source v_{R2} into Fig. 3. v_{R2} has a noise voltage PSD of $4kTR_2$. Solving the symmetric network,

$$v_1 + v_4 = -v_{R2}\frac{C_2}{C_1 + C_2 + j\omega R_2 C_1 C_2}, \tag{2}$$

$$v_2 + v_3 = v_{R2}\frac{C_1}{C_1 + C_2 + j\omega R_2 C_1 C_2}. \tag{3}$$

The equivalent hybrid representation noise sources are then obtained as

$$v_h = \frac{v_1 + v_4}{2}, \tag{4}$$

$$i_h = -g_{m2}(v_2 + v_3) - g_{m1}(v_1 + v_4)$$

$$= \left[\frac{g_{m2}}{C_2} - \frac{g_{m1}}{C_1}\right]C_1(v_1 + v_4). \tag{5}$$

If the f_T of A_1 and A_2 are the same, i.e., $g_{m2}/C_2 = g_{m1}/C_1$, then $i_h = 0$. The base hole noise can be fully described by v_h. As discussed in Section II, $g_{m1}/C_1 < g_{m2}/C_2$, therefore i_h has the same sign as v_h, leading to a positive $\Re(S_{ihvh^*})$.

B. Noise in hybrid representation

For convenience, we define two partition factors

$$\lambda_c = \frac{C_1}{C_1 + C_2} < 1, \lambda_{gm} = \frac{g_{m1}}{g_{m1} + g_{m2}} < 1. \tag{6}$$

Note $\lambda_{gm} < \lambda_c$. As $C_2 << C_1$, we neglect the $\omega R_2 C_1 C_2$ term in both v_h and i_h. The noise due to the left R_2 can be obtained as

$$S_{vh,R2} = < v_h v_h^* > = 4kTR_2[\lambda_c^2/4],$$

$$S_{ih,R2} = < i_h i_h^* > = 4kTR_2(g_{m1} + g_{m2})^2(\lambda_c - \lambda_{gm})^2,$$

$$S_{ihvh^*,R2} = < i_h v_h^* > = 4kTR_2(g_{m1} + g_{m2})(1 - \lambda_c)(\lambda_c - \lambda_{gm}). \tag{7}$$

The right R_2 has the same noise as the left R_2, therefore the two R_2 contribute two times of the noise shown in (7). Now the overall noise can be obtained by adding the contributions of two R_1 and two R_2 in (7) as

$$S_{vh} = 4kTR_{bn},$$

$$S_{ih} = 4kTR_{bn}g_m^2 K_1,$$

$$S_{ihvh^*} = 4kTR_{bn}g_m K_2, \tag{8}$$

284

where

$$g_m = 2(g_{m1} + g_{m2}), \quad R_{bn} = R_1/2 + (R_2/2)\lambda_c^2,$$

$$K_1 = \frac{R_2/2}{R_{bn}}(\lambda_c - \lambda_{gm})^2, \quad K_2 = \frac{R_2/2}{R_{bn}}2(1 - \lambda_c)(\lambda_c - \lambda_{gm}).$$

(8) is our model equation, and has three parameters R_{bn}, K_1 and K_2. According to stochastic physics, the normalized correlation $c \leq 1$ [2], meaning $K_2^2 \leq K_1$. For device simulation, these parameters can be directly extracted from the base hole noise in hybrid representation. Fig. 5 shows the simulated base hole noise in hybrid representation with the new model at V_{BE}=0.90V. Note that $\Re(S_{ihvh*}) > 0$, which

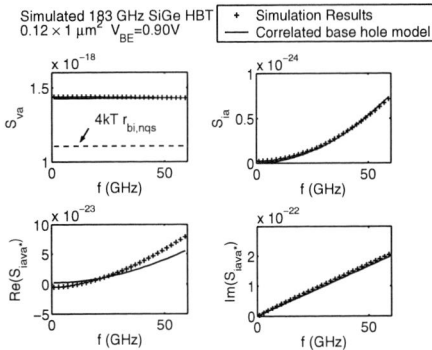

Fig. 6. Comparison of simulation and new model for base hole noise in chain representation at one bias V_{BE}=0.90V.

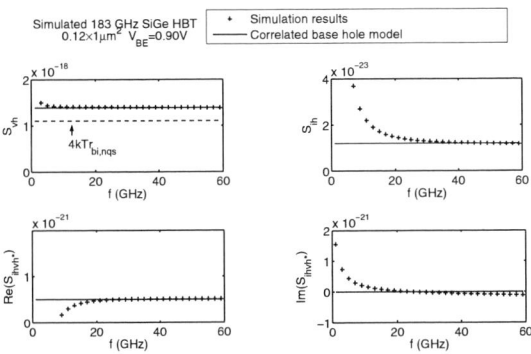

Fig. 5. Comparison of simulation and new model for base hole noise in *hybrid* representation at one bias V_{BE}=0.90V.

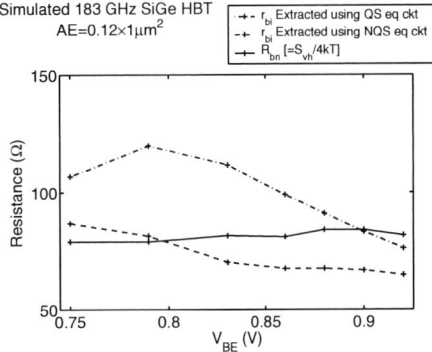

Fig. 7. Comparison between thermal resistances R_{bn} and small signal resistance r_{bi}.

is consistent with $g_{m1}/C_1 < g_{m2}/C_2$. The spikes at low frequencies can be modeled at extra complexity if g_{be} is included in Fig. 3. However, the spikes will disappear in chain representation due to the Y_{11} factor in (1), which decreases as frequency decreases.

Fig. 6 shows the modeling results in chain representation using (1) at V_{BE}=0.90V. The new model correctly models S_{va} and $\Im(S_{iava*})$. Note that the simulated S_{ia} and $\Re(S_{iava*})$ are nonzero at low frequencies. They are zero in the new model because g_{be} was neglected. These low frequency errors are negligible compared to the large value of base electron noise and emitter hole noise. S_{va} and $\Im(S_{iava*})$, however, are correctly modeled. The small error in $\Re(S_{iava*})$ is not important as $\Re(S_{iava*}) \ll \Im(S_{iava*})$.

C. Bias dependence of model parameters

We need to investigate the bias dependence of R_{bn}, K_1 and K_2 because of the unknown bias dependence of λ_c and λ_{gm}. We examined two SiGe HBTs simulated by DESSIS. One has 85 GHz peak f_T and the other has 183 GHz peak f_T.

Fig. 7 shows R_{bn} and r_{bi} obtained using simulation for the 183 GHz SiGe HBT. Two different r_{bi} extracted based on QS and NQS equivalent circuits are shown. The QS r_{bi}, extracted using circle method overestimates R_{bn} at low biases and underestimates R_{bn} at biases around peak f_T. This phenomenon has already been shown in [9]. The NQS r_{bi}, which is more physical, has a value close to R_{bn} at low biases, however, underestimates R_{bn} at high biases. It can be

observed that R_{bn} has a weak bias dependence. Therefore R_{bn} can be modeled as a constant, whose value can be approximated by the r_{bi} value based on NQS equivalent circuit at *low* biases.

Fig. 8 (a) shows the bias dependence of K_1 for the two simulated devices. K_1 is nearly constant around peak f_T for each device. For low biases, S_{ih} is not important due to small g_m, hence the final noise is not sensitive to K_1. Further, the new model is proposed to improve noise modeling for biases before f_T roll off, the K_1 value of peak f_T bias can be used for all biases. Fig. 8 (b) shows the bias dependence of K_2 for the two devices. Again K_2 is nearly constant around peak f_T. Similarly, the K_2 value of peak f_T bias can be used for all biases.

Because of the weak bias dependence of K_1 and K_2, according to (8), S_{ih} and S_{ihvh} go to zero at low biases due to small g_m. The correlation of v_h and i_h affects mainly the biases around peak f_T.

IV. RESULTS

Now we use experimental noise data on a 50 GHz peak f_T SiGe HBT to demonstrate how the new base hole noise model solves the two problems discussed in Section I. The data in [2] are used here. NQS equivalent circuit is used without explicitly separating the fringe region. We first examine the $S_{ic} > 2qI_c$ problem. Fig. 9 shows the impact of the correlation for base hole noise on intrinsic noise ex-

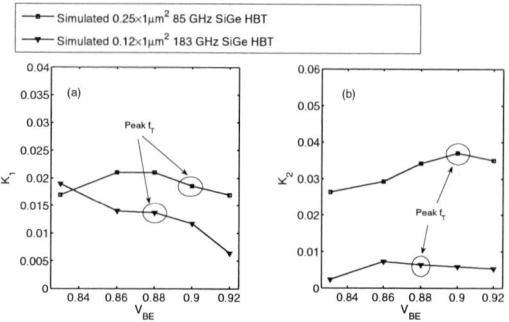

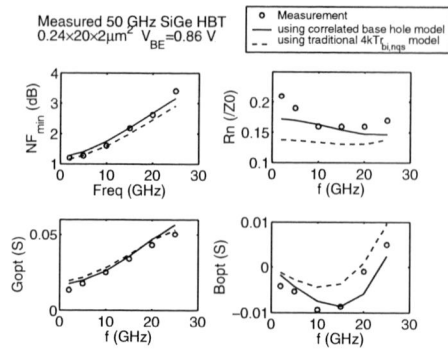

Fig. 8. (a) K_1 extracted for simulated 85 GHz and 183 GHz peak f_T SiGe HBTs. (b) K_2 extracted for simulated 85 GHz and 183 GHz peak f_T SiGe HBTs

Fig. 10. Noise parameter modeling result using the new base hole model and the traditional model.

traction. Only S_{ib} and S_{ic} are shown here. The triangle-up symbols represent the extracted noise using $4kTR_{bn}$ as the base hole noise. Obviously $S_{ic} > 2qI_C$. One can brutally increase R_{bn} to make $S_{ic} = 2qI_C$, this decreases S_{ib} and makes it unphysically negative as shown by cross symbols. However, if we use correlated base hole noise as shown by the triangle-down symbols, S_{ic} can be decreased to $2qI_C$ while S_{ib} is not affected. The frequency dependence of S_{ic}, especially at low frequency, is likely a problem of measurement accuracy [2], and is never observed in simulations.

parameters R_{bn}, K_1 and K_2, all of which can be modeled as bias independent constants. The correlation between two noise sources increases the absolute value of $\Im(S_{icib^*})$. The utility of the model has been demonstrated using noise simulation and experimental data.

ACKNOWLEDGMENTS

This work was supported by SRC under #2003-NJ-1133 and IBM under an IBM Faculty Partnership Award. We thank Yan Cui for the help of DESSIS simulations, D. Sheridan, S. Sweeney, J. Cressler and the IBM SiGe team for their contributions.

REFERENCES

[1] K. Xia et al., *Proc. of the IEEE BCTM*, pp. 180-183, 2005.
[2] K. Xia et al., *IEEE Trans. ED*, 53 (3), pp. 515-522, 2006.
[3] K. M. van Vliet, *Solid state electronics*, 15(10), 1033, 1972.
[4] Y. Cui et al., *Proc. of the IEEE BCTM*, pp. 225-228, 2003.
[5] J. C. J. Paasschens, *IEEE Trans. ED*, 51 (9), pp. 1483-1495, 2004.
[6] Q. Cai et al., *IEEE Trans. MTT*, 45 (12), 1997.
[7] J. Roux et al., *IEEE Trans. MTT*, 43 (2), pp. 293-298, 1995.
[8] H. A. Haus et al., *Proc. IRE*, 48(1), pp. 69-74, 1960.
[9] C. Jungemann et al., *IEEE Trans. ED*, 51 (6), pp. 956-961, 2004.

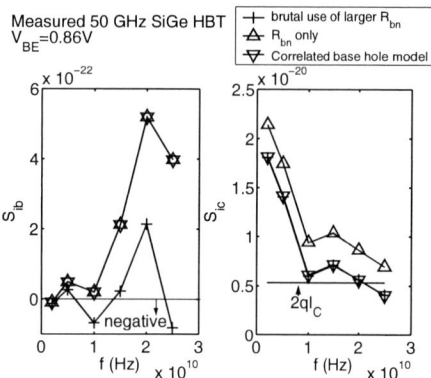

Fig. 9. The impact of the correlation of base hole noise on intrinsic noise extraction using the experimental data of a SiGe HBT with 50 GHz peak f_T.

Fig. 10 shows the noise parameters modeled using the new base hole noise model and the traditional $4kTr_{bi}$ model using experimental data based on NQS equivalent circuit. Van Vliet model for electron noise is used. Modeling of $\Im(S_{icib^*})$ is improved. Modeling of thermal resistance R_n is also improved.

V. CONCLUSIONS

We have explained the impact of fringe BE junction on the base majority carrier noise based on equivalent circuit analysis. The noise is modeled with a noise voltage source and a correlated noise current source in hybrid representation. The resulting new noise model includes three model

Fully coupled dynamic self heating model for power SOI Lateral Insulated Gate Bipolar Transistors

S. Gamage[1], V. Pathirana[2] and F. Udrea[1,2]

[1]University of Cambridge, Electrical Engineering Division, 9, J. J. Thompson Avenue, Cambridge, CB3 0FA UK [2]Cambridge Semiconductor Ltd, St. Andrews House, St. Andrews Road, Cambridge, CB4 1DL, UK

Abstract — **Several vertical IGBT electro-thermal models are currently available on circuit simulators. However, no reliable electro-thermal models have been proposed for the lateral Insulated Gate Bipolar Transistor (LGBT). In this paper, for the first time we present a fully coupled electro-thermal model for a LIGBT structure based on a novel concept recently reported in [1]. The model relies on a systematic study of both isothermal and self heating behavior of the device. It is valid in both steady state and switching conditions. The model is further implemented in the Spice circuit simulator and validated against extensive numerical simulations and experimental data.**

Index Terms — **Power semiconductor switches, Electrothermal effects, Modeling, Insulated gate bipolar transistors.**

I. INTRODUCTION

The lateral Insulated Gate Bipolar Transistor (LIGBT) is a mainstream power device for power ICs and smart power circuits. Today the preferred route for many LIGBT manufacturers (e.g. Toshiba, Hitachi, Camsemi) is Silicon-On-Insulator (SOI) due to the buried oxide layer (BOX) which prevents the charge build-up in the substrate while improving the device isolation. A physics based compact electrical model for a thin SOI LIGBT was presented in [1]. Further in [2] it was extended to model the thick SOI LIGBT. The purpose of this paper is to extend the already proposed LIGBT model to include the electro-thermal interrelations that is valid at transient and switching conditions. Since the main application of the LIGBT is in switching it is expected that the Spice model presented here would be useful to power electronic circuit and system designers. The model has also been implemented in 'Verilog-A' and 'C' and the results will be presented elsewhere.

The silicon on membrane technology (SOM) reported in [3], is one of the most promising approaches to improve the isolation of the LIGBT and concomitantly achieve switching frequencies specific to LDMOS devices rather than bipolar devices. The structure relies on embedding the drift region of the LIGBT in a thin SOI membrane. This achieves perfect isolation, as displacement currents generated by high dV/dts in the anode region are virtually nil due to a minute anode to substrate capacitance. The structure also achieves very high breakdown and low on-state resistance [3]. However in common with other SOI structures thermal management of this class of devices is very important. Hence an electro-thermal model of this structure becomes not only desirable but the only realistic route to describe the static and dynamic performance of the device.

State of the art vertical IGBT electrical models for Spice and Saber circuit simulators are mainly based on the work carried out by Hefner [4] and Kraus [5]. Unfortunately electrical device models on vertical IGBTs are not well suited for the LIGBT due to horizontal carrier dynamics, substrate effects, latch up etc. Some of these models are later extended to include the thermal effects. However in the vertical devices the heat is mainly generated near the top oxide/semiconductor interface and traverses the base vertically before reaching the IGBT package structure. In the SOMs the heat has a complex two dimensional flow in both lateral and vertical directions.

A novel self-heating model is developed by extensive analysis of the self-heating inside the LIGBT using Medici electro-thermal coupling module [7]. The validity of the model in switching is also studied. Then the self-heating model is verified through numerical simulations and experimental data supplied by CamSemi [3], [8].

II. SELF-HEATING MODEL

A. Modelling Approach

The identified modelling scheme is to have two coupled circuits. One circuit is to model the electrical characteristics of the LIGBT with the temperature dependencies. The physical origin and empirical expressions for the temperature-dependent physical properties of silicon are well documented in the literature. The temperature dependent LIGBT parameters are divided into two regions: the base region and the MOS region. These dependencies are given in [1]. The other circuit is to model the thermal network made of thermal resistors and capacitors. The average powers generated by each section of the electrical circuit are taken as flow variables or 'current' sources to the thermal circuit. The resulting average temperatures in the thermal circuit will be fed back to the electrical network. This is also the approach preferred by majority in the context of electro-thermal modelling since it is computationally viable while retaining a good accuracy level [6] [10].

B. Thermal circuit of the device

In this work we attempt to build a self-heating model for the SOM LIGBT, which is shown on Fig. 1. This structure is based on a thin silicon LIGBT membrane consisting of a thin silicon layer (0.1-0.5 μm) on top of a thin BOX. (1-5 μm). Fig. 2 shows a 3D picture of the temperature distribution when the gate is at 5V and drain is at 10V.

When the drain voltage is low the device will be in linear region and the highest temperature peak will appear at the drain end due to the potential drop at the p+/n junction and air membrane underneath which blocks the heat dissipation. When the device enters saturation at high drain voltages (or low gate voltages) there is another temperature peak appearing at the gate end due to the high electric field at the pinch off region. To account for both effects a novel model based on two heat sources is developed.

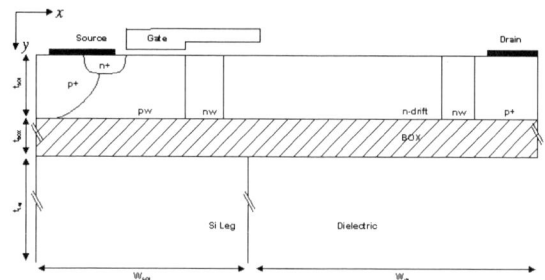

Fig. 1 The silicon on membrane LIGBT device structure

The first heat source is accounting mainly for the power generated in the channel and accumulation layer while the second heat source is accounting for the power generated at the drain junction and in the drift region (i.e. base of the pnp bipolar transistor). The magnitude of these heat sources would depend on the generated power in each of these regions. The two port self-heating network captures accurately the behaviour of the LIGBT both in the linear and saturation regions. While here we specifically demonstrate its use for the SOM LIGBT, it is worthwhile noting that two node concept for self heating is also valid for other LIGBTs such as those based on traditional SOI substrate [11].

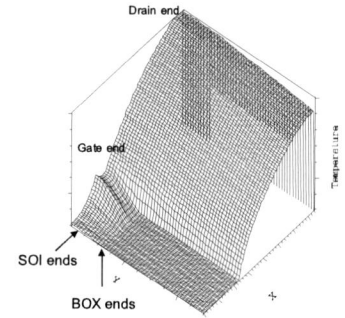

Fig. 2 Temperature distribution inside the membrane LIGBT at V_g=5V and V_d=10V

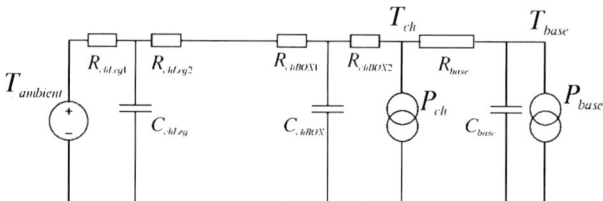

Fig. 3 Thermal circuit for the silicon on membrane LIGBT

Given that two nodes are required, one for the channel and another for the p+/n drain junction and base region, the equivalent circuit shown in Fig. 3 is proposed. There is no significant thermal flow to the right hand side of the base heat source as there is no power dissipation path to that side. The lateral thermal resistances around the source area within the top silicon are relatively small compared to the base and the leg thermal resistances; hence these can be ignored. There will be no power source in the vertical heat dissipation path since there is no power generation in the leg area. The total thermal resistance of the leg would be the series combination of the thermal resistance of the top Si, BOX and substrate silicon. The theoretical equations to calculate the thermal circuit parameters are given in [11].

Also the resistors under steady state conditions can be extracted though numerical simulation. This procedure also acts as a verification of the validity of the two port network. This procedure is discussed at lengths in [11]. Here we present an alternative calculation method of the base thermal resistance which is valid for dynamic switching case as well.

Assuming the thermal resistance at each point is constant

$$\delta T(x) = \delta R(x) P(L) \int_{x}^{L} pdf(x) dx \qquad (1)$$

where $pdf(x)$ is the power per unit distance along x. $dT(x)$, $dR(x)$ and $P(L)$ are the infinitesimal temperature change at point x, the infinitesimal thermal resistance change at point x and total base power at $x=L$ at the end of the base. Hence integrating (1) wrt x and taking the average in x direction gives

$$T_{ave} = \left[\frac{1}{2} - \frac{1}{L} icdf_{ave} \right] R(L) P(L) \qquad (2)$$

where $icdf_{ave}$ is the averaged cumulative power density function in x direction. This can be easily calculated from numerical simulations using the normalized voltage drop across the device into total current across the device, since this represents an approximation for the cumulative power density function ($cdf(x)$). Integrating this wrt x and taking the average gives $icdf_{ave}$. Thermal resistance of the base $R(L)$ is calculated using the theory as given [11].

B. Transient turn-off behavior

Fig. 4 depicts the temperature distribution inside the device at different time points in a switching cycle. The highest peak in the temperature at turn off is seen at the source side of the base where the gate field plate ends. This is due to the highest electric field in the blocking mode appearing there in switching.

The 2D temperature distribution at $t=6e-7s$ and $t=1.12006e-2s$, which are equivalent repetitive points at switching steady state, are shown in Fig. 5. The temperature distribution across device at start up (Fig. 5(a)) in switching is very different to steady state behaviour as shown in Fig. 2. This is mainly due to very large thermal capacitance under the SOI which includes the leg and the dielectric. While this is being charged the thermal flow is essentially 2D and differ from the model assumption we have derived. However we are mainly interested in switching steady state assuming the frequency of switching remains static throughout the operation of the device. Hence as shown in Fig. 5(b) the shape of thermal flow is similar to the steady-state case as in Fig. 2. Hence we expect our model to be valid in switching steady state. The thermal resistance for this case can be calculated from (2) with the aid of numerical simulation of one cycle since the normalised power distribution ($cdf(x)$) does not change significantly with the

switching cycle.

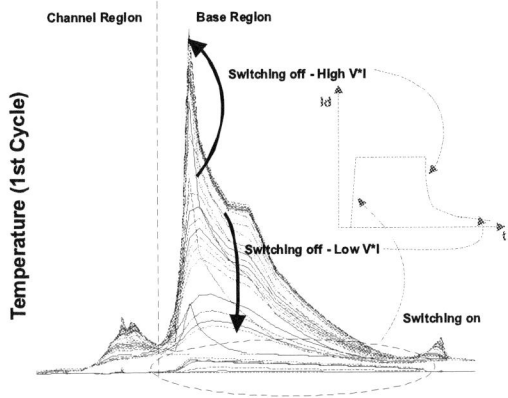

Fig. 4 The internal temperature distribution across the device in switching using numerical simulation. Shown here is the first cycle

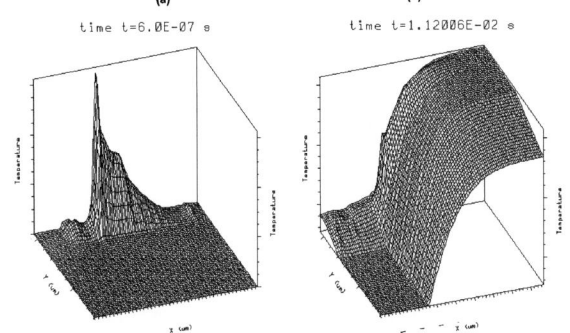

Fig. 5 The internal temperature distribution at (a)t=6e-7s and (b) 1.12006e-2s for current switching of 0.125A with 10us period at 5% duty cycle. The blocking voltage is 560V.

III. SELF-HEATING MODEL RESULTS

The membrane devices fabricated by Camsemi [3][8] were measured to verify the model against the experimental data.

The on state comparison between the model and the measurements are shown in Fig. 6. Furthermore the internal temperatures of these test structures were obtained from embedded calibrated diodes in the drain and source. This is shown in Fig. 7. The matching of currents and internal temperatures are very good. Fig 8 depicts the matching between the channel and drain region average temperatures of the model compared to simulations of current of 0.125A switching at 10us period with 5% duty cycle. The blocking voltage is 560V. Note that switching transient temperatures in this case is not matched very well. This is because the 2D effects in switching transient case (Fig. 5(a)) are not considered in the compact model. However the model matches well towards the switching steady state (Fig 5(b)) where the compact mode assumptions are more accurate.

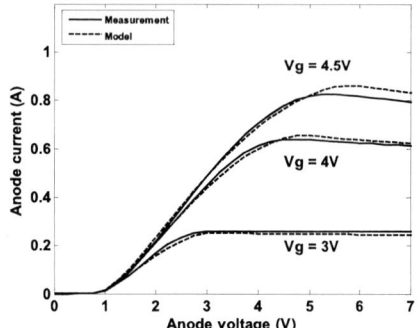

Fig. 6 On state measurements

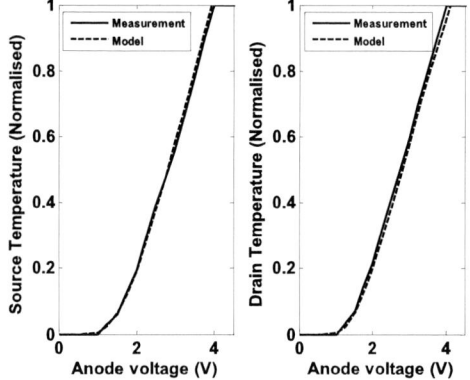

Fig. 7 Internal temperature measurements using embedded calibrated diodes

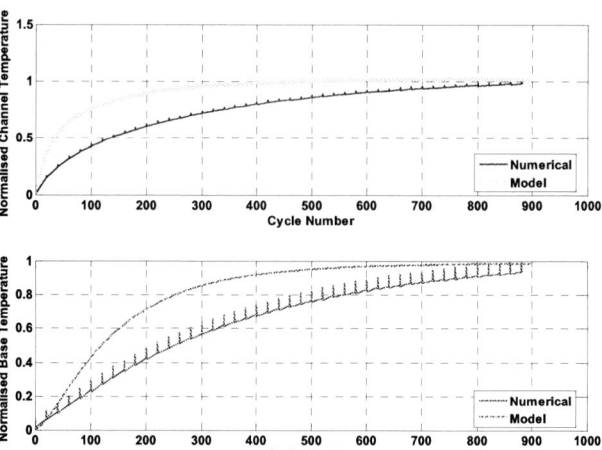

Fig. 8 Simulated internal average channel and drain temperatures compared against the model in switching. A current of 0.125A switched at 10us period with 5% duty cycle. The blocking voltage is 560V

IV. CONCLUSIONS

In this paper a systematic analysis and study of the thermal behaviour of a LIGBT in silicon on membrane technology was presented. The proposed model can simulate both the isothermal and self-heating effects. For the first time we proposed a two port self-heating thermal circuit to model the self-heating inside the LIGBT with much improved accuracy in steady-state and more importantly in switching conditions. The developed electro-thermal model results are compared against numerical simulations as well as experimental data and found to be in very good agreement.

ACKNOWLEDGEMENT

The authors wish to acknowledge the financial support of the European Union 6th framework project ROBUSPIC for this work and device structures and measurements from Camsemi.

REFERENCES

[1] E. Napoli, V. Pathirana, F. Udrea, G. Bonnet, T. Trajkovic, G. Amaratunga, "A compact model for thin SOI LIGBTs: description, experimental verification and system application", in *Proc. 17th International Symposium Power Semiconductor Devices and ICs*, 2005, pp. 95-98

[2] V. Pathirana, S. Gamage, E. Napoli, F. Udrea, "A Complete Isothermal Model for the Lateral Insulated Gate Bipolar Transistor on SOI technology", *IEEE Tencon*, Nov 2005, Australia

[3] F. Udrea, T. Trajkovic and G. A. J. Amaratunga, "Membrane High Voltage Devices – A Milestone Concept in Power ICs", *IEDM*, 2004

[4] A.R. Hefner, D Blackburn, "An analytical model for the steady-state and transient characteristics of the power insulated-gate bipolar transistor", *Solid State Electronics*, vol. 31 , no. 10, pp. 1513–1532, 1988

[5] R. Kraus; K. Hoffmann, "An analytical model of IGBTs with low emitter efficiency" in *Proc. 5th International Symposium Power Semiconductor Devices and ICs*, 1993, pp. 30–34.

[6] Allen R. Hefner, Jr., "A Dynamic Electro-Thermal Model for the IGBT", *IEEE Transactions on Industry Applications*, vol. 30, vol. 2, pp. 394–405, Mar/Apr 1994

[7] Medici user guide, Fremont CA, Avant Corporation, 2003

[8] F. Udrea and G.A.J. Amaratunga, US patent 6,703,684

[9] S. M. Sze, *Physics of Semiconductor Devices*. 2nd ed., New York: John Wiley & Sons, 1981.

[10] C. Anghel, R. Gillon and A. M. Ionescu, "Self-Heating Characterization and Extraction Method for Thermal Resistance and Capacitance in HV MOSFETs," *IEEE Electr. Dev. Lett.*, vol. 25, no. 3, pp. 141–143, March 2004.

[11] S. Gamage, V. Pathirana and F. Udrea, "Fully coupled dynamic self heating model for power SOI Lateral Insulated Gate Bipolar Transistor," Accepted for publication in *IEEE Trans. Elec. Dev.*, 2006

A Broadband and Scalable On-Silicon-Chip Inductor Model

for Varying Substrate Resistivities

J. C. Guo and T. Y. Tan

Dept. of Electronics Engineering, National Chiao Tung Univ., Hsinchu, Taiwan
Tel: +886-3-5131368, Fax: +886-3-5724361, E-mail: jcguo@mail.nctu.edu.tw

Abstract — **A broadband and scalable model is developed to accurately simulate on-chip inductors of various dimensions and substrate resistivities. The broadband accuracy is proven over frequencies up to 20 GHz, even beyond resonance. A new scheme of RLC networks is deployed for spiral coils and substrate to account for 3D eddy current, substrate return path, and spiral coil to substrate coupling effects, etc. The 3D eddy current is identified as the key element essential to accurately simulate broadband characteristics. A key element (R_P) introduced in our model can successfully accounts for the conductor loss due to eddy current arising from magnetic field coupling through substrate return path. This broadband and scalable model is useful for RF circuit simulation. Besides, it can facilitate on-chip inductor design and optimization through physics-based model parameters relevant to lossy substrate.**

Index Terms – **Inductor, broadband, scalable, eddy current**

I. Introduction

On-chip inductor model becomes one of the most challenging topics for Si-based RF IC design due to difficulties in broadband accuracy, scalability, and physical interpretation of complicated EM coupling through semiconducting substrate by lumped elements. Many publications reported improvement on the commonly adopted π-model [1]-[2]. However, limited bandwidth to few GHz remains an open issue. A two-π model was proposed to improve the accuracy beyond self-resonance frequency f_{SR} [3]. Unfortunately, this two-π mode suffers a singular point above f_{SR} [4]. Besides, the complicated circuit topology with double element number will add difficulty in parameter extraction and greater time consumption in circuit simulation [5]. In our previous work, a T-model of new circuit scheme was created to realize three primary features, i.e. broadband accuracy, scalability, and a parameter extraction flow capable of automation by Agilent IC-cap or other extraction tools [6]. However, there remained two major points deserving deeper study for enhancement. One is the precise matching of broadband S-parameters beyond resonance and another is the prediction of broadband features associated with varying substrate resistivities (ρ_{Si}). In this work, an improved T-model is devised to enhance the broadband accuracy beyond resonance. The earlier resonance suffered by larger

spiral coils generally reveals lower f_{SR}, maybe far below 20 GHz. It brings a challenge to most of lump element models and even to EM simulators to achieve precise matching with measurement beyond resonance. In this work through 3D EM simulation (Ansoft HFSS) we identify that substrate eddy current effect is actually a 3D coupling behavior rather than a simplified planar feature. The 3D eddy current effect plays a major role in frequency response beyond resonance for the on-chip inductors. In this improved T-model, three branches of parallel RL elements are deployed to emulate the 3D eddy current effect. All the equivalent circuit elements are kept constants independent of frequencies and can be expressed by a closed form derived from circuit analysis. We believe that this scalable inductor model can effectively improve RF circuit simulation accuracy over broadband and facilitate the design optimization using on-chip inductors.

II. Technology & Characterization

Spiral inductors of square coils were fabricated by 0.13μm BEOL technology with 8 layers of Cu and FSG IMD (k=3.8). The top metal of 3μm Cu was used to implement the spiral coils of width fixed at 15 μm and inter-coil space at 2μm. The inner radius is 60 μm and outer radius is determined by different coil numbers, N=2.5, 3.5, 4.5, and 5.5 for this study. The physical inductance achieved at sufficiently low frequency are around 1.96~8.66 nH corresponding to N=2.5~5.5. S-parameters were measured by using Agilent network analyzer up to 20 GHz and open de-embedding was carefully done to extract the intrinsic characteristics for model parameter extraction.

III. Improved T- Model's Equivalent Circuit & Model Parameter Extraction Flow

Fig.1(a) illustrates the circuit schematics of the improved T-model for on-chip inductors. The major enhancement over our original work [6] is the deployment of three branches of parallel RL elements using a T-shape configuration to emulate the 3D eddy current effect. The idea comes from our study through 3D EM simulation (Ansoft HFSS). We identify that substrate eddy current effect is actually a 3D coupling behavior rather than a simplified planar feature. Two branches of parallel RL in series with C_{ox1}/C_{ox2} account for the eddy current component normal to the substrate plane and the other RL element in series with substrate RC represents the eddy current in the plane. The 3D eddy current effect plays a major role in frequency response beyond resonance for the on-chip inductors. In this way, two RLC networks of four physical elements for each are linked through $C_{ox1,2}$ in series with parallel RL to account for the EM coupling between the spiral inductors and lossy substrate underneath. Fig.1(b) indicates the block diagrams derived by circuit analysis theory to extract the physical circuit elements as proposed. Z_1 represents the RLC network for spiral inductor and Z_4 is another one representing lossy substrate. Z_2 and Z_3 consisting of $C_{ox1,2}$ and parallel RL as proposed for normal component of eddy current act as the coupling path between Z_1 and Z_4. The circuit scheme is further transformed to Fig.1(c) to correlate with Y-parameters from 2-port measurement. As a result, all the physical elements composing the model can be extracted by the flow as shown in Fig.2. The details of extraction for the circuit elements in two primary RLC networks, i.e. Z_1 and Z_4 and the physical properties can be referred to our previous work [7]. Regarding the extraction of new elements added in this improved T-model, i.e. $L_{sub1}//R_{loss1}$ and $L_{sub2}//R_{loss2}$, the first step is to extract C_{ox1} and C_{ox2} from Z_2 and Z_3 under very low frequency provided that $\omega L_{sub1}//R_{loss1}$ and $\omega L_{sub2}//R_{loss2}$ are negligibly small impedances compared to $1/\omega C_{ox1}$ and $1/\omega C_{ox2}$, respectively. After the extraction of C_{ox1} and C_{ox2}, the four new elements (L_{sub1}, R_{loss1}) and (L_{sub2}, R_{loss2}) can be derived easily from Z_2 and Z_3 under very high frequency. It is noted that R_P is a critical element introduced in our model to simulate the spiral conductor loss and Q degradation before resonance. Skin effect caused by eddy current in the coil metal arising from magnetic field coupling through the substrate return path is proposed as the origin responsible for the conductor loss.

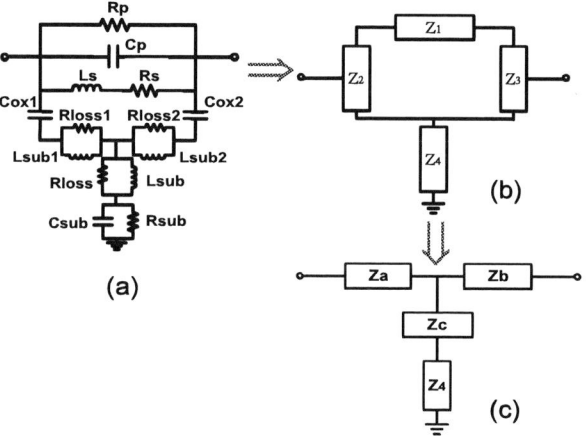

Fig.1 Improved T-model (a) equivalent circuit schematics, (b) and (c) schematic block diagram for circuit analysis

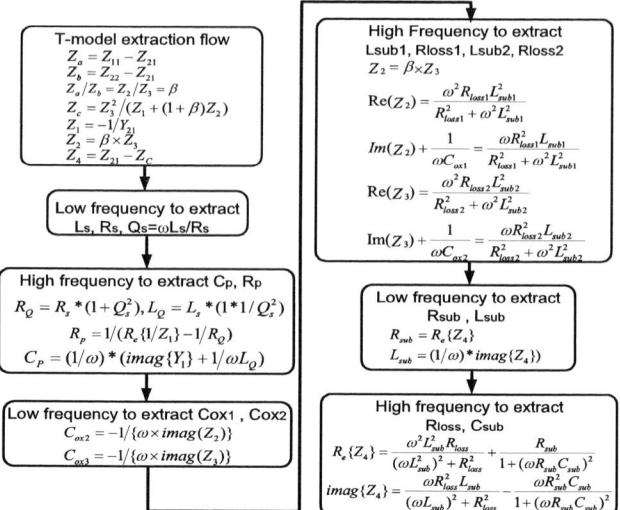

Fig.2 Improved T-model parameter formulas and extraction flow chart

IV. Broadband Accuracy and Scalability

The improved T-model has been extensively verified by comparison with measurement in terms of S-parameters (S_{11}, S_{21}), $L(\omega)$, $Re(Z_{in}(\omega))$, and $Q(\omega)$ over wideband up to 20 GHz. The scalability is justified by various dimensions of coil numbers, N=2.5, 3.5, 4.5, and 5.5. Broadband accuracy is justified by good match with measurement in terms of the mentioned performance parameters. Figs.3(a)-(d) indicate the comparison for magnitude and phase of S_{11} between the model and measurement. As for S_{21}, the results and comparison are presented in Figs.4(a)-(d) for magnitude and phase respectively. Excellent match is achieved for all coil numbers even beyond resonance, which happened at $f_{SR}<<20GHz$ for larger coil numbers (N=3.5, 4.5, 5.5). It is an obvious improvement over the original T-model and even better match is achieved as compared to EM simulation before dedicated calibration. More extensive verification has been done by comparison of three key parameters for spiral inductors, i.e. $L(\omega)$ ($=Im(Z_{in}(\omega)/\omega)$, $Re(Z_{in}(\omega))$, and $Q(\omega)$. $Q(\omega)$ is the quality factor defined by $\omega L(\omega)/Re(Z_{in}(\omega))$. All three parameters are frequency dependent that is critically related to the spiral conductor loss and Si substrate loss. Figs.5(a)-(d) illustrate the excellent fit to the measured $L(\omega)$ and $Re(Z_{in}(\omega))$ by the improved T-model for all inductors operating up to 20GHz. The transition from inductive to capacitive mode at $f > f_{SR}$ is accurately reproduced. Besides, the model can exactly capture the full band behavior of $Re(Z_{in}(\omega))$ even beyond resonance such as the dramatic increase prior to resonance, peak at resonance, and then sharp drop after the peak. Eventually, $Q(\omega)$ is the most critical parameter governing RF IC performance such as power, gain, and noise figure, etc. Fig.6 reveals the excellent match with the measured $Q(\omega)$ over broadband of 20GHz. The exact fit to the peak Q and capture of full band behavior for all coil numbers suggests the advantage of the improved T-model compared to the existing π-model.

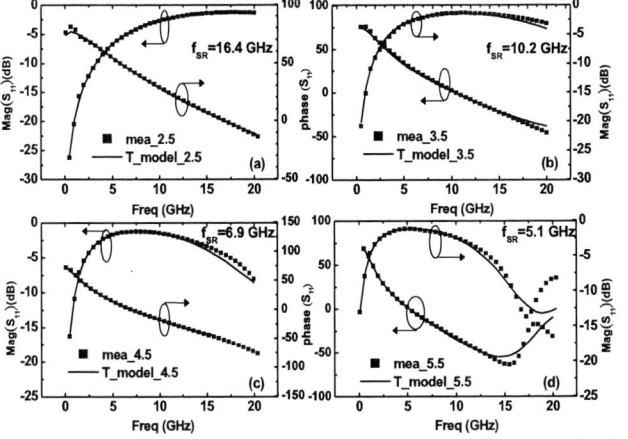

Fig. 3 Comparison of improved T-model and measured S_{11}(mag, phase) for inductors. Coil numbers (a) N=2.5 (b) N=3.5 (c) N=4.5 (d) N=5.5

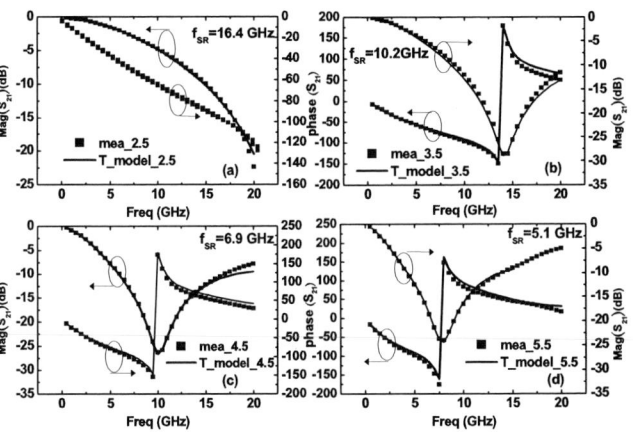

Fig. 4 Comparison of improved T-model and measured S_{21}(mag, phase) for inductors. Coil numbers (a) N=2.5 (b) N=3.5 (c) N=4.5 (d) N=5.5

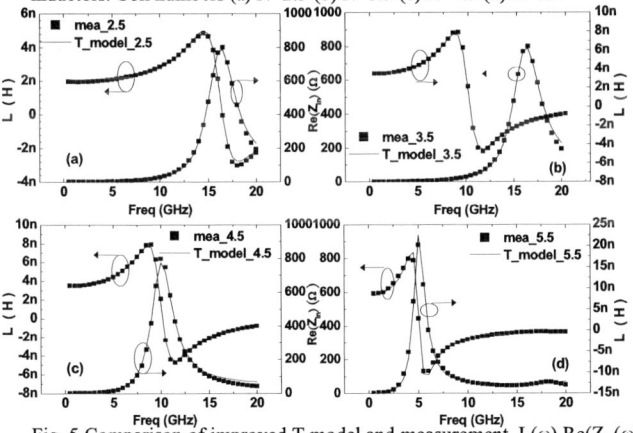

Fig. 5 Comparison of improved T-model and measurement, $L(\omega), Re(Z_{in}(\omega))$ (a) N=2.5 (b) N=3.5 (c) 4.5 (d) N=5.5

Another important feature is the good scalability w.r.t. dimension for all model parameters. Figs.7(a)-(d) present good match with a linear function of coil numbers for each

model parameter in the spiral coil's RLC network, i.e. L_S, R_S, R_P, C_P, and $C_{ox1,2}$. Figs.8(a)-(d) indicate the excellent fit by linear function for substrate network involved model parameters, C_{sub}, $1/R_{sub}$, L_{sub}, $L_{sub1,2}$, R_{loss} and $R_{loss1,2}$. The proven scalability suggests that this T-model can be used for pre-layout simulation and optimization.

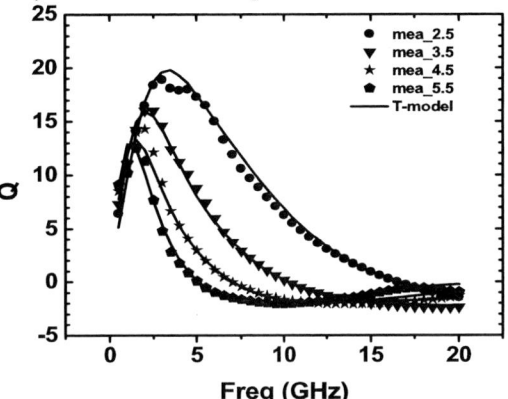

Fig. 6 Comparison of improved T-model and measurement for $Q(\omega)$ for various coil numbers (N=2.5, 3.5, 4.5, 5.5)

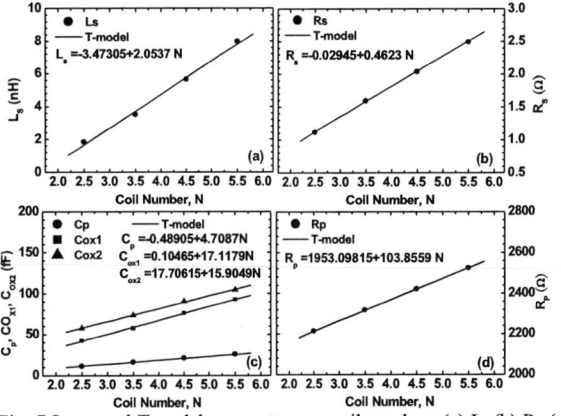

Fig. 7 Improved T-model parameters vs. coil numbers (a) L_S (b) R_S (c) C_P, $C_{ox1,2}$ (d) R_P

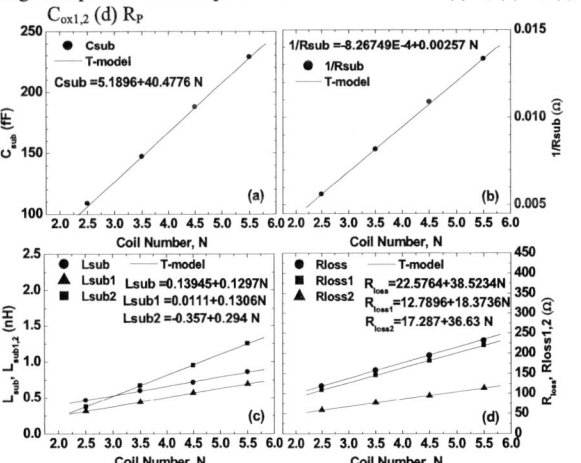

Fig. 8 Improved T-model parameters vs. coil numbers (a)C_{sub} (b) $1/R_{sub}$ (c) L_{sub}, $L_{sub1,2}$ (d), R_{loss}, $R_{loss1,2}$

V. Operation Modes of Varying Substrate Resistivities

ADS momentum simulation with extensive calibration is conducted to predict the broadband characteristics under varying ρ_{Si}. Fig.9 indicate good match between ADS momentum, measurement, and T-model in terms of S_{11}, S_{21}, $L(\omega)$, $Re(Z_{in}(\omega))$, and $Q(\omega)$ for inductors on standard substrate of $\rho_{Si}=10\Omega$-cm. Three operation modes such as TEM, slow-wave, and eddy current [7] corresponding to wide range of ρ_{Si} (0.05~1KΩ-cm) are reproduced. Figs.10(a)-(d) present four key parameters, i.e. Q_m, f_m, f_{Lmax}, and f_{SR} as function of ρ_{Si}. Q_m and f_m are the maximum Q and corresponding frequency. f_{Lmax} is the frequency for maximum L. Interesting result is identified in region of $\rho_{Si}=0.5$~10Ω-cm where f_{SR} drops monotonically with reducing ρ_{Si} while Q_m reveals a hump vs. ρ_{Si}. The drop of f_{SR} and increase of Q_m suggest that the spiral coil is getting into slow-wave mode. As for high resistivity region of $\rho_{Si}>10\Omega$-cm, f_{SR} saturates to a maximum while Q_m increases continuously with ρ_{Si}. This region is so called TEM mode, which favors inductor operation with high Q attributed to suppressed resonance in substrate of dielectric property. Regarding the very low resistivity region of $\rho_{Si}<0.5\Omega$-cm, f_{SR} saturates to a minimum and Q_m drops drastically. The spiral coil is driven into eddy current mode. In the following, improved T-model parameters are extracted from the simulated S-parameters under various ρ_{Si} to verify if the model parameters can reflect the physical properties responsible for the three modes of operation. Figs.11(a)-(b) indicate how the resistive elements (R_P, R_{sub}, R_{loss}, $R_{loss1,2}$) and inductive elements (L_{sub}, $L_{sub1,2}$) vary with varying ρ_{Si}. Quite interestingly, R_P just follows exactly the same trend as that of Q_m vs. ρ_{Si} with a hump in slow-wave mode while the others show monotonic increase with ρ_{Si} in slow-wave and TEM modes and near saturation in eddy current mode. All four capacitances (C_P, C_{sub}, C_{ox1}, C_{ox2}) in Figs.11(c)-(d) demonstrate monotonic increase with reduction of ρ_{Si} in slow-wave mode, saturation in TEM mode while different behaviors in eddy current mode. The larger capacitances associated with lower ρ_{Si} in slow-wave mode just play a major role responsible for the drastic drop of f_{SR} with reducing ρ_{Si}. The result supports an important point that R_P, a new element introduced in our T-model is the key parameter to explicitly guide substrate engineering for on-Si-chip inductor to achieve maximum Q_m.

Conclusion

A broadband and scalable model has been developed to accurately simulate on-chip spiral inductors operating up to 20 GHz. Physics-based model parameters enable this model applicable for three operation modes under varying substrate resistivities. The results suggest that this T-model can facilitate Si-based RF IC design using on-chip inductors

Acknowledgement : The authors would like to thank the support from NSC under Grant 94-2220-E009-018 and CiC

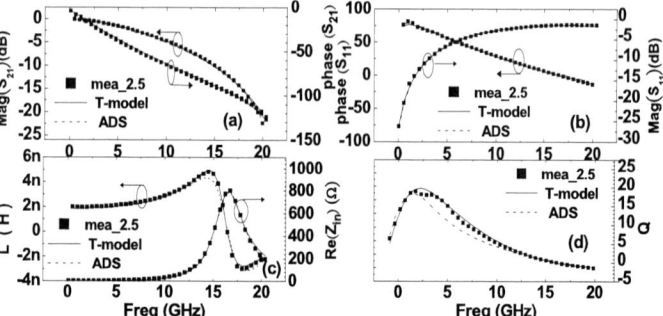

Fig.9 Comparison between ADS momentum simulation, measurement, and improved T-model for on-chip inductor (a) S_{11}(mag, phase) (b) S_{21}(mag, phase) (c) $L(\omega)$, $Re(Z_{in}(\omega))$ (d) $Q(\omega)$

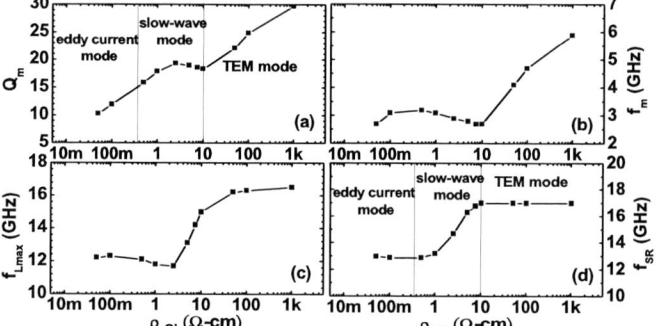

Fig.10 (a) Q_m (b) f_m (c) f_{Lmax} (d) f_{SR} under varying ρ_{Si} (0.01~1KΩ-cm) predicted by ADS momentum simulation.

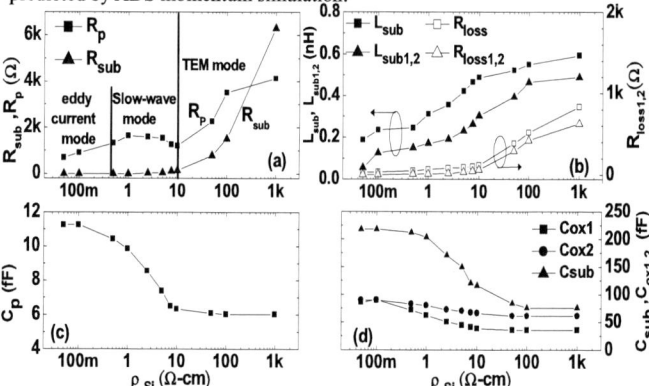

Fig.11 Improved T-model parameters under varying ρ_{Si} (a) R_{sub}, R_P (b) L_{Sub}, $L_{Ssub1,2}$, R_{loss}, $R_{loss1,2}$ (c)C_P (d) $C_{ox1,2}$, C_{sub}

References

[1] C.P. Yue, et al., "Physical Modeling of Spiral Inductors on silicon" IEEE Trans. on Electron Devices, vol.47, pp.560-568, 2000

[2] Min Park, et al., "The detailed analysis of high Q CMOS-compatible microwave spiral inductors in silicon technology," IEEE Trans. on Electron Devices, Vol.45, pp.1953-1959, 1998

[3] Yu Cao, et al., "Frequency-independent equivalent-circuit model for on-chip spiral inductors," IEEE JSSC, Vol. 38, pp.419 – 426, 2003

[4] Fujishima, et al.,"Accurate subcircuit model of an on-chip inductor with a new substrate network," VLSI Symp. on Circuits, Technical Digest, 2004 , pp.376 – 379

[5] J. Gil, et al., "A Simple Wide-band On-chip Inductor Model for Silicon-Based RF Ics," IEEE Trans. on Microwave Theory and Technique, Vol.51, pp.2023-2028, 2003

[6] J.C. Guo, et al., " A Broadband and Scalable Model for On-chip Inductors Incorporating Substrate and Conductor Loss Effect," in IEEE RFIC Symp. Digest, 2005, pp.593-596

[7] H. Hasegawa, et al., IEEE MTT-Vol.19, pp.869-881, 1971

Author Index

A

Ahlgren, D..49
Ancey, P...186
Aniel, F..251
Arnaud, Caroline................................122
Asbeck, Peter M..................................231
Aufinger, K...61
Avenier, G....................................53, 243

B

Babcock, Je.......................................100
Badets, F..211
Bafleur, M..150
Bajolet, A...211
Barbalat, B...................................243, 251
Bécu, S..259
Bellini, Marco......................................41
Belot, D..130
Bennett, Herbert S.................................65
Benoit, D..259
Besse, P...150
Besset, C..259
Bilgen, Halim.....................................122
Biondi, T..207
Blalock, Benjamin J..........................72, 88
Blanc, JP..211
Bock, J..61
Boguth, S...61
Boissonnet, L..53
Bonnard, A..259
Borel, S.......................................196, 259
Borot, G...251
Bouillon, P..53
Bruyère, S...259
Burghartz, J. N....................................191

C

Cai, Jin..41
Cai, Will...126
Campos, R..25
Carchon, G..57
Carusone, A.C......................................223
Chai, Francis.......................................247
Chakravorty, A....................................279
Chalvatzis, T.......................................223
Chanemougame, D...............................196
Chanlo, C..17, 21

Chanlo, Christophe 275
Chantre, A............53, 142, 235, 243, 251
Chataigner, E.......................................211
Chen, Suheng...................................72, 88
Chen, Tianbing.................................41, 100
Cheung, Tak Shun D............................203
Chevalier, P..........53, 142, 235, 243, 251
Choi, L. J...57
Coronel, P...196
Costa, Julio...65
Cottrell, Peter..65
Crémer, S...259
Cressler, John D. ..29, 41, 72, 80, 88, 170
Cui, Yan......................................88, 170

D

D'Alessandro, V. 33
Dahlström, M..................................49, 174
Dahlström, M.E....................................255
Dai, Foster F...................................84, 138
Davis, Paul...158
Dawdy, Morgan...................................247
de Graauw, A.J.M. 17, 21
de Graauw, Anton................................275
De Paola, F. M......................................33
de Vreede, L.C.N. 1
Decoutere, S..57
Dehan, M..57
Dekker, R..178
Delpech, P....................................211, 259
den Dekker, A. 21
Devriendt, K..57
Dickson, T.O.......................................223
Dijkhuis, J.......................................17, 21
Ding, H..49
Dohmen, J.J..108
Dunn, J..49
Dunn, J.S..255
Dupuis, O...57
Dutartre, D. 53, 196, 251

E

Eisenstadt, William R. 215
Endo, Koichi..118

Author Index

F

Farcy, A. 259
Feng, Zhiming 162
Fox, Robert M. 215
Fresina, Mike 154
Fukase, Shin-ichi 263

G

Gamage, S. 287
Garg, A. 223
Gaucher, B.P. 255
Gendron, A. 150
Geynet, B. 243
Gheorghe, M. 271
Ghibaudo, G. 45
Giraudin, JC. 211
Giry, Alexandre 122
Gordon, M. 223
Grens, Curtis M. 29
Groves, Rob 92
Guillaumet, S. 259
Guo, J. C. 291
Gustat, Hans 80

H

Harrison, S. 196
Hazneci, A. 223
Heeres, R.M. 9
Heinemann, Bernd 80
Hernandez, A. Sibaja- 57
Herricht, J. 279
Hijzen, E. 57
Hildreth, Jill 247
Hopper, P.J. 146
Horton, Brian 158
Huang, W. Margaret 65
Hurkx, G.A.M. 25
Hurwitz, Paul 126
Huszka, Z. 104

I

Immorlica, Anthony A. 65
Irmscher, M. 191
Italia, A. 207
Itoh, Nobuyuki 263

J

Jaeger, Richard C. 84, 138
Jagueneau, T. 211
Jarry, P. 13, 130
John, Jay P. 247
Johnson, J.B. 255
Jos, H.F.F. 9
Joseph, A. 49
Joseph, A.J. 174, 255
Joseph, Alvin J. 29, 88
Joseph, Alvin 170
Judong, F. 53, 243, 251
Judong, Fabienne 122

K

Kakani, Vasanth 84, 138
Kerhervé, E. 13, 130
Khater, M. 49
Kikuta, Makoto 263
Kirchgessner, Jim 247
Knapp, H. 61
Krakowski, Tracey L. 100
Kretz, J. 191
Krithivasan, Ramkumar 72, 80, 88
Kunnen, E. 57
Kuo, Wei-Min Lance 80

L

Lachner, R. 61
Lai, B. 223
Laskin, E. 223, 235
Lee, Zachary 126
Leray, P. 57
Letzkus, F. 191
Leyssenne, L. 13
Li, Hao 247
Li, Xiangtao 80
Li, Ying 162
Linten, D. 57
Liu, P. 223
Liu, Q. 49
Liu, Yun 100
Lok, P. 17
Long, John R. 203
Loo, R. 57
Lu, Yuan 72, 80
Lukashevich, D. 61

Author Index

M

Mahmoudi, Reza 275
Malladi, R.M. 174
Manceau, J.-P. 259
Marin, M. 259
Masuoka, Hideaki 263
Matsushita, Kenichi 118
Maxim, A. 239, 271
Maxim, Adrian 76
McAndrew, Colin C. 96
Meister, T.F. 61
Menner, Matt......................................247
Metzger, Andre G. 231
Monfray, S. 196
Monroy, Agustin 122
Moreira, C. P. 130
Morgan, Dave 247
Morita, Makoto 263
Mourier, Jocelyne 122
Mueller, Jan-Erik 65
Muhonen, Kathy 154
Muller, Dorothée.................................122

N

Nagata, Minoru 263
Najafizadeh, Laleh 72, 88, 170
Nakagawa, Akio..................................118
Nakamura, Kazutoshi...........................118
Neri, Bruno 219, 267
Newton, Kim M. 174
Ng, G..223
Nicolson, S. T. 142, 223, 235
Niu, Guofu 88, 162, 166, 170, 283
Nolhier, N. 150

O

Onge, S.A. St. 255
Orner, B. A...............................49, 255

P

Paasschens, J.C.J................................108
Pache, Denis......................................122
Palestri, P. 25
Palmisano, G. 207
Pan, Jun 162
Pananakakis, G. 45
Panko, D...25

P (continued)

Pantel, R. 243
Pathirana, V.......................................287
Payet, F. 196
Perrot, C. 259
Perrotin, A. 53
Pflanzl, W. 104
Pham, J.M. 13
Pijper, R.M.T. 108
Piontek, A. 25, 57
Pothiawala, A......................................49
Pourchon, F. 243
Pramm, S. 17, 21
Pramm, Skule 275
Proust, M..251

Q

R

Racanelli, Marco 65, 126
Ragonese, E. 207
Rascoe, J.S. 255
Rassel, R. M. 49, 255
Rauber, B. 53
Raya, C. 243
Renaud, P. 150
Resnick, D...191
Reuter, Ralf 134, 247
Revil, N. 45
Reynolds, S.K. 255
Richard, C. 251, 259
Rinaldi, N. 33
Rossato, C. 211
Ruat, M. 45
Rubaldo, L...................53, 243, 251
Rueda, Hernan....................................247
Ruijs, Lennart....................................275

S

Sadovnikov, Alexei...............................100
Saguin, F. 243, 251
Saias, Daniel 13
Sakalas, P. 279
Salamero, C. 150
Sautreuil, B. 142, 235
Schafer, H. 61
Schroter, M. 279
Schwartzmann, T. 243
Scuderi, A. 207

Author Index

Seebacher, E. 104
Segura, N. 259
Selmi, L. 25
Shahramian, S. 223
Sheridan, David C. 162
Shi, X. P. 57
Shi, Yun 170
Shichijo, Hisashi 65
Skotnicki, T. 196
Smith, D. 239
Steeneken, P.G. 17
Strachan, Andy 100, 112
Suzuki, Fumito 118
Szelag, Bertrand 122

T

Takahashi, Morio 118
Talbot, A. 196
Tan, T. Y. 291
Tang, K.A. 142
Tchoketch-Kebir, L. 223
Teeter, Douglas A. 154
ten Dolle, H.K.J. 17, 21
ten Dolle, Henk Jan 275
Thomas, M. 259
Turinici, C. 271

U

Udrea, F. 287
Ulaganathan, Chandradevi 72, 88

V

van Bezooijen, A. 17, 21, 275
van den Oever, L.C.M. 9
van der Heijden, M.P. 1
van der Toorn, R. 108
Van Huylenbroeck, S. 57
van Roermund, Arthur H.M. 275
van Straten, F. 17
van Straten, Freek 275
Vandelle, B. 53, 243, 251
Vanhoucke, T. 25
Vashchenko, V.A. 146
Vleugels, F. 57
Voinigescu, S. P. 142, 223, 235
Von Bruns, S. 174
Vytla, R.K. 61

W

Wagner, Lawrence 92
Walter, K. 174
Wan, Ava 92
Wang, Jing 92
Weitzel, Charles E. 65
Widay, David 154

X

Xia, Kejun 166, 283

Y

Yao, T. 223
Yasuhara, Norio 118
Yau, K.H.K 142, 223
Yin, Yi 134
Yoon, Jangsup 215
Yu, Xuefeng 84
Yuryevich, O. 223

Z

Zerounian, N. 251
Zhao, Bin 65
Zheng, Jie 126
Zhu, Chendong 41, 72, 88, 170
Zito, Domenico 219, 267
Zwingman, Robert 126